NUTRITION AND FOOD SCIENCE
Present Knowledge and Utilization

VOLUME 3
Nutritional Biochemistry and Pathology

NUTRITION AND FOOD SCIENCE
Present Knowledge and Utilization

NUTRITION AND FOOD SCIENCE
Present Knowledge and Utilization

VOLUME 3
Nutritional Biochemistry and Pathology

Edited by

Walter Santos, Nabuco Lopes,
J. J. Barbosa, and Dagoberto Chaves
Brazilian Nutrition Society
Rio de Janeiro, Brazil

and

José Carlos Valente
National Food and Nutrition Institute and
Associação Brasiliense de Nutrologia
Rio de Janeiro, Brazil

PLENUM PRESS · NEW YORK AND LONDON

Library of Congress Cataloging in Publication Data

International Congress of Nutrition, 11th Rio de Janeiro, 1978.
  Food and nutrition programs and policies.

  (Nutrition and food science; v. 1)
  Includes index.
  1. Nutrition policy — Congresses. 2. Food supply — Congresses. 3. Food
relief — Congresses. I. Santos, Walter II. Title. III. Series.
TX359.I57 1978   362.5   79-27952
ISBN 0-306-40344-7 (v. 3)

Proceedings of the Eleventh International Congress on Nutrition, organized by
the Brazilian Nutrition Society, held in Rio de Janeiro, Brazil, August 27—
September 1, 1978, and published in three volumes of which this is Volume 3.

© 1980 Plenum Press, New York
A Division of Plenum Publishing Corporation
227 West 17th Street, New York, N.Y. 10011

Printed in the United States of America

FOREWORD

The Brazilian Society of Nutrition, through the present publication, brings to the attention of the world scientific community the
works presented at the XI INTERNATIONAL CONGRESS OF NUTRITION which,
promoted by this Society and under the sponsorship of the International Union of Nutritional Science, was held in the city of Rio de
Janeiro from August 27th to September 1st, 1978.

The publication, edited by Plenum Publishing Corporation, is
titled "Nutrition and Food Science: Presented Knowledge and Utilization" and appears in three volumes, under the following titles and
sub-titles:

Vol.   I - FOOD AND NUTRITION POLICIES AND PROGRAMS
              - Planning and Implementation of National Programs
              - The role of International and Non-governmental Agencies
              - The role of the Private Sector
              - Program Evaluation and Nutritional Surveillance
              - Nutrition Intervention Programs for Rural and Urban Areas
              - Mass Feeding Programs
              - Consumer Protection Programs

Vol. II - NUTRITION EDUCATION AND FOOD SCIENCE AND TECHNOLOGY
              - Animal and Vegetable Resources for Human Feeding
              - Food Science and Technology
              - Research in Food and Nutrition
              - Nutrition Education

Vol. III - NUTRITIONAL BIOCHEMISTRY AND PATHOLOGY
              - Nutritional Biochemistry
              - Pathological and Chemical Nutrition
              - Nutrition, Growth and Human Development

It is hoped that this publication may prove useful to all those who are interested in the different aspects of Nutrition Science.

Editorial Committee:

Walter J. Santos
J.J. Barbosa
Dagoberto Chaves
José Carlos Valente
Nabuco Lopes

PREFACE

The XI INTERNATIONAL CONGRESS OF NUTRITION — XI ICN —, promoted by the INTERNATIONAL UNION OF NUTRITIONAL SCIENCES — IUNS — , was carried out by the BRAZILIAN NUTRITION SOCIETY — BNS — , in the Convention Center of the Hotel Nacional, in the city of Rio de Janeiro, Brazil, from August 27th to September 1st, 1978.

Taking place for the first time in the southern hemisphere, the XI ICN received the collaboration and participation of several international agencies, including the World Health Organization (WHO), the Pan American Health Organization (PAHO), the United Nations Children's Fund (UNICEF), the Food and Agriculture Organization (FAO), the International Fund for Agricultural Development (IFAD), the United Nations Educational, Scientific and Cultural Organization (UNESCO), the World Food Program (WFP) and the World Food Council (WFC).

The meeting had a multi-disciplinary character, with the participation of professionals and students from the different sectors related to the field of food and nutrition, and was characterized by the interest it arose, which was demonstrated by the presence of 5,026 participants from 92 countries, and the presentation of more than 1,200 scientific papers.

During the period prior to the Congress, the different committees, aside from the regular work such as receiving requests for registration and selection of the scientific papers to be presented, also developed an intensive activity of divulgation, both in Brazil and in the other countries, by means of written, verbal and televised press, distribution of circular letters, and the publication of a monthly, bilingual (Portuguese and English) Information Bulletin.

In addition, there were several preparatory meetings, not only in Brazil but also in the exterior, in which members of the Executive Committee participated.

In this Congress, besides the scientific activities other events were considered such as:

- Parallel Meetings promoted by scientific entities related to food and nutrition, trough a formal request to the Congress Executive Committee;

- Intensive Courses about different subjects, especially about Diet Therapy and physiopatology;

- The V International Congress of Paraenteral Nutrition, promoted by the International Society of Paraenteral Nutrition.

We feel the need to clarify that the works were not grouped in accordance to the various events of the Congress, which were organized by the classifications as per argument elaborated with the objective of providing the specialists with correlated matters in each one of the three volumes of the Proceedings.

Since many of the works were not presented in the English language, the Brazilian Society of Nutrition had them translated to the referred idiom, trying, as much as possible, to homogenize the scientific nomenclature of these works.

Lastly we would like to render account for two facts: one, in respect to the exclusion of works related to Paraenteral Nutrition, originated by the fact that the Society never received the respective original works, although these were insistently requested. The other, in reference to the choice made by Plenum Publishing Corporation for the editing and distribution at international level of the Proceedings, given the international tradition of this Publishing House in works of this nature.

Hopefully, the Proceedings will be usefull to all those who are interested in the different aspects of the science of Nutrition.

Editorial Committee:

Walter J. Santos
J.J. Barbosa
Dagoberto Chaves
José Carlos Valente
Nabuco Lopes

CONTENTS

I. NUTRITIONAL BIOCHEMISTRY

III.   NUTRITION, GROWTH AND HUMAN DEVELOPMENT

I. Nutritional Biochemistry

THE METABOLISM OF PROTEIN DURING PREGNANCY

D.J. Naismith

Department of Nutrition

Queen Elizabeth College, London, W8 U.K.

The control of amino acid metabolism is partly exercised by hormones that regulate the activities of rate-limiting enzymes in the tissues. During pregnancy the amounts of the hormones secreted may be profoundly changed, patterns of secretion may differ at different stages of pregnancy, rates of degradation and excretion may by altered, and normal metabolic effects may be modulated or even suppressed as a result of hormone antagonisms. This reorganization of the normal female endocrinology is to a large extent initiated by the emergent endocrine function of the placenta. Hormones produced by the cooperative enterprise of the placenta and the foetal adrenals and liver, participate in or even appropriate the roles of the maternal pituitary, the ovaries, adrenal cortices and the pancreas. Thus the foetus is directly involved in the maintenance of pregnancy, a cardinal feature of which is the assurance of an adequate nutrient supply.

Metabolism in pregnancy is essentially anabolic; energy and nitrogen balances both are positive throughout gestation (Hytten & Leitch, 1971; Calloway, 1974). The need for an assured energy supply is of paramount importance for normal growth and development of the foetus, and adjustments in the utilization of both fat and carbohydrate permit the foetus to re-direct the use of available energy from furthering maternal storage to satisfying its own increasing demands (Freinkel, 1964; Grumbach, Kaplan, Sciarra & Burr, 1968; Strange & Swyer, 1974). Intervention experiments in which the effects of dietary supplements of protein and energy have been compared have established beyond doubt that energy rather than protein is the major determinant of foetal growth (Lechtig, Habicht, Yarbrough, Delgado, Guzman & Klein, 1975). The same conclusion has been reached in studies on the rat (Naismith, Richardson &

3

Ritchie, 1976). Nevertheless the acquisition of amino acids by the growing feotus appears also to be safeguarded by the action of hormones secreted by the foeto-placental unit.

Studies on protein metabolism during pregnancy in women have, of necessity, been confined largely to measurements of N-balance. Far from shedding light on the nature of protein utilization, these studies have served only to confuse the picture with gross over-estimates of nitrogen retention (e.g. Macy & Hunscher, 1934). The results of more recent balance studies have, however, revealed an interesting paradox. Calloway (1974) observed little difference in the rate of N retention between early and late pregnancy, although the rate of N accretion by the conceptus shows an eight-fold rise between the second and last quarters of pregnancy (Hytten & Leitch, 1971). We have also failed to detect a significant trend in N-balance during pregnancy in women (Naismith and Emery, unpublished results). An explanation for this surprising finding may be found in the results of animal studies.

From analyses of the bodies of rats at different stages of gestation, a picture has been built up of the changes in gross body composition that occur during pregnancy (Naismith, 1966; 1969). During the first two weeks of gestation, the 'anabolic phase', when accretion of N by the conceptus is insignificant, food intake is increased, and a substantial store of protein is laid down in the maternal muscles. During the third and final week, the period of rapid foetal growth, the protein reserve is withdrawn despite a further rise in food intake. This 'catabolic phase' occurs irrespective of the protein intake of the mother, indicating that it is under hormonal rather than dietary control. The use of protein stored in early pregnancy to subsidize the rising cost of foetal protein synthesis in late pregnancy was unequivocally demonstrated in phased dietary supplementation experiments (Naismith & Morgan, 1976). Feeding extra protein to marginally-nourished rat dams for 5 days in early pregnancy promoted substantial increases in protein content and cellularity of the tissues of their offspring. The biphasic character of protein metabolism in pregnancy may thus play a major role in moderating the effects of chronic malnutrition on foetal growth and development. Involving as it does a continuous retention of N, and an internal re-distribution of N, it may also account for the failure of N-balance experiments in women to reveal a consistent upward trend in N-retention with time.

A further adaptation to pregnancy in the rat is the more efficient use of dietary protein. When N-balance was measured in pregnant rats pair-fed with littermate virgin controls, N retention was found to rise progressively in the pregnant animals (Naismith & Fears, 1971). The improvement in protein utilization that occurs in pregnancy could result either from the intervention of an anabolic hormone, or from the reduced secretion of a catabolic hormone.

Involvement of the glucocorticoids might be inferred from their known metabolic effects. These hormones are known to induce protein break-down both in the rat (Silber & Porter, 1953) and in man (Sprague, Mason & Power, 1951), and the plasma concentration has been shown to fall markedly in early pregnancy in both species (Martin & Mills, 1958; Rosenthal, Slaunwhite & Sandberg, 1969; Naismith & Fears, unpublished results). The identification of an anabolic agent was, however more difficult. Progesterone, the most abundant steroid hormone produced during pregnancy, seemed the most obvious candidate for this role. In the rat the function of the placenta is primarily luteotrophic, and secretion of progesterone by the corpus luteum persists. In man, however, synthesis by the ovary is overtaken by the placenta by the end of the second month, and production rises steadily until term (Csapo, Pulkkinen, Ruttner, Sauvage & Wiest, 1972). Surprisingly, progesterone has been claimed by Landau and Lugibihl (1961; 1969) to have a catabolic effect in man, although their choice of subjects for study was far from ideal — healthy young men, men with Addison's disease and post-menopausal women. Since there was no relevant information for the rat, Naismith and Fears (1972) studied the effects of administering progesterone to virgin adult female rats. After 10 days of treatment, the activities of two enzymes that regulate the rate of amino acid deamination and of urea synthesis in the liver were measured. Both the enzymes, alanine aminotransferase and argininosuccinate synthetase are known to be responsive to corticosteroids. In addition blood was analysed for urea and for corticosterone. When compared with untreated ani-mals, the rats that had received progesterone showed a fall in the activity of both enzymes in the liver and in the concentration of urea in the plasma. The adrenal glands were lighter in weight, and the level of circulating corticosterone was greatly reduced. Pro-gesterone appeared therefore to exert an anabolic effect by sup-pressing the activity of the adrenal cortex. Similar changes have been noted in the rat at mid-pregnancy (Naismith & Fears, 1971). The rising secretion of progesterone in early pregnancy could there-fore account for the enhanced efficiency of amino acid utilization and the build-up of a maternal protein store.

During the latter half of pregnancy in the rat and in women, secretory activity of the adrenal cortex appears to be high (Burke & Roulet, 1970; Naismith & Fears, unpublished results). It has been suggested that the rise in corticosteroid secretion is induced by the upsurge in oestrogen production by the foeto-placental unit, and is mediated by way of the pituitary-adrenal axis (Kitay, 1963). Yet the depression in activity of alanine aminotransferase and argininosuccinate synthetase was found to persist in the rat until the end of pregnancy. But not all enzymes that are corticosteroid-inducible behave in this anomalous way. Tryptophan pyrrolase, the enzyme which initiates the degradation of tryptophan to the nicotinic acid derivative NAD and to CO_2 has been found to fall in activity in early pregnancy, but then to rise to four times its original value by the end of pregnancy (Naismith and Morgan, 1975).

The inverse relationship found between the activities of argininosuccinate synthetase and alanine aminotransferase on the one hand, and of tryptophan pyrrolase on the other could provide an explanation, at the cellular level, for the biphasic nature of protein utilization during pregnancy.

Tryptophan, in comparison with the other amino acids, is present in the lowest concentration by far in dietary proteins, in tissue proteins and in the amino acid pools (Munro, 1970). The tissue content is carefully regulated, and Munro has suggested a unique role for tryptophan in the control of protein synthesis.

A dynamic equilibrium exists between free tryptophan in the tissue amino acid pools and protein-bound tryptophan. Excess tryptophan entering the pools is normally catabolized to NAD, CO_2 and other substances by a common pathway regulated by the activity of tryptophan pyrrolase. During pregnancy, the picture changes. Tryptophan is withdrawn from the amino acid pools by the placenta, but on a diet providing an excess of tryptophan (protein), this alone would not account for the breakdown of maternal muscle protein during the 'catabolic phase'. However, the marked increase in tryptophan pyrrolase activity that occurs in late pregnancy could deplete the pools of tryptophan, and thus disturb the equilibrium between free and bound tryptophan in favour of tissue catabolism. Amino acids released in this way would escape further degradation because of the depressed activity of the enzymes controlling amino acid catabolism and urea synthesis, and so augment the supply for nourishment of the fetus. There is some evidence to support this hypothesis. Naismith and Morgan (1975) found that the rise in tryptophan pyrrolase activity in the liver of pregnant rats was accompanied by a fall in the concentration of free tryptophan in their muscles. Furthermore, when the normal increase in tryptophan pyrrolase activity was suppressed, through feed-back inhibition, by supplementing the diet with nicotinic acid during the last week of pregnancy a significant reduction in the weight, protein content and cellularity of the offspring occurred.

It would appear that this modification of tryptophan metabolism is not confined to the pregnant rat. An elevated urinary excretion of metabolites of tryptophan and of nicotinic acid has been reported in women during the second and third trimesters of pregnancy (Brown, Thornton and Price, 1961; Hernandez, 1964). It was suggested by Rose and Braidman (1971) the the altered hormone environment of pregnancy was responsible for promoting this increase in tryptophan degradation, by raising the activity of one or more of the enzymes concerned in the tryptophan-nicotinic acid ribonucleotide pathway. On theoretical grounds, they argued that tryptophan pyrrolase would be the most likely enzyme to be involved.

In order to reconcile the changes in activity of these three enzymes that occur in pregnancy with the antagonistic roles of progesterone and corticosterone, as described above, it was necessary to postulate that tryptophan pyrrolase must respond differently to the action of the two hormones, and must be controlled independently.

An experiment was designed to test this theory (Naismith and Morgan, unpublished results). Female rats, selected in littermate triplets, received injections of corn oil (the vehicle used for the hormones), corticosterone alone, or corticosterone with progesterone. This last treatment was intended to simulate the endocrine balance prevailing in late pregnancy. Alanine aminotransferase and tryptophan pyrrolase were estimated in the livers after 10 days. The activity of alanine aminotransferase was increased by 25 per cent in the animals which received corticosterone alone, but was decreased by 14 per cent when progesterone was given simultaneously. Quite a different pattern of behaviour was found with tryptophan pyrrolase. Corticosterone administration caused a three-fold rise in enzyme activity, as was noted in late pregnancy but in this instance progesterone had no moderating influence whatsoever. A similar observation has been made in women taking the estro-progestogen type of oral contraceptive (Rose and Braidman, 1971). The progestogenic component, similar in structure to progesterone, was found not to affect the high urinary excretion of tryptophan metabolites induced by the estrogenic component. This selective antagonism displayed by progesterone towards the induction of hepatic enzymes by the glucocorticoids would permit the release of amino acids from maternal muscle protein (the 'catabolic phase') but would inhibit further degradation in the liver, as would occur in the non-pregnant animal.

The description I have given of the modification of protein metabolism in pregnancy is undoubtedly an over-simplification. The role of other hormones will, in time, be clarified. For example placental lactogen, a hormone that closely resembles growth hormone in chemical structure and in physiological properties, is believed to be the main antagonist to the action of insulin in late pregnancy (Grumbach, Kaplan, Sciarra & Burr, 1968) and so would reinforce the catabolic state by depressing the uptake of amino acids by muscle tissue. A clear picture does, however, emerge of the foeto-placental unit as a pluripotent endocrine organ adjusting the maternal metabolism to ensure its independence of a vulnerable maternal food supply.

BIBLIOGRAPHY

Brown, R.R., Thornton, M.J. & Price, J.M. (1961), J. Clin. Invest. 40, 617.

Burke, C.W. & Roulet, F. (1970). Brit. Med. J. (i), 657.
Calloway, D.H. (1974). In: "Nutrition and Fetal Development", (M. Winick, ed.), Wiley, New York, p. 79.
Csapo, A.I., Pulkkinen, M.O., Ruttner, B.; Sauvage, J.P. & Wiest, W.G. (1972). Am. J. Obstet. Gynec. 112, 1061.
Freinkel, N. (1964). Diabetes. 13, 260.
Grumbach, M.M., Kaplan, S.L., Sciarra, J.J. & Burr, I.M. (1968). Ann. N.Y. Acad. Sci. 148, 501.
Hernandez, T. (1964). Fed. Proc. 23, 136.
Hytten, F.E. & Leitch, I. (1971). "The Physiology of Human Pregnancy", 2nd ed., Blackwell, Oxford.
Kitay, J.I. (1963). Endocrinology, 72, 947.
Landau, R.L. & Lugibihl, K. (1961). Rec. Prog. Horm. Res. 17, 249.
Landau, R.L. & Lugibihl, K. (1967). Metabolism, 16, 114.
Lechtig, A., Habicht, J.P., Yarbrough, C., Delgado, H., Guzman, G. & Klein, R.E. (1975). "Proc. IXth Int. Congr. Nutr.", Mexico. Vol. 2, A. Chavez, ed.; Karger, Basel, p. 44.
Macy, I.G. & Hunscher, H.A. (1934). Am. J. Obstet. Gynec. 27, 878.
Martin, J.D. & Mills, I.H. (1958). Clin. Sci. 17, 137.
Munro, H.N. (1970). "Mammalian Protein Metabolism", Vol. 4, H.N. Munro, ed., Academic Press, New York, p. 299.
Naismith, D.J. (1966). Metabolism, 1, 582.
Naismith, D.J. (1969). Proc. Nutr. Soc. 28, 25.
Naismith, D.J. & Fears, R.B. (1971). Proc. Nutr. Soc. 31, 8A.
Naismith, D.J. & Fears, R.B. (1972). Proc. Nutr. Soc. 31, 79A.
Naismith, D.J. & Morgan, B.L.G. (1975). Proc. Nutr. Soc. 34, 27A.
Naismith, D.J. & Morgan, B.L.G. (1976). Br. J. Nutr. 36, 563.
Naismith, D.J., Richardson, D.P. & Ritchie, C.D. (1976). Proc. Nutr. Soc. 35, 124A.
Rose, D.P. & Braidman, I.P. (1971). Am. J. Clin. Nutr. 24, 673.
Rosenthal, N.E., Slaunwhite, W.R. Jr. & Sandberg, A.A. (1969). J. Clin. Endocr. 29, 352.
Silber, R.H. & Porter, C.C. (1953). Endocrinology. 52, 518.
Sprague, R.G., Mason, H.L. & Power, M.H. (1951). Rec. Prog. Horm. Res. 6, 315.

PROTEIN TURNOVER IN MALNUTRITION,

OBESITY AND INJURY

J.C. Waterlow

Dept. of Human Nutrition - London School of Hygiene

and Tropical Medicine - London, U.K.

During the last ten years there has been a great increase in interest in trying to find out the rate at which protein is synthesized and broken down in the body under different conditions. In animals we can make measurements on individual tissues, but this is very difficult in man, and we have to content ourselves with estimates of the overall rate of turnover in the whole body. Although all the methods for measuring total protein turnover are open to theoretical objections, I think that all the same useful comparative results can be obtained, even though the absolute values may be in error.

There are two main methods of estimating total protein turnover. The first, which I call for short the plasma method, relies on the specific activity of a labelled amino acid in the plasma. Usually ^{14}C-amino acids have been used, e.g. lysine, tyrosine, leucine, but one study has been published with ^{15}N-labelled lysine. The second group of methods, which I call end-product methods, depends on the labelling of urinary urea or ammonia with ^{15}N. The amino acid most commonly used is ^{15}N-glycine, because it is readily available and cheap. This introduces some difficulties because glycine has rather special metabolic pathways.

The plasma and end-product methods rely to a very large extent on different assumptions, and therefore if we are interested in absolute values it is important to see how far they agree. Table 1 summarizes published results obtained by these methods in adult human subjects. ^{14}C seems to give somewhat higher results than ^{15}N, but the agreement is not bad. I think we can conclude that the normal rate of protein turnover in adult man is about 3.5 - 4.5 g/kg/day — i.e. about four times the dietary intake.

TABLE 1

RATES OF PROTEIN SYNTHESIS IN ADULT HUMAN SUBJECTS
MEASURED WITH RADIOACTIVE OR ^{15}N-LABELLED AMINO ACIDS

AUTOR	SUBJECTS No.	SUBJECTS AGE	AMINO ACID	SYNTHESIS RATE, g/kg/day mean ± SD	CV[3]
Steffee et al. (1976)[1]	6	20–25	^{15}N–glycine	3.0 ± 0.3	10
Halliday & McKeran (1975)	5	31–46	^{15}N–lysine[2]	3.54 ± 1.06	30
Crane et al. (1977)	11	20–65	^{15}N–glycine	3.83 ± 0.73	19
Winterer et al. (1976)					
Males	4	65–72	^{15}N–glycine	3.18 ± 0.71	22
Females	5	69–91	^{15}N–glycine	2.25 ± 0.37	16
Golden & Waterlow (1977)	6	66–91	^{15}N–glycine	3.31 ± 0.27	8
Golden & Waterlow (1977)	6	66–91	^{14}C–leucine	2.67 ± 0.71	26
O'Keefe (1974)	4	52–75	^{14}C–leucine	3.52 ± 0.99	28
James et al. (1976)	6	30–65	^{14}C–tyrosine	4.60 ± 0.77	17

[1] Subjects on 1.5 g protein per kg per day.
[2] Measurements on plasma. In all other studies with ^{15}N, measurements were made on urine with urea as end–product.
[3] Coefficient of variation.

A problem in comparing methods or comparing groups studied by the same method is that of the variability between individuals. Young and his group at MIT obtained a coefficient of variation of only 7% in their studies on young adults, but, as the table shows, in our work the variability seems to be twice as great, perhaps because the groups were less homogeneous. There is very little information about the variability of protein turnover in the same subject at different times, because few comparisons have been made under identical conditions. Figure 1 shows results obtained in an obese patient when the measurements were repeated several times on the same diet. These results give some indication of the confidence which can be attached to a single measurement.

I have mentioned these difficulties of measurement and interpretation, because account must be taken of them in considering the results now to be presented.

MALNUTRITION

In children in a normal nutritional state the rate of protein turnover is 6-7 g protein/kg/day, which is about twice that found in adults. When they are malnourished the rate of turnover, per kg body weight, is reduced, and while they are recovering and growing very rapidly it is increased above normal levels. Figure 2 shows the turnover rates in a small group of malnourished, recovering and recovered children in whom the measurements were made at maintenance levels of protein and energy intake.

The effects of food intake on protein synthesis and turnover are not entirely clear. In normally nourished children, at constant energy intake, the level of protein in the diet did not affect the turnover rate. However, we know from results on adults reported elsewhere in this Congress that if a subject is studied continuously over 24 hours the rates of protein synthesis and oxidation were much higher during the day, when food was being supplied, than during the night when there was no food. Studies of this kind have not been done on children, but measurements of oxigen consumption may have some relevance to this problem. Brooke and Ashworth measured the increased oxygen consumption after a meal in malnourished, recovering and recovered children. In malnutrition this post-prandial increase was quite small. In recovering children the size of the increase was linearly related to the rate of growth. This suggested the hypothesis that a meal may provoke a burst of protein synthesis, which requires an increase in oxygen consumption. This would fit in with the results on adults referred to above.

The Jamaica group has measured the rate of total protein synthesis in recovering children during the rapid phase of growth, eating ad lib a diet of constant composition. These children vary in their food intake, rate of weight gain and rate of nitrogen re-

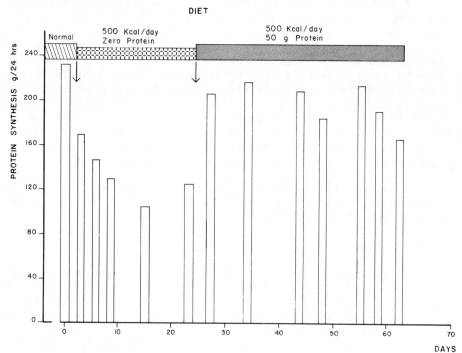

Fig. 1 - Variability of protein turnover in the same subject at different times.

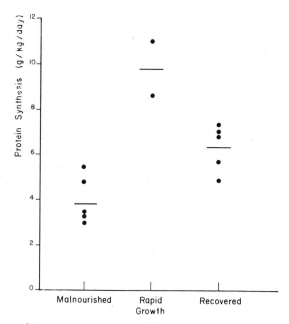

Fig. 2 - Protein turnover rates in small group of malnourished, recovering and recovered children.

tention. In general, the greater the food intake the higher the
rates of growth and of protein synthesis. Figure 3 shows that the
synthesis rate increases with increasing food intake, whereas there
is little change in breakdown rate. In fact, the slope of breakdown versus intake is not significantly different from zero. In
these studies we cannot disentangle the effects of energy intake
and of protein intake. Figure 4 shows the relationship between the
rate of protein synthesis and the rate of protein gain or protein
deposition. At maintenance, when there is no gain, the   rate of
protein synthesis is 4.6 g/kg/day, which is similar to the rate found
in malnourished children. The regression equation shows that above
maintenance, for each gram of protein gained 1.37 g protein are synthesized. This represents a remarkably efficient process. If we
suppose, from the known amounts of ATP and GTP required, that the
energy cost of peptide bond synthesis is 0.85 kcal/g, then the actual
cost of protein deposition in the organism would be 1.37 x 0.85 =
= 1.16 kcal/g. To this, of course, must be added the energy value
of the protein laid down, which may be taken as 5.4 kcal/g. Thus in
conventional terms the energetic efficiency of protein synthesis
would be 5.4/(5.4 + 1.16) = 80%.

        We can draw some general conclusions from these studies on
children. The reduced rate of protein turnover in malnutrition is
accompanied by a reduction in the resting oxygen consumption. Thus
the vital processes seem to be at a low ebb. We do not know whether
this process ever becomes irreversible. Secondly, once the turning
point is past, and rapid growth begins, net deposition of protein
is produced by an increase in the rate of synthesis and not by a
fall in the rate of breakdown. This finding is perhaps surprising,
but it fits in with results obtained from measurements of muscle
protein turnover in the rat. Whenever growth is rapid, the rate of
muscle protein breakdown is increased. This is found in the weanling
rat, as compared with the adult; in refeeding after undernutrition;
and in muscle hypertrophy provoked by nerve section. Situations of
this kind illustrate very clearly how little we know about the mechanisms which regulate rates of synthesis and breakdown.

OBESITY

        It is difficult to compare rates of protein turnover in normal
and obese subjects, because of the problem of choosing a suitable
basis of reference. In our Unit we do not have means of measuring
lean body mass, except for creatinine output. Our impression so
far is that on a normal diet the rate of protein turnover in obese
people is much the same as in the non-obese. When the energy intake
is reduced, the effect on protein turnover depends on whether or
not protein is provided. Table 2 compares the rate of turnover on
a reducing diet (500 kcal/day) with that on a normal diet (2000 kcal/
day). When the low energy diet provided 50 g protein, there was no

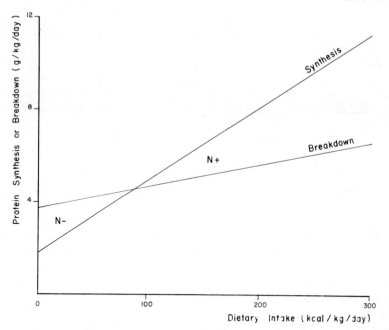

Fig. 3 - Relationship between the rates of protein synthesis
          and breakdown and dietary intake.

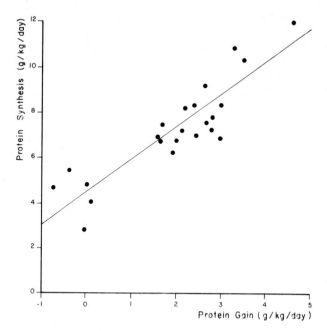

Fig. 4 - Relationship between rate of protein synthesis
          and the rate of protein gain.

TABLE 2

EFFECT OF A LOW ENERGY DIET (2.1MJ) WITH OR WITHOUT
PROTEIN (50g) ON PROTEIN SYNTHESIS IN OBESE PATIENTS

DIET	No. OF SUBJECTS	SYNTHESIS RATE FINAL/INITIAL x100 Mean and range
+ Protein	6	99.5 (71-124)
0 Protein	4	49.5 (40-66)

Measurements with ^{15}N glycine, end-product ammonia.

effect on the rate of protein synthesis. When it contained no
protein the synthesis rate was halved. These results were first
obtained with ^{15}N-glycine, but they have been confirmed by studies
with L-^{14}C-leucine. Therefore we believe that this quite dramatic
difference is not an artefact. Figure 5 shows that the fall in
protein synthesis was accompanied by a reduction in the rate of
amino acid oxidation. These responses to a change in protein and
energy supply are quite rapid, as is evident from Figure 1. As we
report elsewhere in this meeting, when measurements are made continuously over a 24 hour period, rates of protein synthesis and oxidation are much lower during the night, without food, than during the
day when food is provided.

From the point of view of treatment it is clearly beneficial
to provide protein, in order to maintain synthesis as well as to
minimize nitrogen loss. If at a very conservative estimate we take
the energy cost of protein synthesis to be 0.85 kcal/g, an extra
200 g protein synthesized per day (Figure 5) would require 170
kcal/day. This is not a negligible amount.

INJURY

Injury presents a very interesting problem in protein metabolism. How do we explain the so-called 'catabolic' loss of nitrogen? The negative nitrogen balance could result either from an
increase in protein breakdown or from a block in synthesis. The
answer to this question might have quite important implications
for treatment as well as for a better understanding of the mechanism
of injury.

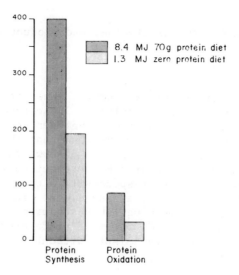

Fig. 5 – Effect of dietary protein and calorie restriction
        on protein synthesis and oxidation in obesity.

TABLE 3

EFFECT OF ABDOMINAL OPERATIONS ON RATES OF
PROTEIN SYNTHESIS AND BREAKDOWN

Continuous infusion with $1-^{14}C$-leucine

4 patients : mean ± SEM

g/kg/day	Before	After
Synthesis	3.52 ±0.49	3.21 ±0.53
Breakdown	3.63 ±0.96	4.04 ±0.68

(From O'Keefe 1974)

TABLE 4

PROTEIN TURNOVER IN 11 PATIENTS BEFORE AND AFTER
ELECTIVE ORTHOPAEDIC OPERATIONS

	Before	After
N balance, g/day	-0.52 ±1.31	-7.51 ±4.5
Synthesis g protein/kg/day	3.83 ±0.73	2.94 ±0.83
Breakdown g protein/kg/day	3.88 ±0.66	3.66 ±0.65

Data of Crane et al., 1977.
Measurements by intermittent dosage of ^{15}N glicine,
  end-product urea.

Our approach has been to select patients who were to undergo
elective surgical operations, and to measure rates of protein synthesis before and after operation. Preliminary studies with
^{14}C-leucine in four patients undergoing abdominal operations showed
a small fall in synthesis, with no evidence of an increase in protein breakdown (Table 3). Similar results were obtained in a series
of patients undergoing elective orthopaedic operations, in whom
measurements were made with ^{15}N-glycine (Table 4).

These findings certainly suggest that the major problem is
not an increase in breakdown but a block in synthesis. However,
none of these subjects was in very large negative N balance. We
need to know what happens in patients who, as a result of burns or
severe trauma, are in negative balance to the extent of 20-30 g N/
day. However, it is hard to study patients of this kind. I think
it is entirely possible that in very severe injury there is both
a block in synthesis and an increase in breakdown. Again we can
draw an analogy with changes in muscle protein turnover in the rat.
When the rat is starved, the first effect is a decrease in the rate
of muscle protein synthesis, but if starvation is continued, after
2-3 days the rate of breakdown rises. A rise in the excretion of
creatine and of methylhistidine in injured subjects is further
evidence of an increase in breakdown rate.

In conclusion, all we can say at present is that studies of
protein turnover in different pathological states are in their
infancy. Since the rate of growth of studies in this field seems
to be very rapid, there will probably be a great deal more to
report at the next Congress.

METABOLISM OF LONG CHAIN FATTY ACIDS

WITH THE EMPHASIS ON ERUCIC ACID

Rolf Blomstrand

Department of Clinical Chemistry, Huddinge Univ. Hospital

Karolinska Institutet, Stockholm, SWEDEN

A joint FAO/WHO expert consultation on the Role of Dietary Fats and Oils in Human Nutrition was held in Rome in September 1977. A major topic discussed was the health implications of the brassica derived oils and the partially hydrogenated marine oils[1].

Guidelines regarding the use of such oils and fats for human consumption were given and the background to these recommendations will be discussed.

Oils from the seed of different Brassica varieties is produced and consumed in many countries. Before 1970 all commercially used rape seed oils contained 30-50% of erucic acid (cis - ∇ 13 - doco- senoic acid). Through plant breeding of rapeseed the content of erucic acid and eicosenoic acid has been lowered to close to zero in different varieties of the seed.

Since the 1940s reports in the literature have revealed that feeding of large amounts of rapeseed oil high in erucic acid to laboratory animals and swine causes growth retardation and changes in the heart, adrenals and liver. The above observations did not bring about any actions as far as consumption in humans was con- cerned until the second half of 1970. By this time there was evi- dence not only that feeding of rapeseed oil high in erucic acid to rats and several other animals was consistently followed by lipi- dosis and fibrosis of the myocardium, but also that it interfered with mitochondrial function in the rat.

CHEMICAL CHARACTERISTICS OF <u>BRASSICA</u> DERIVED OILS

The most characteristic feature of oils from Brassica spp. is
the presence of erucic acid, cis - Δ 13 - docosenoic acid, which
amounts to 20 to 55% of the total fatty acids. There are differences
between <u>B</u>, <u>napus</u> and <u>B</u>. <u>campestris</u> as well as between different cul-
tivars of these species. In addition to erucic acid, the oils con-
tain about 10% of <u>cis</u> - Δ 11 - eicosenoic acid. These two long
chain fatty acids are not present in other vegetable oils used for
human consumption. In addition, the oils contain linolenic, linoleic and oleic acid.

Oils with a high level of erucic acid contain a low proportion
of oleic acid; when the erucic acid decreases the oleic acid increases, while the proportions of the other fatty acids vary only
little.

The saturated fatty acids of the oils are palmitic and stearic
acids; which together are present at a level of about five per cent.
This proportion is unusually low in comparison with other vegetable
oils.

By plant breeding, it has been possible to reduce the level
of erucic acid to a low level, and a substantial part of the rapeseed
crop is now made up of these improved varieties. The fatty acid
composition of these oils, in rounded off values, are as follows:

Palmitic acid	3%
Stearic acid	2%
Oleic acid	50-65%
Linoleic acid	20%
Linolenic acid	10%
Eicosenoic acid	0-10%
Erucic acid	0-5%

Apart from the change in fatty acid composition, no changes in
the chemical composition of the oil have been introduced, as far as
is known.

BIOLOGICAL EFFECTS OF ERUCIC ACID

In essence, these studies have proven that <u>oils high in</u> erucic
acid cause a retardation of growth in experimental animals in comparison to other fats and oils. Further, heart and red skeletal
muscle show intrafibrillar fatty infiltration that in the heart can
be made visible within a few hours after administration of rapeseed
oil to young rats. On prolonged feeding the amount of fat in the
heart decreases, without disappearing completely, and other changes
are seen.

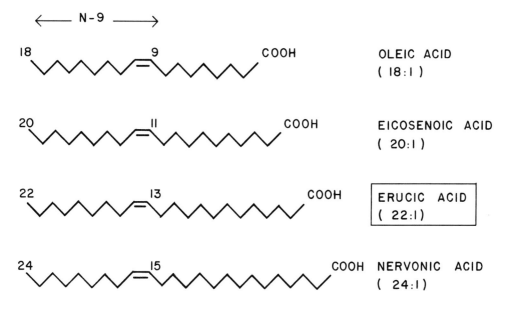

Fig. 1 - Fatty acids of the oleic acid (n-9) family.

Myocardial changes are characterized by cellular infiltration, death of contractile tissue and replacement of this by scar tissue.

Concomitant to the fat infiltration a decrease of the glycogen stores in the heart has been found, indicative of a shift in metabolism upon feeding conventional rapessed oil. Other chemical findings are changes in the proportions of different phospholipids and of the fatty acid composition of mitochondrial membranes.

The fat in the heart and the red skeletal muscle is mainly present in the form of triglycerides, in which erucic acid is a substantial part of the fatty acids.

Also, the adrenals accumulate fat in the cortex when the rat is fed conventional rapeseed oil, the fat being present mainly as cholesteryl erucate.

Extensive studies of the series of events in the heart have been performed in the rat, and agreement between different laboratories on the effects caused by oils high in erucic acid is good. The male rat seems more seriously affected than the female.

The fat infiltration of the heart described above has been induced in the rat by feeding fats that contain synthetic triglycerides of erucic acid. Prolonged feeding of synthetic triglycerides containing erucic acid has also been proven to cause death of contractile tissue and its replacement by scar tissue[2, 3, 4].

METABOLISM OF ERUCIC ACID

Digestion and absorption

The digestibility of traditional rapeseed oil has been studied in several species. From the early work of Deuel et al.[5] it was apparent that the traditional rapeseed oil had an unusual slow rate of absorption in the rat. It was concluded that the long chain fatty acids of the rapeseed oil were responsible for its digestibility. Carrol[6] noted that the absorption of fatty acids decreased with increasing chain length and was 85% for oleic acid, 55% for erucic acid and 25% for nervonic acid.

As early as 1918 Holmes[7] fed 82 g of rapeseed oil to each of four humans and obtained an average coefficient of digestion of 98.8%. Deuel et al.[8] have later confirmed these high absorptions in human beings.

In our laboratory we[27] have shown that dietary erucic acid is absorbed and transported unchanged mainly in the chylomicrons tri-

glycerides of the human thoracic duct. A minor part is transported
in the very low density lipoproteins. The incorporation of erucic
acid in different lipid classes of chylomicrons and VLDL is somewhat
different from that of oleic and palmitic acid.

Mitochondrial oxidation

Dietary rapeseed oil with high content of erucic acid results
in fat accumulation in the hearts of rats, ducks, guinea pigs, rabbits and hamsters. Because the deviations in fat metabolism are
primarily observed in heart and skeletal muscle, in which a major
function of the fatty acids is to supply energy to the muscle cells,
the possibility has been discussed that erucic acid is less efficiently oxidized and therefore accumulates in the triglycerides.

As suggested by Swartouw[9] the inhibitory effect of erucic acid
might be exerted on any of the following steps in the combustion
of fatty acids by mitochondria: fatty acid activation, acyl-CoA
transfer across the mitochondrial inner membrane, beta-oxidation
of the activated fatty acid to acetyl-CoA and, finally, the complete
oxidation of active acetate to CO_2 and water.

Studies on the ability of isolated mitochondria from rat heart
to oxidize substrate and to syntesize ATP led to considerable controversy. A depressed production of energy as first reported by Houtsmuller[10] was not subsequently confirmed[11, 12, 13, 14].

The translocation of erucic acid through the mitochondrial membrane involves its carnitine derivative but Gumpen and Norum[15] found
no difference in the relative amounts of free carnitine to acyl-
carnitine in the rat heart. Christophersen and Bremer[16, 17] found
that oxygen uptake was lower in the presence of erucylcarnitine than
with palmitylcarnitine. They suggested that erucyl-CoA might have an
inhibitory effect on the beta-oxidation of other fatty acids, which
could explain the accumulation of triglycerides in the heart of rats
fed rapeseed oil. Recently Heijkenskjöld and Ernster presented
results[18] which indicated that erucic acid might interfere with mitochondrial beta-oxidation at the site of acyl-CoA-dehydrogenase.
Inhibition in the oxidation of palmitylcarnitine, in the presence
of erucylcarnitine (without ATP) was accompanied by a reduction in
flavoprotein. The oxidation of palmitylcarnitine was reduced by
more than 50%, even with levels of erucic acid as low as 1.4% in
the diet. Korsrud et al.[19] have studied the activity of acyl-CoA-
dehydrogenase with substrates of varying chain lengths. They found
a progressive decrease from oleyl-CoA to eicosenoyl-CoA to erucyl-
CoA. The activity of this first step of beta-oxidation was therefore lowered by each additional 2-carbon on the substrate.

Christiansen, Christophersen and Bremer[20, 21, 22] presented results that indicate that erucoylcarnitine competitively inhibits

the oxidation of other acylcarnitines in the heart mitochondria.
They have also compared the inhibitory effect of erucoylcarnitine
with the effects of carnitine esters of long-chain monounsaturated
fatty acids occurring in hydrogenated marine oils and received similar results. They did not find any significant difference in the
ability of the beta-oxidation system of the heart and liver mitochondria to handle very long chain monosaturated acylcarnitines.

INCORPORATION OF ERUCIC ACID INTO MEMBRANE PHOSPHOLIPIDS

Considerable study has been given to the role of unsaturated
fatty acids in mitochondrial function[23]. There is evidence that
changes in the fatty acid composition are fundamental to metabolic
and physical differences in mitochondria. Recently a close correlation between the alteration of cardiolipin and mitochondrial
ATP-activity has been reported, indicating the existence of specific
association between this enzyme and cardiolipin, independently of
other phospholipids[24]. With a high dietary level of erucate the
cardial triglycerides contained more than 50% erucic acid and the
phospholipids approximately 7%[25]. Maranesi et al.[26] found 2.5% erucic
acid in the phospholipid fatty acids of rats fed high erucic acid
rapeseed oil.

Blomstrand and Svensson[27,28] studied the influence of rapeseed
oil and trierucate on the distribution of different phospholipids
in the rat heart mitochondria. In the experiment with 9.8% erucic
acid in the diet, there was a tendency to an increased concentration
of phosphatidylcholine and a decreased concentration of phosphatidylethanolamine. The concentration of cardiolipin was mainly unchanged
in all of these experiments.

In our studies in rats we found that the fatty acids of phosphatidylcholine, phosphatidylethanolamine and cardiolipin were all influenced by the diet. Erucic acid seems to have a specific affinity
to be incorporated in the cardiolipin of the rat heart. The isolated
cardiolipin from rat heart mitochondria was found to contain 12%
erucic acid. Of great interest is that there is a decrease in the
linoleic acid (18:2) content from 82-47%.

The incorporation of erucic acid into phosphatidylethanolamine
and phosphatidylcholine causes a corresponding decrease in the arachidonic acid (22:4) content.

Beare-Rogers et al.[29] found that in experiments with rapeseed
oil high in erucic acid, the intact rat heart incorporated only low
levels of erucic acid into the major phospholipids, but the heart
did increase its concentration of sphingomyelin and the content of
erucic acid in that fraction.

An alteration in the composition of cardiac phospholipids in rats supplied with erucic acid has been confirmed by several laboratories.

New born rat heart cells in culture became appreciably modified in membrane constituents when subjected to erucic acid. In such heart cells the fatty acids of phosphatidylcholine contained 9% erucic acid.

It is also possible that mitochondrial membrane fatty acid transitions affect ATP-utilization via alterations in membrane associated ATP translocase transporting ATP from the inner membrane matrix compartment directly to the active sites of creatine phosphokinase thereby affecting the complex control of oxidative phosphorylation.

The exact relationship between mitochondrial functional changes and transitions in membrane fatty acid composition induced by erucic acid and other long-chain fatty acids still remains unknown.

It is tempting to suggest that the long-chain fatty acids might interfere with the prostaglandin synthesis in the myocardium. Numerous critical investigations are still necessary for the characterization of the underlying biochemical regulatory mechanisms which are influenced by long-chain fatty acids in the diet.

CONCLUSIONS

Reports from several laboratories indicate that a short-term dietary intake of brassica oils with a high percentage of erucic acid (22:1, n-9) causes transient diffuse myocardial lipidosis (fat deposition of the myocardium) in several animal species. The accumulation of triglycerides in the heart of the rat reaches a peak after about one week and falls to almost normal levels by four weeks, despite continued feeding of the dietary fat. The accumulation of triglycerides in the heart is directly proportional to the amount of erucic acid (22:1, n-9) in the diet.

The possibility has been discussed that erucic acid (22:1, n-9) is less efficiently oxidized and therefore accumulates in the form of triglycerides in the heart cells. Experimental data indicate that erucoyl carnitine competitively inhibits the oxidation of other acyl carnitines in the heart mitochondria. The fatty acid compositions of the main membrane phospholipids of rat heart mitochondria, i.e. phosphatidylcholine, phosphatidylethanolamine, sphingomyelin and cardiolipin, are influenced by dietary erucic acid (22:1, n-9).

Long-term feeding of high erucic acid brassica oils induces
focal necrotic lesions with reactive cellular infiltration leading
to fibrotic changes in the rat heart muscle. It has been reported
that in the rat, erucic acid (22:1, n-9) per se is a cardiopatho-
genic agent and can induce cardiac necrosis and fibrosis.

Studies of low erucic acid brassica oils show that they permit
growth in the rat to the same extent as do other fats and oils.

REFERENCES

[1] FAO. "Dietary Fats and Oils in Human Nutrition". FAO and Nutri-
    tion Paper 3, Rome, 1977.
[2] Aaes-Jørgensen, E. Nutritional value of rapeseed oil, in: "Rape-
    seed", Appeqvist-Ohlson, ed., Elseirer Publ. Co., 1972, p. 301.
[3] Vles, R.O. Nutritional aspects of rapeseed oil, in: "The Role of
    Fats in Human Nutrition", Vergroesen, A.J., ed., Academic Press,
    1975, p. 433.
[4] Beare-Rogers, J.L. Docosenoic acids in dietary fats, in: "Progress
    in the Chemistry of Lipids and other Substances", Pergamon
    Press, 1976.
[5] Deuel, H.J., A.L.S. Cheng, and M.G. Morehouse, J. Nutr. 35: 295,
    1948.
[6] Carroll, K.K. J. Biol. Chem. 200: 287, 1953.
[7] Holmes, A.D., U.S. Dept. Agr. Bull. No. 687, 1918.
[8] Deuel, H.J., Johnson, R.M., Calbert, C.E., Garner, J. and Thomas,
    B. J. Nutr. 38: 369, 1949.
[9] Swarttouw, M.A. Biochim. Biophys. Acta 337: 13, 1974.
[10] Houtsmuller, U.M.T., Struijk, C.B. and van der Beek, A. Biochim.
    Biophys. Acta 218: 564, 1970.
[11] Beare-Rogers, J.L. and Gordon, E.R. Lipids 11: 287, 1976.
[12] Cheng, C. and Pande, V.S. Lipids 10: 335, 1975.
[13] Dow-Walsh, D.S., Mahadevan, S., Kramer, J.K. and Sauer, F.D.
    Biochim. Biophys. Acta 396: 125, 1975.
[14] Kramer, J.G.K., Mahadevan, S., Hunt, J.R., Sauer, F.D., Corner,
    A.H. and Charlton, K.M. J. Nutr. 103: 1696, 1973.
[15] Gumpen, S.A. and Norum, K.R. Biochim. Biophys. Acta 316: 48, 1973.
[16] Christophersen, B.O. and Bremer, J. FEBS Lett. 23: 230, 1972.
[17] Christophersen, B.O. and Bremer, J. Biochim. Biophys. Acta
    280: 506, 1972.
[18] Heijkenskjöld, L. and Ernster, L. Acta Med. Scand. Suppl. 585:75,
    1975.
[19] Korsrud, G.O., Conacher, H.B.S., Jarvis, G.A. and Beare-Rogers,
    J.L. (unpublished data).
[20] Christiansen, R.Z., Christophersen, B.O. and Bremer, J. Biochim.
    Biophys. Acta 280: 506, 1972.
[21] Christiansen, R.Z., Christophersen, B.O. and Bremer, J. Biochim.
    Biophys. Acta 388: 402, 1975.
[22] Christiansen, R.Z., Christophersen, B.O. and Bremer, J. Biochim.

      Biophys. Acta 487: 28, 1977.
[23]Haslam, J.M. Biochem. J. 123: 6, 1971.
[24]Santiago, E., Lopez-Moratalla, N. and Segovia, J.L. Biochem.
      Biophys. Res. Comm. 53: 439, 1973.
[25]Beare-Rogers, J.L., Nera, E.A. and Craig, B.M. Lipids 7: 548,
      1972.
[26]Marenesi, M., Biagi, P.L. and Turchetto, E. Boll. Soc. Ital.
      Sper. 48: 1203, 1973.
[27]Blomstrand, R. and Svensson, L. Acta Med. Scand. Suppl. 585: 51,
      1975.
[28]Blomstrand, R. and Svensson, L. Lipids 9: 771, 1974.
[29]Beare-Rogers, J.L. "Modification of Lipid Metabolism", Academic
      Press, New York, 1975.
[30]Clatinin, M.T. J. Nutr. 108: 273, 1978.
[31]Saks, V.A. et al. Arch. Biochem. Biophys. 173: 34, 1976.

LIPID COMPONENTS IN NUTRITION

M.F. Nesterin

Nutrition Institute

Moscow, USSR

A modern tendency in nutrition in most developed countries is characterized by an increase of fat in the diet. Consequences of such a situation appear to be associated not with energy imbalance but with the specific action of fat upon the organism.

Tests on rats were performed to study bodily responses to differences in the quantity and qualitative peculiarities of diets. The general characteristic of these diets is presented in Table 1.

Animals maintained on a diet of 50% fat became obese. A clear relationship of the degree of obesity to fat qualities tested was observed. As Fig. 1 shows the most significant manifestations of obesity were recorded when butter was the source of fat. Intensive fat accumulation was proved by the analysis of body chemical composition. In rat weanlings fed a high-butter diet fro 3 months, later transferred to a control diet ad libitum (30% caloric value by lard and sunflower seed oil) higher body weights were observed (Fig. 2), leading to the conclusion that the quantity and qualitative peculiarities of alimentary fat have a role in occurrence of experimental obesity. Study of biological oxidation in liver mitochondria in rats fed diets with qualitatively-different fats revealed certain peculiarities. The rate of oxygen consumption added with endogenic substrates and succinate was lower in animals fed partially-hydrogenated sunflower seed oil or mustard oil; as Table 2 shows, the coupling of respiration to phosphorylation was changed only with mustard oil in diets (the respiratory reduction in the presence of uncouples DNP).

Our joint studies with N.V. Lachneva showed that qualitative peculiarities of fat produced no effect upon the rate of substrate

TABLE 1

CHARACTERISTICS OF RAT DIETS

SOURCE OF FAT IN DIETS	DIET CONTENT (Caloric Value %)			LINOLIC ACID CONTENT (CALORIC VALUE %)	Polyunsaturated fatty acids / Saturated fatty acid ratio
	Protein	Fat	Carbo-hydrates		
Lard and sunflower seed oil	20	30	50	4.25	0.75
Lard and sunflower seed oil	20	50	30	5.60	0.75
Butter	20	30	50	0.90	0.12
Butter	20	50	30	1.40	0.12
Sunflower seed oil	20	30	50	16.00	10.4
Sunflower seed oil	20	50	30	27.00	10.4
Mustard oil	20	30	50	4.09	5.2
Lineseed oil	20	30	50	2.98	6.4
Partially-hydrogen-ized sunflower seed oil	20	30	50	2.27	0.60
Rapeseed oil	20	50	30	4.31	5.58

All diets contained proper levels of vitamins and minerals.

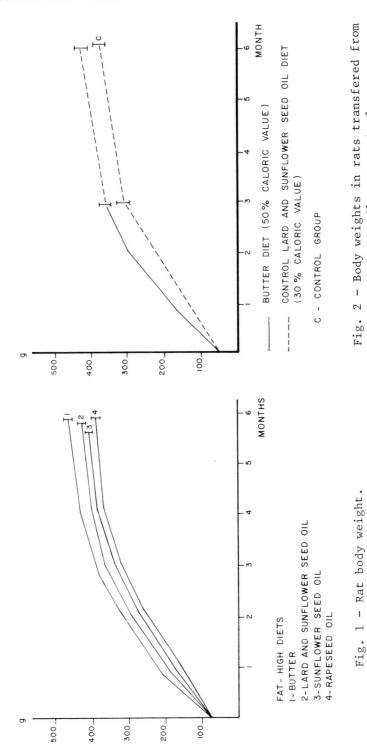

BUTTER DIET (50 % CALORIC VALUE)

CONTROL LARD AND SUNFLOWER SEED OIL DIET
(30 % CALORIC VALUE)

C - CONTROL GROUP

Fig. 2 - Body weights in rats transfered from
         a butter diet to control one.

FAT- HIGH DIETS
1- BUTTER
2- LARD AND SUNFLOWER SEED OIL
3- SUNFLOWER SEED OIL
4- RAPESEED OIL

Fig. 1 - Rat body weight.

TABLE 2

RATE OF OXYGEN CONSUMPTION BY LIVER MITOCHONDRIA
IN RATS FED VARIOUS FATTY PRODUCTS
(M O_2/1mg of protein/min = $25^{\circ}C$ = 6, m±)

FATTY PRODUCTS	MITOCHONDRIA	ADDITION TO INCUBATION MEDIUM	
		Succinate	Dinitro-phenol
Lard + sunflower seed oil	41,2 ± 2,3	83,0 ± 8,9	225 ± 8,2
Partially hydro-genated sun-flower seed oil	22,8 ± 5,0*	61,2 ± 5,4	196 ± 10,0
Mustard oil	17,4 ± 1,0**	48,7 ± 2,3	150 ± 11,0

* - p 0,05
** - p 0,01

hydroxylation by liver microsomes, amidopyrine dimethylation in particular. An activation of the microsomal system of foreign substance
detoxification was observed with a rise of total fat in diets. The
metabolic response of the body to the action of hormonal factors also
depends on fat levels in diets. In daily administration of 5 mg of
DES, lipid content in the serum and liver differed in animals kept on
a normal and high-fat diet (Fig. 3).

Qualitative peculiarities of fat in diets produce an influence
upon cholesterol content in blood, liver and bile (Fig. 4). It suggests a determining role of the alimentary factor in the regulation
of metabolic processes in the body. Joint investigations with L.G.
Panomareva showed that both fatty products of lower biological value
(rapeseed oil) and high-quality fats (butter) in imbalanced proportion favoured an increase of animal sensivity to stress (Fig. 5).

New data on the significance of edible fat qualities in lipid
plastic function on the level of membranes in erythrocytes, throm-
bocytes, mitochondria and endoplasmic reticulum were obtained. The
fitness of alimentary fats to requirements of the organism in an essential spectrum of fatty acids for synthesis of membrane lipids
might be estimated by means of metabolic efficiency coefficient
(MEC). Lower MEC values characterize limited biological efficiency

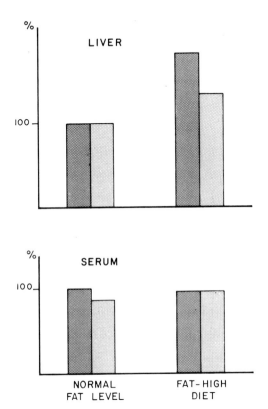

Fig. 3 - Total lipid content.

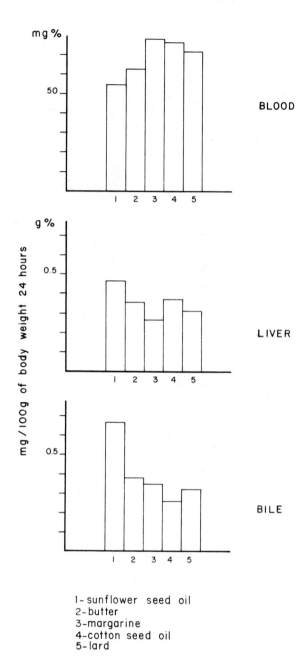

I- sunflower seed oil
2- butter
3- margarine
4- cotton seed oil
5- lard

Fig. 4 - Blood, liver and bile cholesterol content in rats
         fed balanced qualitatively - different fats -
         included diets.

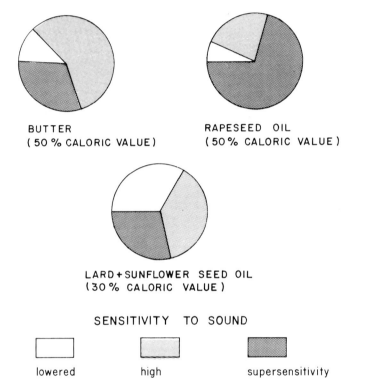

Fig. 5 – Sensitivity of rats fed various fatty products
          to sound irritant.

TABLE 3

MEC VALUES OF VARIOUS MEMBRANE LIPIDS IN
RATS FED QUALITATIVELY-DIFFERENT FATS

M E C	FATTY PRODUCTS		
	Lard and sunflower seed oil	Rapeseed oil	Partially hydrogenated sunflower seed oil
Mitochondria	3.90	1.54	1.12
Endoplasmic reticulum	3.72	1.43	1.02
Erythrocyte membranes	2.60	1.32	0.98
Thrombocyte membranes	2.82	1.48	1.06

of products (Table 3). Cited data indicate a complex dependence of
metabolic processes both on quantity and qualitative peculiarities
of fats in diets. It appears reasonable to search for effective
criteria for assessing the   influence of the fatty component in
diets upon the body.

LIPID REQUIREMENTS DURING PREGNANCY AND LACTATION

M.A. Crawford, M.D.

Dpt. Bioch, Nuffield Lab. of Comparative Medicine

Regents Park - London N.W.1, U.K.

INTRODUCTION

Nutrition concepts have been dominated by considerations of
protein and body growth. Lessons from comparative studies tell us
that the difference between man and other animals is not concerned
with body growth but with the specialisation of the brain. For
example, the rhinocerus at the age of four years has a body weight
of about one ton yet its brain only weighs a few hundred grams. By
contrast the body size of the four year old human child is small
yet its brain weighs about 1.5 kg and has virtually completed all
its growth. Hence it seems to us that different biological principles may be involved in body growth on the one hand and brain
growth on the other. In this context it is interesting that whilst
protein is quantitatively the most important component of muscle
and minerals of bone, lipids are the more important in the brain
and nervous system.

There are two different types of lipid in the body: (i) storage,
(ii) structural. The storage fats are mainly triglycerides whilst
the structural lipids are mainly composed of phospholipids and some
cholesterol. Whilst the fatty acid composition of the storage fats
changes quickly in response to changes in dietary fats, the structural lipids are much more resistant to change and show a species
& tissue specific pattern. Comparison of different animal species
have shown that although the liver and adipose tissue lipids were
variable in composition and may reflect food selection practices,
the composition of the brain, by contrast, was remarkably similar
regardless of species or wide differences in natural food selection
patterns[1]. In particular, the essential fatty acids in the brain
phosphoglycerides are mainly as the long chain derivatives of the

parent acids. For example, brain ethanolamine phosphoglycerides may contain about 22-26% of their fatty acids as docosahexaenoic acid (22:6, n-3) derived from α-linolenic acid (18-3, n-3); the long chain derivatives of linoleic acid accounts for about the same proportion. At the same time linoleic & α-linolenic acids together would only amount to 1% of the total. This pattern of brain fatty acids was found to be common to all 45 species so far studied[1].

Disturbance of this normal fatty acid pattern of the brain by the use of extreme dietary deprivation of essential fatty acids, during brain growth, has been shown in a number of laboratories to lead to defects[2, 3, 4, 5].

We have been interested in understanding how the brain specific fatty acid pattern is accummulated. Using radioactive isotopes, we and Pascaud in Paris (paper 122 in this symposium) have shown that the initial reports that linoleic acid did not cross the placenta were incorrect. In fact there is a concentration gradient across the placenta which reduces the linoleic acid and increases the arachidonic acid on the foetal side[6]. That is, linoleic acid is either metabolised by the placenta or arachidonic acid is preferentially transferred. Indeed the concentration of linoleic acid diminishes and that of its long chain polyunsaturated derivatives increase in the phosphoglycerides from the maternal liver, to placenta, foetal liver and finally the foetal brain[6]. The same events take place with regard to α-linolenic acid. We have termed this sequence a process of biomagnification and it explains the high content of the long chain polyunsaturated derivatives of linoleic and α-linolenic acids in the brain.

An important point emerging from this evidence is that linoleic acid is obviously not converted directly to arachidonic acid: the conversion rate appears to be slow. Hassam [7, 8] has shown in the developing rat pup in vivo, that only about one molecule in thirty of linoleic acid was converted to arachidonic acid for the developing brain over the first 24 hours. In addition, it appears that in vivo the rate of conversion is limited by the first desaturation reaction which is in agreement with in vitro studies by Dr. Brenner in the Argentina and Dr. Sprecher in the U.S.A.

This slow rate of conversion is important because much of the physiological activity of the essential fatty acids, linoleic and linolenic acids, can be attributed to their conversion to their long chain derivatives which are used in cellular structures and for the synthesis of prostaglandins[9]. It is not surprising that the long chain polyunsaturated fatty acids are more potent than linoleic acid for example, in curing EFA deficiency and have a more pronounced effect on platelet aggregation. The higher activity[10] of the long chain derivatives is important when one comes to consider the sig-

nificance of what may appear to be small amounts in foods such as
human milk.

The main bulk of brain development in the human takes place
during foetal growth and the first two years of life[11]. About 70%
of the brain cells divide before birth. We have found that the most
active period for incorporation of the long chain derivatives of the
essential fatty acids is during fetal growth and during cell division.
Post-natally, the emphasis changes quickly in the human to myelination (i.e. towards making connections between the cells). Whilst
the fatty acids in the cellular lipids are highly unsaturated those
of myelin contain more of the long chain saturated and unsaturated
types such as lignoceric (C24:0) and neruonic (24:1) monoacids.

POST NATAL CONSIDERATIONS

The infant obtains its nourishment from human milk which has a
highly complex mixture of fatty acids in its glycerides. Milk lipid
accounts for 50-60% of the dietary energy and triglyceride is its
major component. Phospholipids are also present at about 0-29/100g
but they have a different fatty acid composition to the triglycerides.
In the triglycerides the total essential fatty acid component amounts
to about 5/6% of the total dietary energy of the milk; this total
includes the long chain EFA derivatives, the sum of which amounts
to about 1% of the energy.

In fact, in human milk the total EFA content is approximately
equivalent to the protein which occupies about 6% of the dietary
energy. As only part of the protein is present as essential amino
acids this means the total EFA content of human milk is greater than
the total essential amino acids. These quantitative relationships are
not generally appreciated because the EFA content is thought of in
the context of the total fat in comparison with which it appears relatively small. However, it has to be remembered that milk fat is
acting as the major energy source of milk as well as providing essential fatty acids.

Cow's milk by contrast provides about three times the amount
of protein; its mineral content is much higher and EFA content lower
than human milk. Assumedly the high protein and mineral content of
cow's milk is appropriate to the rapid rate of body growth of the
calf. Indeed on a comparative basis human milk has a relatively low
protein concentration. The lowest concentration of protein in milks
that we have studied have been all in the primates.

There has been some debate about the consistency of composition
of human milk lipid. However, our own studies in 7 different countries reported earlier at the Conference suggest that milk fatty
acid composition does appear to change with time. In a study of
10 mothers over 9 months of lactation we found that the phospho-

glycerides were initially rich in the long chain polyunsaturated
essential fatty acids; these, however, diminished in proportion,
but the long chain saturated and monounsaturated fatty acids tended
to increase with time[12]. That is, the composition of the milk seemed
to change in a manner consistent with the postnatal change in emphasis from cell division to myelination.

ENERGY AND EFA REQUIREMENTS IN PREGNANCY

It is estimated that the growing brain of the neonate utilises
some 70% of the dietary energy. Consequently, energetic considerations will be especially important. Discussion of the energetic
aspects of early development again focuses attention on lipids and
also enables us to obtain some idea of the requirements for essential
fatty acids.

Measurements by Hytten and Leicht[13] showed that in well nourished mothers the normal gain in protein including the foetus,
expansion of maternal blood volume, uterine growth and development
of the mammary gland was about 900g; the foetal gain included in
this figure was about 400g of protein. Total maternal fat deposition was however, about 4.5 kg. If one averages the gain in protein
and fat over the whole pregnancy in terms of dietary energy they
amount to a daily gain of 20 kcal (0.08 MJ) of protein and 120 kcal
(0.5 MJ) of lipid.

However, if we break these data down into their components and
express them as the velocities or rates of gain with time throughout
the pregnancy we find that the foetus does not really grow rapidly
until the last half of pregnancy. Foetal growth is preceeded by
the expansion in blood volumes, placental and uterine growth and
deposition of fat. These 'support' systems develop early in pregnancy, well ahead of the main phase of foetal growth. Assumedly
the lipid stores provide for the energetic (and EFA) requirements
when the foetal growth rate is maximal near the end of pregnancy
and when maternal food intake may diminish.

In addition, the maternal lipid deposition also provides for
about one third of the energy cost of milk production for the first
100 days of lactation[14].

On the basis of this information we can obtain an idea of the
essential fatty acid requirement for the mother if we assume that
the composition of the fat deposited in pregnancy should be similar
to that which will be produced in the milk; such data as is available from maternal adipose tissue biopsy studies in pregnancy would
support this assumption[15]. The total gain in EFA should then be
between 600 to 650 g which in terms of energy represents an average
increase of 1% of the non-pregnant woman's dietary energy. However,

this calculation concerns the parent essential fatty acids, linoleic
and linolenic acids; one also has to take into account the presence
of the long chain polyunsaturated EFA derivatives, the slow rate of
conversion of linoleic and linolenic and the small amount metabolized
into their long chain derivatives. Taking these factors into account
it is likely that additional 0.5% of the dietary energy would be required.

CONSIDERATIONS IN LACTATION

During lactation when maternal lipid stores have been mainly
used, the demand for dietary energy is about 745 to 875 Kcal/day (or
3.1-3.7 MJ per day)[14], which is much higher than during pregnancy.

Similar calculations based on the mean fatty acid composition
of human milk (from 261 mothers from five different countries)[10]
indicate that to replace the essential fatty acids secreted in the
milk would require an additional 2-4% of the dietary energy as EFA
over and above the basal EFA requirement. It was felt best to express these figures as a range because of the uncertainty in calculating the equivalent of the long chain EFA derivatives in terms
of linoleic and linolenic acids[10]. Whilst there is evidence that
the same general principles regarding the low efficiency of conversion and limitation by the delta 6 desaturase apply to man there
is obviously uncertainty about the quantitative relationships.

If we take the FAO/WHO recommendation of a baseline EFA requirement of 3% of the dietary energy[10] then a total 4.5% and 5-7% of
the dietary energy would be used in pregnancy and mature lactation[10].

CONCLUSION

It is of special interest that the problem of undernutrition
or 'malnutrition' is increasingly being recognized as a function of
dietary energy; it has been estimated that in India some 70% of the
malnutrition is due to an energy deficit (not to specific protein
deficiency) and the remainder due to a total shortage of food[10].
Much of the problem seems to be in the fact that many common carbohydrate or protein rich foods contain much water and bulk and it
can be difficult with these to meet the energy demands of the young
growing child and the pregnant or lactating mother. Dietary fats,
on the other hand have the highest energy density of all food types.
100 g of boiled rice needs only 10 to 13 g of fat added to it to
double its energy value. This may well mean that if one is interested in tackling the 'malnutrition problem that dietary fats could
have an important part to play quantitatively to meet energy requirements and qualitatively to meet essential fatty acid requirements.

The evidence we have presented indicates that problem of essential fatty acid requirements is more complex than previously assumed. One not only has to deal with two different parent fatty acids with essential fatty acid functions, but also their long chain derivatives. It is the long chain derivatives which appear to be used mainly in cell structures and are the precursors for prostaglandin synthesis. During pregnancy the 'biomagnification' process for the essential fatty acids in favour of the foetus, and the early, substantial store of lipid, offer powerful mechanisms for protecting foetal development and ensuring adequate initiation of lactation. This, in turn, is a powerful protective mechanism for the early period of brain growth. However, if these protective mechanisms fail for any reason then experience shows that the consequences in terms of brain retardation or handicap are likely to be permanent and life long. Because this period of early development is so important to the future of the child and to national health it seems to us that this is a period in the life cycle to which very special attention should be given.

REFERENCES

[1] Crawford, M.A., Casperd, N.M. and Sinclair, A.S. (1976). Comp. Biochem. Physiol. 54B: 395.
[2] Paoletti, H. and Galli, C. (1972). Lipids, Malnutrition and the Developing Brain. CIBA Symposium, Elsevier, Associated Scientific Publishers, Amsterdam.
[3] Sinclair, A.J. and Crawford, M.A. (1973). Brit. J. Nutr. 29: 127.
[4] Sun, G.Y. and Sun, A.Y. (1974). J. Neurochim., 22: 15.
[5] Lamptey, M.S. and Walker, B.L. (1976). J. Nutr. 106: 86.
[6] Crawford, M.A., Hassam, A.G., Williams, G. and Whitehouse, W.L. (1976). Lancet 1: 452.
[7] Hassam, A.G., Sinclair, A.J. and Crawford, M.A., (1975). Lipids, 10: 417.
[8] Hassam, A.G., Crawford, M.A., (1976). J. Neurochem. 27: 967.
[9] Van Dorp, D.A., Beerlhuis, R.K., Mugteren, D.H. & Vonkeman, H. (1964). Nature, Lindon, 203: 839.
[10] FAO (1977). Dietary fats & oils in human nutrition, Food and Nutrition Paper No. 3., FAO/WHO expert consultation, Rome.
[11] Dobbing, J. (1972), Malnutrition and the developing brain, in: "Lipids". CIBA Symposium, Elsevier, Associated Scientific Publishers, Amsterdam.
[12] Crawford, M.A., Hall, B., Laurance, B.M. and Muanhambo, A. (1976) Curr. Med. Res. Op. 4, (Suppl. 1), 33.
[13] Hytten, P.E. & Leicht, I. (1971). "The Physiology of Human Pregnancy". Blackwell Scientific Publications, Oxford, London.
[14] WHO (1964). Nutrition and Pregnancy and Lactation. Technical Report, Series No. 302, WHO, Geneva.
[15] Robertson, A.F. and Sprecher, H. (1968). Acta Pediat. Scand., Suppl. 183.

THE PLASMA LIPOPROTEIN SYSTEM IN MAN

Barry Lewis

Prof. Dept. Chem. Path. Met. Disorders

St. Thomas Hospital Medical School, London S.E. 1 U.K.

In the past decade there have been rapid advances in characterizing the plasma lipoproteins; their lipid and apoprotein content have been well defined, and their structure, including the nature of lipid-protein interactions have in part been identified. The metabolic relationships between the lipoproteins, their physiological roles and the regulation of their synthesis and degradation are the subjects of vigorous research. Some of this work suggests that the conventional, operational definitions of lipoproteins (very-low density, low density and high-density classes, VLDL, LDL and HDL, and chylomicrons) are no longer adequate to the components of this dynamically interrelated system. These names will however be retained in this review.

CHYLOMICRON METABOLISM

Chylomicrons are synthesized in small intestinal mucosal cells, and contain triglyceride chiefly, not entirely of dietary origin. They also bear dietary and reabsorbed endogenous cholesterol and fat-soluble vitamins. Their size appears to reflect the rate of fat absorption; they become larger, and more triblyceride-rich, during absorption of a high-fat meal than when intake of fat is smaller. Apolipoprotein B is an obligatory component of the protein moiety of chylomicrons. They also contain small amounts of apo E and apo C, derived from plasma, but acquire more apo C by transfer from HDL. Of great potential importance is the recent demonstration that the chylomicrons of rat lymph have newly synthesized apolipoprotein A — I as their major polypeptide. In the fasted rat, the rate intestinal synthesis could contribute about one quarter of the circulating mass of apo A-I; and during fat absorption its production increases considerably. Within the circulation, chylomicron apo + A-I is probably transferred to HDL during catabolism of chylomicrons, and is conceivably a major source of HDL protein.

The initial site of chylomicron degradation is the capillary bed of extrahepatic tissues, notably skeletal and cardiac muscle and adipose tissue. The apo C-II moiety (largely aquired from HDL) activates the endothelium bound enzyme lipoprotein lipase, which clears the triglyceride component of chylomicrons. The lipolysis of chylomicron triglyceride occurs as a series of reactions between which particles re-enter the circulation. A population of primary particles and of chylomicrons progressively depleted of triglyceride is thus produced.

In the course of this process, the volume of the triglyceride-containing core of chylomicron particles is reduced. Some of the surface material (including apoprotein C and E, phospholipids and unesterified cholesterol) is consequently made redundant. Normally these components are transferred to HDL. The partly metabolised chylomicron particles remain in the circulation, retaining their apolipoprotein B, their cholesteryl ester and 15% or more of their triblyceride; these are chylomicron "remnants", a term which has been criticised but not superseded. In laboratory rodents and presumably in man these particles are taken up by the liver; the process is probably mediated by specific high-affinity receptors on hepatocyte plasma membranes.

While it is possible that chylomicron apo-B-containing remnants are in part converted to LDL, such conversion is not necessary to account for quantitative observations on LDL synthesis in man.

The physical properties of chylomicron remnants resemble those of smaller VLDL particles and these physiologically-distinct entities cannot be separated by currently available methods.

VERY-LOW DENSITY LIPOPROTEINS

These triglyceride-rich particles, smaller and denser than chylomicrons are chiefly secreted by the liver. The contained triglyceride is thus of endogenous origin. Normally it is obtained by reesterification of free fatty acid derived from plasma and initially released from adipose tissue. When an exceptionally high carbohydrate diet is provided — in acute experiments — triglyceride fatty acid is partly produced de novo from carbohydrate. A small proportion of plasma VLDL is of intestinal origin, as shown by its presence in intestinal lymph; this source contributes perhaps 20% of VLDL turnover, the contained triglyceride fatty acids being largely of dietary origin. VLDL triglyceride is therefore chiefly but not entirely endogenous. Whether the term "intestinal VLDL"

or "small chylomicrons" is used is a semantic nicety of no great
importance. As pointed out above, the VLDL density class in plasma
includes chylomicron remnant particles.

VLDL varies widely in particle size, from 30-80 mm. Within
this range are the primary particles and their metabolic products,
for VLDL undergoes progressive depletion of its triglyceride content by the action of lipoprotein lipase, exactly analogous with
the clearance of chylomicron — borne dietary triglyceride. Similarly
some of its surface components, notably apo C, apo E, lecithin and
some free cholesterol transfer to HDL. It has therefore been assumed
that primary VLDL is the large C. 80 mm, Sf 100-400 group of particles, and the products are progressively smaller. This is partly
but not entirely true: kinetic studies by Steiner et alia provide
specific activity-time curves for VLDL subclasses which demonstrate
that newly-secreted VLDL is not only in the Sf 100-400 range but
includes smaller particles as well.

When VLDL triglyceride synthesis is increased experimentally
the particle size distribution shifts towards a preponderance of
larger, less-dense particles.

The preferred substrates of lipoprotein lipase are chylomicrons;
at equal triglyceride concentration the rate of lopolysis by this
enzyme decreases with decreasing particle size. Chylomicrons and
VLDL compete for this final common path of triglyceride removal and
pronounced increase in VLDL concentration results in impairment of
chylomicron catabolism. Normally the fractional catabolic rate of
chylomicrons is very much faster than that of VLDL.

In addition to lipoprotein lipase, a triglyceride lipase is
present in the liver, probably at the surface of hepatacytes. No
physiological role has been established for this enzyme, but recent
findings suggest a possible site of action in lipoprotein metabolism.
By contrast with lipoprotein lipase, hepatic lipase has a marked
substrate preference for smaller triglyceride-bearing particles and
has only minor activity against chylomicrons. These differences
would be predicted to result in selective initial metabolism of
chylomicrons and larger VLDL in extrahepatic tissues, while their
metabolic products, the chylomicron and VLDL 'remnants' would undergo
further triglyceride hydrolysis in the liver.

INTERMEDIATE DENSITY LIPOPROTEIN

Ongoing lipolysis of VLDL yields still smaller particles of
density greater than that of VLDL. They are best known as intermediate density lipoprotein (IDL)-isolated between 1.006 and 1.019
g/ml. Their concentration is normally low. Originally they were
termed LDL_1. These particles contain all the apo B and most or all

of the cholesteryl ester present in their precursors, the primary
VLDL. Although the usual operational classification makes them
appear a distinct entity, in a functional sense they are part of the
spectrum of VLDL catabolic products or remnants. Because much of
the triglyceride has been removed (also some phospholipid and apo C)
they are relatively enriched in cholesteryl ester. When these particles accumulate in plasma they lead to increased levels of both
triglyceride and cholesterol.

In the rat, functional hepatectomy leads to retention of remnant
particles in plasma. They are also the preferred physiological substrate for hepatic lipase.

Recently Turner, Hazzard and I have investigated the metabolic
fate of VLDL and IDL in man, by measuring their concentrations in
arterial and in hepatic venous blood. The trans-splanchnic arterio-
venous difference in apo B concentration was determined in multiple
replicate samples by the tetramethyliver precipitation procedure,
in six normolipidaemiac individuals to date. There was evidence
of net secretion of large VLDL (Sf 100-400) and of net extraction
both of VLDL of Sf 20-100 and of IDL. Hence 'remnant' catabolism
in man also appears to take place in the splanchnic bed.

LOW DENSITY LIPOPROTEIN METABOLISM

LDL is the final lipoprotein produced in the course of VLDL
catabolism. (Other catabolic products are discussed in the next
section.) The original evidence for this was obtained by Gitlin
more than 20 years ago, and was based on the use of VLDL labelled
in the apoprotein moiety. More recently Levy, Eisenberg and their
colleagues confirmed that VLDL-apo B is converted via IDL - apo B
to LDL - apo B in man and the rat. This precursor product relationship is also true for the cholesterol component of these lipoproteins. The subsequent metabolism of LDL is extravascular. Because
the fractional rate of LDL catabolism is far slower than that of
VLDL. The response of VLDL concentrations to dietary manipulation
is more rapid than that of LDL. There is evidence that this takes
place in extrahepatic tissues though this conclusion awaits un-
ambiguous confirmation. LDL is catabolised intracellularly, after
internalization by cell surface receptors which are homeostatically
regulated.

QUANTITATIVE ASPECTS OF VLDL-IDL-LDL
METABOLISM IN MAN

Nicoll, Janus, Sigurdsson, Wootton and I have quantified the
processes in this pathway in normal and hyperlipidaemic subjects.
Most data has been based on apo B specific activity measurements,

and on the use of deconvolution integral, a form of non-compartmental analysis. Eisenberg has calculated that apo B is quantitatively conserved as the VLDL particle is remodelled to IDL and finally to LDL - apo B metabolism is therefore a valid marker of the metabolism of the entire particle. The main findings are summarised in this section.

In normolipidaemic man, about 50% of VLDL 1 apo B is converted to LDL - apo B. At least 90% of VLDL - apo B reaches the IDL - LDL pool. VLDL - apo B appears to be quantitatively converted to IDL - apo B. There is no evidence that LDL or IDL are directly secreted; they are derived entirely from the catabolism of VLDL (and conceivably from chylomicrons). As discussed earlier, liver cells possess high-affinity receptors mediating uptake of remnant particles, so that a basis for direct extravascular metabolism of IDL exists. In the rat, following injection of VLDL labelled with radioiodine in the apoprotein moiety, most isotope is recoverable from the liver and some 10% is converted to LDL; the relative magnitude of these two pathways shows marked species differences, man showing greater conversion of VLDL to LDL, the guinea pig being intermediate. Thus the rate of LDL synthesis is influenced both by the rate of VLDL synthesis and by the percentage of VLDL and its products which is catabolized to LDL.

There is indirect evidence, then, that the liver is responsible for the final conversion of partly metabolized VLDL to LDL. Possibly the hepatic lipase plays a role in this process. On the other hand, it has recently been shown that when VLDL is incubated with purified lipoprotein lipase in a simple in-vitro system, a major product is a particle resembling, though not identical with authentic LDL. This finding reopens the possibility that conversion of VLDL to LDL could conceivably be mediated entirely by this extrahepatic enzyme.

Patients with primary hypertriglyceridaemia may be considered, on the basis of current data, in four groups. Those with endogenous hypertriglyceridaemia most often have an increased VLDL synthetic rate which correlates with plasma VLDL concentration. But such patients are clearly heterogeneous. When they are grouped into genetically-defined categories it appears that increased levels of VLDL due to familial combined hyperlipidaemia result from overproduction of this lipoprotein; familial endogenous hypertriglyceridaemia (so far studied in smaller numbers) appears to result from a defect in VLDL catabolism.

When overproduction of VLDL occurs, the concentration of LDL may be low, normal or high. Low levels of LDL are largely the result of increased fractional catabolic rate of this lipoprotein. The dissociation between synthetic rates of VLDL and of its products LDL is largely a consequence of the direct extravascular catabolism of VLDL and its immediate catabolic products. Catabolism of IDL has already been referred to. In endogenous hypertriglyceridaemia, an average of 50% of VLDL - apo B is metabolized extravascularly and does not reach the IDL-LDL pool. As a result, overproduction of VLDL does not necessarily lead to over-production of LDL.

In some circumstances changes in VLDL and LDL synthesis appears to be coupled. When levels of both lipoproteins are increased due to the mutant gene for familial combined hyperlipidaemia, the synthetic rates of VLDL - apo B and LDL - apo B are consistently elevated. When plasma levels of VLDL and LDL are produced by substitution of a diet rich in saturated fat (P:S = 0.25) with one with a P:S ratio of 1.8 - 1, the synthetic rates of both VLDL and LDL decrease.

A less common variant of combined hyperlipidaemia is remnant or Type III hyperlipoproteinaemia or broad β-diesase. The fractional catabolic rate of β - VLDL and IDL is abnormally prolonged and there is evidence also for direct synthesis of IDL.

Familial hypercholesterolaemia in the heterozygous state is associated with a 50% reduction in the number of cell-surface LDL receptors in mutant fibroblasts in tissue culture, a defect also demonstrable in other cell lines. These receptors mediate LDL uptake from extravascular fluid, and the consistently-reduced fractional catabolic rate of LDL in familial hypercholesterolaemia may reasonably be attributed to this receptor deficiency. However the elevated plasma LDL concentration is not purely the result of impaired removal. Myant and his colleagues have shown in homozygous patients with familial hypercholesterolaemia that some circulating LDL is not derived from VLDL. In heterozygotes we find that up to 50% of LDL can be calculated as the difference between total LDL - apo B turnover and the rate of VLDL - apo B conversion to LDL; this direct secretion rate is linearly related to plasma LDL concentration in heterozygotes.

HIGH DENSITY LIPOPROTEIN METABOLISM
AND CENTRIPETAL TRANSPORT OF CHOLESTEROL

Peripheral cells appear to acquire a major proportion of their cholesterol requirements by receptor-mediated uptake of LDL from extracellular fluid. This cholesterol is ultimately derived from diet or from synthesis in the liver or small intestine. In addition, almost all cells have the capacity to synthesize cholesterol, though this potential is, evidently, largely repressed in the steady state.

If such extrahepatic tissues can acquire and/or synthesize cholesterol it follows that they msut also be able to get rid of this sterol in order to maintain a constant cholesterol content. A little cholesterol is lost from the body in desquamated epidermal

cells, and some is utilized for steroid hormone synthesis; but only
the liver can excrete substantial amounts of cholesterol. It is the
liver, too, which is the major site of cholesterol catabolism, the
excreted end products being the bile salts.

There is, therefore, an obligatory need for a centripetal transport mechanism by which cholesterol is carried from peripheral tissues
to the liver. Ten years ago Glomset postulated that this pathway
comprised HDL and the closely-associated enzyme lecithin: cholesterol
acyltransferase (LCAT).

Substantial indirect evidence has been adduced in support of
this concept, though a great deal of uncertainty remains. HDL serves
as a receptor for cholesterol released from Erlich ascites tumour
cells in tissue culture. In normal man, the size of the two major
pools of exchangeable cholesterol bear a significant negative correlation with plasma HDL concentration. Storage of cholesterol in
reticulo-endothelial cells is a feature of Tangier disease in which
normal HDL is absent from plasma.

Some discordant findings have also been reported. In the rat,
there is evidence that the flux of cholesterol in LDL (presumably
reflecting centrifugal cholesterol transport) is several fold greater
than that of cholesterol in HDL; if so HDL could not be the major
vehicle for centripetal transport. Very recently it has been suggested, on the basis of in vitro studies with cholesteryl ester-
enriched cardiac cells in tissue culture, that the plasma fraction
of density >1.21 (i.e. depleted of HDL, LDL and VLDL) is an effective
receptor for cholesterol.

The role of HDL in centripetal lipid transport is the subject
of intensive current research. At the same time the origin of HDL,
and regulation of its metabolism are under renewed scrutiny.

Under appropriate circumstances the perfused rat liver releases
HDL into the perfusate, in addition to VLDL. The HDL appears in
nascent form, as disc-shaped particles chiefly containing apo E,
apo A and apo C, unesterified cholesterol and lecithin. Such particles are normally converted to 'mature' HDL under the influence of
LCAT which esterifies cholesterol by transfer of a fatty acyl residue
from lecithin within the HDL molecule.

More recently it has become evident that at least some of the
components of HDL could — at least in theory — be derived from surface material of triglyceride-rich lipoproteins during the catabolism
of these particles by lipoprotein lipase. After injection of VLDL
(labelled with ^{32}P) into rats, transfer of lecithin to the HDL density class occurs rapidly if lipolysis is stimulated by heparin.
When labelled VLDL is exposed to lipoprotein lipase by perfusion
through the rat heart and its products examined much of its free

cholesterol and apo C, and some lecithin and sphingomyelin are recoverable from the HDL density range; in this range electron microscopy reveals disc-like particles reminiscent of nascent HDL.

In extension of these studies Eisenberg has incubated VLDL and
lipoprotein lipase with the HDL_3 subclass of HDL. Surface components
of VLDL appear to combine with HDL material, the product being similar
to, but not identical with HDL_2.

As discussed above, apo A-1 is synthesized in the small intestine and an appreciable proportion of plasma apo A-1 may enter plasma
with chylomicrons; it is probable this apoprotein is transferred to
HDL within the circulation. Hence there is substantial if indirect
evidence that much HDL material would be derived from VLDL and chylomicrons during their intravascular catabolism. Two findings argue
against the simplistic concept that HDL is derived entirely as a
product of these lipoproteins however. One is the demonstrable release of nascent HDL by the perfused rat liver. The other is the
occurrence of HDL, albeit abnormal in composition, in plasma in
abetalipoproteinaemia, a condition charecterized by failure to synthesize VLDL, chylomicrons and LDL.

DIETARY VARIABLES AND LIPOPROTEIN METABOLISM

While there is abundant evidence that plasma lipoprotein levels
are influenced by the intake of several nutrients, the mechanism of
these effects is poorly documented. The association between obesity
and elevated levels of VLDL has been studied qualitatively by the
IV fat tolerance test: decreased removal of VLDL appears to be present in hypertriglyceridaemic obese subjects. Reduction of LDL and
VLDL levels by isocaloric substitution of polyunsaturated fatty acids
for saturated fatty acids is the most widely used therapy for hyperlipidaemia; yet its mode of action has aroused controversy for twenty
years. Recent studies have suggested that the synthetic rates of
VLDL - apo B and of LDL - apo B are reduced when the dietary P:S
ratio is increased. An acute increase in the proportion of energy
derived from carbohydrate leads to temporary elevation of VLDL
levels, seemingly involving enhanced synthesis and impaired removal.
The reduction in HDL concentration during this "carbohydrate induced
lipaemia" is associated with accelerated catabolism of HDL from its
plasma compartment.

CHEMISTRY AND METABOLISM OF VITAMINS

G. Brubacher

Dept. of Vitamin and Nutritional Research

F. Hoffmann-La Roche & Co., Ltd., Basel/SWITZERLAND

In contrast to the other nutrients which are treated within the frame of the symposia on basic nutrition, vitamins do not belong to a homogenous group of chemical compounds like proteins, lipids and minerals. Therefore, it will not be possible to discuss general points of view in the chemistry and metabolism of vitamins. On the other hand, it is not possible to treat in a sophisticated manner thirteen different chemical classes within the short time which we have at our disposal. Therefore, we have to make a choice.

I have chosen topics of actual interest which are in some connection with the problem of assessment of nutritional status and the question of daily allowances.

Of the thirteen vitamins which we know, the mode of action of ten has been elucidated or is about to be elucidated. The eight vitamins of the so-called vitamin-B complex, vitamin B_1, vitamin B_2, vitamin B_6, vitamin B_{12}, vitamin PP, folic acid, biotin and pantothenic acid act as part of a co-enzyme or as a prosthetic group of an enzyme; vitamin K probably has also co-enzymatic functions, and vitamin D is a hormone-like substance, whereas the mode of action of vitamins A, C and E is still unknown. The knowledge of the mode of action could give us a better estimate of the daily dietary allowance of a nutrient and would give us the possibility to develop better methods to assess the nutritional status of the vitamin in question. Therefore, from the point of view of the aim of our symposium, more research work should be done to elucidate the mode of action of these three vitamins.

We can assume that the chemical properties of a compound are essential for its biological action. Therefore, knowledge of these chemical properties is the prerequisite for all further research.

Since we have no time to discuss the chemistry of all the three
remaining vitamins I have chosen only vitamin C to go a little deeper
into this matter. This does not mean that vitamins A and E should
be excluded from further discussion.

Concerning the metabolism of vitamins in man in the last few
years research has switched from mainly qualitative aspects to
quantitative statements. This again allows better estimates of the
daily dietary allowance and better judgement of the different methods
for assessment of the nutritional status. I have, therefore, included a lecture on the quantitative aspects of vitamin metabolism.

Now we come to the problem of the practical utilization of basic
knowledge of chemistry and metabolism of vitamins in respect to the
problems of assessment of nutritional status and the question of the
daily dietary allowance. Here research is not concerned so much
with the development of new and better tests as with questions of
standardization and of feasibility in field work. Finally, the
results must be correctly interpreted. Men are not rats! Of the
whole range of problems which stem from this fact, only two are
stressed: the problem of interrelationships between different nutrients, and the problem of inborn error of vitamin metabolism. This
latter problem will be sure to gain even more importance in the
future since with modern medical treatment more children with inborn
errors of metabolism reach the child-bearing age, so that the percentage of persons with inborn error of metabolism will increase
in the general population. Therefore, this part of the population
should no longer be neglected in our estimate of the daily dietary
allowance of a certain nutrient.

QUANTITATIVE ASPECTS OF VITAMIN METABOLISM

Anders Kallner M.D.

Dept. of Clinical Chemistry, Huddinge Univ. Hospital

Karolinska Institutet, S-141 86 - Huddinge, SWEDEN

Usually, quantitation of vitamin metabolism aims at determining the requirement of the vitamin for man or animal. It is anticipated that during the course of its action, the vitamin is metabolized. Even if this is largely so, it may not be the whole truth and at least theoretically a substantial amount of vitamin may be absorbed, distributed into various pools and finally excreted as such after having exerted its physiological functions.

Vitamins are a heterogenous concept both from a chemical physiological and pharmacological point of view. By convention we subdivide the vitamins into fat soluble and water soluble vitamins. Besides the properties apparent from the names a justifiable generalization may be that the animal body can store relatively large amounts of fat soluble vitamins with a long half life whereas body-pools of water soluble vitamins are comparatively small and above all have a relatively rapid turnover rate.

Originally, the action of a vitamin was described as curing one disease or participating in one or a limited number of biochemical reactions or types of biochemical reactions. This is still the case for some vitamins whereas the majority of vitamins seem to participate in a number of various reactions. It is conceivable that if the vitamin action becomes fully known on a molecular level, one should be able to establish the total need of the vitamin. Quantitative metabolic studies on the vitamins will eventually also enhance a clarification of qualitative modes of action of the compound. For many vitamins we have to admit that the biochemical actions or physiological functions are not fully known. It has often been revealed that several compounds of more or less different chemical structure can exert similar vitamin-like effects.

It is virtually impossible task to make a general review of
the quantitation of vitamin metabolism. The following presentation
will therefore be limited to recent approaches to determine pool
sizes of a fatsoluble vitamin, vitamin A, and a kinetic determination
of the turnover rate and pool sizes of a watersoluble vitamin, vitamin C.

Vitamin C deficiency is still a major nutritional problem in
many parts of the world (for a recent review, the reader is referred
to [1]). The reason for this deficiency is the limited sources of
vitamin A in the diets and the most prominent deficiencies are found
in growing children.

The first physical signs of vitamin A deficiency is usually
said to be hyperkeratosis and other skin manifestations as well as
night blindness to various degrees. Assessment of the vitamin A
status of humans and animals is usually based on the determination
of the vitamin A concentration in plasma. It is well-known, however, that the plasma vitamin A level is correlated to the amount
of vitamin A in the main store — the liver — only in extreme hypo-
vitaminoses. This is shown in Fig. 1[2]. On depletion of vitamin A
in rats, the liver vitamin A stores decrease continuously whereas
vitamin A plasma level remains quite constant until the liver stores
have been almost totally depleted. On repletion a rapid increase
in liver stores and plasma occurs but the latter does hardly rise
above the initial level.

Probably the best assessment of vitamin A status would be obtained by the analyses of liver biopsy material. Although this
can be carried out only in special cases, surveylike studies have
been carried out in several countries (as reviewed in [1]). These
studies have revealed an incidence of 20–30% of cases with levels
of less than 40 ug vitamin A per gram of liver. This is the accepted
borderline with respect to vitamin A reserves.

In a classical and very fundamental study Sauberlich et al.[1]
showed that depletion of vitamin A may take from less than one to
more than two years until physical signs appear, even if the subject
is kept on a diet with negligible contents of vitamin A or less than
20 ug retinol per day.

By repletion of the subjects Sauberlich and coworkers could
estimate the daily metabolism of retinol. It was found that repletion of the vitamin A as judged from the change in plasma required the administration of about 600 ug retinol/day or about
1200 ug of β-carotene per day (Fig. 2).

Depletion — repletion studies of this type require an outstanding cooperation of subjects and a very long time. It is under-
standable that the amount of data does not allow any statistical

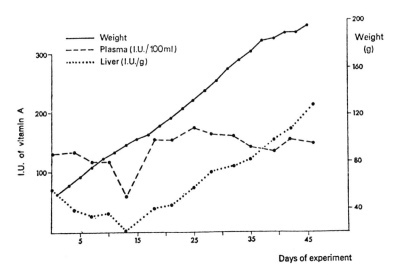

Fig. 1 - Correlation between Vitamin A levels in plasma and liver.
(From Bausch and Rietz, 1977)

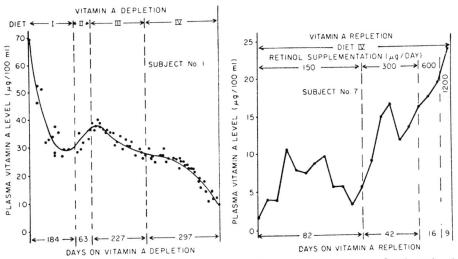

Fig. 2 - Changes in plasma Vitamin A levels in states of Vitamin A
depletion and repletion. (From Sauberlich et al., 1974)

evaluation although they certainly indicate a substantial inter-
individual variation.

Rietz, Wiss and Weber at Hoffman-La Roche in Basel have attempted another approach to determine the vitamin A body pool[3] in animals
and ultimately in humans. It has previously been shown[4] that a dose
of vitamin A is homegenously mixed with the pre-existing vitamin A
body pool. A constant excretion rate indicating an equilibrium is
obtained after about 3-7 days in the rat and[3, 5] after about 14-17
days in man[2] (Fig. 3). The main store of vitamin A is the liver
where the vitamin is stored with a considerable half-life. In the
Rietz study, the assumption was made that it would be possible to
calculate the amount of vitamin A stored in the body by measuring
the specific activity in plasma, after equilibration, in relation
to the amount of tracer dose given. In a series of experiments
with this isotope dilution technique in rats[3], sheep and piglets[2]
the method was proven successful by comparison with the post-mortem
determination of vitamin A in the carcass or the liver. An inaccuracy is introduced in this method by an experimental factor des-
cribing how much of the administered dose that (Fig. 4) reaches
the body pools. This factor was found to approach 0.5 in rat and
sheep but lower values were obtained in cattle and depleted cattle[2].
In order to avoid the use of radioactive tracers the application
of stable isotopes (deuterium) was worked out and shown to give
equally good results.

It might be difficult to define an experimental model which
allows post-mortem or even biopsy analyses to be carried out in
humans. Therefore the magnitude of the experimental factor in
humans cannot be proven as easily as in animals and the calculation
of human body pools remains uncertain with this method. However,
the use of a factor 0.5 in the only study of a human subject, which
has been published so far, results in a plausible size of the liver
pool[2].

Assumption of an arbitrary experimental factor could be circum-
vented by a careful assay of all excreted label in the urine and
faeces. Thus the used algorithm would be changed (Fig. 5).

In a recent report[6] this approach was tried in a model experiment with rats. Tritium labelled vitamin A at two different levels
was administered orally to one group of animals and parenterally
to another. Faeces and urine were collected and the total body pool
calculated from the specific activity of vitamin A in plasma. It
is interesting to point out that also after intravenous administration a significant amount of radioactivity is excreted in faeces
(Fig. 6). The calculated results were good estimates of the body
pools (Fig. 7).

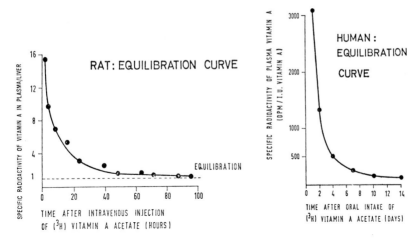

Fig. 3 - Vitamin A equilibrium curves. (From Bausch and Rietz, 1977)

$$\text{Vit A body pool} = \frac{K \cdot \text{Test dose (dpm)}}{\text{Spec. radioactivity of Vit A in plasma (dpm/IU)}}$$

Fig. 4.   Calculation of vitamin A body pool using an experimental
          factor. K = experimental factor.
          (From Bausch & Rietz, 1977)

$$\text{Vit A body pool} = \frac{\text{Test dose} - \text{totally excreted amount (dpm)}}{\text{Spec. radioactivity of Vit A in plasma (dpm/IU)}}$$

Fig. 5.   Calculation of vitamin A body pool.
          (From Bausch et al., 1978)

	ORAL APPLICATION		INTRAVENOUS APPLICATION	
	High dose	Low dose	High dose	Low dose
Excreted in urine	8.4 1.7	10.1 1.4	7.6 0.9	7.8 0.6
Excreted in faeces	17.3 2.0	20.3 1.0	11.6 1.6	11.0 0.6
Remaining in the body	74.2 3.0	69.6 2.2	80.8 1.6	81.2 0.9

Fig. 6. Percentage of radioactive dose excreted in urine and faeces during 7 days and remaining in the body 7 days after administration of labelled vitamin A acetate. (From Bausch et al., 1978).

Evidently it is possible to use this type of isotope-dilution in the estimation of the vitamin A body pool. It is likely that this model could be used for humans as well.

One has to be careful, however, in using radioactive isotopes for this purpose because of the slow turnover and thus the relatively large exposure of the subject to radioactivity and it would be feasible to use stable isotopes, i.e. deuterium instead. The experimental model requires the assay of the total amount of excreted label in the given dose. This might cause technical problems when stable isotopes are used since the natural abundance of deuterium containing compounds in urine and faeces is not negligible.

Quantitative determination of the metabolism of vitamin C or ascorbate has been approached by similar techniques as have been used for other vitamins, i.e. vitamin A. Due to the nature of vitamin C and the fact that ascorbate is synthesized endogenously in the majority of animals, model experiments have usually been conducted in the guinea-pig. This species, like many primates and man, requires an adequate dietary supply of ascorbate.

Qualitatively the metabolism is to a certain degree different in different species. In man, the main metabolites of ascorbic acid are oxalic acid, dehydroascorbic acid, 2,3-diketogulonic acid and ascorbate-2-sulfate (Fig. 8). Practically no conversion of ascorbic acid to respiratory carbon dioxide occurs in man in contrast to guinea-pig[7] and negligible amounts of radioactivity is recovered from faeces after administration of ascorbic acid-1-^{14}C[8]. In man the main excretory pathway of ascorbic acid and its metabolites is thus the urine and an almost complete recovery of labeled material has been obtained from this source[9].

Baker et al.[9] administered ascorbic acid-1-^{14}C to three healthy volunteers and estimated the poolsize to between 2 and 3 g from determinations of the specific activity of plasma ascorbate. The turnover rate was expressed as a half-life of about 20 days on ascorbate intakes of about 100 mg per day. Later, in depletion — repletion studies[10-12] it was found that the pool size was remarkably constant at about 1500 mg and that the body pools decreased with about 3% of the existing pool per day. This was interpreted that in steady state, with normal pools, the quantitative metabolism of ascorbate is about 45 mg per day. During the depletion stage the blood level of ascorbate fell until a body pool size of about 300 mg was reached when it levelled off. At this pool size clinical signs of scurvy were observed. A comprehensive review of the research and results on recent vitamin A requirements was published by Irwin and Hutchins in 1976[13].

Quantitative determination of vitamin C metabolism can also be obtained by a pharmacokinetic approach. In this technique different compartmental models are tested against the experimental data. Use

Rat No.		Vitamin A body pool (IU)		$\frac{\text{calculated pool}}{\text{analyzed pool}} \times 100$
		Analyzed	Calculated	
high dose	1	6472	5994	92.6
	2	5346	6151	115.1
	3	6617	7187	108.6
	4	6306	6879	109.1
	5	7394	7192	97.3
	6	6772	6497	95.9
				mean ± SD 103.1 ± 8.2
low dose	7	3483	3643	104.6
	8	3276	3507	107.1
	9	3879	4089	105.4
	10	3178	3217	101.2
	11	3163	3654	115.5
	12	3529	3189	90.4
				mean ± SD 104.0 ± 7.5

Fig. 7. Comparison of analyzed and calculated vitamin A body pool in rats with oral application of labelled vitamin A acetate. (From Bausch et al., 1978)

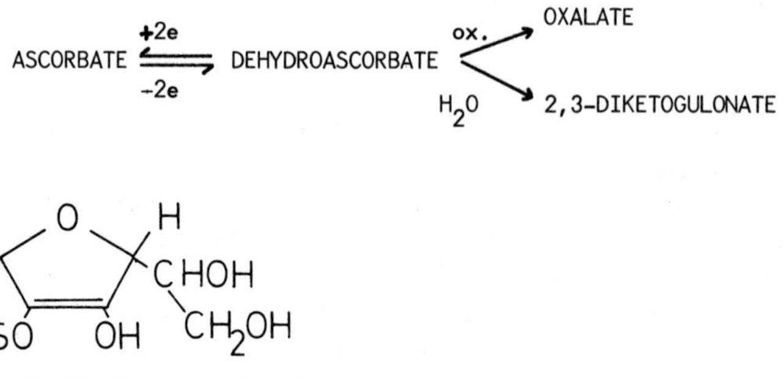

ASCORBATE - 2 - SULFATE

Fig. 8 - Main metabolites of ascorbate.

of the model which fits the data best permits estimation of, among other parameters, pool-sizes and turnover rates.

In a series of studies[14, 16] we determined quantitative parameters of ascorbate metabolism in man.

Non-smoking volunteers were given a restricted diet with respect to vitamin C containing food-stuff. The subjects were divided into four cathegories, given 30, 60, 90, and 180 mg of ascorbate daily in a steady state. After an equilibration time of three weeks, a known amount of ascorbic acid-1-^{14}C was given orally (Fig. 9). The subjects were kept on the diet and supplementation of ascorbate for an additional two weeks after administration of the labelled dose. The total radioactivity in plasma, the plasma concentration of ascorbic acid, and the excreted amount of activity in the urine, totally and as ascorbic acid, were determined. The experiment was finished by a period of high ascorbate intake (4x1000 mg). This serves the dual task of limiting the radiation exposure and improving the accuracy of estimating the recovered amount of radioactivity.

A rough estimate of the compartmental model can be achieved from the shape of the disappearance curve of radioactivity from plasma (Fig. 10).

In these experiments a three-compartment model (Fig. 11) was postulated. It consists of a central compartment (compartment number 1) into which dietary ascorbate is assumed to be absorbed and from which it is eliminated unchanged. Two peripheral compartments are in equilibrium with the central compartment. Compartment number two is the shallow compartment where the metabolism is considered to take place whereas number three is considered a deep pool and a mere storage pool.

Typical examples of the time course of excretion of radioactivity in urine are shown in Fig. 12. It can be seen that the relative amount of radioactivity excreted as ascorbic acid is high in the subject supplemented by 180 mg but it is interesting to point out that also in the low-dose subject (supplemented by 30 mg) a certain amount of radioactivity is excreted (Fig. 13) as labelled ascorbic acid.

The model permits the calculation of the apparent volume of distribution. The resulting volumes are unphysiologically large (Fig. 14) and since the amount of bound ascorbic acid in plasma is known to be low[17], a pronounced tissue binding must be postulated.

The renal turnover is affected by the glomerular filtration rate, the plasma concentration and the tubular reabsorption. As seen in Fig. 15, a steep increase in the renal turnover (T_{REN}) occurs

3 weeks   2 weeks   1 week

EQUILIBRATION PERIOD   EXPERIMENTAL PERIOD   HIGH DOSAGE PERIOD

Restricted diet

Plasma ascorbic acid
  (mg/100 ml)   (1–¹⁴C) ASCORBIC ACID

Zero levels of radioactivity
  Plasma
  Urine

Ascorbic acid intake (mg/day)   Carrier with· C laben and
  1 x 30   intake for first 12 h (mg/h)
  2 x 30   1.25
  2 x 45   2.50
  4 x 45   3.75
                                   7.50

                                 Blood sampling
                                 Plasma ascorbic acid (mg/100 ml)
                                 Plasma radioactivity (dpm/ml)

                                 Urine collection
                                 Cumulative excretion (dpm)

                                 Excretion as (1–¹⁴C) ascorbic acid (dpm)

Fig. 9. Study of ascorbate metabolism in non-smoking volunteers.

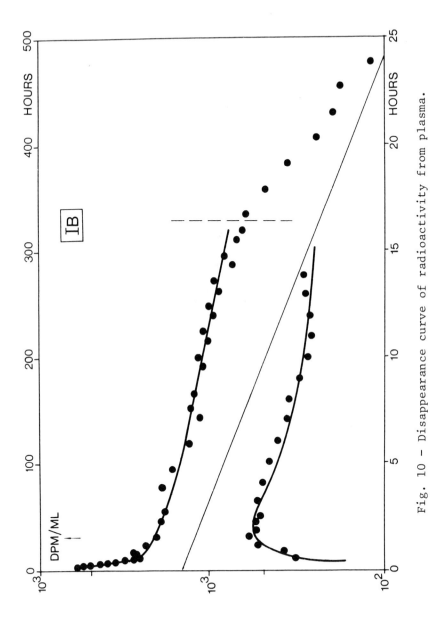

Fig. 10 - Disappearance curve of radioactivity from plasma.

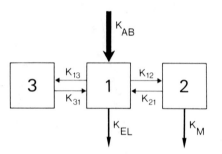

Fig. 11 - Three compartment model.

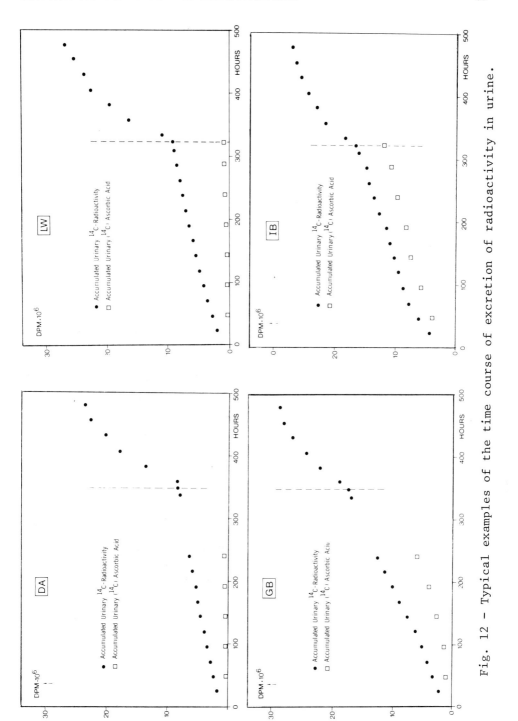

Fig. 12 - Typical examples of the time course of excretion of radioactivity in urine.

Initials	Dosage ascorbic acid mg/day	Percentage of total urinary ^{14}C-radioactivity as $(1-^{14}$C) ascorbic acid
ST, ED. DA, PS	1 x 30	6.6 ± 1.1
JO, NI, RC, LW	2 x 30	20.3 ± 7.2
SO, GB, BW	2 x 45	34.1 ± 6.3
JA, TU, CL, IB	4 x 45	61.7 ± 2.5
ED, NI, SO, TU	4 x 250	82.4 ± 1.8
ED, NI, SO	4 x 500	87.9 ± 3.8
DA, RC, GB, CL, PS, BW, LW, IB	2 x 1'000	87.0 ± 2.1

Fig. 13. Urinary excretion of $(1-^{14}$C) ascorbic acid as percentage of total urinary $(^{14}$C) radioactivity for different levels of daily dosages of unlabelled ascorbic acid (± SEM) after single oral administration of $(1-^{14}$C) ascorbic acid to man.

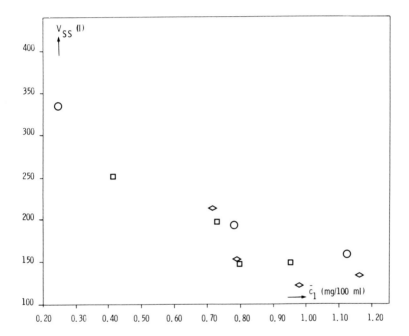

Fig. 14 - Apparent volume of distribution V_{ss} in dependence
on plasma steady concentration.

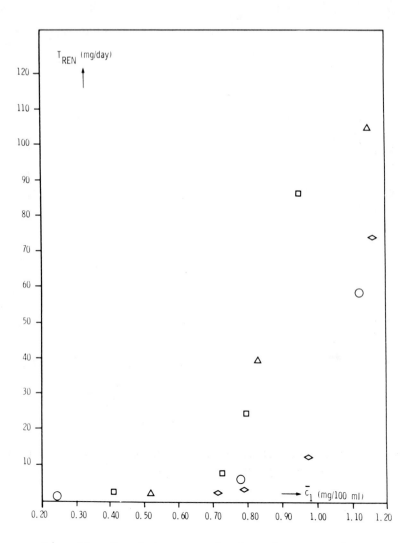

Fig. 15 - Renal turnover in dependence on plasma
steady state concentration

at a plasma concentration of about 0.7 mg/100 ml. This threshold
value indicates a saturability of the reabsorption process.

The experimental procedure with a restricted diet allows for
an inaccuracy in the determination of the daily intake of vitamin
C. Therefore, the total turnover (T_{TOT}) was calculated from kinetic
data. As shown in Fig. 16 a good agreement exists between calculated
C_1 and T_{TOT} and the relation between $\bar{C}_1$ and the intake in controlled
studies. The relation between metabolic turnover (T_{MET}) and T_{TOT}
is shown in Fig. 17. The metabolism of ascorbate appears to be a
saturable process as well and it reaches a plateau at a T_{TOT} of
about 60 mg/day. Our model also allows the calculation of the total
body pool of ascorbic acid. Depending on the dose, the total body
pool approaches about 20 mg/kg body-weight when T_{TOT} is above
60 mg/day. This would be in agreement with earlier estimates of the
body-pool at saturation. It is interesting to see that the size of
pools 2 and 3 behave differently on changing the total turnover
(described as $\bar{C}_1$ in Fig. 18). Thus the shallow pool is sensitive
to the total turnover, whereas the deep pool is relatively stable.
In depleted subjects pools 1 and 2 are more or less emptied and
the turnover rate, which was determined in depleted subjects as
described previously, may refer to pool 3 only. The turnover obtained in that way may therefore be an underestimate of the total
turnover in steady state.

Based on the pharmacokinetic study it can be calculated that
at a total turnover, T_{TOT}, of about 60 mg/day the metabolic turnover is saturated. At about the same order of magnitude of total
turnover the renal tubular reabsorption is saturated and the total
body pool approaches 20 mg/kg or about 1500 mg for a normal adult.
If these data were to be used for estimation of required daily oral
intake one also would have to consider that the gastrointestinal
absorption is less than 100%[18] and that an estimate of the variation
in a population has to be made.

It is evident from the presently available information that
saturation of a metabolic turnover should occur at the same level
as an optimal or desirable total turnover. In the case of ascorbate,
for instance, which participates in redox systems, it can be well
imagined that unmetabolized ascorbate has in fact participated in
physiological processes. On the other hand it may also be possible
that the excretion of ascorbic acid or its metabolites is a process
of detoxication of an exogenous substance.

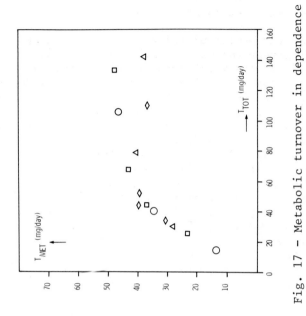

Fig. 17 – Metabolic turnover in dependence
on total turnover.

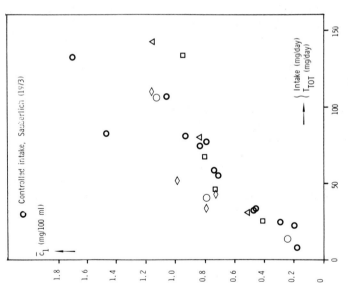

Fig. 16 – Plasma steady state concentration in
dependence on total turnover and on
controlled intake respectively.

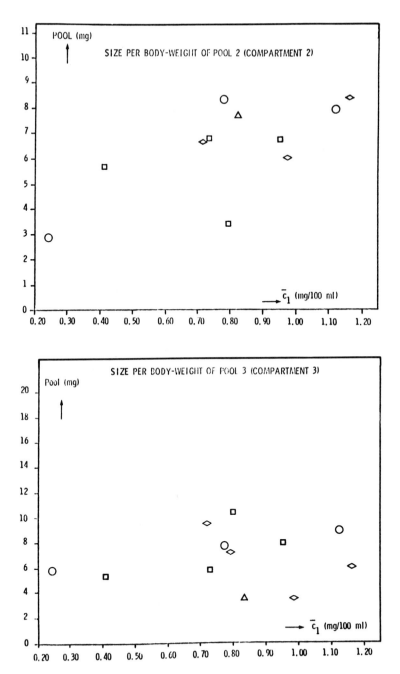

Fig. 18 - Sizes per body weight of pools 2 and 3 in dependence
on plasma steady state concentration.

72

A. KALLNER

REFERENCES

[1] Sauberlich, H.E., Hodges, R.E., Wallace, D.L., Kolder, H., Canham, J.E., Hood, J., Raica Jr., N. & Lowry, L.K. Vitamins and Hormones 32: 251 (1974).

[2] Bausch, J. & Rietz, P. Acta Vitamin. Enzymol. (Milano) 31: 99 (1977).

[3] Rietz, P., Wiss, O. & Weber, F. Vitamins and Hormones 32: 237 (1974).

[4] Rietz, P., Vuilleumier, J.P., Weber, F. and Wiss, O. Experientia 29 (2): 168 (1973).

[5] Varma, R.N. & Beaton, G.H., Con. J. Physiol. Pharmacol. 50: 1026-1037 (1972).

[6] Bausch, J., Rettenmaier, R. & Horning, D. 4th Fat Soluble Vitamin Group Meeting, Leeds 7-8 April (1978).

[7] Tolbert, B.M., Downing, M., Carlson, R.W., Knight, M.K. & Baker, E.M. Ann. N.Y. Acad. Sci. 258: 48 (1975).

[8] Horning, D. "World Review of Nutrition and Dietetics", Karger, Basel 23: 225 (1975).

[9] Baker, E.M., Saari, J.C. & Tolbert, B.M. Am. J. Clin. Nutr. 19: 371-378 (1966).

[10] Baker, E.M., Hodges, R.E., Hood, J., Sauberlich, H.E. & March, S.C., Am. J. Clin. Nutr. 22: 549-558 (1969).

[11] Baker, E.M., Hodges, R.E., Hood, J., Sauberlich, H.E., March, S.C. & Canham, J.E. Am. J. Clin. Nutr. 24: 444-454 (1971).

[12] Sauberlich, H.E. Ann. N.Y. Acad. Sci. 258: 438-450 (1975).

[13] Irwin, M.I. & Hutchins, B.K. J. Nutr. 106: 821-879 (1976).

[14] Kallner, A., Hartmann, D. & Hornig, D. Fed. Proc. 36: 1177 (1977).

[15] Kallner, A., Hartmann, D. & Hornig, D. Nutr. Metab. 21 (Suppl. 1): 31-35 (1977).

[16] Kallner, A., Hartmann, D. & Hornig, D. Am. J. Clin. Nutr. Accepted for publication.

[17] Zetler, G., Seidel, G., Siegers, C.P. & Iven, H. Europ. J. Clin. Pharmacol. 10: 273 (1976).

[18] Kallner, A., Hartmann, D. & Hornig, C. Internat. J. Vit. Nutr. Res. 47: 383-388 (1977).

INBORN ERRORS OF VITAMIN METABOLISM

V.B. Spirichev

Nutrition Institute

Moscow - USSR

Discovery of vitamins helped to reveal the nature of such severe diseases as scurvy, beri-beri, pernicious anemia, etc. Wide vitamin utilization for preventive and therapeutic purposes led to the fact that in most developed countries diseases of vitamin deficiences have become rare and the problem of such deficiences is rather socio-economic than scientific. Although the development of balanced nutrition favoured disappearance of alimentary avitaminosis there appeared in literature the description of diseases with clinical signs similar to avitaminosis but occuring in normal vitamin status.

Such troubles of hereditary origin occur frequently in children. Sometimes a disease is fully or partially treated by constant administration of respective vitamins in doses exceeding physiological requirements by 100-1000 times as much. Such cases were termed vitamin-dependent states.

In states termed vitamin-resistant ones the administration of vitamin high levels eliminated no manifestations involving severe affections and fatal cases.

In the 30's against the background of wide vitamin D utilization for prevention and treatment of rickets cases of vitamin D-resistant rickets unresponsive to normal vitamin D levels were discovered.

In 1954 Hunt described infant convulsions stopped by vitamin B_6 administration in doses 10-50 times exceeding the physiological requirement.

This case and similar observations helped Hunt to launch a conception of pyridoxine-dependent convulsion syndrome (Hunt, A.D. & Stokes, I., 1954). Later pyridoxine-responsive anemia and other inborn diseases with increased vitamin requirements were described (Harris, J.W. & Whittington, R.M., 1956).

Further biochemical investigations revealed that the origin of these diseases is accounted for by the inherited defects of vitamin metabolism or their interaction with specific transport and enzyme proteins in metabolic processes.

Data on vitamin metabolism and their mechanism of action show that no vitamin fulfills its functions in metabolic processes in primary form as it occurs in food. Before fulfilling its functions any vitamin should undergo certain transformations assisted by sepcial transport, enzyme and receptor proteins.

Fig. 1. presents basic stages of vitamin metabolism.

Stage 1 - absorption in the intestine. For most vitamins this stage is realized by special transport systems that 'catch' occasional vitamin molecules and provide their transfer through the intestinal mucosa. As a classic example we can cite the well-known intrinsic factor that binds Vitamin B_{12} in the stomach and provides its transfer to the specific receptor site of entherocyte apical membrane. The cause of Addison anemia is proved not to be Vitamin B_{12} deficiency but malformation of the intrinsic factor. The intrinsic factor is one of the proteins necessary for the utilization of vitamins in the body. At present we know certain carrier proteins participating in absorption of vitamins, their transport to deposition and utilization sites. Transport of absorbed vitamin B_{12} to the tissues is realized by special transport proteins: transcobalamins I and II. Vitamin D transport — by means of special protein — transcalciferin. Vitamin A transport is realized by recently-discovered retinol-binding protein. In the lack of this protein retinol would fail to be transported from the liver to the cells.

The next stage is vitamin conversion into an active form. It is known that water-soluble vitamins perform their functions in the form of coenzymes or prostetic groups.

In recent years active forms of some fat-soluble vitamins have been discovered: 1,25-dihydroxycholecalciferol, an active form of Vitamin D; retinylphosphate, an active form of Vitamin A.

The conversion of vitamins into coenzyme and other active forms is effected by means of special enzymes and enzyme systems.

However vitamin coenzymes perform their functions in metabolic processes not by themselves but in cooperation with respec-

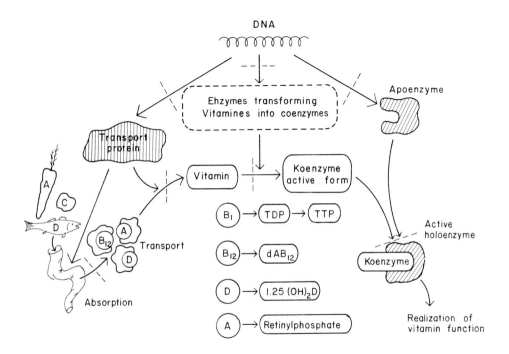

Fig. 1 - Vitamin metabolism and its inborn errors.

tive proteins-apoenzymes. It also concerns active forms of fat-soluble vitamins: e.g. 1,25-dihydroxycholecalciferol functions depend on the normal functioning of protein systems in regulation of calcium and phosphorus absorption in the intestine, their reabsorption in kidneys and mobilization from bone tissue.

The question arises why nature, while freeing the man and animals from a large biochemical work in vitamin synthesys, has however retained some final chemical stages for a vitamin to undergo before it can fulfil its functions in the body. One could suppose that this retention under the control of the organism is a useful evolutionally-determined mechanism aiming at maintaining homeostasis and metabolic equilibrium despite short-term fluctuations of vitamin intake.

No maintenance of homeostasis would be possible if vitamins and coenzymes possessed a catalytic activity of their own and if the rate of biochemical reactions changed depending on varied vitamin intake.

It can be seen on 1,25-dihydroxycholecalciferol formed in kidneys and controlled by a feedback mechanism assisted by parathyroid hormone.

This regulation provides Ca homeostasis and constant Ca levels in plasma despite variations in Ca and Vitamin D intakes and changes in physiological status.

These stages in vitamin metabolism controlled by the regulatory systems of the organism are the necessary condition for the realization of vitamin's functions and could be regarded as an important mechanism of integration and regulation of metabolism. However this mechanism would cause some trouble.

Since any vitamin could fit in metabolism after certain preparatory stages, any disorder in the chain of its absorption, transport and activation would result in a vitamin failing to perform its functions. As one of the causes we consider the losses or defects in genetic information encoding the synthesis of certain protein.

Let's look at the scheme (Fig. 1). Any protein is encoded in the genetic apparatus of the cell. Any mutation concerning respective sites of genetic information could disturb the synthesis or structure of such proteins involved in vitamin metabolism and functions. In consequence the vitamin fails to perform its function despite its proper intake.

We can cite as an example many such diseases with pathogenesis revealed in recent years.

Leigh's disease or inborn subacute necrotizing encephalomyelopathy (SNE) is associated with disorders in thiamine metabolism (Fig. 2). This disease is characterized by basic neurologic signs of vitamin B_1 deficiency but even high doses of this vitamin produce no therapeutic effect and the disease involves fatal cases in the firts months of life. According to J.R. Cooper and J.H. Pincus (1972) the disease is caused by disorders in thiaminetriphosphate (TTP) synthesis. The cause of disorders is associated with the presence in tissues and body fluids of anomalous glycoprotein inhibiting ATP-TDP-phosphotranspherase that synthesizes TTP from TDP and ATP.

Besides SNE inherited disturbances of TDP coenzyme functions are described: intermittent ataxia with disorders in pyruvate dehydrogenase activity and thiamine-responsive 'maple syrup urine' disease. In the last case the defect concerns TDP-dependent system of branched chain α-keto acid decarboxilation. In both cases the defect is associated with disorders in the structure and properties of TDP-interacted apoenzymes.

Short characteristics of inherited defects in thiamine metabolism and functions are shown in Table 1.

These are descriptions of many inborn diseases determined by genetic defects of pyridoxalphosphate (PALP) coenzyme function (Table 2): pyridoxine-dependent and pyridoxine-resistant forms of homocystin-uria and cystathioninuria, pyridoxine-responsive convulsive syndrome, pyridoxine-responsive anemia and xanturenuria. In such diseases the cause of the metabolic defect is the genetically-determined disturbance in structure and properties of PALP-interacted respective apoenzymes. Sometimes the defect directly destroys the structure of apoenzyme sites responsible for PALP binding.

Kim, V.J. & Rosenberg, L.E. (1973) showed that cystathionine synthase isolated from skin fibroblast of a pyridoxine-responsive homocystinuric patient bound PALP more than 100 times weaker than normal anoenzyme did. In such cases regular administration of pyridoxine high doses with increased PALP concentration in tissues provides better saturation of mutant apoenzyme and a partial correction of the defect. If inactivation of anoenzyme is determined by other structural changes pyridoxine high doses produce no biochemical or clinical effects.

With regard to vitamin B_{12} (Table 3) one should note inborn megaloblastic anemia when as a result of genetic mutation the structure of intrinsic factor is disturbed and it loses its capability of binding vitamin B_{12}.

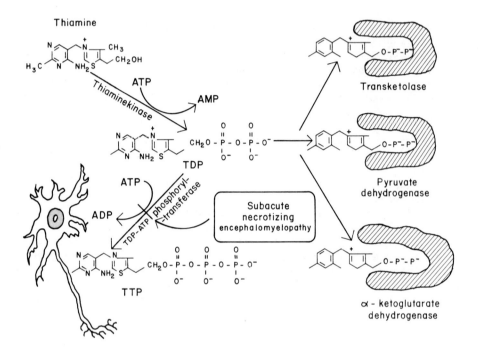

Fig. 2 – Metabolic functions of thiamine and their inborn disturbances.

TABLE 1

INBORN DEFECTS OF THIAMINE METABOLISM AND FUNCTION

DISEASE	DEFECT	TREATMENT
Leigh's subacute necrotizing encephalomyelopathy (SNE) (Cooper, J.R., Pincus, J.H., 1972	TTP biosynthesis	1,5-2,0 g thiamine or thiamine propyldisulfide
Intermittent ataxia (Blass, J.P., Lonsdale, D., 1970)	Pyruvate dehydrogenase	0,3-0,6 g thiamine
Thiamine-responsive megaloblastic anemia (Rogers, L.E. et al., 1968)	?	20 mg thiamine
Thiamin-responsive Maple Syrup urine disease (Lancaster, G. et al., 1974)	Branched-chain keto acid dehydrogenase	Protein restriction, 10 mg thiamine

TABLE 2

INBORN DEFECTS OF VITAMIN B$_6$ METABOLISM AND FUNCTION

DISEASE	DEFECT	TREATMENT
Homocystinuria, pyridoxine-responsive or resistant (Gerritsen, T., Waisman, H.A., 1972)	Cystathionine synthase	250-500 mg pyridoxine, reduction of methionine intake
Cystathioninuria, pyridoxine-responsive or resistant (Frimpter, G.W., 1972)	Cystathionase	25-50 mg pyridoxine
Pyridoxine dependent convulsions (Hinkel, G.K., 1970)	Glutamic acid decarboxylase	10-80 mg pyridoxine
Pyridoxine responsive anemia (Horrigan, D.L., Harris, J.W., 1964)	Defect of Fe utilization in the Hb biosynthesis	5-10 mg pyridoxine, 750 mg tryptophan
Pyridoxine dependent xanthurenic aciduria (Knapp, A., 1977)	Kynureninase	80-120 mg pyridoxine and 50 mg nicotinamide

TABLE 3

INBORN DEFECTS OF VITAMIN B_{12} METABOLISM AND FUNCTION

DISEASE	DEFECT	TREATMENT
Imerslund-Gräsbeck syndrome (Tiganova, I.S., Stoljarova, V.K., 1972)	Selective vitamin B_{12}	100-200 mcg vitamin B_{12} i.m.
Megaloblastic anemia (Katz, M., 1972)	Vitamin B_{12} malabsorption due to biologically inactive intrinsic factor (IF)	pig IF or 1-2 mcg vitamin B_{12} i.m.
Megaloblastic anemia (Hakami, R., Neumann, P.E., 1971)	Inherited transcobalamin II deficiency	2000 mcg vitamin B_{12} i.m.
Inherited deficiency of transcobalamin I (asymptomatic) (Carmel, R., Herbert, V., 1969)	Inherited transcobalamin I deficiency	—
Inherited methylmalonic acidemia: 1. vitamin B_{12} responsive 2. vitamin B_{12} resistant 3. vitamin B_{12} resistant 4 methylmalonic acidemia with homocystinuria (Rosenberg, L.E., 1972; Kang, E.S. et al., 1972)	dAB_{12} biosynthesis apo-methylmalonyc-CoA mutase methylmalonyl-CoA-racemase dAB_{12} and CH_3-B_{12} biosynthesis	1000 mcg vitamin B_{12} protein restriction

In Immerslund-Gräsbeck syndrome intrinsic factor is properly synthesized but the structure of the receptor binding intrinsic factor on entherocyte membrane is disturbed. This defect is also followed by vitamin B_{12} malabsorption and anemia.

Inborn defects of transcobalamine I and II that transfer vitamin B_{12} into the cell are also known.

Inherited methylmalonic acidemia studied by L. Rosenberg (1972) and S.H. Mudd et al. (1970) comprise a special group (Fig. 3). These diseases are caused by a selective disturbance in functions of vitamin B_{12}-dependent enzyme — methylmalonyl Co A — mutase converting methylmalonic acid into succineic acid. One of the forms of methylmalonic acidemia is caused by a disorder in the synthesis of vitamin B_{12} coenzyme form-5'-deoxyadenosyl-cobalamine. In another form the defect concerns the structure of the apoenzyme. In both cases vitamin B_{12} functions are disturbed despite the proper intake of the vitamin.

As far as folic acid is concerned certain forms of megaloblastic anemia and mental retardation caused by malabsorption of folates in the intestine, some defects in their transport into the brain, synthesis and transformations of folic acid coenzymes are described (Lanzkowsky, P. et al. 1969; Arakawa, T., 1970).

Inborn errors in vitamin D metabolism and function are of great practical significance. Detection of the biochemical mechanism of these errors is closely-associated with great successes in the study of vitamin D metabolism attained by E. Kodicek (1974), H.F. de Luca (1974, 1976, 1977), A.W. Norman (1968, 1974), Ch.R. Scriver (1973) a.o.

In pseudo deficient vitamin D-dependent rickets hypocalcemia and defects of bone mineralization are caused by genetically determined malformation of vitamin D hormonal form — 1,25-dihydroxycholecalciferol (1,25-DCC) in kidneys. In inherited hypophosphatemic vitamin D-resistant rickets the formation of 1,25-DCC is not disturbed; the defect concerns the transport system involved in phosphate reabsorption by kidneys. It is followed by an excessive loss of phosphate with urine, hypophosphatemia and severe deformations of the skeleton in spite of normal vitamin D intake (D. Fraser, Scriver, Ch.R., 1976).

Similar types of disturbances are known at present for all vitamins. It allows to single out besides classic avitaminosis and hypovitaminosis a special group of numerous inborn errors of vitamin metabolism and functions.

On the basis of the primary biochemical defect the following classification of these disturbances can be proposed (Table 4) (Spirichev, V.B. & Barashnev, Ju.I., 1977).

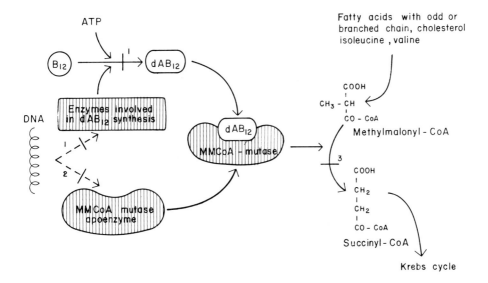

Fig. 3 – Inborn errors of vitamin B$_{12}$ metabolism in methylmalonic acidemiae.

TABLE 4

CLASSIFICATION OF INHERITED DEFECTS OF VITAMIN METABOLISM AND FUNCTION

A. ON THE BASIS OF BIOCHEMICAL DEFECT

1. Inherited defects of vitamin absorption (defect of intrinsic factor, Imerslund-Gräsback syndrome)

2. Inherited defects of vitamin transport (transcobalamin II deficiency, abetalipoproteinemia)

3. Inherited defects of vitamin conversion into active form (vitamin B_{12} responsive methyl-malonic acidemia, SNE, vitamin D dependent rickets)

4. Defects of apoenzyme or another proteins (pyridoxine responsive anemia, vitamin D resistant rickets)

B. ON THE BASIS OF VITAMIN RESPONSIVENESS

1. Vitamin dependent

   - responsive to high doses of vitamin

2. Vitamin resistant

   - unresponsive to high doses of vitamin

The frequency of inborn diseases associated with disorders in metabolism and functions of various vitamins is different and rather high in some cases. For instance homocystinuria appears to be the most frequent inborn error after phenylketonuria Cystathioninuria accounts for 1 case per 10-20 thousands of newborns. Vitamin D-resistant rickets and pyridoxine-responsive anemia seem to be rather frequent too. The frequency of inborn errors of vitamin metabolism and functions appears to depend on the experience of medical practitioners and their ability to diagnose such diseases.

According to Ch. R. Scriver at present disorders in vitamin metabolism are a more serious and urgent problem than classic avitaminosis at least in developed countries.

The severity of clinical manifestations in such diseases varies depending on the character of defect and its metabolic consequences. Rarely such defect has no apparent signs. But more often they involve severe disturbances of physical and mental development and even fatal cases.

Early diagnostics and detection of a primary biochemical defect allows in most cases to suggest pathogenetically determined methods for therapy.

In vitamin-dependent states one should select megavitamin therapy-administration of a respective vitamin in doses 100-1000 times larger than the Physiological requirement.

Vitamin-responsive states include all disorders of vitamin absorption since in such cases respective vitamins could be parenterally administered.

In disturbances of vitamin conversion into coenzyme forms the defect could be corrected if there was some residual enzymatic activity left that could be used more intensively by administrating higher vitamin doses.

The same refers to inborn errors of apoenzyme structure if an interaction of apoenzyme with coenzyme is disturbed. In such cases megavitamin therapy aims at increasing concentration of coenzyme in tissues as is the case in pyridoxine-responsive homocystinuria. This method favours better saturation of mutant protein by coenzyme and rises residual catalytic activity.

A therapeutic effect of high vitamin doses might be caused by the activation of other metabolic compensatory ways too.

As an example of a successful usage of megavitamin therapy we can cite a treatment of pyridoxine-responsive convulsion syndrome, pyridoxine-responsive anemia and homocystinuria by high doses of

vitamin B_6. Table 5 illustrates the results obtained by the treatment of a pyridoxine-responsive homocystinuric patient (Spirichev,
V.B. & Barashvev, Ju. I., 1977). During the first examination of
the patient high urinary excretion of methionine and homocysteine
and mental retardation (IQ=32) were recorded. Four days later pyridoxine treatment (100 mg/day) improved biochemical indices: methionine
excretion decreased, homocystein disappeared. Later on the patient
was treated with 30 mg pyridoxine daily. During the second examination a year later favourable results of treatment were noted:
IQ = 45, normal aminoacid metabolism, darkening of hair and eye-
lashes.

The intake of high vitamin doses in inborn errors of vitamin
metabolism should be regular during the whole life of the patient.
Thus patients with inborn disorders of vitamin metabolism present a
special category of consumers of great quantities of synthetical
vitamin drugs. From this point of view the investigations in the
field of inborn error of vitamin metabolism may be of interest for
the companies producing vitamin drugs.

Special attention should be paid to the problem of the use of
synthetical coenzymes and other active vitamin forms in inborn disorders of their biosynthesis or interaction with proteins.

The usage of synthetical 1,25-DCC and its analogue 1α-hydro-
xycholecalciferol in inborn vitamin D-dependent rickets shows the
successfulness of such an approach in some pathologies.

However such an approach is not applicable in all cases. Most
vitamin B phosphorylated forms: TDP, PALP, vitamin B_{12} coenzyme
etc. penetrate very badly through cellular membranes and are degraded by blood phosphatases. In this connection it is desirable
to create chemically-modified coenzyme forms resistant to phosphatase action and possessing high ability to penetrate into the cell.
It might be obtained by addition to coenzyme molecule of lipophylic
groups capable of penetrating into the cell; their introduction
into liposomes, etc.

We are facing an urgant problem of the creation of TTP forms
freely-penetrating through the hemato-encephalic barrier for the
treatment of SNE and of similar folic acid forms — for prevention
of CNS troubles in inborn defects of folate transport. It's also
desirable to create TDP and PALP forms penetrating into the cell.

Besides developing freely-penetrating drugs a particular attention should be paid to finding new ways of increasing membrane
permeability to coenzymes and rising of their intracellular concentrations by accelerating biosynthesis and decreasing degradation.

TABLE 5

BIOCHEMICAL AND CLINICAL FINDINGS IN A PATIENT WITH
PYRIDOXINE DEPENDENT HOMOCYSTINURIA

The effect of pyridoxine treatment (100 mg per day)

(Spirichev, V. & Barashnev, Ju., 1977)

	BEFORE TREATMENT	AFTER TREATMENT
Urinary homocysteine (mg/24 h) Normal = 0,0	180	0
Urinary methionine (mg/24 h) Normal = 2-7	44	5
I.Q.	32	45

When treatment with high doses of vitamins or their active
forms fails to be effective the basic method for treatment is dietetic correction of the defect. This approach consists in the decreasing intake of food constituents whose splitting is inhibited
and in the enrichment of the diet with compounds whose synthesis is
disturbed. The success of such approach depends on the development
and production of special artificial feed mixtures.

In the future the development of methods for enzymotherapy and
gene engineering will assume greater importance as the most radical
methods for treatment of inborn metabolic errors.

BIBLIOGRAPHY

Arakawa, T. Am. J. Med., 48: 594-601, 1970.
Blass, J.P., Lonsdale, D. et al. Clin. Res., 18: 393, 1970.
Carmel, R. and Herbert, V. Blood, 33: 1-12, 1969.
Cooper, J.R. and Pincus, J.H. J. Agr. Food Chem. 20: 490-493, 1972.
De Luca, H.F. Am. J. Clin. Nutr., 28: 339-345, 1975.
Fraser, D. & Scriver, Ch.R. Amer. J. Clin. Nutr., 29: 1315, 1329,
      1976.
Frimpter, G.W. In: "The Metabolic Basis of Inherited Diseases",

3. ed., 1972, 413-425.

Gerritsen, T. and Waisman, H.A. In: "The Metabolic Basis of Inherited Diseases", 3. ed., 1972, 404-412.

Hakami, N., Neumann, P.E. et al. N. Engl. J. Med. 285: 1163-1170,
1971.

Harris, J.W. and Whittington, R.M. Proc. Soc. Exper. Biol. Med.,
91: 427-432, 1956.

Hinkel, G.K. and Kintzel, H.W. u.a. Schweiz. Med. Wochenschr.,
100: 1152, 1970.

Horrigan, D.L. & Harris, J.W. Adv. Intern. Med. 12: 103-174, 1964.

Hunt, A.D., Stokes, J. et al. Pediatrics, 13: 140, 1954.

Kang, E.S., Snodgrass, P.J. and Gerald, P.S. Ped. Res., 6: 875-879,
1972.

Katz, M. N. Engl. J. Med. 287: 425-430, 1972.

Kim, V.J. & Rosenberg, L.E. Ped. Res., 7: 291, 1973.

Knapp, A. Genetische Stoffwechselstörungen, 2. Aufl., 1977, Jena.

Kodichek, E. Lancet, N 7853: 325-328, 1974.

Lancaster, G., Mamer, O.A. & Scriver, Ch.R. Metabolism, 23: 257-265,
1974.

Lanzkowsky, P., Erlandson, M.E. & Bezan, A.I. Blood, 34: 425-465,
1969.

Mudd, S.H., Edwards, W.A. et al. J. Clin. Invest. 49: 1762-1773,
1970.

Norman, A.W. & Henry, H. Recent Progr. Hormone Res., 30: 431-473,
1974.

Rogers, J.F., Porter, F.S. & Sindbury, J.B. J. Pediatr. 74: 494-
505, 1969.

Rosenberg, L. In: "The Metabolic Basis of Inherited Dieases",
3. ed. 1972, 440-458.

Scriver, Ch.R., Metabolism, 22: 1319-1344, 1973.

Spirichev, V.B. & Barashnev, Ju. I. "Inborn Errors of vitamin metabolism", Moscow, 1977.

Tiganova, I.S. & Stoljarova, V.K. Pediatria (Russ.) 8: 62-65, 1972.

FACTORS AFFECTING THE BIOAVAILABILITY AND METABOLISM

OF VITAMIN A AND ITS PRECURSORS

James Allen Olson

Prof. of Biochemistry and Biophysics
Iowa State University
Ames, Iowa 50011 - U.S.A.

In dealing with the problem of vitamin A deficiency in preschool children, the dietary intake required to satisfy the needs of a child for growth and the prevention of clinical symptoms of vitamin A deficiency should obviously be defined. At first glance, we are seeking a specific number, which appears to be a simple, easily resolvable task. In actuality, it is not a simple task at all, because the number keeps changing as a result of five major factors: 1) the bioavailability of vitamin A and its precursors in the diet, 2) nutrient interaction, 3) the presence, length, and severity of disease, 4) metabolic parameters — and in particular storage efficiency, the turnover of reserves, and recycling efficiency, and 5) statistical considerations. I wish first to consider each of these major factors, and then to suggest a new approach to vitamin A nutriture which might allow simplification of many of the complexities involved in assessing their importance.

The amount of carotenoids and vitamin A in the diet can be fairly well estimated under two major conditions: 1) when essentially no vitamin A is present, e.g., on a post-weaning diet primarily composed of sticky rice or of cassava gruel and molasses, or 2) when abundant amounts of vitamin A or provitamins are ingested, such as liver, red palm oil, papaya, amaranth, or indeed vitamin A supplements. Accurate dietary information becomes most difficult to obtain among children who receive marginal amounts (100-300 μg) of vitamin A or its equivalent. Information obtained by dietary recall is only semiquantitative, even under the best of conditions; local fruits and vegetables often contain quite different concentrations of provitamin A than those indicated in standard dietary tables, the state of maturity and mode of preparation of fruits and vegetables can drastically affect the percentage of provitamins present among total carotenoids, the direct chemical analysis of food together with the professional monitoring of intake over a significant period is both costly and time consuming, and seasonal changes in the diet can be considerable. Now I do not wish to downgrade the general utility of dietary information in assessing many aspects of nutritional status; knowing what children eat is an essential part of baseline information on which nutritional programs should be based. What I am stressing is that the actual intake of carotenoids and vitamin A is particularly difficult to assess among that large group of children who are receiving marginal levels of many nutrients, including vitamin A and carotenoids.

The conversion of carotenoid intake into retinol equivalents, even when the carotenoid intake is accurately known, poses additional problems. The generally recognized equivalencies are 1 μg retinol = 6 μg β-carotene = 12 μg mixed dietary carotenoids. In India 1 μg retinol is equated with 4 μg β-carotene, and in recent controlled studies with adult American males, 2 μg β-carotene, given as a pure substance in oil, gave the same response as 1 μg retinol. But I do not urge that the retinol equivalent of carotenoids be redefined; indeed, it may still be the best overall value to use. The point I am making is that the value is not necessarily reliable under a given set of conditions, and consequently, that estimates of vitamin A intake become even more uncertain.

Quite apart from the problem of estimating the amount of vitamin A and biologically-active carotenoids in the diet, the absorption efficiency must be considered. Happily, vitamin A is very well absorbed under normal conditions, i.e., about 80% or more of the dose, and the absorption efficiency falls only slightly with increasing dosage. Thus, a 200,000 I.U. dose of retinyl palmitate, which has been extensively used in massive dosing programs, is >60% absorbed. Carotene is different; first of all, more fastidious requirements exist for the formation of a suitable micelle for carotenoids; secondly, the absorption rate is much slower than for vitamin A; and thirdly, the absorption efficiency falls rapidly with increasing dosage. Thus, under ideal conditions, about 60% of a dose of β-carotene is absorbed, but it can be much lower.

But let us move on to consider the interaction of other nutrients with vitamin A. We are all aware that nutrients affect each other in many ways, and vitamin A is no exception to this rule. Predominant among nutrient factors which are associated with acute vitamin A deficiency is severe protein energy malnutrition. Although corneal disease due to vitamin A deficiency may occur in moderately well-nourished children, extensive corneal involvement and keratomalacia seem to be mainly, if not invariably, associated with severe PEM. In a biochemical sense, we can cite many enzymes and proteins involved in vitamin A metabolism which are impaired in PEM: carotene cleavage in the intestinal mucosa, the synthesis and release of

plasma retinol binding protein, the formation of the storage lipoglycoprotein of liver, the formation and release of lipolytic and
proteolytic enzymes of the gastrointestinal tract, and many more.
Thus, in PEM vitamin A and carotenoids are less well absorbed, less
well stored as retinyl ester in the liver, and less well transported
to peripheral tissues as a RBP-retinol complex.

But just as PEM markedly affects vitamin A utilization, so also
does vitamin A deficiency impair protein utilization. Of particular
note are the abnormal plasma amino acid patterns, the negative nitrogen balance, and the loss of appetite observed in acute vitamin A
deficiency in experimental animals. Thus, vitamin A deificency and
PEM exacerbate the imbalances caused by each other, thereby setting
in motion a tragic, vicious circle of rapid debilitation. The close
epidemiological relationship among deficiencies of these nutrients,
therefore, is in full accord with biochemical and physiological observations.

Vitamin E and vitamin A also have been linked nutritionally.
In vitamin E deficient experimental animals, vitamin A is less well
absorbed and stored and is more rapidly depleted from the liver.
Interestingly these effects seem to be relatively specific, in that
other anti-oxidants do not replace α-tocopherol. The existence of
frank vitamin E deficiency in children is still uncertain, alghough
severely malnourished children must certainly be marginal in vitamin E nutriture if not acutely deficient. Nonetheless the addition
of vitamin E to a vitamin A preparation given to malnourished children did not significantly enhance vitamin A absorption and storage.
Nonetheless, in view of its anti-oxidant properties, its low cost
and its possible effectiveness in severe cases, the inclusion of
vitamin E in vitamin A preparations is certainly well-warranted.

Other nutrient interrelationships with vitamin A are less well
defined. Anemia due to iron deficiency has been correlated with
vitamin A deficiency epidemiologically, but whether a specific molecular interaction exists is yet uncertain. Similarly, zinc deficiency and vitamin A deficiency produce some similar clinical
signs, such as growth depression and epithelial keratinization.
Although plasma vitamin A levels are depressed in zinc deficiency
and are restored upon supplementation with zinc, zinc deficiency
seems to act indirectly on vitamin A metabolism by reducing tissue
requirements for vitamin A as a result of a reduced growth rate.
Undoubtedly, other interactions exist; calcium excretion increases
in vitamin A deficient animals, for example, although vitamin A
status and urinary calcium excretion have not been correlated in
malnourished children. But as a general principle, we might expect
that any stress — whether nutritional imbalance, infection, harsh
climate, or the like — which impairs the capability of an organism
to function optimally will inevitably increase the need for any
specific nutrient.

Disease is a good case in point, in that the processes of vitamin A absorption, transport, storage or metabolism may be disrupted. In chronic lipid malabsorption syndromes, such as pancreatitis, biliary cirrhosis, sprue, chronic diarrhea and cystic fibrosis, one or several aspects of the absorption process is defective, and consequently, vitamin A together with most lipids is poorly absorbed. Interestingly, a significant part of a massive dose of vitamin A is still taken up in such patients, alghough the efficiency is of course low. The absorption of β–carotene and other carotenoids, which have more fastidious requirements for micelle formation and are normally absorbed at a much slower rate than vitamin A, is particularly affected by malabsorption syndromes.

Diseases, such as hyperthyroidism or those associated with febrile conditions, increase the rate of tissue metabolism and consequently increase the needs for vitamin A. Other energy-draining stresses, such as chronic exposure to cold, have similar effects both on the metabolism of vitamin A and on its required intake.

Because of the importance of the liver and kidney in the storage of vitamin A and in the transport and metabolism of retinol binding protein, diseases of these two organs will naturally affect vitamin A nutriture in a marked way. Parenchymal cell disease, such as hepatitis, will depress the synthesis of apo-RBP, prealbumin and storage lipoglycoprotein, and will therefore influence both the storage of vitamin A and its transport in the plasma. The kidneys are involved both in holo-RBP absorption from the tubules and in the catabolism of apo-RBP. In tubular dysfunction, therefore, much RBP is lost in the urine, and in renal diseases which depress glomerular filtration, holo-RBP levels in the plasma rise dramatically.

Since the physiological role of vitamin A is expressed within cells, the ultimate objective of nutrition is the provision of an optimal amount of a biologically active molecule at its active site within cells at the proper time. Outside of vision, where 11-cis retinal is involved as the chromophore of visual pigments in rod and cone cells, we do not know precisely what the physiological site is for the somatic functions of vitamin A. Nonetheless, a phosphorylated form of vitamin A can serve a function in glycoprotein synthesis, and a protein-bound form may influence the expression of the genome. But whatever its precise function, the presence of vitamin A within target cells requires suitable intracellular transport, recognition of RBP by cells, uptake of vitamin A by cells, transport in the plasma and controlled release from the liver.

To maintain a proper steady-state, vitamin A must also be oxidized, cleaved and excreted. Indeed, the major protection against hypervitaminosis A when large doses of vitamin A are ingested are these same flexible catabolic systems. Finally, conservation mechanisms exist; vitamin A released in the form of glucuronides in the bile is partially reabsorbed in the intestine and recycled back to the liver. Similarly, retinol released as a complex with retinol binding protein is recycled to a significant extent from peripheral tissues, probably as retinyl ester, back to the liver.

Each of these interrelated processes is under subtle metabolic controls about which we still know very little. But in a nutritional sense, the disruption of any one of them can either precipitate a state of deficiency or increase substantially the requirements of vitamin A.

Statistical considerations are of great significance in vitamin A nutriture, and in nutrition generally. Ideally, all members of a society should be well nourished and happy. Unfortunately, such Utopian dreams help us little in dealing practically with current problems of malnutrition. In approaching this issue, current Recommended Dietary Allowances are defined as "the levels of intake of essential nutrients considered to be adequate to meet the known nutritional needs of practically all healthy persons." With the exception of energy, which is expressed as an average requirement for 97.5% of a given group, i.e., two standard deviations from the mean. Unless the distribution curve is well-defined with respect to a given indicator, however, no way exists of estimating the degree of sufficiency of any given intake or the mean requirement.

The R.D.A. for vitamin A for children 0.5-4 years of age is 400 μg retinol equivalents per day. Although this value is rather an arbitrary one, no child receiving that amount of vitamin A, to my knowledge, has suffered any impairment of growth or any sign of deficiency. On the other hand, we know little of the distribution curve for vitamin a intake with respect to any given indicator, and, hence, can make no meaningful estimate of the mean requirement, or of a minimal requirement to prevent corneal disease. Perhaps the effects of nutrient interaction, disease, differences in growth rate, and variations in the metabolic balance of children would make such estimates either impossible or useless. But this matter should probably be probed a bit further, at least to the extent of relating vitamin A intake, insofar as possible, to clinical signs of deficiency.

A new approach, which avoids most of the difficulties cited above, might be taken to the problem of assessing vitamin A status; namely, to define the vitamin A intake needed to maintain the total body stores of vitamin A at a satisfactory steady state level. In infants and small children, a satisfactory level might be set at 20 μg retinol/g liver, or about 10 mg retinol for a 15 Kg child. The replacement rate might then be calculated both with respect to liver turnover and to RBP synthesis. By using an estimated half life of 50 days for liver retinol, the fractional metabolic rate would be 0.014 of the total body pool per day, or 140 μg. If the

efficiency of storage is 50%, the child should ingest 280 µg retinol
per day to maintain the total body pool at 10 mg. At a plasma RBP
concentration of 20 µg/ml, the total plasma RBP pool would amount
to 11.1 mg, which would bind 150 µg retinol. If replaced 2.7 times
daily, retinol used daily for holo-RBP synthesis would be 410 µg.
A significant portion of that retinol, perhaps 70%, would be recycled back to the liver. Thus, with 30% excretion and 50% absorption, 246 µg retinol would be required in the diet daily. However,
if no recycling takes place and the storage efficiency is 50%, 820 µg
retinol should be ingested daily to maintain the steady state. These
calculated dietary intake values of 280, 246, and 820 µg retinol
agree remarkably well with the empirically set R.D.A. of 400 µg
retinol for 0.5-4 year old children.

Determination of the total body stores of vitamin A by the use
of deuterated vitamin A has been accomplished in experimental animals
and has been applied to humans in a few cases. Further refinement of
this method, as well as other non-invasive techniques, might allow
this more kinetic approach to the nutrition of vitamin A to be fruit-
fully applied.

The above calculation is not intended as a definitive expression
of the required intake of vitamin A for young children, but rather
as an example of a different approach to the definition of vitamin A
needs. The example also points out the need to obtain more information about the turnover of vitamin A in the liver and of RBP in the
plasma as a function of age and stress, and to define the extent of
recycling under various conditions. The fact that calculated values
based on the concept agree reasonably well with those obtained empirically bodes well for the use of these procedures in the future.

BIBLIOGRAPHY

NAS. Recommended Dietary Allowances, Eighth revised edition, Food
        and Nutrition Board, National Research Council, National
        Academy of Sciences, Washington, D.C., 1974.

CHEMISTRY AND METABOLISM OF VITAMIN C

Bert M. Tolbert

Dpt. of Chemistry - Univ. of Colorado

Boulder, Colorado 80309 - USA

INTRODUCTION

The nutritional requirement for ascorbic acid is based on a physiological function for this vitamin, which in turn is based on biochemical roles that operate at an enzymic level. This dogma is easy to state, but an abundance of experimental data indicate that our understanding of the physiological and biochemical roles of ascorbic acid are far from complete. Four enigmas in the physiology of ascorbic acid will serve to illustrate my argument: radioautographic studies as well as tissue assay show high levels of ascorbic acid in adrenals, brain, gonads, eye and salivary glands. In these tissues at least 90% of the ascorbic acid is present as free ascorbic acid. Experiments demonstrating a physiological function for these high levels of ascorbic acid are not available except perhaps for the adrenals.

Ascorbic acid has not been observed in E. Coli and other procaryotes, but is present in all animals and plants, and it or some closely related compounds may be present in yeast and fungi. What is unique to the eucaryotes that requires this substance? Eucaryotes that cannot make ascorbic acid and have a nutritional requirement for ascorbic acid now include: primates; guinea pigs; bats; many birds; trout, salmon, bass, carp and probably many other fish; and several insect species.

In higher vertebrates such as man and rat, experimental data indicate that tissue ascorbic levels are maintained by: 1) a facilitated transport system in the intestines; 2) An efficient kidney excretion; 3) An inducible kidney excretion process to increase elimination during periods of high ingestion of ascorbic acid;

4) a facilitated transport system for ascorbic acid into specific tissue which follow saturation kinetics.

In both plant and vertebrate tissue, levels of ascorbic acid vary in response to physiological conditions. A classical example is the increase in tissue levels of ascorbic acid during regeneration of wounds. In plants, ascorbic acid levels are high in growing tips, and rises and falls in a cycle with ripening of fruits. Teleologically it is reasonable to argue that there should be a role for ascorbic acid that is common to both plants and animals. Except for a general oxidation-reduction role, no common function has been proposed. The observations just listed give many indications that ascorbic acid is in some way related to the regulation of growth.

One of the many experimental approaches that can help answer the question of what are the physiological roles of ascorbic acid is to determine the metabolism of the compound, together with tissue and subcellular localization of metabolic sites. Since it is not certain that ascorbic acid is the only active form of this vitamin, metabolic studies have the exciting possibility of discovery of new structures with biological activity, even as has been done in recent years for Vitamin D.

CATABOLIC FATE OF ASCORBIC ACID

In guinea pigs 60 to 70% of a tracer dose of ascorbic acid labeled with carbon-14 in the one position is converted to CO_2 and excreted in the breath. The rest is excreted as water soluble compounds in the urine and less than 1% is present in the feces. This is observed whether the ascorbic acid is given by mouth or by injection. Similar results are observed in the rat, except only 20-30% is oxidized to CO_2 in breath. In man and monkeys rather different results are observed. Burns and later Baker have shown that when small amounts of labeled ascorbic acid are ingested orally in man, no carbon-14 is excreted in the breath or feces and that urine is the only metabolic fate. More recent unpublished work by Johnson, Tolbert and co-workers has shown that when labeled-ascorbate is given to monkey by mouth 20 to 80% of the label may be excreted as CO_2, but when given by intravenous injection less than 1% is excreted as CO_2. These data show that there is a fundamental difference in the metabolism and fate of ascorbic acid in primates as compared to rats and guinea pigs. Rats and guinea pigs have a metabolic pathway for the degradation of ascorbic acid to CO_2 not present in primates. The missing factor is probably the enzyme dehydroascorbic acid lactonase.

There are many data to suggest incomplete absorption of oral ascorbic acid in man. Angel et al have studied ascorbic acid ex-

cretion as a function of intake in a series of weekly regimes. They
find that on ingestion of 1/2 to 3 g ascorbic acid taken in divided
doses with meals, the excretion of ascorbic acid and other DNPH
reactive metabolites is about 1/2 the intake. On the other hand,
unpublished studies by E.M. Baker showed that intravenous infusion
of 3 g of ascorbate into human volunteers resulted in the prompt
excretion of more than 98% of the ascorbate in the urine. Thus if
ascorbic acid is absorbed in man or monkeys, it is all excreted in
the urine, either as metabolites or unchanged ascorbic acid, and
any degradation of ascorbic acid to CO_2 in primates requires an intestinal process, probably a microbiological oxidation.

METABOLITES OF ASCORBIC ACID

One may assume that the pathway leading to CO_2 is not basic
to any essential function of ascorbic acid, but is rather an elimination pathway. Therefore in primates all excretion metabolites of
ascorbic acid are of interest as possible indicators of biological
functions.

A well known metabolite of ascorbic acid in urine is oxalic
acid, derived from the 1 and 2 carbons of ascorbic acid by a C_2/C_3
cleavage process. It is a minor metabolite of ascorbic acid, accounting for 5-10% of the excretion products. The precursors and
catabolic pathways for oxalate are unknown.

Ascorbate-2-sulfate has been shown to be present in the urine
and tissue of a variety of animals — man, rat, trout, brine shrimp
cysts and others. It is a vitamin in fish, but in reasonable doses
does not cure or prevent scurvy in guinea pigs or monkeys. In
mammals it is both rapidly eliminated after parenteral administration
and poorly absorbed on oral ingestion. The defect seems to be one
of absorption, both cellular and intestinal, for an enzyme with
ascorbate sulfatase activity is present in mammalial tissue. A biological role for ascorbate sulfate is not known. In fish and brine
shrimp cysts, it may serve as a storage form of ascorbic acid since
it is stable to air oxidation. It does not seem to be a sulfating
agent. It may be mainly an elimination product created by a lack of
substrate specificity of the enzyme that sulfates catechols. Ascorbic acid is similar in size and polarity to the catechols.

2-O-methylascorbic acid is present in the urine of rats. Again
it is a minor metabolite, formed by action of the enzyme that methyl-
ates catechols.

The major urinary metabolites of ascorbic acid have not been
identified. Studies on the urine of monkeys, rats and guinea pigs
given labeled ascorbic acid show that there is a large number of
metabolites — 3-4 major ones and 9-12 significant ones. These can

be separated by various means. DEAE sulfate column chromatography
of urine from monkeys given [1-^{14}C]-ascorbic acid show a number of
peaks and rather clearly show ascorbate sulfate and at least seven
other peaks.

In our laboratory Ron Harkrader has developed the following
scheme for the partial isolation of ascorbic acid metabolites from
monkey urine. Most of the labeled metabolites of urine may be extracted from lyophylized urine with absolute ethanol-3% trichloro-
acetic acid. Ethanol is a better solvent for these metabolites
than acetone or diazone and trifluroacetic acid is a better acidi-
fying agent than formic acid. After lead acetate precipitaion of
the extract the solution is chromatographed on a Dowex-1-formate
column using a formic acid gradient. Individual fractions were
assayed both for carbon-14 and by the DNPH reaction which is fairly
specific for the ene-diol lactone structure of ascorbic acid. The
major group of metabolites is not resolved by this procedure, but
the data show that many of the metabolites may be DNPH reactive.
When the major peak from the Dowex-1 column was rechromatographed
on Dowex-50 using water elution, the carbon-14 and DNPH reactive
material eluted on the front, indicating that the metabolites do
not contain basic groups. It is both interesting and significant
that as the purification continued, the carbon-14 labeled fractions
continue to give a positive DNPH assay. We know these compounds
are not ascorbic acid nor dehydroascorbic acid. They elute at
entirely different locations and are chemically stable. Thus the
DNPH assay for ascorbic acid in urine and tissue has much metabolite interference. In recent months we have been able to convert
some of these metabolite fractions into volatile trimethylsilyl
derivatives for gas chromatography and mass spectrometry. The major
peak from paper chromatography is resolved by gas chromatography
into 3 major peaks. Further identification of this material is
not complete. Analysis of human urine using the separation scheme
described above and DNPH assay give yield and chromatographic results similar to those for monkey urine.

We believe it will be easy to devise abbreviated separations
and assays for these metabolites of ascorbic acid in urine. Such
procedures will offer important new analytical tools to the nutritionists who desire to study the effects of environmental and nutritional factors on ascorbic acid metabolism.

The salivary glands are high in ascorbic acid, and we have
examined saliva from rats given [1-^{14}C] ascorbic acid for labeled
compounds. Saliva was deproteinated by ultra filtration and chromatographed on a DEAE-formate column using a formate gradient.
The saliva contained no ascorbic acid, consistent with the recent
results of Feller published in 1975. It contained one major metabolite. Calculations using the radioactivity in this peak together
with dilution factors for the size of the ascorbate pool in rats

showed that there was so little material in this peak or in saliva
that it could not have been detected by DNPH assay. Thus although
the salivary glands are high in ascorbic acid, neither ascorbic
acid nor any metabolite of ascorbic acid is excreted in saliva in
significant amounts.

Harkrader, Tolbert, Johnson and Joyce have studied the metabolism of ascorbic acid labeled either with carbon-14 or tritium
on the 6-carbon of ascorbic acid. From monkeys injected with [6-^{3}H] -
'ascorbic acid it was found that about 45% of the tritium had been
released from the ascorbic acid metabolites and appeared in water
of urine; this result was confirmed by periodate degradation which
showed that about 45% of the carbon-14 containing metabolites no
longer had a primary alcohol group at the C-6 carbon. This indicates
that C-6 oxidation is an important metabolic process for ascorbic
acid. One urinary metabolite degrades as if it were a 5-Keto-6-
carboxy derivative of ascorbic acid. These studies suggest extensive
involvement of the side chain of ascorbic acid in its metabolism,
and raise the question whether there could not be side chain altered
forms of ascorbic acid with specific biochemical or co-enzymic functions.

In plants Loewus infers a 5-keto intermediate. In ascorbate
metabolism he observes C_2-C_3 cleavage of the chain to give oxalate.
He also finds C_2-C_3 cleavage to give tartaric acid from carbons C_3
to C_6, suggesting a C-6 oxidized intermediate. There is evidence,
then, that ascorbic acid metabolism in plants and in vertebrates
have many features in common. The retention of these pathways in
such diverse biological systems suggest that they may fulfill important metabolic roles.

RECENT DEVELOPMENTS IN THE CHEMISTRY OF ASCORBIC ACID

Chemical derivatives of the side chain of ascorbic acid are
of metabolic and nutritional interest. The lability of the ascorbate
ring makes it difficult to do chemistry on this molecule, but new
ascorbic acid derivatives with modified side chains are being made.

Stuber and Tolbert have prepared saccharoascorbic acid by a
procedure involving protecting and deprotecting the ascorbate ring.
We have prepared it by the method shown. It is not a major urinary
metabolite, although it could be present at the few percent level.

Prof. Paul Seib has found that in concentrated sulfuric acid
ascorbic acid is rapidly sulfated at the 6-carbon with a minor product of ascorbic acid 5-sulfate. It is remarkable that the ascorbic
acid lactone ring is stable in concentrated sulfuric acid. In iso-
lating these compounds Prof. Seib observed that the barium salt of

the 5-sulfate decomposed to give a new product, which was shown to
be a 4,5-dehydrate compound. It is a very good reducing agent and
quite stable at pH 2 and pH 7, but decomposes at pH 11.

The 6-halo-6-deoxy derivatives of ascorbic have been patented
as antioxidants.

Ascorbate-2-phosphate is also stable in water at neutral pH.
It is a powerful inhibitor of Arylsulfatase A with a Ki of 1μM.
It is also an effective blood preservation agent. It and other
phosphates may have biological roles. I believe that the catabolism
of ascorbate probably proceeds through an activated molecule, and
ascorbate-6-phosphate is the most likely candidate for this intermediate.

ASCORBATE-MONODEHYDROASCORBATE-DEHYDROASCORBIC ACID

The reversible oxidation sequence of ascorbate-monodehydroascorbate-dehydroascorbic acid has long been considered central to
the physiological role of ascorbic acid. Studies on the monodehydro-
ascorbate ion have been made using pulsed electron irradiation and
spectroscopy. Laroff has shown that in the radical intermediate
the unpaired electron is spread over a highly conjugated tricarbonyl
system and that the same radical is produced from pH 1 to 13. Thus
it is a strong acid. The radical is relatively stable and reacts
only slowly with oxygen.

The commonly written structure of dehydroascorbic acid as a
tricarbonyl is an unlikely formula. Studies by Pfeilstiker on the
monomeric form of dehydroascorbic acid using NMR show that the compound is a lactone, probably hydrated. A better name would be
desirable.

Dehydroascorbic acid reductase has been purified from carp by
Yamamoto and from spinach leaves by Foyer and Halliwell. Both
enzymes catalyzed the reaction using glutathione as the reactant.
Km values for dehydroascorbic acid, 1/2 mM corresponding to 5-10
mgd As/100 ml, are high compared to blood dehydroascorbate levels
so we still are unsure how the As/d As ratio is maintained.

The spontaneous hydrolytic decomposition of dehydroascorbic
acid at pH 7 has been studied by Valisek and is sufficiently rapid
that this should be a significant pathway for ascorbic acid metabolism. But in primates this is not so. This suggests that either
tissue levels of dehydroascorbic acid are very low in primates,
or that free dehydroascorbic acid is not present. The measurement
of dehydroascorbic acid in biological samples is subject to many
chemical questions and interpretation of such data must be guarded.

TRANSPORT AND CONTROL OF ASCORBIC ACID METABOLISM

Not only the metabolism of ascorbic acid but also the control of metabolism is of interest to nutrition. Little is known concerning the direct control of ascorbate catabolism in higher animals. This is unfortunate because the question of rate of use of ascorbic acid is central to the controversy of whether supplementary ascorbic acid can be beneficial in disease or environmental stress.

At present the primary control of ascorbic acid metabolism appears to be mediated through facilitated transport process. Two species, man and guinea pig, have active transport systems for ascorbate asborption in the ileum. In contrast the rat and hamster which biosynthesize ascorbic acid have passive transport systems. In man there is a Na^+ dependent active transport system plus a simple diffusion absorption that predominates at higher mucosal ascorbic acid concentrations. Thus absorption is very efficient at low intakes of ascorbic acid and increasingly poor as stomach levels of ascorbic acid increase.

The upper level of ascorbic acid in the blood is in turn limited by a clearance system in the kidney. If this level is exceeded by excessive intestinal absorption for some days, the efficiency of the kidney clearance improves, probably by raising the level of one or more inducible proteins.

Spector and Lorenzo have shown that transport of ascorbic acid into the central nervous system of rabbits satisfies a Michaelis-Menten model with an active saturable carrier. Transfer of ascorbic acid into the eye is also a facilitated saturable process. It is probable that such controlled facilitated transport occurs for other tissue showing high levels of ascorbic acid.

Such control, first of intestinal absorption, then of blood levels and finally of tissue transport serves to maintain and to sharply limit maximum levels of ascorbic acid in any tissue, and consequently, metabolic rates. In guinea pigs on a scorbutigenic diet ascorbic acid levels in the brain are substantially maintained in the face of severe blood level depletion, while high intakes do not markedly increase brain ascorbate levels. Thus control of brain metabolism of ascorbic acid is probably under primary control of the transport system.

There are several experimental approaches to demonstrate changes in ascorbate metabolism as a function of environmental or medical stress. One is to measure tissue levels of ascorbic acid, and indeed Hume, Wilson and others can demonstrate blood and leucocyte changes after wounds, illness, cold, and cancer. Such changes are observed for adrenal ascorbic acid but these experiments do not demonstrate changes in the rate of metabolism, for the effect could be merely a redistribution of the body pool of ascorbic acid.

The current experiments on pharmacokinetics of ascorbic acid by Kallner, Hartmann and Horning are an ideal approach to a method that could demonstrate metabolic changes under stress.

We are working to identify and establish assay methods for the major urinary metabolites of ascorbic acid. Measurement of the excretion rate of these compounds can be used to establish changes in metabolic rate. If one can establish that there are significant increases in excretion rate of ascorbic acid metabolites during disease or environmental stress one would surely establish a fundamental rational of increasing intake during such episodes.

That ascorbic acid catabolism does increase during certain forms of environmental stress is already suggested or shown by many literature reports. In a 1978 report by Nishie, a number of nitrosamines were shown to dramatically increase dinitrophenylhydrazine reactive material in urine of rats. Most of this reactive material is probably ascorbate metabolites. Hodges observed a dramatic increase in the rate of excretion of labeled ascorbic acid metabolites during a period of psychological stress in an experimental human scurvy subject. Many other experiments showing a decrease in tissue levels of ascorbic acid after medical or environmental stress are also indicative of increased metabolism.

I think it is very likely that in the next few years either the pharmacokinetic approach of Kallner or the metabolite approach we are interested in can be used to establish clearly significant increases in ascorbic acid catabolism under special environmental conditions.

METABOLISM OF PYRIDOXINE IN RIBOFLAVIN DEFICIENCY

Mahtab S. Bamji

National Institute of Nutrition

Hyderabad-500007, INDIA

Epithelial lesions of the mouth have been reported to respond to treatment with either riboflavin or pyridoxine. Earlier we had hypothesized that these lesions may be due to cellular deficiency of pyridoxal phosphate (PLP), since the enzyme pyridoxaminephosphate oxidase (PPO) is a flavo protein. Subsequent studies showed that PPO activity is markedly reduced in tissues of riboflavin deficient rat and man[1]. The in vivo conversion of parenterally administered pyridoxine to PLP was also markedly impaired in riboflavin deficient humans[2]. Despite this, the concentration of PLP in tissues of riboflavin deficient rat and blood of deficient humans was normal[1].

To investigate the possibility of a riboflavin independent pathway for PLP synthesis, metabolism of ^{14}C pyridoxine was studied in the livers of ad lib-fed control, pair-fed control and riboflavin deficient rats. The rats were killed at 3 or 5 minutes and 10 minutes after parenteral injection of ^{14}C pyridoxine. Pyridoxine metabolites were separated by ion exchange chromatography[3].

The concentration of phospohrylated derivates of pyridoxine (pyridoxine phosphate and PLP) was markedly lower in pair-fed control and deficient rats compared with ad lib-fed controls, the lowest concentration being in the deficient animals. This was an unexpected finding. Pyridoxal was a major metabolite in the deficient and the pair-fed control groups at 3 minutes, suggesting oxidation of pyridoxine to pyridoxal (a reaction not known to need riboflavin in mammalian tissues) in rat liver. However, due to impaired phosphorylation, this pathway cannot augment PLP synthesis in riboflavin deficiency.

In a separate experiment, pyridoxal kinase and pyridoxal

phate phosphatase activities were examined in control and riboflavin deficient rats. Pyridoxal kinase activity was reduced in pair-fed control but not in riboflavin deficient rats compared with ad lib fed control rats. Thus the impairment in phosphorylation in the two groups cannot be attributed to a common mechanism. In riboflavin deficient rats the substrate ATP may become limiting and impair phosphorylation, whereas in food restriction lower kinase activity may impair phosphorylation.

Earlier we had reported that the cellular requirement of PLP may be higher in riboflavin deficiency due to increase in some PLP enzymes[4]. The metabolic blocks described would limit the supply of additional coenzyme and produce relative deficiency of PLP.

Hypothesis: Actual or relative PLP deficiency may reduce lysyl oxidase activity[5] and affect the formation of collagen cross links. A weak collagen support may be aetiological in the development of epithelial lesions seen in riboflavin and/or pyridoxine deficiency.

REFERENCES

[1] Lakshmi, A.V. and Bamji, M.S. Br. J. Nutr. 32: 249 (1974).
[2] Lakshmi, A.V. and Bamji, M.S. Nutr. Metab. 20: 228 (1976).
[3] Contractor, S.F. and Shane, B. Biochem. Biophys. Acta 230: 127 (1971).
[4] Lakshmi, A.V. and Bamji, M.S. Indian J. Biochem. Biophys. 12: 136 (1975).
[5] Murray, J.C. & Levene, C.I. Biochem. J. 167: 463 (1977).

CLASSIFICATION OF BIOLOGICAL EFFECTS

OF TRACE ELEMENTS

Stanley C. Skoryna, M.D., Ph.D. (Biol.),

Sadayuki Inoue, Ph.D. and M. Fuskova, B.Sc. (Nutr.)

St. Mary's Hospital Centre and Gastrointestinal

Research Laboratory, McGill University, Montreal, Canada.

In 1912 Gabriel Bertrand[1] who can be regarded as the father
of trace element research promulgated a theory which now is known
as the Bertrand's Law "A plant cannot live with a deficiency while
an excess is toxic". Since trace elements cannot be synthetized
by living organisms but must be absorbed from soil and water in order
to enter the food chain of animals and man, Bertrand's law, which
has been extended to all species, has acquired general significance.
Trace elements research has revealed numerous pathological conditions
in animals and man where deficiency or excess plays a significant
etiological role and, needless to say, the industrialized society
has to cope with environmental problems due to toxic trace elements
such as lead and cadmium. According to the definition of essentiality of trace elements, fifteen trace elements are known to be essential to life. These include Arsenic, Cobalt, Copper, Chromium,
Fluorine, Manganese, Molybdenum, Nickel, Iodine, Iron, Selenium,
Silicon, Tin, Vanadium and Zinc. Boron is known to be essential
to life of higher plants, due to its requirement in flavonoid synthesis. When concentrations of trace elements in man are compared with those in seawater and on the earth's crust it becomes
evident that only Iodine concentration is higher in man, when compared to the environment; the significance of this finding being
unknown (Table 1). In some cases, such as Manganese, Iron and Silicon,
there is a tremendous difference between the environmental and
tissue concentrations; in other cases, such as Arsenic and Zinc, the
difference is small. Schroeder[2] has pointed out that from paleo-
biological point of view, the environment was probably deficient in
some trace elements, such as Selenium, Molybdenum and Iodine, in
Jurassic time, while excessive in others, such as Iron, Manganese,

105

TABLE 1

CONCENTRATION OF ESSENTIAL TRACE ELEMENTS IN SEA WATER ON THE EARTH'S
CRUST AND IN MAN (AFTER SCHROEDER; H.A., TIPTON, I.H., AND NASON, A.P.,
J. CHRON. DIS., 1965)

	SEA WATER % p.p.b.	EARTH'S CRUST % p.p.m.	HUMAN BODY	
			p.p.m.	mg/70kg SM
Arsenic	1.6–5.0	2	< 1.4	< 100
Cobalt	0.1	23	< 0.04	± 3
Copper	1.0–25	45	1.4	100
Chromium	1.0–2.5	200	< 0.09	< 6
Fluorine	1300	700	P	P
Manganese	0.7–1.0	1000	0.3	20
Molybdnum	12 –16.0	1	< 0.07	5
Nickel	1.5–6.0	80	< 0.14	< 10
Iodine	50	0.3	0.43	30
Iron	3.4	50000	57	4100
Selenium	4.0	0.09	P	P
Silicon	10–4000	277200	P	P
Tin	3.0	3	0.43	30
Vanadium	2.4–7.0	110	0.3	20
Zinc	9.0–21.0	65	33	2300

P = Present, but amounts not estimated.

TABLE 2

CONCENTRATION OF MAJOR ELEMENTS IN SEA WATER ON THE EARTH'S CRUST
AND IN MAN (AFTER SCHROEDER, H.A., TIPTON, I.H. AND NASON, A.P.
J. CHRON. DIS., 1965)

	SEA WATER %	EARTH'S CRUST %	HUMAN BODY	
			%	g/70kg
Chlorine	1.9	0.02	0.15	105
Sodium	1.05	2.83	0.15	105
Magnesium	0.135	2.09	0.05	35
Sulfur	0.0885	0.52	0.25	175
Calcium	0.04	3.63	1.5	1050
Potassium	0.038	2.59	0.2	140
Strontium	0.0008	0.045	0.0002	0.140
Phosphorus	0.000007	0.118	1.0	700

Manganese, Chromium and Fluorine and that homeostatic mechanisms have gradually developed in the Silurian period. Further evolution led to development of highly specific homeostasis in mammals, which prevented intake of trace elements present in excessive amounts in the environment, such as Iron, and Chromium, while protecting themselves from deficiency of Iodine and Selenium[2].

One of the problems related to levels of intake of trace elements, which developed in the industrial society, is the deficiency created by processing of foods. Sugar refining, polishing of rice and production of patent flour deprive man of such essential trace elements as Chromium, Copper, Manganese, Molybdenum and Strontium, which remain in the residue. As a result, animals receive high amounts of trace elements in ingredients, such as mill feeds, while man is left with what has been termed by Schroeder[3] as "empty calories". The bulk minerals (Table 2), calcium, magnesium, phosphorus, sodium and potassium were lost up to 60-84.7%; trace elements loss was estimated to be in the range of 40-88.5%; the most significant losses occurring in chromium, zinc and manganese, which are concentrated in germ and bran[3]. Unfortunately, present requirements for a balanced diet do not contain provisions for analysis of food ingredients for such essential trace elements as chromium, copper, manganese and molybdenum, although it would be relatively easy to establish standards. According to data collected by Gormican[4] essential trace elements are deficient in hospital diets. Murphy, Page and Watt[5] reported analyses of 300 type A school lunches and these were, according to accepted criteria, deficient in chromium, manganese and copper[5]. To give just one example, chromium, which is considered essential for glucose and cholesterol metabolism and has been demonstrated by Mertz[6] to control the Glucose Tolerance Factor, is deficient in most North American diets. It is not necessary in the context of this presentation to review the principles of action of trace elements within the framework of nutritional requirements, since abundant literature is available on the subject[7]. Suffice to say that when the concept of the effects of trace elements is extended to different intake levels, the magnitude of the problem becomes apparent.

While a great deal is known about the deficiency and toxic effects of trace elements, the pharmacological action represents largely a big question mark and it is only in recent years that detailed studies were made possible by popularization of accurate methods of analysis of trace elements, such as atomic absorption spectrophotometry and neutron activation analysis. Generally speaking, much more is known about the effects of trace elements in animals than in human subjects. It can be predicted that clinical significance of trace elements levels in chronic diseases such as cancer, diabetes and atherosclerosis will become a major area of clinical investigation in the next few years. One aspect which requires special consideration is the requirement for trace elements in parenteral nutrition, which cannot be designated as "total" unless deficiencies in trace elements intake are corrected.

Venchikov[8] extended Bertrand's concept to animal species
and proposed "zoning" of the effects of trace elements into areas
of deficiency, physiological intake, pharmacological action,
toxicity and lethal dosage. The objective of this classification
is to eliminate presently existing confusion with regard to the
observations made in different species, using different routes of
administration and with respect to interactions among trace
elements and between the trace and bulk element[9]. The "zone
limits" are set on the basis of available data: the effects may
relate to a particular system, a function, or reflect a general trend.
The information presently available on some of the trace elements
including stable Sr^{2+} is frequently insufficient to construct such
a biological scale and "zone limits" will undoubtedly change as
new data become available. The need for standardization of
reporting on effects of trace elements is one of the urgent tasks
of current nutrition research[9]. In addition to Cr^{3+}, Mn^{2+}, stable
Sr^{2+} belongs to the group of "newer" trace elements, newer only
because of the renewed interest in their effects and the realization of their significance in physiological functions and in
pathological conditions. Pharmacological effects of trace elements
can be examplified by studies on the effects of Sr^{2+}. Perhaps
insufficient attention was paid in the past to the fact that effects
of Sr^{2+} intake or body content should be investigated within the
framework of trace element levels rather than levels increased to
those of Ca^{2+}; no other trace element can totally replace a bulk
element in physiological function. Sr^{2+} occupies a special
position among trace elements, since it has the highest concentration in sea water, is ubiquitously distributed in relatively
large amounts in the geosphere and has been found in all organisms
analyzed[10]. Sr^{2+} interacts with Ca^{2+}, but its absorption is not
dependent on it, a point of physiological significance to be
emphasized. Another reason for delay in the investigation of
effects of stable Sr^{2+} in human subjects, was probably the confusion
existing in the public mind between the effects of radioactive
strontium, a product of atomic fission, and the stable strontium,
a naturally-occurring trace element. It has not been shown that
stable Sr^{2+} is essential to life, although its essentiality was
suggested on geochemical basis by the late Henry Schroeder[10].
The most convincing evidence of the effects of increased intake of
stable Sr^{2+} in human subjects was published by McCaslin and Janes[11]
in 1959 who have demonstrated both the safety and beneficial
effect of stable Sr^{2+} in treatment of osteoporosis. Curzon and
Spector[12] and several other authors reported the anti-cariogenic
action of stable Sr^{2+} associated with higher intake in drinking
water.

Our own studies on stable strontium began in 1965 with a
reversal of a simple question: if the radioactive strontium produces
bone tumours[13] what would be the effect of non-radioactive strontium
on soft tissues. Studies were carried out in male adult rats of
R.V.H. strain weighing 300-350 grams. Stable strontium was added

TABLE 3

CONCENTRATION AND RELATIVE CONCENTRATION OF Sr^{++} TO Ca^{++}
IN RATS KEPT ON STABLE STRONTIUM SUPPLEMENTATION

MEAN VALUES IN 6 ANIMALS	CONTROL GROUP CaCl$_2$ 0.17% in water			STABLE STRONTIUM SUPPLEMENTATION SrCl$_2$ 0.34% in water		
	Sr^{++} μg/g	Ca^{++} μg/g	Relative Concentration Sr:Ca	Sr^{++} μg/g	Ca^{++} μg/g	Relative Concentration Sr:Ca
LIVER	.149 ± .02	106.0 ± 2.6	1: 709	5.98 ± 1.2	117.8 ± 5.7	1:20
HEART	.261 ± .04	125.6 ± 11.5	1: 482	10.71 ± 1.3	107.8 ± 9.2	1:10
LUNGS	.315 ± .09	458.3 ± 71.0	1:1451	16.32 ± 1.7	584.6 ± 79.0	1:36
BRAIN	.505 ± .12	339.0 ± 97.0	1: 671	10.93 ± 1.3	188.8 ± 50.0	1:17
WHITE MUSCLE	.226 ± .02	185.7 ± 13.8	1: 821	10.71 ± 3.1	174.6 ± 11.3	1:16
RED MUSCLE	.235 ± .01	162.0 ± 6.0	1. 689	9.77 ± 1.8	177.8 ± 11.5	1:18
ADRENALS	.737 ± .12	193.6 ± 27.2	1: 263	11.54 ± 4.9	256.0 ± 26.7	1:22
KIDNEY	.328 ± .05	261.6 ± 12.3	1: 797	23.66 ± 3.0	225.0 ± 9.6	1:10
BONE	65.92 ± 11.3	247433.0 ± 37293.0	1:3753	11116.0 ± 1928.0	239319.0 ± 10564.0	1:22

to drinking water as 0.34% solution of $SrCl_2$ for a period of
6 months. A control group receiving an equimolar amount of
$CaCl_2$ (0.17%) was kept. Sr^{++}, Ca^{++} and Mg^{++} were determined
in serum and 8 organs: liver, lungs, heart, skeletal muscle,
brain, adrenals, kidney and bone, using Atomic Absorption Spectrophotometry with Heated Graphite Atomizer. No change in weight or
any other untoward effects of stable strontium were observed in
these animals. Organ concentrations recorded after 6 months of
Stable Strontium Supplementation are shown in Table 3. It is
evident from this table that concentration of stable strontium increased markedly. No significant changes were found in the organ
content of calcium and magnesium, except for lowering of calcium
content in bone; the Sr:Ca ratio changed from 1:3753 to 1:21.

Group of rats were subjected to exhaustive exercise using a
vertical motor-driven treadmill, for 60 minutes (Table 4).
Animals were decapitated immediately following the exercise and
a "red" skeletal muscle specimen divided into 1 mm^3 pieces was
immersed quickly in a 3% glutaraldehyde in 0.1 M sodium cacodylate
buffer (pH 7.4). After post-fixation, thin sections were prepared and observations were made on a Philips EM 301 microscope
after staining with lead citrate. One hundred mitochondria were
counted in each animal. The results were tabulated in
groups according to disruption of cristae structure and severity
of matrix swelling: 1) Intact cristae; 2) Partially intact
cristae ($\geqslant 50\%$); 3) Loss of structure and matrix swelling. The
results, tabulated in Table 5, indicate marked differences between
control group and those animals receiving Stable Strontium Supplementation. In animals receiving stable strontium supplementation, the cristae structure was well preserved in 69.8% of mitochondria, while in the control group a great majority of mitochondria (70.2%) exhibited loss of structure and cloudy swelling.
If mitochondrial swelling and loss of cristae structure are taken
as an indication of cellular injury, though reversible, these
results have wide-ranging application, because of basic significance in preservation of mitochondrial structure for energy metabolism of the cell. Only limited literature is available on the
subject of electronmicroscopic changes following exhaustive
exercise. Banister et al.[14] observed swelling of cardiac mito-
chondriae and King and Gollnick[15] found disruption of cristae
structure and degeneration following exhaustive exercise. Taylor
et al.[16] extended these studies further by making simultaneous
observations on rate of oxidative phosphorylation and the morphological changes taking place in cardiac mitochondria in exhausted
pigs. Taylor et al.[16] concluded that impairment of mitochondrial
structure induced by exercise diminishes significantly rates of
oxidative phosphorylation when compared to resting animals.

Mitochondrial uptake of calcium and strontium in vitro has
been demonstrated in several organs[17] and considerable literature
exists on the significance of divalent cations in mitochondrial
electron transport systems. If the zoning principle is applied

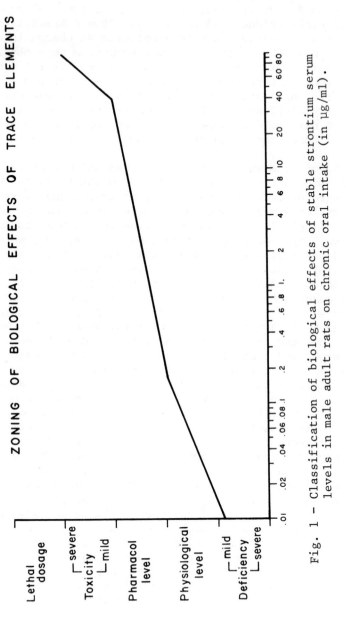

Fig. 1 – Classification of biological effects of stable strontium serum levels in male adult rats on chronic oral intake (in μg/ml).

TABLE 4

CONCENTRATION AND Sr^{++}/Ca^{++} RATIO IN THE RED MUSCLE
OF RATS KEPT ON STABLE STRONTIUM SUPPLEMENTATION

MEAN VALUES IN 6 ANIMALS	Sr^{++} µg/g	Ca^{++} µg/g	RELATIVE CONCENTRATION Sr^{++} : Ca^{++}	Sr^{++}/Ca^{++} RATIO
CONTROL GROUP RESTING $CaCl_2$ 0.17% in water	.235 ± .01	162.0 ± 6.0	1 : 689	.002
CONTROL GROUP AFTER EXERCISE	.234 ± .02	214.3 ±17.8	1 : 916	.001
STABLE STRONTIUM SUPPLEMENTATION RESTING $SrCl_2$ 0.34% in water	9.77 ±1.88	177.8 ±11.5	1 : 18	.055
STABLE STRONTIUM SUPPLEMENTATION AFTER EXERCISE	10.41 ±1.67	205.0 ±18.4	1 : 20	.051

TABLE 5

ALTERATIONS IN MITOCHONDRIAL STRUCTURE   IN RAT
SKELETAL (RED) MUSCLE   AFTER EXHAUSTIVE EXERCISE

Six animals	EXTENT OF LOSS OF CRISTAE STRUCTURE (per 100 mitochondrial profiles)		
	IC	PIC	LOS
Rat 1	13	14	73
2	11	23	66
Control 3	14	21	65
Group 4	13	7	80
Cacl 0.17% 5	7	14	79
solution in 6	17	25	58
drinking water			
Av	12.5	17.3	70.2
Rat 1	80	7	13
2		12	23
Stable			
Strontium 3	71	13	16
Supplementation			
Group 4	56	25	19
SrCl 0.34% in 5	77	10	13
drinking water 6	70	17	13
Av	69.8	14.0	16.2

IC = Intact cristae
PIC = Partially intact cristae ($\geqslant 50\%$)
LOS = Loss of structure
Av = Average

to the biological effects of strontium administration in rats
(Figure I) serum levels in order of 0.02-0.2 µg/ml can be
considered within physiological limits. Values between 1-20 µg/ml
would fall into pharmacological range. Although stable strontium
has low toxicity and no lethal effects of oral administration have
been reported, serum levels over 50 µg/ml should be considered
toxic due to displacement of calcium by strontium in the bone,
which occur only if very high doses of stable strontium are used[18].

CONCLUDING COMMENTS AND SUMMARY

Trace elements research has revealed numerous pathological
conditions in animals and man, where deficiency or excess plays a
significant role. Attention is drawn to deficiencies occurring
in trace element intake following processing of foods, such as
polishing of rice, refining of sugar and production of patent flour.

In view of recent trends in classification of biological
effects of trace elements it is proposed that the concept of
"essentiality to optimum health" be considered as a separate entity
from "essentiality to life". In this context the dietary intake
of some of the trace elements such as Zinc, Chromium or stable
Strontium should be increased beyond the level required for a simple
correction of deficiency. Naturally, the toxic effects should be
taken into consideration. For instance, marked increase in Zinc
intake may not be toxic directly but produce side effect by
depressing copper levels; nevertheless, higher Zinc intake may be
recommeneded in certain conditions, as demonstrated by findings of
the Denver Head Start Programme[19]. Higher Chromium intake may be
useful in diabetics due to its effect on Glucose Tolerance Factor[6].
Studies of our group at McGill University and St. Mary's Hospital
Centre[20] are used as an example of pharmacological effects of stable
strontium. Preliminary findings indicate that stable strontium
exerts a protective effect on mitochondrial structure under conditions in which extent of mitochondrial swelling exceeds the physiological limits for adaptation to configurational changes. Attention
is also drawn to relatively high organ levels of stable strontium
in normal rats, when compared to serum levels, as if all strontium
in the environment was required by mitochondria. Calcium and
Magnesium levels do not change following dietary supplementation
of stable strontium. Using relatively small doses of stable
strontium (when compared to 0.5%, 1% or 1.5% used by several
investigators), no toxic effects were observed. In bone, stable
Strontium replaces Calcium: when 0.34% of $SrCl_2$ was added to
drinking water of rats, the Sr:Ca bone ratio changed from 1:1753
to 1:21. However, as demonstrated by Weber et al.[18] very high
doses of strontium produce excessive loss of bone calcium, which
is probably responsible for the so-called "strontium rickets".

REFERENCES

1 Bertrand, G. Proc. Int. Cong. Appl. Chem., 28: 30, 1912.
2 Schroeder, H.G. J. Chron. Dis. 18: 217-228, 1965.
3 Schroeder, H.E., Balassa, J.J., Nasam, A.P. and Tipton, I.H.
   Essential trace metals in man. J. Chron. Dis., 15: 941, 1962;
   19: 545 & 1007, 1966; 20: 179 & 869, 1967; 21: 815, 1968;
   23: 227, 1970; 25: 562, 1971.
4 Gormican, A. J. Am. Diet. Assoc., 56: 397-403, 1970.
5 Murphy, E.W., Page, L. and Watt, B.K. J. Am. Diet. Assoc.,
   58: 115, 1971.
6 Mertz, W. Physiol. Rev., 49: 163, 1969.
7 Underwood, E.J. "Trace Elements in Human and Animal Nutrition",
   545 pp., Academic Press, 1977.
8 Venchikov, A.I., Vop. Pitan., 19: 3-11, 1960.
9 Skoryna, S.C., Tanaka, Y., Wellington Moore, Jr., and Stara, J.F.
   Trace Elem. Environ. Health, 6: 3-11, 1973.
10 Schroeder, H.A., J. Chron. Dis., 25: 491-517, 1972.
11 McCaslin, F E. and Janes, J.M. Proc. Mayo Clinic, 34: 329-334,
   1959.
12 Curzon, M.E.J. and Spector, P.C. In: Handbook of Stable
   Strontium". S.C. Skoryna, Ed., Plenum Publ., 1980.
13 Skoryna, S.C. and Kahn, D.C. Cancer, 12: 306-322, 1959.
14 Banister, E.W., Tomanek, R.J. and Cvorkov, N. Amer. J. Physiol.,
   220: 1935-1940, 1971.
15 King, D.W. and Gollnick, P.B. Amer. J. Physiol., 218: 1150-1155,
   1970.
16 Taylor, P.B., Lamb, D.R. and Budd, G.C. Europ. J. Appl. Physiol.,
   35: 111-118, 1976.
17 Carafoli, C., Crompton, M., Malmström, K., Siegel, E., Saltzmann, M.
   Chiesi, M. and Affolter, H. pp. 535-551, "Biochemistry of Membrane Transport", G. Semenza and F. Carafoli, ed., Springer Verlag,
   1977.
18 Weber, C.W., Doberenz, A.R., Wyckoff, R.W.G. et al.
       Poultry Sc., 47: 1318-1323, 1968.
19 Hambidge, K.M., Walravers, P.A., Brown, R.M., et al. Amer. J.
   Clin. Nutr., 29: 734-738, 1976.
20 Skoryna, S.C. and Fuskova, M. In: Handbook of Stable Strontium.
   S.C. Skoryna, ed., Plenum Publ., 1980, pp.

ACKNOWLEDGEMENT

     This work was supported by grants from Department of Health
and Welfare, St. Mary's Hospital Foundation, Rideau Institute of
Canada and The Stuart Foundation, Montreal, Canada.

ARE DIETARY GUIDELINES FOR TRACE ELEMENTS NEEDED?

THE DEVELOPMENT OF PROVISIONAL RECOMMENDED ALLOWANCES IN THE USA

Walter Mertz, M.D.

Nutrition Institute - U.S. Dept. of Agriculture

Beltsville, MD 20705 USA

History has shown that the consumption of a diet well balanced
with regard to the major nutrients and derived from a variety of
foods furnishes an adequate supply of the essential micronutrients,
including trace elements. Imbalances have existed and continue to
exist in unusual geochemical environments and in situations where
natural catastrophies, economic or medical reasons, restrict the
quantity of food choices. Since the beginning of the twentieth
century the reliance on a "balanced" diet as an adequate source for
all micronutrients has become somewhat questionable. The consumption
of refined, partitioned foods and of new food analogues has increased
in industrialized societies and completely new foods are manufactured
and consumed in large quantities. The impact of these changes on
the health and trace element status of the affected populations is
not well defined. Although severe trace element deficiencies are
extremely rare in the United States and other industrialized countries, marginal intakes of essential trace elements, in some cases
resulting in an impaired physiological function are being described
with increasing frequency. Some part of the public expresses serious
doubts about the adequacy of the modern diet and resort to self supplementation with vitamins and trace elements ranging from zinc to
selenium which are freely available in the health food stores.

The Recommended Dietary Allowances of the U.S. National Academy
of Sciences prior to 1974 contained a recommended allowance only for
iron and iodine. The 1974 Edition added zinc with an allowance of
15 mg/day for adults. The periodic table shows that there are many
more trace elements for which there is evidence of an essential
role, at least in animal species. For many of those, for example,
vanadium, nickel, and arsenic, there is a total lack of data relating
to a human requirement. For others, notably fluorine, chromium,

117

manganese, copper, selenium, and molybdenum, an approximate estimate
of safe, adequate intakes can be made. The Committee on Recommended
Dietary Allowances is proposing that "Provisional Allowances" for
these six trace elements be included in the next revised edition.
Although this proposal had been discussed in scientific meetings in
the USA it should not be considered as final as it must be acted on
by the National Academy. The Provisional Allowances are proposed
as ranges of intakes that are considered adequate and safe. The
concept of ranges, rather than of one particular intake is based on
two arguments. First, the healthy organism can maintain desirable
tissue concentrations and trace element dependent functions over a
whole range of intakes. When desirable concentrations or functions
are reached, regulation of intestinal absorption or excretion or of
both will tend to maintain these levels, in spite of higher intakes
as indicated by the plateau of the curve. When this range is exceeded by excessive intakes the first signs of toxicity set in as
shown by the declining function. The range of safe intakes varies
from one nutrient to another; this will be discussed later.

The second argument for the range concept is based on our knowledge of dietary interactions. The biological availability of many
trace elements depends on the chemical form in which they are present
in foods. Iron is the best example, but other elements, such as
cobalt, copper, selenium, and chromium, have shown similar dependence
of biological availability on chemical form at least in animal experiments. Biological availability and therefore the requirement
for a trace element also depend on other mineral elements in the
diet. The antagonism between copper and molybdenum is well known
from animal and human experiments but it is only one example of a
very large number of known mineral interactions. Finally, organic
constituents of the diet have a strong influence on availability of
several trace elements. The best defined effects are those of fiber,
phytate, vitamin C, an unknown factor in animal meat, but there are
many more natural compounds and drugs that affect the bioavailability
of trace elements. The range concept takes into account the efficiency of the organism's homeostatic regulation and the diverse influences on biological availability of dietary interactions. It
must be understood that the concept does not imply deficiency or
toxicity of intakes outside of the suggested range. It does imply
however an increased risk for deficiency or toxicity when intakes
are consistently low or high.

Let us now look at these six elements individually in the following lines, I must emphasize that the data to be discussed are
valid for the United States but not necessarily for other countries
with different nutritional habits and different geochemical environments. The syndrome of copper deficiency in man is well described
with the main features of anemia, neutropenia, and bone changes.
Experimental copper deficiency in animals affects the integrity of
large arteries and, in rodents, increases serum cholesterol levels.

Copper deficiency has been described in children with protein energy
malnutrition and also in a few isolated cases of patients on total
parenteral nutrition. The amount of copper required to maintain
balance in adults on typical American diets is close to 1.5 mg,
perhaps between 1.5 and 2 mg. The copper intake from self-selected
or institutional diets as determined by modern methods is consistently below 2 mg in the USA, with the average very close to only
1 mg/day. Dietary copper toxicity does not appear to be a problem
within wide ranges of intakes, but the figures that we just presented
suggest the possibility of the existence of marginal deficiencies
that have not yet been recognized. Alternatively, it is possible
that balance studies overestimate the real requirement by not taking
into account the possibility of long-term adaptation of the organism
to lower intakes. These questions can be resolved only by additional
research. With regards to the situation with manganese, deficiency
in free-living human subjects is unknown, and only one case of experimental deficiency has been described at intakes of less than
.35 mg. Equally unknown is toxicity of dietary manganese. All balance studies have observed equilibrium or accretion at intakes of
around 2.5 mg, an amount easily obtained from almost all diets, as
manganese is widely distributed in foods. Thus it appears that this
element presents no great nutritional problems in the United States
and most probably not in other countries. The proposed ranges of
the provisional allowances for copper are 2-5 mg/day, and for manganese 2.5-5 mg/day for adults. The recommended intake for other
age groups are arrived at by either direct experimentation or by
extrapolation. Selenium deficiency is widespread in animal species
but is not known to exist in human subjects. The selenium intake
is very low in certain countries and can be extremely low in children
with inborn errors of metabolism requiring highly purified diets,
but no deleterious effects to health have been reported. Thus, a
human selenium requirement can only be extrapolated from the fact
that in all mammalian animal species investigated the selenium requirement is amply met by diets furnishing 0.1 ppm. This can be
translated into an estimate of adequate selenium intake by an adult
person of 50-75 µg. The habitual intake in the United States ranges
from this amount to more than 200 µg, but this is not so in all countries. The range of safe intakes for selenium is smaller than that
for other trace elements; this cautions against indiscriminate use
of selenium supplements which are now quite fashionable in some
parts. With the wide circulation of foods within the United States
selenium does not appear to present a nutritional problem. The
health consequences of either extremely low or extremely high selenium intakes in certain areas of the world are unknown at the
present time and require much additional research for clarification.
The trace element molybdenum is similar to selenium in that it
presents a major problem in animal nutrition but that human deficiencies are unknown. Balance studies indicate a requirement of
approximately 0.15 mg/day for the adult, this requirement is easily
met by typical Western diets. Adverse effects of excessive molyb-

denum intakes have been reported in human subjects living in some
molybdenum-high provinces of the Soviet Union, where daily intakes
of several milligrams were associated with a goutlike syndrome. But
even more moderate intakes, at around one-half of a milligram, are
known to adversely affect the metabolism in human subjects of another essential trace element, copper. Considering the uncertainty
of the adequacy of our copper intake, it appears wise to avoid excessive molybdenum exposure. The provisional recommendations for
selenium and molybdenum are 50-200 µg of selenium for adults and
150-500 µg of molybdenum. As stated before these intakes should
be easy to obtain in the United States, but they may present difficulties in regions where the concentration of these elements in
the environment is abnormally low or high. The next element fluorine, is not yet universally recognized as essential, although its
beneficial effects on dental health are beyond any doubt and growth
effects have been demonstrated in experimental animals under certain
conditions. Recommendations for this element are presented because
of its very narrow range of safe intakes and the wide differences
of fluoride levels in local environments. A protective effect
against caries in children is observed when the total intake from
diet and water exceeds 1 mg/day. The first undesirable effects,
namely mottling of the teeth enamel, occur with intakes of more
than 4 mg in children and higher doses in adults. Pronounced toxicity, fluorosis, is observed however only with much greater intakes.
The desirable range of fluoride intake giving protection against
caries is not always obtained in the United States, because a major
contribution to it comes from the local water supply which varies
in its fluoride content. Fluoridation of local water supplies to
1 ppm insures a desirable dietary intake and is considered safe in
areas with a moderate climate. Chromium deficiency is known in experimental animals and in human subjects; the signs consist in an
impaired effectiveness of insulin, impaired glucose tolerance, and
a disturbed fat metabolism. Chromium deficiency has been described
in malnourished children in several areas; in association with aging
and repeated pregnancies, and it can be aggravated by diabetes.
The absolute requirement is in the order of 1 µg/day in adults. In
view of the low intestinal absorption of chromium a dietary intake
of 50-100 µg is necessary. This intake is not always provided by
normal diets and there are foods with extremely low chromium contents. Oral toxicity of trivalent chromium has never been observed
in human subjects or in experimental animals receiving excessive
doses in their diet. The implementation of the provisional allowances for chromium is difficult at the present time because of analytical problems. However, methods for chromium analysis in foods
have been developed and are being used to produce the necessary
nutrient composition information. The proposed provisional recommendation for fluorine for adults is 1.5-4 mg/day for the total
intake from diet and drinking water and correspondingly lower for
the younger age groups. This range is believed to have a measurable
impact on the prevention of tooth decay. At the same time it is

low enough to avoid the danger of mottling of the enamel. The proposed range of chromium intake is from 50-200 µg/day with correspondingly lower values for the younger age groups.

   This part of the discussion was directed to one of the important
endeavors of trace element research namely human requirement. The
next part deals with a subject that is equally important, namely
our increasing awareness of interactions that influence the biological availability of trace elements. The RDA Committee is proposing a new method of evaluating dietary iron which takes into
account influences of chemical form of the iron compounds and influences from other dietary constituents on the biological availability. The implementation of the recommended allowance of 18 mg
of iron for women of child-bearing age has always been very difficult, because the typical American diet provides no more than
6 mg/1000 kcal. The proposed approach takes into account the new
concept of two iron pools, heme and non heme which evolved through
the collaboration of scientists in many countries, including Venezuela, and which can be summarized as follows. The dietary iron
that is in form of the heme complex, amounting to a little less
than half of the total in meat, poultry, and fish is consistently
better absorbed than the rest the non heme iron. The absorption
of non heme iron, that is the dietary iron in all other foods, in
iron supplements and in contamination, is considerably less and it
depends strongly on influences from other dietary ingredients. Of
the many constituents that influence the absorption of non heme
iron either positively or negatively, two are standing out by their
strong effects: Vitamin C and a yet unidentified agent in animal
meats. These two factors, either alone or together, greatly increase the absorption of dietaty non heme iron, provided that they
are present in the iron-containing meal. Conversely, the absorption of non heme iron will be poor when these enhancing factors are
absent in the meal and are consumed either between meals or with one
providing little dietary iron. The enhancing effect of ascorbic
acid is considerable although the amounts needed are greater than
those required to prevent scurvy. However, these intakes are perfectly feasible, particularly in tropical countries.

   In an attempt to provide for the proper evaluation of absorbable
dietary iron and also to provide means for the planning of proper
diets, the RDA Committee is proposing the following scheme. First,
each meal has to be evaluated individually and its content of heme
iron and non heme iron has to be calculated. As a general approxi-
mation, 40% of the iron in meat, poultry, and fish, can be assumed
to belong to the heme category. For this amount an absorption factor
of 23% is used. All the rest of the meal's iron content is treated
as non heme iron, with one of three absorption coefficients, depending on the nature of that meal. A low availability meal contains
less than 25 mg vitamin C and less than 30 mg of meat; a low absorption factor of 3% of the non heme iron is used. A medium availabi-

lity meal is defined as containing more than 25 mg vitamin C or more than 30 mg of meat; and the absorption factor of the non heme iron is increased to 5%. A high availability meal contains more than 75 mg ascorbic acid or more than 90 mg of meat, or both meat and vitamin C in excess of 30 mg and 25 mg respectively. For this an absorption factor of 8% is used. The sum of the non heme iron and heme iron is the total absorbable iron in this meal. The total absorbable iron of the whole day can be calculated from the individual meals and related to the requirement of absorbable iron for various age groups.

IRON INTAKE AND IRON ABSORPTION

Miguel Layrisse

Universidad Central de Venezuela and Instituto Venezolano

de Investigaciones Científicas - Caracas, VENEZUELA

Accumulating evidences demonstrated that the amount of iron potentially available from food depends not only from the amount of iron intake but from the nature of that iron and the composition of the meal in which it is consumed. Thus, the total iron content of a diet is a poor indicator of the amount of iron that is really absorbed and utilized. It is essential, therefore, to know the mechanism that takes place at the lumen of the gut which regulates the different patterns of iron absorption.

Several methodological procedures have been assayed during this century to estimate the dietary iron absorption. However, it was only by the introduction of radioactive iron in Biology, and more specifically the biological incorporation of radioiron in the food[1], that is, during the life of plants and animals, that students count with a ground to perform appropriate basic research.

The studies on iron absorption from food with this new methodology have provided progressive knowledge on the mechanisms which regulate absorption of dietary iron. The first stage of these studies was the determination of iron absorption from a single food. It showed in general that iron from animal food, specially meat, is absorbed three or four times more than vegetable iron. The exceptions to this rule were eggs, milk and milk derivatives in which the percentage of iron absorption is similar to that found in vegetable food[2, 3]. Studies on animal food such as meat, fish and liver have shown that iron absorption depends on the proportions in which heme, ferritin and hemosiderin are present. Iron absorption from purified heme is poor (about 4%)[4], but increases about three times when it is ingested as hemoglobin[2, 5] and even more when it is ingested as hemoglobin and myoglobin from meat[6]. Iron from purified ferritin and hemosiderin are also poorly absorbed, about 2%, but absorption increases several folds when they are incorporated into the liver and meat, though still below the level of absorption of heme iron under the same conditions[7, 8].

The second stage of knowledge with this new methodology was obtained by measuring the food interaction effect on iron absorption at the lumen of the gut[9]. Foods biosynthetically labeled with radioactive iron and administered first separately and then together at 15 day intervals and in the same individual have shown that iron absorption from two vegetable foods is very similar when they are administered together[10] and a similar absorption pattern is found when one vegetable is mixed with an iron salt as either ferric or ferrous compound[11, 12, 13, 14]. Absorption from vegetable food iron is about double when given together with meat, liver or fish[6, 9, 15, 16]. This absorption increases progressively according to the amount of ascorbic acid administered[17].

The effect of vegetables and chelating agents on iron absorption from meat, fish and liver depends on the proportion in which heme, ferritin and hemosiderin are present in the food. Absorption of heme iron is not affected by vegetables and chelating agents but absorption from ferritin and hemosiderin is affected. This explains why vegetable foods do not affect iron absorption from meat[6], but do affect iron absorption from fish and liver[15, 16].

These results provided the background to build the concept of two iron pools in term of iron absorption: heme and non-heme. Iron compounds belonging to the same iron pool show the same percentage of absorption when they are administered together. The heme iron pool is formed by hemoglobin and myoglobin and its absorption is not affected by other foods or chelating agents[12]. The non-heme pool is formed by vegetable foods, eggs and iron salts, its absorption is enhanced by combined administration of meat, liver or ascorbic acid, and reduced by desferrioxamine[11, 12, 13].

These results provided also the basic knowledge for the extrinsic label method to determine the absorption from the non-heme and heme iron pools. Iron absorption from non-heme iron is measured by mixing 0.1 mg or traces of labeled iron, such as ferric chloride, with one or more foods biosynthetically labeled with different radioactive iron. The ratio of absorption from the extrinsic labeled represented by the iron salt and that from the intrinsic labeled represented by the natural food iron is close to the unit, indicating that absorption from extrinsic label accurately reflects absorption from the non-heme pool[11, 12, 13].

With the same method 0.1 mg of labeled rabbit hemoglobin iron mixed with meat reflects the absorption from heme iron[12]. It is also possible to measure accurately the iron absorption from ferritin

and hemosiderin using a small amount of purified labeled rabbit ferritin[8].

    As practically the total iron compounds present in foods belong
to the heme and non-heme pools, this new methodology has permited
to determine the total iron that is absorbed from a meal and from
a daily diet[18, 19]. Studies carried out with different diets have
shown that the total iron intake does not reflect the amount of iron
which could be absorbed and that enhancing absorption foods such as
meat and fruits are paramount to obtain an adequate iron absorption
from a meal.

    This method was used to measure three Venezuelan diets consumed
by low or lower-middle class groups[18]. Two diets, Andes and Central,
consisted of vegetable foods eaten in each meal, plus milk drunk
once or twice and a small piece of meat eaten at lunch in the Andes
diet and lunch and supper in the Central diet. The third diet was
that of people living on the coast and consisted mainly of maize
and fish at each meal accompanied by fruit at supper. These studies
showed that milk, cheese and butter did not improve absorption from
vegetable food iron. Absorption of 3-4 mg of vegetable iron was
increased about two folds by 50 mg of meat, about three fold by
100 g of fish and five fold by 100 g of fish and 150 g of papaya
containing 66 mg of ascorbic acid. Absorption of heme iron accounts
for 42% and 62% of the total iron absorbed from Central and Andes
diet, while fish accounts for about 10% of the total iron absorbed
from the coast diet.

    Since ferric and ferrous salts belong to the non-heme iron pool,
the extrinsic label method can also be used to measure iron absorption from iron-fortified food[14, 20, 21]. The studies on iron absorption carried out in the last eight years demonstrated that the iron
absorption from salts used to fortify foods enter into the non-heme
pool and that increasing doses of iron from 0.1 to 60 mg mixed with
vegetable foods are absorbed to the same extent as the native food
iron. According to these results, food vehicles such as wheat,
maize and rice inhibit considerably the absorption of iron fortification.

    Although some of the participants to this symposium will speak
specifically on iron fortification, I would like to devote the last
part of my presentation to our studies on iron fortification specially those concerned with the characteristics of the Fe(III)-EDTA
as iron fortification[22, 23].

    Six food vehicles have been tested for iron absorption: refined
sugar, sugar cane syrup, sweet manioc, milk, wheat and maize. The
absorption ratio of 3 mg of iron as ferrous sulfate mixed with each
vehicle to the same amount of ferrous sulfate given alone showed
that either refined sugar or sugar cane syrup practically do not

TABLE 1

ABSORPTION FROM IRON AS FERROUS SULFATE
MIXED WITH VARIOUS FOOD VEHICLES

| VEHICLES | IRON ABSORPTION (%) | | ABSORPTION RATIO |
| | A | B | A/B |
	Ferrous sulfate (3-5 mg Fe) + food vehicle	Ferrous sulfate (3-5 mg Fe) given alone	
Refined sugar	31.6	33.2	.95
Sugar cane syrup	30.3	36.9	.82
Sweet manioc	12.4	35.2	.35
Milk	7.0	27.8	.25
Wheat	7.7	49.3	.16
Maize	2.7	32.9	.08

contain absorption inhibiting substances while wheat and maize contain strong inhibiting substances. It can be expected that rice and beans behave the same as maize. It is interesting to notice that sweet manioc has a relatively small amount of inhibiting substances.

Table 2 shows the iron absorption from Fe(III)-EDTA tested with 6 diffefent food vehicles and compared with the absorption from ferrous sulfate administered with the same food vehicles. The original results were calibrated by multiplying the observed absorption by the ratio of the composite mean absorption from the reference dose of iron ascorbate of all individuals and the mean absorption from the reference dose of iron ascorbate for the given study. It can be observed that while the mean absorption from ferrous sulfate varies from 2% to 32% according to the food vehicle mixed with the salt, the absorption from Fe(III)-EDTA remains practically the same, ranging from 8 to 13%. These results indicate that while iron absorption from ferrous sulfate is very sensitive to the inhibiting absorption substances present in food vehicles, the absorption from Fe(III)-EDTA complex is only slightly affected by the presence of such substances in vegetable food or milk.

Adding to the characteristics of Fe(III)-EDTA, iron formed by this complex exchanges completely with vegetable iron in the lumen of the gut and the absorption from both extrinsic and intrinsic food iron are the same and higher than that expected from other iron salt used as iron fortification. According to this information wheat and maize are not adequate food vehicles when they are fortified with ferrous sulfate and are administered with a diet poor in enhancing absorption foods, since a large amount of the salt would be necessary to meet the normal iron requirement. However, if Fe(III)-EDTA is used, a small amount, such as 10 mg of iron per day, would be sufficient in most instances to cover the physiological needs.

The amount of information on iron absorption from food and from iron fortification which has been accumulated in the last ten years permits to estimate on solid basis, the daily availability of iron from the diet in a population and the amount of iron fortification that is needed daily to prevent iron deficiency in most subjects. Following our studies in this field, we started to coordinate a collaborative study sponsored by the United Nations University and the Instituto Venezolano de Investigaciones Cientificas to determine the iron availability from the most common South American diets and establish the base line for iron fortification studying the food vehicles and iron salts suitable for one or various countries, according to the food habits and the technological facilities. This program will start this month with laboratories located in four countries and it is expected that others will be incorporated in the collaborative study during the first year of work.

TABLE 2

IRON ABSORPTION FROM Fe(III)-EDTA COMPLEX AND FERROUS SULFATE MIXED WITH
VARIOUS FOODS AND CALIBRATED ACCORDING TO THE ABSORPTION FROM THE REFERENCE DOSE

FOOD ENRICHED	IRON ABSORPTION (%)		
	Fe(III)-EDTA (3 or 5 mgFe)	Ferrous sulfate (3 or 5 mg Fe)	Composite mean absorption from reference dose (3 mg Fe)
Refined sugar	8.4	30.0	
Sugar cane Syrup	8.4	25.8	
Milk	13.1	7.9	31.5
Sweet manioc	12.8	11.8	
Wheat	11.5	4.9	
Maize	8.2	2.0	

REFERENCES

[1] Moore, C.V. and Dubach, R. Trans. Ass. Am. Phychs. 64: 245-256, 1951.

[2] Layrisse, M., Cook, J.D., Martínez-Torres, C., Roche, M., Kuhn, I.N. and Finch, C.A., Blood, 33: 430-443, 1969.

[3] Layrisse, M. and Martínez-Torres, C. "Progress in Hematol." Vol. VI, pp. 137-160, 1971.

[4] Conrad, M.E., Benjamin, B.I., Williams, H.L. and Foy, A.L. Gastroenterology 53: 5-10, 1967.

[5] Turnbull, A., Cleton, F. and Finch, C.A. J. Clin. Invest. 41: 1897-1907, 1962.

[6] Martínez-Torres, C. and Layrisse, M. Am. J. Clin. Nutr. 24: 521-540, 1971.

[7] Layrisse, M., Martínez-Torres, C., Renzy, M. and Leets, I. Blood, 45: 689-698, 1975.

[8] Martínez-Torres, C., Renzi, M. and Layrisse, M. J. Nutr. 106: 128-135, 1976.

[9] Layrisse, M., Martínez-Torres, C. and Roche, M. Am. J. Clin. Nutr. 21: 1175-1183, 1969.

[10] Layrisse, M. and Martínez-Torres, C. Clin. Haemat. 2: 339-352, 1973.

[11] Cook, J., Layrisse, M., Martínez-Torres, C., Walker, R., Monsen, E. and Finch, C.A., J. Clin. Invest. 51: 805-815, 1972.

[12] Layrisse, M. and Martínez-Torres, C., Am. J. Clin. Nutr. 25: 401-411, 1972.

[13] Björn-Rasmussen, E., Hallberg, L. and Walker, R.B., Am. J. Clin. Nutr. 25: 317-323, 1972.

[14] Layrisse, M., Martínez-Torres, C., Cook, J.D., Walker, R. and Finch, C.A., Blood, 41: 333-352, 1973.

[15] Martínez-Torres, C., Leets, I., Renzi, M. and Layrisse, M. J. Nutr. 104: 983-993, 1974.

[16] Martínez-Torres, C., Leets, I. and Layrisse, M. Arch. Latin. Nutr. 25: 199-210, 1975.

[17] Layrisse, M. Proc. Third Conference Hoechst. Excerpta Medica 1975.

[18] Layrisse, M., Martínez-Torres, C., González, M. Am. J. Clin. Nutr. 27: 152-162, 1974.

[19] Björn-Rasmussen, E., Hallberg, L., Isaksson, B. and Arvidsson, B. J. Clin. Invest. 53: 247-255, 1974.

[20] Cook, J., Minnich, V., Moore, C.V., Rasmussen, A., Bradley, W.B. and Finch, C.A. Am. J. Clin. Nutr. 26: 861, 1973.

[21] Sayers, M.H., Linch, S.R., Charlton, R.W., Bothwell, T.H., Walker, R.B. and Mayet, F. Brit. J. Haemat. 28: 483-495, 1974.

[22] Layrisse, M. and Martínez-Torres, C. Am. J. Clin. Nutr. 30: 1166-1174, 1977.

[23] Martínez-Torres, C., Romano, E.L., Renzi, M. and Layrisse, M. J. Clin. Nutr. (in press).

THE FUNCTIONAL EVALUATION OF VITAMIN STATUS WITH

SPECIAL ATTENTION TO ENZYME-COENZYME TECHNIQUES

Myron Brin, Ph.D.

Director, Clinical Nutrition – Roche Research Center

Nutley, New Jersey 07110 USA

INTRODUCTION

Following the discovery that there were essential dietary nutritional factors such as vitamins which if not consumed in adequate
quantities would result in severe clinical deficiency diseases, and
that these diseases could be cured by repletion with the essential
nutrients, it soon became evident that the public health problem of
nutritional adequacy should be managed by techniques of preventive
medicine. Accordingly, an estimate of the vitamin status of the
population was needed. With that information one could judge the
necessity for intervention programs.

Based upon the large number of cases of pellagra etc. in the
United States in the 1920's and 1930's, and compounded by the need
to satisfy the nutritional requirements of military forces in the
far-flung world during the Second World War, the United States in
1942 took steps to assure that its populations would be appropriately
nourished with the known nutrients by formulating a bread enrichment
program[1]. Since bread was a basic food, the addition of thiamine,
niacin and riboflavin and iron would assure a basic foundation for
nutrient adequacy.

However, additional nutrients have been discovered since 1942
when the bread enrichment program was initiated. These included
vitamin B_6, folacin, vitamin B_{12}, etc. A recent review of the adequacy of bread enrichment for the current era by the NAS/NRC Food
and Nutrition Board has indicated that the enrichment of cereal
grains should be broadened to encompass 6 vitamins and 4 minerals
and that the enrichment should be done at the mill rather than at
the product manufacturing site, in order to assure uniform and ex-

tensive distribution of a minimum nutrient level in all foods made
of grain flours[1, 2]. Accordingly, in order to judge the adequacy
of current intervention programs it is necessary to continuously
monitor the nutritional status of the target population.

A recent issue of the American Journal of Public Health entitled
"Nutritional Assessment of Health Programs" presented four modes for
the evaluation of nutritional status[3] (Table I). The demographic
mode is a technique for use where very little specific information
is available on individuals but one must rely upon the eating habits,
ethnic background and socioeconomic status of the people being evaluated. This technique is used for a gross evaluation of the status
of a relatively homogeneous ethnic or cultural group. The second
mode is that of dietary history. If managed accurately, this mode
can reveal a great deal of information about nutritional status.
One can query about food groups consumed, the specific menus for
each meal in the course of a day or a week, the amount of money
spent on food, snacking habits, etc. The limitation of this mode
is the need to question the person very carefully and to judge the
credibility of the response. Furthermore, the person who is of low
socioeconomic status may be embarrassed to indicate that he cannot
afford high quality food, the alcoholic may not admit to drinking
rather than eating and, therefore, will fabricate an imaginary diet,
or the obese person may not admit to being unable to diet, etc. It
therefore requires a high level of discretion on the part of the
interrogator in order to gain accurate information. The third mode
to evaluate nutritional status is physical examination. Here there
are objective and subjective findings. For instance, height, weight
and age can be measured accurately as one can also get objective
evaluations of the skin, the hair, the teeth, etc. These give us
an overall insight into clinical nutritional status. Physical
examination, however, cannot reveal a condition of marginal deficiency[4]. In the marginal deficiency state, for instance, there
is modified immune response, serious adverse affects on behavior
and impaired ability to neutralize drugs and environmental chemicals, which in themselves increase vitamin needs (Table II)[5, 6, 7].
The physician would be hard pressed to reveal these inadequacies
by simple physical diagnostic techniques.

The fourth mode is the laboratory evaluation for vitamin
status as shown in Table I and in more detail in Table III. The
traditional way to evaluate nutritional status by laboratory analysis has been to determine vitamin levels in physiological fluids
such as blood and urine. For certain of the nutrients such as
vitamins A, D, E, B_1, B_2, B_6, folacin, B_{12}, and biotin, there are
useful methods for these analyses. In some cases another component
rather than the vitamin is measured. For instance, one may measure blood calcium or 1,25-dihydroxyvitamin D^8 for vitamin D, and
for vitamin B_6 one may measure pyridoxic acid excretion or pyridoxal phosphate in plasma. Also in some cases blood analyses are

TABLE I

FOUR MODES TO EVALUATE NUTRITIONAL STATUS

1. Demographic

2. Dietary History

3. Physical Examination

4. Laboratory Analysis

TABLE II

ADVERSE EFFECTS OF MARGINAL VITAMIN DEFICIENCY

1. Impaired Immune Response

2. Adverse Effects on Behavior (MMPI)

3. Reduced Neutralization of Drugs and
    Environmental Chemicals

TABLE III

LABORATORY EVALUATION FOR VITAMIN STATUS

1. Chemical Status

    a. Blood
    b. Urine

2. Functional Status

    a. Systemic

        1) in vivo
        2) in vitro

    b. Enzyme-Coenzyme (Activation)

---

preferred such as for vitamins A, E, B_6, folacin, B_{12} and biotin, while for others the urinary excretion levels are preferred such as for vitamin B_1 and B_2. The blood and urine assays are to a great degree presented in the manual for Nutrition Surveys of the Interdepartmental Committee for National Development[9]. Additional methods are available for pyridoxal phosphate[10], folacin[11], vitamin B_{12}[12], and biotin[13].

In the early 1960's, as a consequence of our work with vitamins B_1 and B_6, we coined the phase "functional methods to evaluate nutritional status"[14] because we recognized that if we could measure the activity of a vitamin in a physiological functional mode such as an enzyme activity, we could then better relate actual nutritional status to the optimum condition. If so, we should be better equipped to differentiate between a clinical deficiency state, a marginal deficiency state and a state of normalcy. This concept has since been adopted by many investigators and extended to other vitamins beyond our own observations.

In Table IV are outlined the functional methods to evaluate vitamin status. There are systemic methods by which the total body response to a metabolic load involving the function of an enzyme or group of enzymes is evaluated either by an in vivo loading test or the manipulation of a biopsy tissue in vitro. Secondly, there are also the in vitro enzyme-coenzyme activation tests such as we initiated for thiamine. We separated the functional methods into systemic as well as enzyme-coenzyme groups. For instance, as a systemic functional test, dark adaptation[15] is employed for the evaluation of vitamin A status. This is an in vivo measure. Similarly, other in vivo systemic tests are the tryptophan and methionine load tests for vitamin B_6 by which the person is given a few grams of tryptophan and/or methionine and the excretion of xanthurenic acid or cystathionine, respectively, are measured in urine[16, 17]. For folacin[18], there is a histidine load test following which the excretion in urine of FIGLU (formiminoglutamic acid) excretion is measured. The histidine load test is also useful for vitamin B_{12}[19]. In addition, methylmalonate excretion can be measured in urine to assess vitamin B_{12} adequacy. For biotin one can perform a loading test with propionic acid[21].

There are also totally in vitro systemic functional tests. A typical example is the erythrocyte peroxidative hemolysis test for vitamin E[22] or the modification which we developed by using the Fragiligraph[23]. For vitamin K there is the measurement of prothrombin time which is a standard clinical procedure.

The employment of specific enzyme-coenzyme relationships, however, lends another dimension to the evaluation of nutritional status. By the use of these techniques one not only measures the activity of an enzyme requiring a vitamin but also the activity

TABLE 4

FUNCTIONAL EVALUATION OF VITAMIN STATUS

VITAMIN	SYSTEMIC		ENZYME-COENZYME
	In vivo	In vitro	
A	Dark adaptation		
E	---	RBC peroxidative Hemolysis	---
K	---	Prothrombin time	---
B_1	---	---	TPP-transketolase (RBC)
B_2	---	---	FAD-GSH reductase (RBC)
B_6	Tryptophan methionine load	---	B_6P-aminotransferase (RBC)
Folacin	Histidine load (FIGLU)	---	---
B_{12}	Histidine load (FIGLU) Valine load (methylmalonate)	---	---
(Biotin)	---	---	(Pyruvate carboxylase)
(Co-Q_{10}	---	---	(Co-Q_{10}-succinic deH$_2$)

level of the apo-enzyme and the proportion of the apo-enzyme which
is not saturated with coenzyme. This proportion of unsaturation,
which we call the "coenzyme-effect" gives us a measure of extent of
insufficiency for the vitamin which exists in the person's biochemical system. Let us illustrate this for thiamine. Early studies
in our laboratories with intact red blood cells demonstrated that
methylene blue activated the pentose phospate pathway in nonnu-
cleated human red blood cells[24]. Recognizing that the isolation
a few years earlier of the enzyme transketolase, independently by
Horecker and by Racker, revealed that the isolate contained thiamine
pyrophosphate, we embarked on studies to determine whether the metabolism of the pentose phosphate pathway in the erythrocyte was modified in thiamine deficiency. Studies in intact rat erythrocytes,
with the use of specifically carbon labeled glucose molecules, showed
that the transketolase activity of these cells was gradually depleted
as the thiamine deficiency became more severe[25]. Furthermore, it
was shown that significant effects were found within the period of
4 to 7 days at a time when the rats were growing normally and appeared in perfect health. This suggested that the method could
measure marginal or preclinical deficiency. Coincidentally, we had
occasion to apply the method to alcoholic individuals showing the
signs of Wernicke's encephalopathy and found that the method confirmed a biochemical thiamine deficiency in these subjects who
responded clinically to administered thiamine[26]. Subsequently it
was appreciated that the intact red cells and the use of radioactive
substrate were not suitable for routine analysis because the cells
retained their active level of metabolism for only about 12 hours
on the one hand and radioactive substrate measurement instrumentation
was not generally available during that era. Therefore, an alternate
mode of assay was desirable. Accordingly, a hemolysate assay was
developed[27, 28] which eliminated the need for radioactive substrate
and permitted the freezing of hemolyzed red cells for periods up
to 3 to 5 months without major loss of activity. The hemolysate
assay was applied to human subjects maintained on thiamine low
diets[29, 30] and some interesting observations are shown in Figure 1.
Here we note that urine thiamine reached minimum levels in about
10 days with very little change between 10 days and 49 days. On
the other hand the TPP-effect in the transketolase assay increased
gradually until it exceeded 15% in 10 to 12 days and continued to
rise to 35% in about 30 days. In other words, urine thiamine remained constant once it reached its low level, while the TPP-effect
continued to rise. This was the first indication in human subjects
that the use of the enzyme-coenzyme functional test permitted us
to differentiate between a marginal deficiency state and an acute
deficiency since the urine thiamine level did not permit us to determine which end of the depletion period we were at.

These observations permitted us to delineate a sequence of five
stages in the development of a thiamine vitamin deficiency[14] (Table
V). In doing so, we were able to designate the first three stages

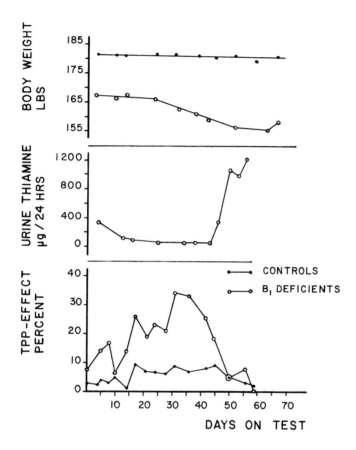

Fig. 1 - Effects of thiamine deficiency in man on body
weight, urinary excretion of thiamine, and TPP
effect. The group consumed approximately 190μg
thiamine daily and was depleted for six weeks
before being given a supplement (30).

TABLE 5

THE DEVELOPMENT OF VITAMIN DEFICIENCY

Sequence	Deficiency Stage	Symptoms and Comments
1.	Preliminary	Depletion of tissue stores (Due to diet, malabsorption, abnormal metabolism, etc.). Urinary excretion depressed.
2.	Biochemical	Enzyme activity reduced due to coenzyme insufficiency, urinary excretion negligible.
3.	Physiological	Loss of appetite with reduced body weight, insomnia or somnolence, irritableness, adverse MMPI scores. Reduced drug metabolism.
4.	Clinical	Exacerbated nonspecific symptoms plus appearance of specific deficiency syndrome.
5.	Anatomical	Clear specific syndromes with tissue pathology. Death ensues unless treated.

as comprising the "marginal deficiency" state to which we referred to early in this paper[4].

Subsequently, we were able to demonstrate that the level of transketolase activity in erythrocytes could be controlled by dietary thiamine[31] and that by those response curves one could assay for thiamine in foods. Also, the transketolase test was independent of deficiencies of protein, pyridoxine and riboflavin[31]. We were able to show that brain was the last tissue to be depleted for transketolase activity in thiamine deficiency[33], that pyrithiamine had greater effectiveness in reducing brain transketolase activity than oxythiamine[34], that sorbiton and penicillin administration to rats in low quantities resulted in increased gastrointestinal synthesis of thiamine[35], and that the administration of amprolium, a poultry coccidiostat, confirmed the resultant thiamine deficiency which occurred as measured by the transketolase assay[36]. The use of this enzyme technique also showed that thiamine is depleted in fasting adult women in a period of 4 days[37], suggesting the need for daily intake for this vitamin.

The data presented in Figure 1[30] on the rate of change of the TPP-effect in human subjects, and in Figure 2[31] which is a scatter diagram for values of hexose formation and the TPP-effect in 210 subjects, permitted us to suggest that normal values for TPP-effect are below 15% while those between 15 and 25% comprise the marginal deficiency state. Since individuals with Wernecke's encephalopathy (the alcoholic form of beriberi) generally had TPP-effect values of 25% or higher we designated that range the severe deficiency of clinical deficiency range. The TPP-effect transketolase assay is now recognized to be the most sensitive test for thiamine adequacy[32].

While a large number of modifications of our basic TPP-effect transketolase assay have been published[38-47] no concerted attempt has been made to develop criteria of thiamine adequacy for these other assays. Rather, the original criteria for thiamine adequacy have been adopted for virtually all assays, although it would be desirable and accurate to develop criteria of adequacy for each assay modification.

The TPP-effect transketolase assay has been used to evaluate nutritional status in countries worldwide and for thiamine adequacy in a variety of nutritional as well as other disease conditions[48, 58]. It has been used in the field by the ICNND[59] with success (minor modifications in the cirteria for adequacy were suggested) and the concept has been adapted for other enzyme assays for riboflavin, pyridoxine, biotin and coenzyme Q as will be discussed.

Additional studies demonstrated that young erythrocytes had higher levels of transketolase activity and that this also applied to young thiamine deficient erythrocytes[60]. An automated assay for erythrocyte transketolase activity has been developed[61].

PYRIDOXINE

Many years ago we observed that pyridoxine deficiency markedly depressed the activities of the glutamic alanine and glutamic as-partic aminotransferases in duck heart, by 50 and 30%, respective-ly[62]. In view of our successful development of an enzyme-coenzyme assay for thiamine adequacy based upon the transketolase enzyme in erythrocytes as described earlier in this paper, we undertook to determine whether it was feasible to measure pyridoxine adequacy by the use of the enzyme-coenzyme principle. We soon observed that serum aminotransferases were less subject to change than tissue enzymes in rats exposed to fasting and cortisone[63] but that when rats were fed pyridoxine deficient diets the alanine and the as-

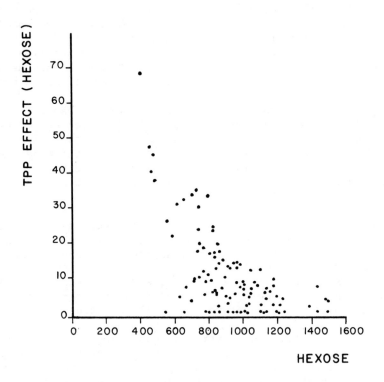

Fig. 2 - A scatter diagram of the transketolase assay values
for hexose formation and TPP effect (hexose) for
two-hundred and ten subjects. A clear relationship
is seen between these for TPP effect (hexose) values
in excess of 15% (p < 0.001). (32).

partic transaminase enzymes in plasma were quickly depressed 85% and 63%, respectively[64, 65]. These data are shown in Table VI. Although the activities of both enzymes were rapidly restored in deficient rats by the administration of pyridoxine by injection or by feeding, the addition of pyridoxal phosphate to the serum _in vitro_ only restored activity partially to normal and this was not a constant observation. Although the activity of the serum enzyme with and without pyridoxal phosphate was not a very reliable method to evaluate nutritional status, the effects of the deficiency on the enzyme activity per se, was useful in determining the bioavailability of vitamin B_6 in foodstuffs[66]. Furthermore, it was soon realized that serum enzymes and particularly the aminotransferases were highly variable depending upon the physiological state of other tissues in the body such as heart, liver and muscle, since they are released in hepatitis, myocardial infarction, and various myopathies, respectively.

Accordingly, efforts were made to determine the usefulness of the erythrocyte aminotransferases to assess pyridoxine adequacy. It was noted (Table VII[67, 14]) that after 18 days of pyridoxine deficiency there was about a 60% reduction in the erythrocyte alanine aminotransferase activity which was restored only partially to normal by the addition of pyridoxal phosphate. Accordingly, the absolute enzyme activity in the erythrocyte was more useful to asess nutritional status than the reactivation with pyridoxal phosphate in the rat. Furthermore it was observed that when deficient rats were treated with vitamin B_6 the plasma enzyme reverted to normal within 6 days while the erythrocyte enzyme did not for up to 3 weeks (Figure 3).[67]

It remained for other investigators to confirm our findings on the effects of pyridoxine deficiency on aminotransferase activity in human erythrocytes and serum[68, 69, 71] and to extend this to leucocytes[69]. It has been shown that the erythrocyte enzymes have greater usefulness to study pyridoxine status than the serum enzymes in man[69, 70]. The pyridoxal phosphate effect has value in determining human pyridoxine status although the values are much more valiable than the TPP-effect for thiamine deficiency. The B_6P-effect for the glutamic oxalacetic aminotransferase is usually no more than 50% and that for the pyruvate aminotransferase rarely more than 25%. The criteria for normalcy, which have been suggested by the U.S. Army Medical Nutrition Laboratory, are a B_6P-effect of less than 1.5 for the oxalacetic aminotransferase and less than 1.25 for the pyruvate transaminase[70].

Various methods of assay are available for these pyridoxine enzymes, although much variation is observed between different laboratories[66-67-72-75]. While hemolysates stored for aminotransferase activity are readily stable in the freezer, activity is often lost by thawing and refreezing.

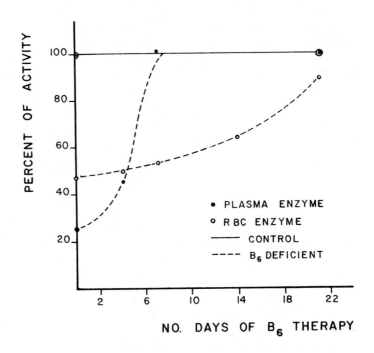

Fig. 3 - A diagrammatic presentation of the difference in
         rate of recovery to normal activity of plasma and
         erythrocyte (RBC) alanine transaminase activities
         in pyridoxine-deficient rats, following placement
         on the pyridoxine-adequate diet. (14).

TABLE 6

EFFECT OF PYRIDOXINE DEFICIENCY ON PLASMA
TRANSAMINASE IN YOUNG RATS[a] (14)

GROUP	NUMBER OF RATS	ALANINE ENZYME		ASPARTIC ENZYME	
		No addition (units)	Depression[b] (%)	No addition (units)	Depression[b] (%)
Control	13	378 ± 74^{c}	---	831 ± 211^{c}	---
Deficient	6	55 ± 24^{c}	85	333 ± 87^{c}	63

[a]Rats were on test 2-4 weeks at time of assay. Data are presented as mean ± S.E.
[b]Percentage of depression calculated as (Control - Deficiency)/control x 100.
[c]$P < 0.01$

TABLE 7

EFFECT OF PYRIDOXINE REPLETION ON PLASMA AND HEMOLYSATE
TRANSAMINASES IN B_6-DEFICIENT RATS*

(14)

Group Injected	Alanine Transaminase Activity (Units)	
	Plasma	Hemolysate
Control		
Saline	234.9 ± 21.5	723.8 ± 23.0
Pyridoxine	241.2 ± 15.1	741.2 ± 36.9
Deficient		
Saline	68.6 ± 17.2	332.8 ± 17.8
Pyridoxine	250.8 ± 15.8	397.4 ± 28.7

* Groups of 7-8 rats. Na pyruvate produced in hg/ml sample/hr.
Data are means ± S.E.

RIBOFLAVIN

Following the observation by Glatzle et al. that glutathione
reductase can serve as an enzymatic test for the detection of a
riboflavin deficiency[76] other workers simultaneously confirmed and
extended this observation[77-80]. The test system measures the reduction of oxidized glutathione as it is coupled to the oxidation
of NADPH to NADP. This can be done spectrophotometrically and the
FAD-effect can be calculated for the erythrocyte enzyme. The FAD-
effect for normal subjects is about 1.0 while values of 1.2 or higher
are considered to be reflections of inadequate riboflavin
status[78, 81, 82]. Not only is the erythrocyte glutathionine reductase
activity useful in evaluating riboflavin status, but it has been
used to reflect interactions with vitamin B_6 metabolism[83, 84]. It
has been observed that glutathione reductase activity serves as a
more sensitive procedure for assessing riboflavin status than does
urinary riboflavin measurement[78, 80].

The enzyme is readily measured on small quantities of blood
and is relatively stable and therefore can be collected and analyzed
at a later date.

The interrelationship between riboflavin intake and urinary
excretion versus erythrocyte glutathione reductase activity is shown
in Figure 4[85]. We notice that on a reduced riboflavin intake, urinary riboflavin reaches almost a minimal value in two weeks while
the FAD-effect of "EGR activity coefficient" as shown in the figure

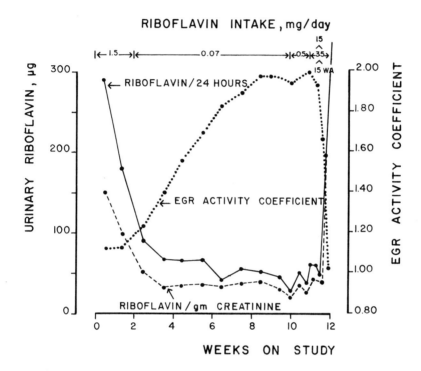

Fig. 4 – Relationship of riboflavin intake to urinary
riboflavin intake to urinary riboflavin excretion
and erythrocyte glutathionine reductase activity
coefficients. Mean values for six young adult
males. (85)

continues to increase regularly for a period of 8 weeks before it
levels off. Accordingly, just as in the case of the TPP-effect for
thiamine, the FAD-effect for riboflavin has the potential of revealing more quantitatively the marginal versus the acute deficiency
state.

COENZYME Q

While there might be some controversy as to whether Coenzyme Q
is a vitamin per se, it certainly is a normal component of animal
and human tissues and is an essential coenzyme in metabolism. It
has been shown that Coenzyme Q_{10} is the form necessary in human
subjects[86]. The enzyme-coenzyme principle was developed for coenzyme Q for Folkers' group in association with the activity of
the coenzyme with succinate dehydrogenase[87]. Although used primarily in tissue preparations the assay system can be specifically
applied to blood systems by using it in human leucocyte mitochondrial preparations[87].

BIOTIN

While biotin nutriture has not been of major concern in human
nutrition to this point, it has been of commercial interest in chickens, turkeys and fish[88-90]. Recently a blood carboxylase assay has
been used as a criteria of biotin status in chickens and turkeys as
a functional test[91]. Perhaps this enzyme[92] is worthy of further
investigation in leucocytes and/or platelets as formed elements in
blood of human subjects.

DISCUSSION AND CONCLUSIONS

The enzyme-coenzyme functional tests for vitamin adequacy, as
performed on the formed elements in blood, usually the erythrocyte,
afford the opportunity to evaluate nutritional status in a quantitative functional way in order to differentiate a marginal from a
severe clinical vitamin deficiency. This technology, although slow
in developing for all vitamins, has a sound solid biochemical foundation and is highly useful at this time for thiamine and riboflavin.
For pyridoxine more effort has recently been placed upon plasma pyridoxal phosphate levels due to the high variation of the B_6P-effect
of the aminotransferases in erythrocytes and/or plasma, and the fact
that the administration of high levels of vitamin B_6 is reported to
induce the levels of apoenzyme thereby making the interpretation of
the enzyme-coenzyme functional assay difficult.

Not all vitamins function as coenzymes (the fat-soluble vitamins
and vitamin C in particular) and therefore enzyme functional assays

are less feasible for these at this time. By the same token although
the remaining B vitamins do have coenzyme activity for specific biochemical reactions, they have not in all cases been demonstrated in
blood or other available fluids or cells from subjects. Fortunately,
for those others such as vitamin B_{12} folacin and B_6, the levels in
plasma are more meaningful than the functional assays, whether systemic, in vitro or in vivo. Therefore, one does have the opportunity
to carry out functional assays for most of the vitamins, whether of
the enzyme-coenzyme variety or of the systemic variety.

In the current era when clinical vitamin deficiencies are rare
at least in the developed countries, but biochemical measurements
show that large portions of populations are below optimal, it is
a challenge to define what state exists for the nutriture of the
population. The Ten State Survey of the United States has declared
this is a state of being "at risk" for the development of nutritional
deficiency[93] and the Market Basket Survey of the USDA has suggested
that about 50% of all diets in the United States are below optimum[94].
Accordingly, the nutritional status of populations must be monitored
constantly. In the United States there is the HANES (Health and
Nutrition Examination Survey Program) which is an ongoing program
and which again reveals large groups of the population inadequate
for vitamins C, A, B_6, calcium, and iron, etc.[95].

If, in fact, it is desirable for public health purposes to show
all portions of the population within the normal range, one must
then intervene to accomplish this. A primary procedure to do so
would be to develop and implement massive education programs so
that people will be more highly informed about their food and hopefully modify their behavior to eat protective foods. A second procedure would be to improve the enrichment or fortification of foods
with micronutrients so that people who choose food unwisely and
consume an inordinate amount of cereal grains which when processed
ordinarily are relatively low in nutrient density, but would through
proper enrichment would gain a higher and broadened nutritional status. If food enrichment were implemented appropriately it would contributed very greatly to improving the nutritional status of all populations of the world. In some cases where there is a wheat and corn
economy and the grains are centrally processed this can be done
through grain. In other cases where sugar is the centrally processed
commodity, such as in certain countries of Central America, This has
shown to be possible to be accomplished through its fortification
with vitamin A. There are current programs to develop MSG as a delivery system in countries where the per capita of consumption is in
excess of 3 grams per day. In other words if the motivation exists
to improve micronutrient or vitamin status in population groups,
modes of delivery can be identified by which this can be accomplished, at a cost of fractions of pennies per day. Even a multi-
vitamin supplement can be made available for as little as 1 or 2
cents per day if this is the only last resort available to accomplish

the public health purpose, and certainly this would be cheaper than
delivering supplemental nutrition through food at currently inflated
costs.

Conceived by us 25 years ago, the functional evaluation of nutritional status by enzyme-coenzyme relationships has contributed
a great deal to our knowledge about metabolism of vitamins and the
evaluation of nutritional status of population groups.

REFERENCES

[1] U.S. Federal Register. Title 7, Chapter IX, Part 1404, 7: 11105,
    Issue #255, 1942.
[2] NAS/NRC Food and Nutrition Board, "Proposed Fortification Policy
    for Cereal Grain Products", NAS, Washington, D.C., 1974.
[3] Christakis, G., ed., Nutritional assessment in health programs.
    Am. J. Pub. Hlth. 63, Suppl., November 1973.
[4] Brin, M. Dilemma of marginal vitamin deficiency, in: "Proc.
    9th Intl. Congr. Nutr., Mexico", Chavez, A., Bourges, H.,
    and Basta, S., eds., pp. 102-115, Karger, 1975.
[5] Brin, M. Drugs and environmental chemicals in relation to vitamin
    needs. In: "Nutrition and Drug Interrelationships", Hathcock,
    J.N. and Coon, J., eds., pp. 131-150, Academic Press, N.Y.,
    1978.
[6] Kumar, M. and Axelrod, A.E. Cellular antibody synthesis in
    thiamine, riboflavin, biotin, and folic acid deficient rats.
    Proc. Soc. Exp. Biol. Med. 157: 421-423, 1978.
[7] Brin, M. Examples of behavioral changes in marginal vitamin
    deficiencies in the rat and man. Proc. U.S.-Japan Coop. Sci.
    Progr. on "Behavioral Effects of Energy and Protein Deficits",
    Brozek, J., ed., PAHO, Washington, D.C., November 30-Decem-
    ber 2, 1977 (in press).
[8] Eisman, J.A., Hamstra, A.J., Kream, B.E., and DeLuca, H.F. A
    sensitive, precise, and convenient method for determination
    of 1,25-dihydroxy vitamin D in human plasma. Arch. Biochem.
    Biophys. 1766: 235-243, 1976.
[9] Interdepartmental Committee on Nutrition for National Defense,
    "Manual for Nutrition Surveys", 2nd ed., U.S. Gov. Print. Off.,
    Washington, D.C., 1963.
[10] Rose, C.S., Gyorgy, P., Butler, N., Andres, R., Norris, A.H.,
    Shock, N.W., Tobin, J., Brin, M., and Spiegel, H. Age dif-
    ference in vitamin B_6 states of 617 men. Am. J. Clin. Nutr.
    29: 847-853, 1976.
[11] Herbert, V. Biochemical and hematological lesions in folic acid
    deficiency. Am. J. Clin. Nutr. 20: 562, 1967.
[12] Boddy, K. and Adams, J.F. The long-term relationship between serum
    Vitamin B_{12} and total body vitamin B_{12}. Am. J. Clin. Nutr.
    25: 395, 1972.
[13] Association of Vitamin Chemists, "Methods of Vitamin Assay",

3rd ed., pp. 245-255, J. Wiley, N.Y., 1966.

[14] Brin, M. Use of erythrocyte in functional evaluation of vitamin adequacy. In: "The Red Cell", 1st ed., Bishop C. and Surgenor, D., eds., pp. 451-476, Academic Press, N.Y., 1964.

[15] Dowling, J.E. and Wald, G. Vitamin A dificiency and night blindness. Proc. Nat. Acad. Sci. 44: 648, 1958.

[16] Luhby, A.L., Brin, M., Gordon, M., Davis, P., Murphy, M., and Spiegel, H. Vitamin B_6 metabolism in users of oral contraceptive agents. I. Abnormal urinary xanthurenic acid excretion and its correction by pyridoxine. Am. J. Clin. Nutr. 24: 648-693, 1971.

[17] Park, Y.K. and Linkswiler, H. Effect of vitamin B_6 depletion in adult man on the excretion of cystathionine and other methionine metabolites. J. Nutr. 100: 110, 1970

[18] Luhby, A.L. and Cooperman, J.M. Folic acid deficiency in man and its interrelationship with vitamin B_{12} metabolism. Adv. Metab. Disord. 1: 263, 1964.

[19] Nixon, P.F. and Bertino, J.R. Interrelationships of vitamin B_{12} and folate in man. Am. J. Med. 48: 555, 1970.

[20] White, A.M. and Cox, E.V. Methylmalonic acid excretion and vitamin B_{12} deficiency in the human. Ann. N.Y. Acad. Sci. 112: 915, 1964.

[21] Murthy, P.N.A. and Mistry, S.P. Biotin. Progr. Fd. Nutr. Sci. 2: 402-455, 1977.

[22] Horwitt, M.K., Harvey, C.C., Duncan, G.D., and Wilson, W.C. Effects of limited tocopherol intake in man with relationship to erythrocyte hemolysis and lipid oxidations. Am. J. Clin. Nutr. 4: 408, 1956.

[23] Brin, M. and Danon, D. Some new developments in the functional evaluation of vitamin E and thiamine nutritional status. J. Sci. Ind. Res. 29: 338-344, 1970.

[24] Brin, M. and Yonemoto, R.H. Stimulation of the glucose oxidative pathway in human erythrocytes by methylene blue. J. Biol. Chem. 230: 307-317, 1958.

[25] Brin, M., Shohot, S.S. and Davidson, C.S. The effect of thiamine deficiency on the glucose oxidative pathway in rat erythrocytes. J. Biol. Chem. 230: 319-326, 1958.

[26] Wolfe, S.J., Brin, M., and Davidson, C.S. The effect of thiamine deficiency on human erythrocyte metabolism. J. Clin. Invest. 37: 1476-1784, 1958.

[27] Brin, M., Tai, M., and Ostashever, A.S. Thiamine deficiency and erythrocyte hemolysate metabolism, Fed. Proc. 18: 518, 1959.

[28] Brin, M., Tai, M. Ostashever, A.S., and Kolinsky, H. The effect of thiamine deficiency on the activity of erythrocyte hemolysate transketolase, J. Nutr. 71: 273-281, 1960.

[29] Brin, M., Vincent, W., and Watson, J. Human thiamine deficiency and erythrocyte transketolase. Fed. Proc. 21: 468, 1962.

[30] Brin, M. Erythrocyte transketolase in early thiamine deficiency. Ann. N.Y. Acad. Sci. 98: 528-541, 1962.

[31] Brin, M. Erythrocyte as a biopsy tissue for functional evaluation

of thiamine adequacy. J. Am. Med. Assn. 187: 762-766, 1964.

[32] Brin, M. Functional evaluation of nutritional status: thiamine.
In: "Newer Methods in Nutritional Biochemistry", 3: 407-445,
Academic Press, N.Y., 1967.

[33] Brin, M. Effects of thiamine deficiency and oxythiamine on rat
tissue transketolase. J. Nutr. 78: 179-183, 1962.

[34] Brin, M. The differential effect of pyrithiamine and oxythiamine
on rat brain transketolase activity. Abstr. Am. Chem. Soc.,
Div. Biol. Chem. 146th Ann. Mtg., Denver, Colorado, 1964.

[35] Brin, M. The effects of penicilline, D pencillamine and D sor-
bitol on erythrocyte transketolase activity in thiamine deficient rats. Toxic. Appl. Pharmacol. 6: 631-637, 1964.

[36] Polin, D., Wynosky, E.R., and Porter, C.C. Amprolium: studies
on thiamine deficiency in laying chickens and their eggs.
J. Nutr. 76: 59, 1962.

[37] Haro, E.N., Brin, M., and Faloon, W.W. Fasting in obesity:
thiamine deficiency as measured by erythrocyte transketolase
changes. Arch. Int. Med. 117: 175-181, 1966.

[38] Dreyfus, D.M. Clinical application of blood transketolase determinations. N. Eng. J. Med. 267: 596, 1962.

[39] Brin, M. Clinical applications of transketolase assays. In:
"Methods in Enzymology", Vol. IX, pp. 506-514, Academic Press,
N.Y., 1966.

[40] Brin, M. Transketolase (sedoheptulose-7-phosphate:D glyceral-
dehyde-3-phosphate dihydroxyacetonetransferase, EC 2.2.1.1)
and the TPP-effect in assessing thiamine adequacy. In:
"Methods in Enzymology", 18: 125-133, 1970.

[41] Brin, M. Transketolase. In: "Methods in Enzyme Analysis",
Bergmeyer, H.U., ed., 2: 703-709, 1974.

[42] Warnock, L.H. A new approach to erythrocyte transketolase
measurement. J. Nutr. 100: 1057, 1970.

[43] Schouten, H., Van Eps, S., and Stryker Boudier, A.M. Transketolase in blood. Clin. Chim. Acta. 10: 474, 1964.

[44] Upjohn, D.R., Dohm, G.L., and Ziporin, Z.Z. An enzyme method
for the assay of transketolase activity in the red blood
cell. J. Nutr., 1973.

[45] Chang, Y.H. and Ho, G.S. Erythrocyte transketolase activity.
Am. J. Clin. Nutr. 23: 261, 1970.

[46] Gubler, C.J. and Johnson, L.R. Enzyme studies in thiamine deficiency. In: "Thiamine Deficiency, Ciba Study Group nº 28",
pp. 54-66, 1967.

[47] Bruns, F.H., Dunwald, E., and Noltman, E. III. Quantitative
bestimmung von sedoheptulose-7-phosphat and einige eigenshaften
der transketolase der erythrocyten un des blutserums. Bioch.
Z. 330: 497-508, 1958.

[48] Dibble, M.V., Brin, M., McMullen, E., Peel, A., and Chen, N. Some
preliminary biochemical findings in junior high school children
in Syracuse and Onondaga County, New York. Am. J. Clin. Nutr.
17: 218-239, 1965.

[49] Brin, M., Dibble, M.V., Peel, A., McMullen, E., Bourquin, A., and

Chen, N. Some preliminary findisgs on the nutrition status of the aged in Onondaga County, N.Y. Am. J. Clin. Nutr. 17: 240-258, 1965.

[50] Dibble, M.V., Brin, M., Thiele, V., Peel, A., Chen, N. and Mc-Mullen, E. Nutritional status evaluation in older age subjects with comparisons between Fall and Spring. J. Am. Ger. Soc. 15: 1031, 1967.

[51] Thiele, V., Brin, M., and Dibble, M.V. Nutritional status evaluation in negro migrant workers in Kings Ferry, N.Y. Am. J. Clin. Nutr. 21: 1229-1238, 1968.

[52] Konstirren, A., Louhija, A., and Hartel, G. Blood transketolase in assessment of thiamine deficiency in alcoholics. Ann. Med. Exp. Biol. Benn. 48: 172, 1970.

[53] Tripathy, K. Erythrocyte transketolase activity and thiamine transfer across human placenta. Am. J. Clin. Nutr. 21: 739, 1968.

[54] Brubacher, G., Haenel, A., and Ritzel, G. Transketolaseactivat, thiaminausscherdung under bluttethiamingehalt bein menschen zur beurterlung der vitamin B_1-versorgung. Int. J. Vit. Nutr. Res. 42: 190, 1972.

[55] Tanphaichitr, V., Vimokesant, S.L., Dhanamitta, S., and Valiasevi, A. Clinical and biochemical studies of adult beriberi. Am. J. Clin. Nutr. 23: 1017, 1970.

[56] Akbarian, M. and Dreyfus, P.M. Blood transketolase activity in beriberi heart disease. J. Am. Med. Assn. 203: 77, 1968.

[57] Burgener, M. and Jurgens, P.G. Thiamine excretion and transketolase activity in chronic alcoholism and Wernicke's encephalopathy. Ger. Med. Mon. 12: 396, 1967.

[58] Coon, W.W. and Bizer, L.S. Subclinical thiamine deficiency in post-operative patients. Surg. Gynecol. Obst. 121: 37, 1965.

[59] Interdepartmental Committee on Nutrition for National Defense, Union of Burma Nutrition Survey Report. U.S. Gov. Print. Off., Washington, D.C., May 1963.

[60] Brin, M. The effects of cell age and thiamine on erythrocyte transketolase activity. J. Vitaminol. 15: 338-339, 1969.

[61] Stevens, C.O., Sauberlich, H.E., and Long, J.L. An automated assay for transketolase determinations. In: "Automation in Analytical Chemistry", Medical Publ., N.Y., 1968.

[62] Brin, M., Olson, R.E., and Stare, F.J. Metabolism of cardiac muscle pyridoxine deficiency. J. Biol. Chem. 210, 435-444, 1954.

[63] Brin, M. and McKee, R.W. Effects of X-irradiation, nitrogen, mustard, fasting, cortisone and adrenolectomy on transaminase activity in the rat. Arch. Biochem. Biohpys. 61: 384-389, 1956.

[64] Brin, M. and Tai, M. Pyridoxine deficiency and serum transaminases. Fed. Proc. 17: 472, 1958.

[65] Brin, M., Tai, M. Ostashever, A.S., and Kolinsky, H. The relative effects of pyridoxine deficiency on two transaminases in the growing and the adult rat. J. Nutr. 71: 416-420, 1960.

[66] Brin, M., Ostashever, A.S., Tai, M., and Kalinsky, H. Effects

of feeding X-irradiated pork to rats on pyridoxine nutrition as reflected in the activity of plasma transaminase. J. Nutr. 75: 35-38, 1961.

[67] Albert, D.J. and Brin, M. Comparison of serum and erythrocyte hemolysate systems. Fed. Proc. 19: 321, 1960.

[68] Cheney, M., Sabry, Z.I., and Beaton, G.H. Erythrocyte glutamic pyruvic transaminase activity in man. Am. J. Clin. Nutr. 16: 337, 1965.

[69] Raica, N. Jr. and Sauberlich, H.E. Blood cell transaminase activity in human vitamin B_6 deficiency. Am. J. Clin. Nutr. 15: 67, 1964.

[70] Sauberlich, H.E., Canham, J.E., Baker, E.M., Raica, N. Jr., and Herman, Y.F. Biochemical assessment of the nutritional status of vitamin B_6 in the human. Am. J. Clin. Nutr. 25: 629, 1972.

[71] Driskell, J.A. Vitamin B_6 status of the elderly. In: "Human Vitamin B_6 Requirements", pp. 252-256, National Academy of Sciences, Washington, D.C., 1978.

[72] Reitman, S. and Frankel, S. A colorimetric method for the determination of serum glutamic oxalacetic and glutamic pyruvic transaminases. Am. J. Clin. Pathol. 28: 56, 1957.

[73] Karmen, A. A note on the spectrophotometric assay of glutamic oxalacetic transaminase in human blood serum. J. Clin. Invest. 34: 131, 1955.

[74] Wroblewski, F. and LaDue, J.S. Serum pyruvic transaminase in cardiac and hepatic disease. Proc. Soc. Exp. Biol. Med. 91: 569, 1956.

[75] Giusti, G., Ruggiero, G., and Cacciatore, L. A comparative study of some spectrophotometric and colorimetric procedures for the determination of serum glutamic oxalacetic and glutamic pyruvic transaminase in hepatic disease. Enzymol. Biol. Clin. 10: 17, 1969.

[76] Glatzle, D., Weber, F., and Wiss, O. Enzymatic test for the detection of a riboflavin deficiency. Experientia 24: 1122, 1968.

[77] Beutler, E. Effect of flavin compounds on glutathione reductase activity: in vivo and in vitro studies. J. Clin. Invest. 48: 1957, 1969.

[78] Bamji, M.S. Glutathione Reductase activity in red blood cells and riboflavin nutritional status in humans. Clin. Chim. Acta 26: 263, 1969.

[79] Flatz, G. Population study of erythrocyte glutathione reductase activity. I. Stimulation of the enzyme by flavin adenine dinucleotide and by riboflavin supplementation. Humangenetik 11: 269, 1971.

[80] Tillotson, J.A. and Baker, E.M. An enzymatic measurement of the riboflavin status in man. Am. J. Clin. Nutr. 25: 425, 1972.

[81] Glatzle, D., Korner, W.F., Christellar, S., and Wiss, O. Method for the detection of a biochemical riboflavin deficiency. Int. J. Vit. Nutr. Res. 40: 166, 1970.

[82] Sauberlich, H.E., Judd, J.H., Nichoalds, G.E., Broquist, H.P. and

Darby, W.J. Application of the erythrocyte glutathione reductase assay in evaluating riboflavin nutritional status in
a high school student population. Am. J. Clin. Nutr. 25: 756,
1972.

[83] Krishnaswamy, K. Erythrocyte transaminase activity in human
vitamin B_6 deficiency. Int. J. Vit. Nutr. Res. 41: 240, 1971.

[84] Krishnaswamy, K. Erythrocyte glutamic oxalacetic transaminase
activity in patients with oral lesions. Int. J. Vit. Nutr. Res.
41: 247, 1971.

[85] Sauberlich, H.E., Dowdy, R.P., and Skala, J.H. "Laboratory Tests
for Assessment of Nutritional Status". CRC Press, Inc.,
Cleveland, 1974.

[86] Nakamura, R., Littarru, G.P., and Folkers, K. A new enzymic
assay for human deficiencies of coenzyme Q_{10}. Int. J. Vit.
Nutr. Res. 43: 526-536, 1973.

[87] Folkers, K. Relationship between coenzyme Q_{10} and vitamin E.
Am. J. Clin. Nutr. 27: 1026, 1974.

[88] Hegsted, D.M., Mills, R.C., Briggs, D.M. Elvehjem, C.A., and
Hart, E.B. Biotin in chick nutrition. J. Nutr. 23: 175-179,
1942.

[89] Patrik, H., Boucher, R.V., Dutcher, R.A., and Knandel, H.C.
Prevention of perosis and dermatitis in turkey poults. J.
Nutr. 26: 197-204, 1943.

[90] Castledine, A.J., Cho, C.Y., Slinger, S.J., Hicks, B., and
Bayley, H.S. Influence of dietary biotin level on growth,
metabolism and pathology of rainbow trout. J. Nutr. 108: 698-
711, 1978.

[91] Whitehead, C.C. and Bannister, D.W. Blood pyruvate carboxylase
activity as a criterion of biotin status in chickens and
turkeys. Br. J. Nutr. 39: 547-556, 1978.

[92] Bannister, D.W. and Whitehead, C.C. Presence of pyruvate carbo-
xilase in the blood of the domestic fowl, and an assay procedure. Int. J. Biochem. 7: 619-624, 1976.

[93] Center for Disease Control. Ten State Nutrition Survey in the
U.S., U.S. Department of Health, Education and Welfare,
Atlanta, 1972.

[94] U.S. Dept. of Agriculture. Dietary Levels of Households in the
U.S., Spring 1965, USDA Publ. #ARS62-17, 1968.

[95] National Center for Health Statistics. Health and Nutrition
Examination Survey (HANES), Health Resources Administration,
Washington, D.C., 1973, 1974.

II. Pathological and Clinical Nutritional

ADAPTATION TO SEVERE PROTEIN DEFICIENCY

Hisato Yoshimura, Katsuharu Kubo and Noriko Tanaka

Dept. of Physiology, Hyogo College of Medicine

Nishinomiya - JAPAN

In recent studies on dynamic aspect of protein metabolism, there are culminating evidences which indicate that the muscle protein is rather labile and its turnover rate is very sensitive to low protein intake. In an attempt to clarify the physiological role of muscle protein in severe protein deficiency, by taking account of its largest mass in the body, the authors designed a series of experiments with pregnant rats fed protein free diet. Much protein is required in pregnancy for development of reproductive organs including foetus. Thus the rats fed on protein free diet suffer from severe protein deficiency which results in reabsorption of foetus and the pregnancy is interrupted. By injecting subcutaneously 4 mg progesterone and 0.5 μg esterone daily, the pregnancy can be maintained until parturition. With the pregnant rats thus treated, their body weight and protein contents of various organs were compared with the control non-pregnant rats fed on protein free diet which started on the same day with the pregnant. The comparisons were made at the early (around the 7th day), the middle (around 12th day) and the last stage (after 16th day) of pregnancy.

By feeding the protein-free diet, the body weight decreased to about the same extent with the pregnant rat fed protein-free diet. By comparing the organ weight of the rats, it was revealed that the reduction of muscle mass was larger in the pregnant rat than in the control. The difference almost corresponded to the increase of weight of reproductive organs including foetus caused by pregnancy. Fig. 1 illustrates the difference of nitrogen content in various organs at various stages of pregnancy which was calculated by comparing the pregnant rats and their control non-pregnant rats. Remarkable differences between the two groups are as follows: the nitrogen retention appears in the muscle and liver in the middle

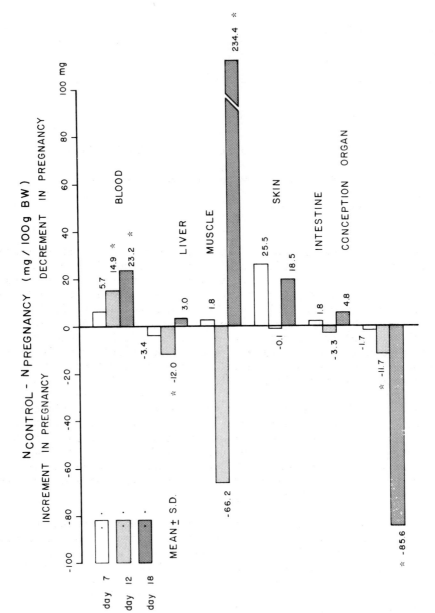

Fig. 1 – Difference of N. content in various organs of rats at various stages of pregnancy maintained on protein free diet.

stage of pregnancy, while the retention of protein in reproductive
organs (including foetus) and the reduction of muscle protein markedly appear in the last stage of pregnancy. Thus it is suggested
that the skeletal muscle reserves the protein in the middle stage
of pregnancy and discharges its nitrogen in the last stage when the
reproductive organs retain the building materials liberated from the
muscle to develop the foetus. In an attempt to confirm this dynamic
aspect in protein metabolism in pregnancy, ^{15}N ammonium citrate
(1.01 mEq ^{15}N/rat) was injected intraperitoneally at various stages
of pregnancy, and then all the urine was collected successively until
72 hours after injection, when the rats injected were killed and
accumulations of ^{15}N in muscle, liver and in the whole body were
determined, ^{15}N being measured by emission spectrometric method.
Results are shown in Fig. 2. It is seen that ^{15}N accumulates in
muscle, liver, in whole body at the middle stage of pregnancy, while
the accumulations in pregnant rats decrease in muscle and liver at
the last stage as compared with the control. In order to analyze
the site of accumulation of ^{15}N and its rate of incorporation in the
protein of muscle and liver, ^{15}N atom percent excess in N. of protein
and N. of protein fractions separated by ultracentrifuge was measured
successively at various stages after intraperitoneal injection of
^{15}N lysine (0.511 mEq ^{15}N/rat) on the 7th day of pregnancy. As seen
in Fig. 3, ^{15}N atom percent excess in liver protein fractions i.e.
supernatant soluble protein (SUP) and microsome protein (MS) gradually decrease in parallel, while those in muscle protein fractions
change differently among SUP, MS and crude contractile protein (CP).
In the early period after injection, ^{15}N incorporation is highest
in MS fraction, and it decreases remarkably after a week, i.e. in
the middle stage of pregnancy and attains the lowest level in its
last stage. ^{15}N incorporations to the fractions of SUP and CP are
rather low in the early stage of pregnancy, but increase in the
middle and last stage. ^{15}N atom percent excess of CP, the largest
component of muscle protein falls at the latest stage of pregnancy.
Patterns of changes of ^{15}N incorporation in various fractions of
muscle protein indicate that the rate of ^{15}N accumulation changes
specifically in various fractions of muscle protein along the time
course of pregnancy, showing dynamic changes in protein metabolism
among various fractions of muscle protein. Thus the change of incorporation rate of ^{15}N in muscle protein is diphasic through the
whole period of pregnancy. These diphasic changes of protein metabolism in pregnancy may be under the hormonal control, and the pregnancy is maintained by mobilizing muscle protein metabolite to the
reproductive organs when the protein cannot be provided from the
diet. Thus the muscle protein plays an important role as protein
reserve in severe protein deficiency at least in pregnancy.

According to the human experiments of Scrimshaw's group on
short-term adaptation to low protein diet, an increase in the catabolic rate of total body protein and also in the synthetic rate appears, and reutilization of endogenous amino acids is accelerated in

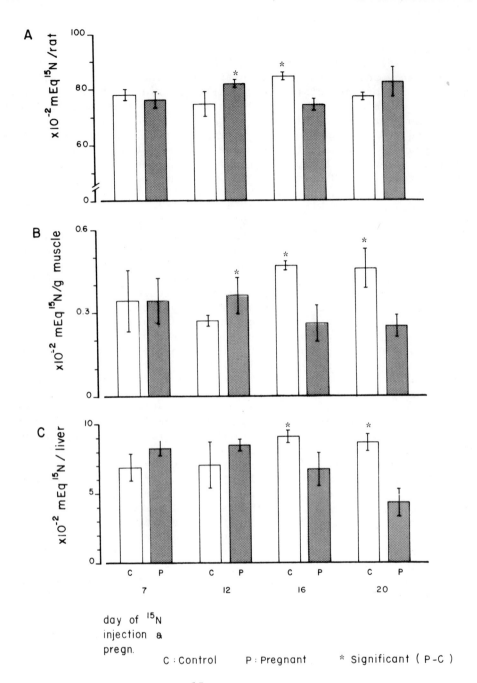

C : Control   P : Pregnant   * Significant (P-C)

Fig. 2 – Rate of ^{15}N accummulation in 3 days after ^{15}N ammonium citrate injection.

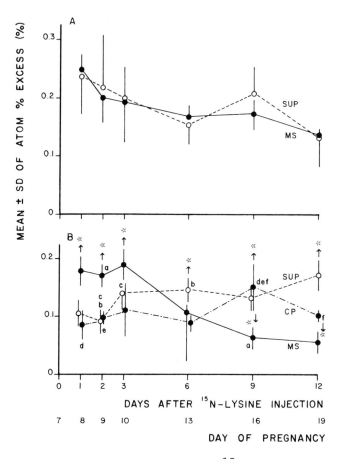

Fig. 3 - Changes in abundance of ^{15}N in the protein
fractions in liver (A) 8_1 muscle (B) of the
protein deficient pregnant rats.

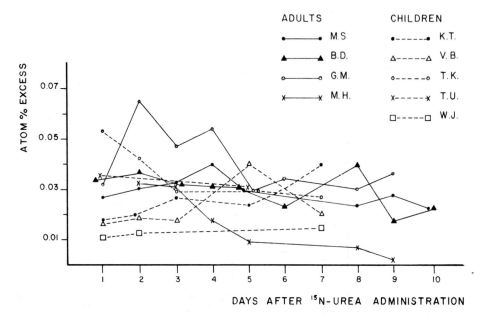

Fig. 4 – P.N.G. ^{15}N incorporated into plasma protein.

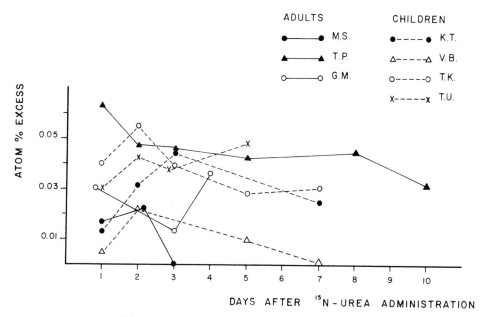

Fig. 5. P.N.G. ^{15}N incorporated into lysin in hydrolysate of plasma
protein.

protein deficiency. In studies with Papua New Guinea (PNG) high-
landers who have been subjected to low protein diet for long time by
taking sweet potato as main food, the authors verified experimentally
a possibility of another type of metabolic adaptation to protein
deficiency. Subjects were ten attendants in Goroka Hospital. The
^{15}N-urea (97 atom % ^{15}N) was taken orally by each subject in a dose
of 20 mg per Kg B.W. Daily urine and morning blood were collected
daily after the administration of ^{15}N urea for about 10 days. The
cumulative ^{15}N excretion, and atom % excess of ^{15}N in plasma protein
as well as NPN in plasma were measured. About 0.04 atom % excess of
^{15}N was found in plasma protein of all the subjects on every morning,
somewhat fluctuating (Fig. 4). ^{15}N atom % excess in NPN in plasma
is higher than that in plasma protein, about 1.0% or so, which fell
to zero 3 days after ^{15}N urea intake. The plasma protein was hy-
drolisated and amino acids were separated from one another on silica
gel plate by chromatography. In about 80% of the samples, ^{15}N was
found in 0.01-0.06 atom % excess in lysine fraction of the hydrolysate as seen in Fig. 5. The lysine is unique among the essential
amino acids in the fact that exchange of amino nitrogen moiety with
nitrogen from other sources does not take place without complete
destruction of the lysine molecule. Therefore ^{15}N-lysine can only
be provided by absorption from the intestine, as ^{15}N-urea could be
converted to ammonia by bacteria in intestinal flora and ^{15}N lysine
might be synthesized there. Other essential amino acids in hydrolysate of plasma protein may originate from those synthesized intestinal flora. Thus the vegetable protein taken by PNG highlander can
be enriched with those essential amino acids, thus contributing to
adaptation to low protein vegetable diet. As the value of percent
excess of ^{15}N in plasma protein is very low with the present subjects
who are hospitalized, further studies are required to verify the
nutritional role of urea recycling by making similar experiments
with subjects in the field where the subjects are living in less
civilized conditions.

THE EFFECT OF MODERATE PROTEIN-ENERGY MALNUTRITION ON

SEVERAL BIOCHEMICAL PARAMETERS IN PRESCHOOL CHILDREN

F. Delpeuch, A. Cornu, and Ph. Chevalier

24 Rue Bayard

75008 Paris, FRANCE

INTRODUCTION

In Cameroon most cases of protein-energy malnutrition in infants and children are mild or moderate. These moderate forms are often overlooked because people don't recognize them since they show no characteristic clinical signs. They are also hard to detect because of the lack of appropriate techniques.

Most techniques used to estimate protein-energy malnutrition require simple anthropometric tests. These tests, however, do not allow an early detection of P.E.M. because they are based on bodily alterations and it is obvious that the children's bodily development is already impaired when the measurements are taken. On the contrary, one can assume that food protein-energy deficiencies quickly lead to some alterations in the proteic metabolism detectable in some blood and urine parameters. Actually, few studies are available as to the use of biochemical indicators of mild forms of malnutrition. Most authors describe the biochemical alterations occurring in severe malnutrition states.

In the present study we intend to compare the levels of several biochemical variables in healthy children with those in moderately malnourished children (excluding the severe clinical forms). We have utilized biochemical parameters known to decrease in cases of severe P.E.M.

METHODS

810 children from 1 to 60 months old have been observed in the South-Cameroonian forest area. Their staple diet is characterized by a low intake of poor quality protein sometimes associated with energy deficiency.

The children have been divided into two groups according to their nutritional status which was estimated by means of anthropometric tests.

Group I (control children) is composed of children whose measurements are as follows:

- Weight for age over 80 per cent of the 50th percentile of Harvard standards;

- Weight for height over 90 per cent;

- Arm circumference for age over 85 per cent of Wolanski standards;

- Arm head ratio over 0.290.

The simultaneous use of several weight and arm tests has ensured a better diagnosis of nutritonal status.

Group II (moderately malnourished children) is composed of children with at least one anthropometric test below the above stated thresholds. Their anthropometric deficiencies are moderate and never attain levels associated with severe malnutriton states.

However the means of all anthropometric tests in Group I and II differ significantly (0.001).

Blood samples were taken by venous puncture in the femoral vein. Urine was collected once in the morning. Samples were stored at -20° C before analyses and all were treated in the same controled conditions. The following biochemical parameters were determined:

- Total serum protein by the method of GORNALL;

- Electrophoretic fractions of the serum protein on cellulose acetate strips stained with ponceau red;

- Prealbumin, transferrin, third component of complement by radial immuno-diffusion;

- Urinary hydroxyproline by the method of HABICHT and the index calculated according to WHITEHEAD;

Differences between mean values and significance of correlation coefficients were assessed by STUDENT's test.

RESULTS

The total serum proteins fo not significantly differ from one group to the other (P over 0.05), on the contrary higly significant decreases (0.001) of prealbumin, transferrion hydroxyproline index and the albumin/glubulins ratio appear in the malnourished group; the differences concerning albumin and the third component of complement are less significant (0.01).

Correlation coefficients between each anthropometric test and each biochemical variable were calculated. Although manu coefficients were rela-ively weak, most of them were statistically significant (0.001).

Hydroxyoroline index, transferrin, albumin/globulins ratio and prealbumin proved to be biochemical parameters best related to all anthropometric tests.

The correlation are generally stronger with the arm for age and the ar/head ratio tests than with the weight for age and the weight for height ones.

DISCUSSION

Our results show a slight but significant influence of moderate malnutrition on several biochemical variables.

However, can such results entitle us to draw conclusions as to the use of biochemical variables in the early detection of malnutrition? - Many things must be taken into consideration.

FIRST   - The comparison of the results of our control group with those of other works show that values given as norms can vary considerably. Such variations can be explained by differences between ecological zones studied as well as by different criteria used to define nutritional status.

SECONDLY - Even for the most sensitive parameters the levels recorded in group II were always much higher than those encoutered in severe malnutrition.

Thus it appears that the effects of moderate malnutrition on biochemical parameters can only be revealed in comparison with well nourished children from the same population.

FINALLY - Some recent studies have demonstrated that the interpretation of an isolated biochemical indicator is usually questionable because of the interaction of factors other than malnutrition (parasitism and infections). These interactions and the relatively low variations observed make us think that one of the variables considered in this study can be used alone to ensure a diagnosis of mild malnutrition.

On the contrary, as assessment of nutritional status based on several parameters seems to be more advisable. A reduction of several variables should be considered as a sign of a global metabolism disfunction determined by moderate malnutrition.

Thus albumin/glubulins ratio, prealbumin, transferrin and hydroxyproline index could form a viable set of tests to be used in determining mild forms of malnutrition.

These parameters show higly significant correlations (o.001). These correlations probably reflect a simultaneous decrease of the concentrations of the parameters concerned.

In conclusion, these biochemical tests may be of value for an early detection of malnutrition but only if used together. Comparison of results mus to be made with those of a control group composed of children out of the same population and of the same age. In the future it seems very important to establish normal values of biochemical parameters which are modified by P.E.M.

THE SHORT AND LONG TERM FUNCTIONAL IMPLICATIONS

OF INFANT CALORIE-PROTEIN MALNUTRITION

José O. Mora, M.D.

Inst. Colombiano de Bienestar Familiar, Dir. de Nutrición

Tv. 39, nº 27-01, aptdo. aereo 15609, Bogotá, COLOMBIA

The contention that infant malnutrition carries long term functional implications for the individual thus affecting the human capital, is often used as the most convincing argument to motivate politicians and other decision making people to give high priority and preferential budgetary allocations to nutrition programs[1,2].

Apart from humanitarian reasons, actions to combat infant malnutrition need to be justified on factual evidence on its functional implications. I would like to briefly comment on such evidence with the aim of providing some insights as to what could be reasonably used as scientifically proved facts in favor of nutrition programs as a high priority issue.

I should start by pointing out that usually "infant malnutrition" is operationally defined as the presence of physical growth retardation supposedly associated with calorie-protein deficient diets and, quite frequently, with a high incidence of infectious diseases morbidity. This definition, which excludes specific nutri-

- Supported in part by the Colombian Institute of Family Welfare, the NICHHD Grant No. 3R01-HDo6774-01A1S1, The Ford Foundation Grant No. 740-0348, the German Research Foundation, and the Fund for Research and Teaching, Department of Nutrition, Harvard School of Public Health.

- From the ICBF-Harvard-Giessen Research Project on Malnutrition and Mental Development, Bogota, Colombia.

ent deficiences, refelects the prevalent situation of most children
in developing countries who are affected by global nutritional deficiencies social and biological environment often labeled as the "culture of poverty".

    Implications of infant malnutriton so defined may be arbitrarily
classified as "short term" and "long term".

1. SHORT TERM IMPLICATIONS

    There is abundant evidence on the following short term implica-
tions of infant malnutrition:

(a) By definition, malnutrition is manifested by <u>physical growth</u>
     <u>retardation</u> of easy and reliable recognition, based on the as-
     sumption that attained values lower than the standards of re-
     ference are attributable to environmental rather than to genetic
     factors[3].

(b) Malnutrition seems to <u>increase susceptibility</u> to <u>infectious dis-</u>
     <u>seases</u> affecting either their frequency or, more probably, their
     intensity and duration[4].

(c) Infant malnutrition is associated with <u>high mortality rates</u> to
     such an extent that infant mortality and, particularly, pre-
     -school mortality are used as indirect indicators of malnutri-
     tion in the community. It is to be established whether such
     high mortality rates are attributable to the nutritional condi-
     tion itself or to the high incidence of disease together with
     lower availability of health resources and less efficient use
     of the available services[5].

(d) The short term <u>behavior modifications</u> of malnourished children
     have been recognized and reported by many and, more recently,
     some efforts have been made to quantitate them[6]. Malnourished
     children are apathetic and less active physically, and exhibit
     decreased exploratory drives and lower attention capacity. As
     a consequence, their ability to interact with and take advantage
     of the eventual stimulatory opportunitites provided by the phy-
     sical and human environment is greatly diminished. Furthermore,
     since the child's behavior appears to determine, to a great
     extent, the responsiveness of the human environment, especially
     affecting the mother-child interaction, a vicious circle has
     been described by which the malnourished, apathetic, passive
     child demands little attention from the mother who then limits
     her interaction with him thus increasing his passiveness and
     apathy, and reducing his opportunities for environmental stimu-
     lation. Again, it is not well known to what extent these be-

havioral modifications are casually related to the presence of
disease which is known to lower motivation and physical activity.

In summary, the high incidence of infectious diseases characteristic of the environment in which the malnourished children live
and grow turns to be a confounding factor in the ascertainment of
the short term implications of malnutrition. The issue becomes quite
relevant if it is considered that morbidity is the rule rather the
exception throughout the first years of life in impoverished communities.

2. LONG TERM IMPLICATIONS

The following are the most commonly recognized:

(a) Lower attained growth which has been also labeled as "stunting".
                            Since retardation in height supposedly
     associated with chronic malnutrition seem to be at least parti-
     ally irreversible, the low stature of the population in most
     developing countries has been attributed to environmental influ-
     ences and, particularly, to infant malnutrition[3]. However,
     apart from some implications on the performance in certain
     sports, there does not seem to be evidence of a casual associ-
     ation between low height and physical activity, labor perform-
     ance and intellectual or social competence.

(b) Intellectual retardation The relationship of infant malnutri-
                               tion and intellectual development has
     been explored mostly through cross-sectional studies plenty of
     limitations and, more recently, by means of longitudinal stud-
     ies, most of them carried out in Latin America[7-11].

     Given the practical impossibility to isolate the nutrition fac-
     tor itself from the complex web of interrelated social, economic
     and health factors associated with malnutrition, all the studies,
     even the longitudinal ones, have had and will have to rely upon
     statistical manipulations for the mathematical isolation of
     single factors. Unfortunately, under real life conditions so-
     cial and even some biological phenomenons are not susceptible
     of the reliable quantitation required for the application of
     valid statistical techniques, at least in the current state of
     the art of social sciences research.

     Correlational and multiple regression analyses, for instance,
     have allowed the artificial isolation of the effects of malnu-
     trition, usually defined on the basis of weight and height meas-
     urements, on the child's performance on a variety of psycholo-
     gical tests. In the best of the cases, however, such supposedly
     isolated effects, although statistically significant when using
     relatively large sample sizes, have been of very small magnitude.

The proportion of variance on test scores accounted for by physical growth measurements alone has ranged between 3% and 19% and, in most of the studies, is practically meanigless[9-11].

A major shift has apparently occurred in the relative importance assigned to the mechanisms postulated to explain the eventual two general mechanisms suggested have been:

a) Alterations in CNS development which, when occurring in the so called "critical period" of rapid cell multiplication, may produce structural irreversible changes; and

b) Behavior modifications that limit adequate environmental stimulation, interfering with learning processess in critical periods of child development.

The apparent failures to replicate animal findings on CNS changes associated with malnutrition when species closer to the human are used, e.g. in monkey research, as well as recent studies by Stein and Coworkers on adults supposedly affected by malnutrition in the critical period, and by Winich on early malnourished Korean children adopted by american families, have weakened the possibility that, under the natural conditions in which malnutrition usually occurs, the CNS suffers structural irreversible changes with functional implications[12-14].

On the other hand, social and behavioral scientists have for a long time studied the key influence of environment stimulation and mother-child interaction on the child's intellectual development. Malnourished children are the product of impoverished environments which make them disadvantaged on all aspects of life, particularly in terms of love, care and the necessary environmental stimulation for an adequate cognitive and emotional development. In such conditions malnutrition represent only a part of a syndrome of multiple deprivation, biological and environmental, in which isolation of a single component appears to be not only impossible but also meaningless[15,16].

The compeling need then is for the search and evaluation of comprehensive approaches for the prevention and/or recuperation of intellectual retardation as indicated by the child's performance on relevant measures which, although not yet fully validated in terms of their implications for the social and academic competence in the school or in adulthood, appear to discriminate children from different socio-economic backgrounds in the same culture. Some of those approaches are food supplementation and early environmental stimulation.

Important contributions to the clarification of these issues are coming from our intervention study on nutrition, early stimula-

tion and intellectual development in Bogotá. The basic research
design is a classical two-by-two factorial with nutritional supplementation and early home stimulation as intervention programs given
to a sample of children at risk of malnutrition, as indicated by the
presence of malnutrition in older siblings. Nutritional supplementation was provided since the sixth month of pregnancy, while infant
stimulation started soon after birth; both were given up to 3 years.

In this longitudinal study 450 children who have been closely
followed suffered from any kind of infectious diseases during 285
days of their first two years (about 40% of their life) experiencing
an average of 32 disease episodes with a mean duration of about 9
days. At 18 months, about 60% of these children were classified as
malnourished according to the Gómez anthropometric criterion, even
though their mean daily intake was around 1300 calories and 30 g.
of protein, well above the recommendations. Furthermore, no epidemics of infectious diseases preventable by immunization occurred
during such a period; morbidity was due to upper respiratory infectious (close to 2/3 of the cases), gastrointestinal infectious (about
1/3 of the cases), and eye, ear and skin infections. As expected on
the basis of their mean dietary intake, food supplementation had a
moderate effect on the prevention of malnutrition, whose prevalence
was only 20% lower in the supplemented group, even though there were
highly statistically significant differences in mean growth attained
or incremental[17].

The coincidence of adequate dietary intake, high infectious disease incidence and retarded physical growth would suggest that morbidity may affect physical growth or nutritional status by mechanisms
other than the reduction of dietary intake by loss of apetite.

Intellectual development was measured by different tests, among
them the Griffiths test of mental abilities, which provides five sub-
quotients over the first three years (locomotor, personal-social,
speech and language, eye-hand coordination, and performance), in addition to the general I.Q. The test was previously adapted to the
local culture and standardized. Both interventions had simple and
interaction effects on specific areas of development. Effects of
supplementation were heavily focused on motor development whereas
those of stimulation were mostly centered on language development;
both affected the general I.Q. and, to certain extent, prevented
its downward tendency observed with age. The results strongly suggest that both nutritional and environmental enrichment interventions
may be required to prevent intellectual retardation in disadvantaged
children; they would also suggest that malnutrition and poor environmental stimulation, acting through similar mechanisms, may interact
to depress the general intellectual ability of low income children
by impinging upon specific developmental domains[18, 19].

REFERENCES

[1] Selowsky, M.   Infant malnutrition and human capital formation.
    Presented at the Research Workshop on Problems of Agricultural
    Development in Latin America. Caracas, May 1971.

[2] Selowsky, M.   A note on prescholl age investment in human capital
    in developing countries. Presented at the Workshop on the
    Economics of Education. Washington, Oct. 1976.

[3] Habicht, L.P., et al. Height and weight standards of preschool
    children. How relevant are ethnic differences in growth poten-
    tial? Lancet I (7858): 611, 1974.

[4] Scrimshaw, N.S.:, Taylor, C.E., and Gordon, J.E. Interactions of
    nutrition and infection. World Health Organization Monograph
    Series nº 57. Geneva, 1968.

[5] Puffer, R.R., and Serrano, C.V. Patterns of mortality in child-
    hood. Pan American Health Organization (PAHO) Scientific Publi-
    cation nº 262, Washington, D.C. 1973.

[6] Chavez, A. and Martine C. Nutrition and development of children
    from poor rural areas. V. Nutrition and behavioral development
    Nutr. Rep. Inter. 11:477, 1975.

[7] Scrimshaw, N.S., and Gordon, J.E. (eds.) "Malnutrition, Learning
    and Behavior". MIT Press, Cambridge, Mass. 1968.

[8] Frisch, R.E. Present status of the suposition that malnutrition
    causes permanent mental retardation. Amer. J. Clin. Nutr. 23:
    189, 1970.

[9] De Licardie, E., and Cravioto, J. Language development in survivors
    of clinical severe malnutrition (a longitudinal study). Pre-
    sented at the IX International Congress of Nutrition, Mexico
    City, Sep., 1972.

[10] Mora, J.O., et al. Nutrition, health and social factors related to
    intellectual performance. World Rev. Nutr. Diet. 19: 205, 1974.

[11] Freeman, Klein, R., Kagan, and Yarbrough, C. Relations between
    nutrition and cognition in rural Guatemala. Amer. J. Pub.
    Health 67:233, 1977.

[12] Cheek, Holt, and Mellits, Malnutrition and the nervous system, in:
    Nutrition, the Nervous System and Behavior. Pan American Health
    Organization Scientific Publication, nº 251, 1972.

[13] Stein, Z., Susser, M., Saenger, C., and Marolla, F. Nutrition and
    mental performance. Science 178: 708, 1972

[14] Winick, M., Katchadurian, M.K., and Harris, R.C. Malnutrition and
    environmental enrichment by early adoption. Science, 190:1173.
    1975.

[15] Hegsted, D.M. Deprivation syndrome or protein calorie malnutrition?
    Nutr. Reviews 30:51, 1972.

[16] Elias, M.F. Malnutrition in infancy: one of the many influences on
    human intellectual development. Presented at the Biennial meet-
    ing of the International Society for the study of behavioral
    development. Ann Arbor. Michigan, Aug. 21-25, 1973.

[17] Mora, J.O., Herrera, M., de Navarro, L, Suescun, L., and Wagner, M.
    The effects of nutritional supplementation on physical growth

of children at risk of malnutrition. Submitted for publication,
1978.

[18] Mora, J.O., Clement, J., Christiansen, N., Vuori, L. Ortix, N.,
Wagner, M., and Herrera, M. Nutritional & supplementation,
early home stimulation and child development. Conference on Behavioral Effects of Energy and Protein Deficits, Washington,
D.C., Nov. 29-Dec. 2, 1977.

[19] Mora. J.O., Weber, Herrera, M., Clement, J., Christiansen, N.,
Vuori, L, and Wagner, M. Effects of nutritional supplementation
and early home stimulation on infant's intellectual development
Presented at the XI International Congress of Nutrition, Rio
de Janeiro, Brasil, Aug., 22.Sept. 1, 1978.

THE CHANGING PATTERNS OF PROTEIN-CALORIE

MALNUTRITION IN THE SUDAN

Mahmoud Mohamed Hassan

Nat. Com. for Child Welfare

P.O. Box 8071, New Extension, Khartoum, SUDAN

Protein-calorie malnutrition is one of the most major child health problems in the Sudan. Its magnitude is reflected in the high incidence in the most vulnerable age-groups of 1 to 3 years. During the last 15 years the pattern of the disease is showing significant changes in the clinical, aetiological and prognostic aspects. Since protein-calorie malnutrition is a complex nutritional problem in which multiple causes, especially social, economical, cultural, medical, psychological and ecological factors, play important roles, the changes in pattern of this disease is expected in any developing community where progressive modern changes are taking place.

CHANGING PATTERN IN INCIDENCE

15 to 10 years ago the most predominant nutritional disorder was kwashiorkor, while marasmus was less in significance (Table 1).

The pattern of protein-calorie malnutrition has greatly changed during the last 5 years. The number of cases of marasmus admitted has remarkably increased (Table 2); in addition cases of marasmic-kwashiorkor are being increasingly admitted (Table 3).

CHANGING PATTERN IN AETIOLOGICAL FACTORS

Changing Pattern in Nutrition

Breast feeding – In the past breast feeding was definitely the

177

TABLE 1

NUMBERS OF CASES OF PCM ADMITTED
TO CHILDREN'S DEPARTMENT KHARTOUM HOSPITAL
(1960 - 1964)

YEAR	KWASHIORKOR	MARASMUS
1960	75	12
1961	74	8
1962	59	2
1963	106	-
1964	114	1
TOTAL	428	23

TABLE 2

NUMBERS OF CASES OF PCM ADMITTED
TO CHILDREN'S DEPARTMENT KHARTOUM HOSPITAL
(1969 - 1973)

YEAR	KWASHIORKOR	MARASMUS
1969	15	64
1970	56	19
1971	76	53
1972	49	40
1973	32	18
TOTAL	228	194

dominant method of feeding especially during the first year of life
(Table 4).

Although breast feeding was continued and the infant thrived
during the first 6 months, weight steadily declined because supplementary feeding was insufficient with only a small amount of starch
or lemon juice with sugar given.

TABLE 3

CASES OF PCM ADMITTED
TO CHILDREN's DEPARTMENT KHARTOWN HOSPITAL
(1976)

TYPE OF PROTEIN CALORIE MALNUTRITION	NUMBER OF CASES
Marasmus	58
Marasmic-Kwashiorkor	16
Kwashiorkor	26
TOTAL	100

TABLE 4

DURATION OF BREAST FEEDING IN
BURRI HEALTH CENTRE IN KHARTOUM
(Hassan M.M., 1967)

DURATION OF BREAST FEEDING	PERCENTAGE OF CHILDREN
5 months	97.5
1 year	84
2 years	35

Since dura (Sorghum) is the main national diet in most parts of the country, and it is deficient in certain essential amino-acids, protein-calorie malnutrition may develop if it is solely used as a post-weaning diet (Table 5).

PRESENT TREND OF INFANT FEEDING

The present trend of infant feeding in the Sudan is character-ized by a steady decline of breast feeding especially in urban pop-ulations.

TABLE 5

PERCENTAGE OF PROTEIN IN DURA, WHEAT, MAIZE AND RICE
AND DEFICIENCY OF ESSENTIAL AMINO-ACIDS

TYPE OF GRAIN	gms PROTEINS PER 100 gr wt.	DEFICIENT ESSENTIAL AMINO-ACID
Dura (Sorgum)	8 - 12	1) Tryptophan 2) Methtonine 3) Lysine
Wheat	11.8	1) Lysine 2) Tryptophan
Maize	9.4 - 12	Tryptophan
Rice	7.5	None

The present pattern of breast feeding practice was studied in
170 mothers in the various socio-economic groups in the society
(Table 6). The total durations of breast feeding indicate a definite
deterioration in feeding practice compared to the previous study
(Table 7).

The main factors influencing breast feeding were investigated
in this group of 170 mothers (Table 8). These were:

1) Many mothers are working during the day-time; this inter-
ruption in breast feeding induces the introduction of artificial
feeding which steadily replaced the breast feeding. In the Sudan
the lactating mother is given one hour leave from her work in order
to breast-feed her baby. Although all working mothers are absent
for one hour from her work during the day, the majority cannot utilize it for breast feeding because of distance of residence and
difficulty in transport.

2) Use of contraceptive tablets. Most mothers complain of
steady decline of breast milk soon after they commence contraceptive
tablets. This is a tragic consequence in rural mothers whose infants
depend solely on the breast for their nutrition. The use of contraceptive tablets is getting widely spread especially in the newly

TABLE 6

SOCIO-ECONOMIC GROUPS OF MOTHERS

GROUP OF MOTHER	PERCENTAGE
1. Upper Socio-Economic	6.5
2. Middle Socio-Economic	10.7
3. Lower Socio-Economic	82.8

TABLE 7

TOTAL DURATION OF BREAST FEEDING IN KHARTOUM
(1977)

DURATION OF BREAST FEEDING	PERCENTAGE OF INFANTS
2	94
3	93.4
4	88.7
5	87.6
6	85.2
12	69.9
18	42.3
24	10.6

introduced "family planning clinics", where tablets are issued free of charge.

3) The increasing propaganda for dried milk in daily journals, radio, televisions, posters and charming photographs and free samples in hospitals and private clinics is definitely influencing the public opinion.

4) Pregnancy, especially in the rural community, is a definite reason for abrupt cessation of lactation.

TABLE 8

MATERNAL FACTORS INFLUENCING BREAST FEEDING

NUMBER	FACTOR	PERCENTAGE OF MOTHERS
1	Type of Work	
	1) Doctor	0.5
	2) Teacher	1.0
	3) Nurse	1.0
	4) Labourer	1.0
	TOTAL	3.5
2	Preference for Powdered Milk	4.2
3	Preference of Cow's Milk	5.7
4	Pregnancy	7.3
5	Contraceptive Tablets	9.4
6	Maternal Attitude	26.0
7	Unknown Factor	43.7

5) The mental attitude of the mother is a most important factor especially in the primipara. This is influenced to a large extent by her cultural, educational and socio-economic standard, her understanding and belief of the value of breast-feeding and her satisfaction in its accomplishment. The role of influence of information disseminated in her environment from her mother, relatives, doctors, midwives and nurses before and after delivery is substantial (Table 9).

6) Age of the mother: it has been found that mothers aged 25-30 years were able to feed their children for a peeiod of 1 year to 1 1/2 years more effectively than younger and older mothers (Table 10).

TABLE 9

DURATION OF BREAST-FEEDING IN THE
DIFFERENT EDUCATIONAL GROUPS OF MOTHERS

EDUCATIONAL STANDARD AND PERCENTAGE OF MOTHERS	DURATION OF BREAST-FEEDING		
	6 Months	12 Months	18 Months
University (3.3)	2.1%	Nil	Nil
Secondary (11.9)	10.8%	6.6%	4.4%
Primary (24.5)	26.2%	24.1%	11.0%
Illiterate (60.3)	44.7%	39.0%	27.4%

7) The effect of complete breast-feeding for 4 to 5 mothers
without the addition of any other food or milk was quite significant
in perpetuating the success of a long duration of breast-feeding.

CONCLUSION

In a predominantly muslim community, the incentive for breastfeeding exists. Most mothers are brought up in societies where
breast-feeding has been the normally-accepted feeding practice especially in rural communities.

The importance of breast-feeding and its exact duration have
been well defined in the Holy Guran in three separate statements
fifteen centuries ago.

The First Statement has explained the proper methods of rearing
the infant and establishing harmony and understanding between the
fatehr and the mother in this respect.

I shall endeavour to present the meaning of these statements:

1) "And the mothers shall breast-feed their children for two
   whole years if they wish to complete lactation, and the
   father is responsible for their maintenance and their clothing in a reasonable manner. Nobody should be burdened beyond
   his capacity. No mother should suffer harm on account of
   her child, nor the father on account of his child; and a

TABLE 10

DURATION OF TOTAL BREAST FEEDING
ACCORDING TO AGE GROUPS

MONTHS DURA-TION	M O T H E R S' A G E S				
	20 years	20-24 years	25-30 years	31-40 years	40-50 years
12	3.4%	16.3%	37.3%	16.0%	2.0%
18	0.7%	7.1%	22.9%	11.7%	2.0%

similar duty devolves on the father's heir. But if both desire weaning by mutual consent and after consultation, they shall incur no sin. And if you wish to engage a wet-nurse for your children, there shall be no offense provided that you pay what you promised according to the accepted practice. And keep your duty to Allah and know that Allah is aware of what you do."

2) "And we have enjoyed man to care for his parents; his mother bears him with weakness upon weakness and his weaning occurs after two years. Give thanks to Me and to your parents. To Me is the eventual return."

3) "And we have enjoined man to be good to his parents. His mother bears him with much trouble and she gives birth to him in much pain; and his bearing and weaning take thirty months."

BIBLIOGRAPHY

[1] Hassan, M.M. (1960) Kwashiorkor in Sudanese Children, Gaz. Egypt. Paed. Ass. Vol. VIU, No. 3, 424.
[2] Hassan, M.M. (1967) Protein-Calorie Malnutrition, S.M.J. Vol. 5, No. 4, 168.

INFLUENCE OF MALNUTRITION ON LACTOSE IN CHILDREN
WHO BELONG TO GENETICALLY LACTOSE TOLERANT
OR INTOLERANT TRIBES IN CENTRAL AFRICA

H.L. Vis

Free University of Brussels
Brussels
BELGIUM

Lactose malabsorption was studied in children of Central Africa. Three groups were investigated. The children of groups I and II belonged to a "lactose intolerant" tribe. Their age was below 5 years in group I and above 5 years in group II. The children of group III, aged less than 5 years, belonged to a tribe in which mixing of "lactose tolerant" (Tutsi) and "lactose intolerant" (Bantus) people is important.

All the children were tested for lactose malabsorption by a lactose loading test during an episode of proteo-energetic malnutrition (P.E.M.) and after recovery. The data is summarized in Table 1.

In group III, the children who were found to be "lactose tolerant" after recovery had a height for age which ranged well above that found in the "lactose intolerant" children. The same was true, in the same group, for the children studied during P.E.M. and who had a delayed but normal lactose laoding curve. Pure Tutsi are known to be taller than Bantus at the same age.

A relation may therefore exist between height and lactose tolerance in Central Africa.

Malnutrition induces little alteration of lactose absorption in genetically "lactose tolerant" children.

TABLE 1

STUDY OF LACTOSE MALABSORPTION

	Nº OF CHILDREN	MALABSORPTION (% of cases)	NORMAL ABSORPTION (% of cases)
Group I (age 5 years)			
– during P.E.M.	53	100.0	0
– after recovery	41	55.0	45.0
Group II (age 5 years)			
– during P.E.M	45	100.0	0
– after recovery	58	100.0	0
Group III (age 5 years)			
– during P.E.M.	26	53.8	46.2*
– after recovery	31	61.3	38.7

* delayed N

NUTRITION, INFECTION AND IMMUNE RESPONSE

Panata Migasena

Dept. of Tropical Nutr. & Food Sci., Fac. of Tropical Med.
Medicine, Mahidol Univ.
Bangkok 4, THAILAND

Infectious diseases and malnutrition are two of the most common health problems in children of developing countries. The situation is more severe if these two conditions occur concurrently. The interaction of infection and malnutrition may be considered to be cyclic insofar as one condition is capable of accentuating the other. Not only may malnutrition increase host susceptibility to infection, but infection on the other hand may also precipitate malnutrition, particularly in borderline cases. Altered eating habits, loss of appetite, malabsorption and negative nitrogen balance are often secondary to chronic infections so common in developing countries. For example, diarrhea and measles can frequently aggravate subclinically malnourished infant also frank kwashiorkor.

The interaction between malnutrition and infection may be either synergistic or antagonistic. The outcome of this interaction are either immune response or host response. Most interactions in human and in experimental animals are synergistic. The antogonism occurs not only the nutritional deficiency state in severe but also the microbial agent has obligatory dependence on the metabolism of the host cell. This phenomenon is restricted primarily to infections caused by virus, rickettsia and protozoa. Although the antagnistic effect is well documented in animal studies, it has not been as clearly demonstrated in man. Malaria is perhaps the only example of antagonism thus for documented in the human host. This is probably because in human population nutritional deficiencies are seldom so specific as to be more damaging to the agent than to the host and these deficiencies are usually not as severe as those produced in the laboratory animals.

The outcome of immune response or host response between malnutrition and infection depends on several factors including the type of infection and the age of the patient. The most serious groups are infants and preschool children. Infections that commonly cause death in these malnourished infants and children are herpes simplex, smallpox, measles, miliary tuberculosis and salmonellosis. Infectious caused by staphylococci, Pneumocystic carinii, Plasmodium falciparum and hoockworm are not uncommon.

The addition, malnourished children often respond to infection differently from well-nourished children and some of these difference are:

(1)   When infection spreads in malnourished children, it often does so with the development of gangrene rather than supperation.

(2)   There is a higher tendency for malnourished individuals to develop a gram-negative septicemia.

(3)   An organism which may cause a mild or subclinical infection in a well nourished individual may cause a severe infection or even a fatal one in a malnourished individual.

(4)   There is usually little or no fever associated with infection in malnourished individuals.

The exact mechanisms by which diet may be alter the immune response between host and parasite are poorly understood and rather complex, as both the quality and the quantity of dietary intake not only affect the host defence, but also have a direct influence on the metabolism of the invading organism. Some of the possible mechanisms are:

(1)   Alteration of the host defence which controls the initial invasion of infections agents, in such a way as to facilitate acess to the underlying tissues.

(2)   Interference with the reparative process of the host with a resultant increase in the severity of the disease and retardation of recovery.

(3)   Alteration of the metabolism of the invading agent once it has become established in the tissue.

(4)   Establishment of condition which favour the development of the secondary infection.

The interdependence of malnutrition and infection might be more clearly defined by systematically examining the resistant factors of the host that are altered by changes in nutritional status. There

include apithelial barriers, leukocytic function, inflammatory response, acquired immune response, etc. This symposium will focus primarily on the effects of malnutrition on acquired immunity in human,
but non-specific factors which potentiate the effectiveness of the
acquired immunity, will also be discussed.

The two distinct type of the immune components which are responsible for defence against infections agents and for the development
of other immunological phenomena are humoral and cellular responses.
The humoral immune response is mediated by specific antibodies which
may belong to anyone of the 5 immunoglobulin classes (IgG, IgA, IgM,
IgD and IgE). Antibody is important in defence against infections
caused by extracellular agents, eg. bacterial organisms. Cell-mediated immune response (CHIR) is, on the other hand, mediated by specifically sensitized lymphocytes and plays an important role in defence against infections caused by intracellular agents e.g. mycobac-
teria, viruses, rickettsia and some protozoa. The mechanism by which
malnutrition may unfavourably alter these two immune components needs
to be investigated further, although a clear and unified concept has
began to emerge through findings from recent investigation in this
area.

Humoral Immune Response

Several groups of investigators have evaluated the function of
the humoral immune system in malnourished patients but the results
are still controversial. The integrity of the humoral immune response has often been assessed by either measuring the levels of the
various classes of immunoglobulin in the serum or by observing the
increase of antibody titre following an appropriate antigenic stimulation. This generally agreed that the level of immunoglobulin
in the serum of malnourished individuals are not depressed. In fact,
if any, they are slightly elevated when compared with the age and
sex-matched controls who are well nourished. However, this does
not necessarily imply that the humoral immune system is capable of
responding normally to antigenic stimulation. The normal levels of
immunoglobulins may reflect a competent humoral immune system responding to previous infections because of circulating immunoglobulin
levels at any time present the cummulative results of not only the
present immunological experience, but also of the past experience
as well. A more suitable method for the evaluation of the functional
integrity of the humoral immune system at the time when the children
are hospitalized for the treatment of malnourished condition, is by
observing the magnitude of change in the antibody titre following an
appropriate antigenic stimulation. it was found that, while the
antibody response to some antigens, e.g. diptheria and tetanus tox-
oid, are depressed the response of some other antigens e.g. typhoid
and yellow fever vaccines are normal. Antibody production returns
to normal following dietary treatment and the titres are roughly

correlated with the quantity and quality of food consumption during
rehabilitation. On the other hand, data-from the study of gamma-
-globulin turnover in children with protein-calorie malnutrition
(PCM) is consistent with the notion that the humoral immune system
is PCM children is unimpaired.

     Because malnourished children are particularly prove to infec-
tions that occur at the body surface, e.g. diarrhea and respiratory
tract infections, it is possible that their local immune system which
is independent of the serum antibodies is defective. Information on
this point is limited but the available data suggest that the local
immune system in malnourished children does not function normally.
The system, however, recovers slowly following appropriate dietary
treatment.

     In order to function to the maximum in host defence, certain
types of antibodies must interact with the normal component of plasma
known as complement and it is the activation of the complement system
that is responsible for the killing of certain infections agents.
Children with PCN have a defective complement system. The low circulating complement level in this children returned to normal upon receiving a high-protein diet. However, in some of these children
there is an increase in the turnover of complement, and it is possible that this may be associated with the presence of substances in
the circulation of these children which can activate and destroy complement. Endotoxin seem to be a good candidate because many of the
malnourished children have endotoxaemia at the time of admission to
the hospital. This not unreasonable to postulate that the high tendency for malnourished children to develop a gram-negative septicae-
mia may be associated with a defective complement system in these
children, since these bacteria are more suscetible to the action of
complement than other.

Cell-Mediated Immune Response

     The cell-mediated immune (CHI) system of malnourished children
is definitely impaired. It is highly possible that this function
defect is responsible for the increase susceptibility of these children to certain intracelular infections, e.g. measles, varicella,
tuberculosis. Measles is an example of an ordinarilly mild infection
which, when it occurs in malnourished children, often results in
fatal giant-cell pneumonitis, a condition similar to that frequently
found in patients with acute leukaemia receiving steroid therapy or
other immunosuppresive drug treatment. Results from histological
studies on the lymphoid tissues that are known to be associated with
cell-mediated immune function are consistent with the functional
impairment of the system. These children have atrophic thymus, depletion of lymphocytes in the thymus-dependent area of lymph nodes,
tonsils, appendices, and Payer's patches. The thymolymphatic atrophy
in these children may result from impaired synthesis of proteins

required for lymphopoesis or from the lytic effect of high plasma corticosteroid. The latter situation is a common finding in children with protein-caloric malnutrition.

There are several possible mechanism which can interfere with the expression of CMI function. Host investigators have used delayed cutaneous hypersensitivity reaction to access the integrity of CMI function in malnourished children and, in this reaction, intact inflammatory response in required. It has been observed that PCM children appear to have a more sluggish inflammatory response than well-nourished children. In addition to defective inflammatory response, both the afferent and the efferent limbs of the CMI response are also impaired in most of those children. All of those defects return to normal following appropriate dietary treatment.

CELLULAR IMMUNITY STUDIES

IN MARASMIC INFANTS

Liana Schlesinger

INTA

Casilla 15138, Santiago, CHILE

Malnourished subjects are known to have frequent infections and some of these, such as measles, tuberculosis, typhoid fever, and virus infections, follow a particularly severe course.

Although the high frequency of infections can be explained, in part, by the poor environment in which the patients live, the severity could be due to a deficiency in the host defense mechanisms.

Immunoglobulin levels in kwashiorkor and marasmus are not decreased. Studies of antibody response have yielded conflicting results. On the other hand, several factors point to a defect of cellular immunity in malnutrition. Alterations described include atrophy of the thymus and other lymphatic structures, reduced number of peripheral blood T lymphocytes, and a higher frequency of negative cutaneous delayed hypersensitivy reactions to tuberculin, mumps, candida, trychophytin and other agents known to elicit a cellular immune response. Most of these studies have been performed in infants with kwashiorkor.

In Chile, marasmus is the prevalent form of undernutrition and occurs as a result of premature weaning of infants of low income families to dilute milk formulas.

Marasmic children seldom have signs of specific vitamin or mineral deficiencies. Iron deficiency anemia is infrequent. In Table 1, we show typical values of hematocrit, hemoglobin, transferrin saturation, serum and erythrocyte, folate, vitamin A, carotene and zinc levels of the subjects we studied. Only carotene showed a significant lower level in marasmus as compared to well-nourished infants. These considerations are important when studying the effect of nutri-

TABLE 1

BIOCHEMICAL LABORATORY DATA IN MARASMIC AND CONTROL INFANTS*

	Hb (gr/dl)	TRANSFERRIN SATURATION (%)	SERUM FOLATE (µg/l)	ERITHROCYTE FOLATE (µg/l)	SERUM CAROTENE (µg/ml)	SERUM Vit. A (µg/ml)	SERUM Zn++ (mg/dl)
Marasmic infants n = 14	11.5 ±0.3	14.2 ±1.9	10.1 ±2.5	274 ±46	40[1] ±46	32 ±2	0.080 ±0.005
Control infants n = 27	12.0 ±0.2	15.0 ±1.4	15.7 ±2.2	357 ±31	125 ±14	31 ±2	0.089 ±0.007

Mean ± S.E.M. $p < 0.001$

tional deficiencies on the immune system in view of the alterations
in cellular function that have been reported in cases of iron, vitamin A and folic acid deficiencies.

In this paper I would like to show some results of cellular immunity studies in marasmic infants performed by our group in the last
few years.

We studied 36 children with "severe" marasmus 3 to 18 months of
age. Table 2 summarizes clinical and laboratory admission data of
marasmic patients in the various studies. Most patients were hospitalized and were receiving milk formulas. The majority had a birth
weight over 2.5 kg and were not gaining weight at the time of study.
Serum protein and albumin were within normal levels. Hb concentration
ranged from 9 to 13.1 with a mean of 11.1 gr/dl. Although none of
them was severely infected at the time of study, most were recovering
from acute mild infections. None had received transfusions or had
been immunized during the 2 months prior to the study. In order to
consider the possible role of concomitant infection, both non infected
and infected well nourished children recovering from acute diarrhea
or upper respiratory infections were used as controls.

We performed tuberculin skin tests, senzitization to 2-4 dini-
tro-chlorobenzene (DNCB), lymphocyte proliferation in response to
phytohemagglutinin (PHA) and assayed some of the lymphokines or lymphocyte mediators.

Tuberculin sensitivity was investigated only in patients who had
received a bacillus Calmete-Guerin (BCG) vaccination at birth. In
Table 3 we can observe that only one out of 12 marasmic children had
a positive PPD skin test. Of 52 healthy well nourished infants 46
showed a positive reaction. On the other hand, only two of eight
infected well-nourished infants showed a positive PPD reaction. The
difference between marasmic and healthy control infants was highly
significant (p< 0.01), as was the difference between health and
infected controls (p < 0.02). No difference existed between marasmic
and infected controls.

As shown in Table 4, sensitization to DNBC was performed in 13
marasmic infants. Reactions were strongly positive in eight of nine
healthy well nourished controls, and in all the well nourished infants
studied. In the marasmic patients, the response differed significantly in those sensitized before as compared to those sensitized after
30 days of nutritional rehabilitation. Only one of the nine marasmic
patients sensitized before 30 days of treatment developed a positive
reaction. On the other hand, all three malnourished patients sensitized after 30 days of rehabilitation gave a positive reaction. Four
of the malnourished infants who had a negative reaction were retested
with DNCB 4, 6, 23 and 26 months after initially studied, when they
were nutritionally recovered. Reactions remained negative in all

TABLE 2

CLINICAL AND LABORATORY DATA OF MARASMIC INFANTS

	AGE (months)	BIRTH WEIGHT (kg)	WEIGHT AT TIME OF STUDY (kg)	TOTAL SERUM PROTEIN (g/100ml)	SERUM ALBUMIN (g/100ml)	HEMOGLOBIN (g/100ml)	INFECTION ON ADMISSION[a]
Mean	7.5	3.0	2.4	6.8	4.4	11.1	AD (15) URI (5) UI (3)
Range	3–18	2.1–4.0	2.6–6.8	5.4–7.8	2.9–5.3	9.0–13.1	

[a] AD, acute diarrhea; URI, upper respiratory infection; UI, urinary infection.
Figures in parentheses indicate the number of patients with each kind of infection

TABLE 3

PPD REACTION IN BCG VACCINATED MARASMIC
AND CONTROL INFANTS

GROUP OF INFANTS	Nº OF PATIENTS	PPD POSITIVE	PPD NEGATIVE	p
Healthy controls	52	46	6	
				< 0.01
Marasmic infants	12	1	11	
				< 0.98
Infected controls	8	2	6	

(From: Schlesinger et al., ref. 1)

TABLE 4

DNCB TEST IN MARASMIC AND CONTROL INFANTS

	Nº OF PATIENTS	DNCB POSITIVE	DNCB NEGATIVE	p[a]
Marasmic infants (less than 30 days in the hospital)	9	1	8	
Marasmic infants (more than 30 days in the hospital)	3	3	0	< 0.02
Healthy controls	9	8	1	< 0.05
Infected controls	6	6	0	< 0.05

[a] Values of p in relation to marasmic infants less than 30 days in the hospital

(From: Schlesinger et al., ref. 2).

four. In three healthy controls retested with DNCB 3, 21 and 24 months after sensitization skin reactions remained positive. (Table 5).

TABLE 5

DNBC TEST IN RECOVERED MARASMIC AND
CONTROL CHILDREN 3 to 26 MONTHS AFTER INITIAL STUDY

PATIENT	AGE AT INITIAL STUDY (months)	AGE AT RETESTING (months)	WEIGHT AT INITIAL STUDY (kg)	WEIGHT AT RETESTING (kg)	DNCB REACTION INITIAL	DNCB REACTION RETESTING
Marasmic						
JR	8	14	4.2	8.3	−	−
LF	3	9	3.4	10.0	−	−
MG	12	38	3.4	9.8	−	−
RM	7	30	4.2	10.5	−	−
Controls						
CM	18	42	10.9	17.0	+++	++
XP	21	42	11.5	14.5	+++	+++
LQ	4	7	7.2	9.1	+++	+++

(From: Schlesinger et al., ref. 2).

Lymphocyte proliferation was studied according to the technique of Moorhead, measuring % of blastic transformation to PHA. We observed that mean transformation was over 80% in all groups. There were no significant differences between them.

Leucocyte inhibition factor (LIF) production by leukocytes was measured in 14 marasmic and 27 healthy control infants employing a modification of the technique of Soborg and Bendixen.

LIF production was induced by PHA and PPH, and measured by the cellular migration assay. Results were expressed as an index of migration with added antigen over control without antigen.

Table 6 shows that marasmic children were capable of producing LIF in the same quantities as the control group when stimulated with various dilutions of PHA.

Marasmic PPD (+) infants are capable of producing LIF at levels even higher than PPD (+) controls, as shown in Table 7. PPD (−) marasmic and PPD (−) controls produce similar quantities of LIF.

Interferon production by leukocytes was studied in 9 marasmic infants and 31 healthy controls.

Interferon production was induced in leukocyte cultures stimulated with Newcastle disease virus. Interferon titration was carried out by a plaque lysis inhibition assay in monolayers of Vero cells challenged with bovine vesicular stomatitis virus. Table 8 shows that the interferon titers in well nourished infants ranged from 68 to more than 450 units/ml with a geometric mean of 220/ml. The marasmic infants showed interferon values ranging from 15 to 144 units/ml with a geometric mean of 58 units/ml. The difference between the means was highly significant ($p < 0.01$).

In summary we have found an impaired cellular immune response in marasmic infants evidenced by an alteration of delayed skin reactions. In addition, we have observed that their leukocytes produce less interferon than well nourished infants.

The impaired cutaneous delayed hipersensitivity response agrees with what has been described in other studies in marasmus and kwashiorkor. The proliferative capacity of lymphocytes to PHA that has been found diminished in kwashiorkor seems to be normal in marasmus. Studies in marasmic adults at the Massachusetts Institute of Technology and experimental data in marasmic pigs show, as our studies, a normal response to PHA. Production of LIF, which was found normal by us in marasmus has been found decreased in subjects with kwashiorkor. We are not aware of other reports on interferon production in malnutrition.

TABLE 6

PHA STIMULATED LYMPHOCYTE MIGRATION IN MARASMIC
AND WELL NOURISHED INFANTS

	PHA DILUTIONS		
	1/300	1/200	1/100
Marasmics n = 14	35.8 ± 7.0	51.6 ± 3.8	73.6 ± 4.3
Well nourished n = 27	34.0 ± 5.0	49.8 ± 3.7	73.9 ± 2.7

Mean ± SEM

(From: Heresi et al. 3).

TABLE 7

PPD STIMULATED LYMPHOCYTE MIGRATION
IN MARASMIC AND WELL-NOURISHED INFANTS

	PPD POSITIVE	PPD NEGATIVE	
Marasmics	55.7 ± 5.4	17.3 ± 9.5	p< 0.01
	(n = 4)	(n = 10)	
Well nourished	38.2 ± 5.0	9.7 ± 7.4	p< 0.01
	(n = 13)	(n = 14)	
	p< 0.05	NS	

Mean + SEM

(From: Heresi et al. ref. 3)

Impairment of the immune response and other defensive mechanisms could explain the severe course of infections in undernourished subjects.

TABLE 8

INTERFERON PRODUCTION BY MARASMIC
AND CONTROL INFANTS (U/ml)

	MARASMIC INFANTS (9)	CONTROL INFANTS (31)	p
Mean[a]	58	> 220	< 0.01
Range	15 - 144	68- > 450	

[a] Geometric mean

(From: Schlesinger et al. ref. 2)

REFERENCES

[1] Heresi, G. Saitúa, M.T. and Schlesinger, L. LIF Production in Marasmic Infants. Unpublished data.

[2] Schlesinger, L., Ohlbaum, A., Grez, L. and Stekel, A. Cellular immune studies in marasmic children from Chile: delayed hypersensitivity, lymphocyte transformation and interferon production. in: "Malnutrition and the Immune Response" R.M. Suskind, ed. Raven Press, New York, 1977

[3] Schlesinger, L. and Stekel, A. Impaired cellular immunity in marasmic infants. Am. J. Clin. Nutr. 27: 615, 1974.

NUTRITIONAL ANEMIA WITH SPECIAL REFERENCE TO IRON DEFICIENCY:

RELATION TO OTHER NUTRITIONAL DEFICIENCIES

Ousa Thanangkal, M.D., D.C.H.

Chiang Mai University

THAILAND

Nutritional anemia is a world wide problem, with its highest prevalence in tropical countries. Several studies have revealed the high incidence of nutritional anemia among different countries all over the world[1-27]. Iron deficiency is by far the leading cause of nutritional anemia in all groups. In Thailand the prevalence is high among the whole population, and especially among children and pregnant women.

Nutritional anemia as related to iron deficiency has been presented earlier in this meeting. I shall focus my presentation mainly on the relationship of nutritional anemia to other nutritional deficiencies, with special reference to folate and protein deficiencies.

Let us look at folate deficiency first. Primary dietary folate deficiency is not a major world problem, but it has nevertheless been shown that folate deficiency is prevalent in many population groups in various part of the world[12, 21, 28-31]. The question is, how do we assess the folate status of a population? Herbert in 1967[32] induced dietary folate deprivation in a healthy subject and showed that the serum level below 3 ng/ml in 22 days. At day 49 there was a hypersegmentation of polymorpho nuclear cells. High urine formimino glutamic acid (Figlu) was found after day 95, low level red blood cell folate level was seen on day 123, and macroovalocytosis after day 127. Megaloblastic changes in the bone marrow were found after day 134 and low hemoglobin was seen after 137 days of deprivation. Consequently, a diagnosis of folate deficiency could be established by the low level of plasma or red cell fotate; the increased excretion of urinary Figlu as shown by a histidine loading test; the increased rate of clearance of folate following an intravenous injection of folate; the increased hypersegmentation of polymorphonuclear cells; macrocytic anemia; and the megaloblastic bone marrow finding.

Diagnosis of folate deficiency is difficult, and the reliability
of all these tests also needs careful consideration. Serum folic
acid level reflects recent dietary intake, but red blood cell folate
changes more slowly and reflects the long term dietary deprivation.
Figlu is a metabolic product of histidine which is normally metabolized to glutamic acid by a folic acid dependant reaction. Its
excretion thus indicates folate deficiency. It was shown that B-12
deficiency could result in increased excretion of Figlu. This is
due to impairment of the folic function arising from B-12 deficiency.
Increased urinary Figlu was also found in iron deficiency, suggesting
a defect in the enzyme glutamate formiminotransferase. Vitab 1967[33]
demonstrated low enzyme glumate formiminotransferase activity in the
liver in iron deficient animals. He also found high Figlu excretion
in his patients with iron deficiency anemia. The level came down
after giving iron, but with a low folate diet. Increased clearance
of folate given intravenously is suggestive of low tissue folate.
Klipstein and Lindebaum in 1965[34] reported rapid clearance of folate
administered intravenously in uncomplicated iron deficiency anemia,
which may indicate rapid utilization. Megaloblastic bone marrow was
also produced by vitamin B-12 deficiency. Chanarin et al in 1965[35]
reported the reduction of megaloblastic anemia by the addition of
iron to the diet. This suggests that iron deficiency in the first
instance produces signs of folate deficiency.

Let us now look at the extent of the folate deficiency problem.
There have been several reports from various parts of the world, including developed countries, which indicate folate deficiency among
pregnant women. Karthigani in 1964[36] reported 54% incidence in
Southern India, as shown by megaloblastic bone marrow. Chatterjear[37]
in 1966 studied megaloblastic anemia in Calcutta. He found that 50%
of the anemia was due to folic acid deficiencies; 30% to a combination of both folic acid and B-12 deficiency; and that 20% was caused
by B-12 deficiency alone. From studies in Singapore in 1972, Habbard[38] found that the incidence of macrocytic anemia of pregnancy
was higher in the Indian population of the city than it was in the
Chinese population. In Japan, Tagushi in 1971[39] reported a 15%
incidence of low serum folate levels, but very little anemia responsive to folic acid was observed. In South Africa, a study was conducted by Coleman et al in 1975[40] which found low red cell folate
deficiency were also conducted by our group in Chiang Mai among
pregnant women and children attending the out patient clinic of the
University Hospital who had a level of hemoglobin below 10 mg%. Low
serum and red cell folate was seen among 20% of anemic pregnant wom-
en, but there was no evidence of low serum or red cell folate among
children who had hemoglobin levels below 10 mg%. Megaloblastic
changes in the bone marrow were not observed among these two populations studied (Tables 1 and 2).

TABLE 1

CAUSE OF ANEMIA IN CHILDREN (50)

Iron Deficiency	50%
Folate Deficiency	0%
Vitamin B_{12} Deficiency	0%
Thalassemia	12%
G-6-PD Deficiency	8%

TABLE 2

CAUSE OF ANEMIA IN PREGNANCY (50)

Iron Deficiency	75%
Folate Deficiency	20%
Thalassemia	10%
Vitamin B_{12} Deficiency	3%

Megaloblastic anemia associated with kwashiorkor has been des-cribed from several countries including Africa, India, Egypt and South America[41-46]. Vitab in 1967[33] showed that the administration of iron alone caused megaloblastic bone marrow to disappear. From this, he concluded that iron deficiency (or perhaps protein deficiency) might depress either the activation or synthesis of the enzyme system which splits polyglutamate into the more easily absorbable mono or triglu-tamate forms.

Allen and Whitehead in 1965[46] had indicated that in megaloblas-tosis associated with kwashiorkor, simple folate deficiency alone is neither the whole answer to, not the cause of, abnormal histidine. It results from the block of its metabolism at the urocanic level, which can be improved by treatment with a diet rich in protein. A contradictory result was observed by Vander Weyden et al in 1972[47] and Kamel et al in 1972[43]. They found that megaloblastic bone marrow and hypersegmentation of P.M.M. became more intensified when iron was given without supplementation of folic acid. Arroyave in 1974[48] sta-ted that in Central America, when iron was supplemented, there was a marked drop in the folate level both in plasma and red cells.

In Chiang Mai, we have also looked at the folate status of children with severe protein calorie malnutrition admitted to our metabolic unit. The children were divided into two groups. Group I received no supplementation of iron or folate throughout the study. Group II received 1.5 mg/d of folic acid parenterally, followed by an oral dosage of 0.1 mg/d until de end of the study period. It was shown that serum folate levels, normal on admission fell during the first week. This may have been due to hemodilution. Of the group which received no folate supplementation, 50% had serum folate below 3 ng/ml. Nevertheless, none developed megaloblastic changes in the bone marrow. However, they received no iron supplementation either, which may have caused the changes in the bone marrow. (Fig. 1)

The iron supplementation trial was also conducted among children and non-pregnant women in rural villages. Our study showed that there were no changes in the levels of plasma and red cell folate when iron alone was supplemented to children and non-pregnant women with hemoglobin levels below 8 mg% (Table 3).

Let us now turn to the relationship between nutritional anemia and protein deficiency. It has been described by several authors. Foy and Kondi in 1957[49] found that when patients with iron deficiency anemia were treated with iron alone, their hemoglobin levels rose to 11-12 gm%. Thereafter iron alone produced no additional response until supplemented protein was added to the diet. Sandozai in 1963[50], however, found malnourished children recovered even in the absence of iron supplementation. Allen and Whitehead in 1965[46] found that anemia in kwashiorkor resulted from diminished erythroid activity in bone marrow. This they believed to be the effect of impaired protein metabolism. Latham and Stare in 1967[51] stated that the role of anemia has been underestimated. They found that Tanzanian pregnant women and school boys, heavily loaded with hookworm, will respond to iron therapy, deworming, and other procedure up to a certain level. But the addition of protein is necessary to restore the hemoglobin level to normal. Reddy and Srikantia in 1967 found that when kwashiorkor children were given adequate amounts of protein and calories alone, their hemoglobin levels did not rise as high as those of similar children who had received an adequate diet with supplemented iron. Marzen et al in 1974[52] and Untario and Pitono in 1974[53] have shown that among children there is a definite correlation between the degree of malnutrition and the severity of anemia.

Studies in Chiang Mai found that the prevalence of anemia among school children was higher in rural areas than in the city. Anemia was found in 5% to 11% of school children in the city, as compared with 35% to 80% of school children in rural areas. Iron deficiency is by far the leading cause of anemia in our children and pregnant women.

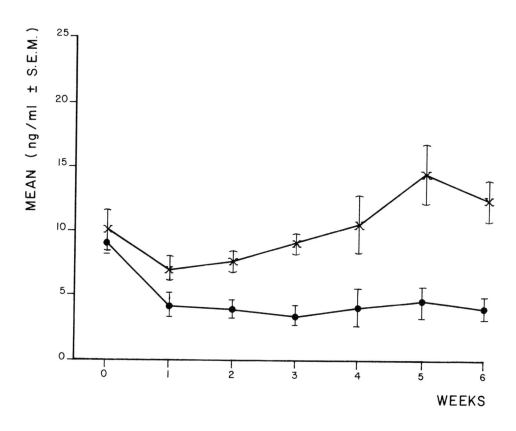

Serum folate changes during PCM treatment in Group I (●) (without folate supplement) and Group II (x) (with folate supplement)

Fig. 1 – Folate status in PCM children.

TABLE 3

HEMATOLOGICAL AND BIOCHEMICAL CHANGES
AFTER IRON SUPPLEMENTATION

GR.	Hgb (gm%)		SERUM IRON (µg%)		% IRON Satn.		SERUM ALBUMIN (gm%)		PLASMA FOLATE (ng/ml)		RBC FOLATE (ng/ml)	
	PRE	POST	PRE	POST	PRE	POST	PRE	POST	PRE	POST	PRE	POST
CHILDREN	6.6 ±0.4	11.8 ±0.4	52.4 ±10.2	56.5 ±8.5	15.8 ±3.8	18.2 ±3.5	3.5 ±0.2	3.7 ±0.1	11.9 ±2.3	18.0 ±1.6	484.5 ±41.4	504.5 ±47.7
WOMEN	6.7 ±0.2	12.3 ±0.4	24.0 ±2.3	79.0 ±9.6	6.6 ±0.9	22.0 ±2.4	3.4 ±0.1	3.8 ±0.1	8.0 ±1.2	14.0 ±1.4	391.0 ±36.2	337.0 ±36.5

Let us now turn to the problem of anemia in severe protein calorie malnutriton. We have conducted two studies to evaluate the effect
of iron and protein supplementation on recovery from anemia among
children with severe protein calorie malnutrition.

In the first study, 70 children with severe protein calorie malnutrition were included in a study to see the effect of supplemented
iron or vitamin E on the hematological response. The design of the
study is shown in Table 4. Al children received adequate treatment
for diarrhea, dehydration and associated infection. All groups
received adequate amounts of protein and calories. The amount administered were gradually increased from 4 gm protein/kg/d and 100
cal/kg/d at the end of the first week, to 180 cal and 6.5 gm protein/
/kg/d at the end of the second week. By the fourth week, their dietary intake had plateaued at about 5 mg protein and 120 cal/kg/d. All
children were divided into four groups. Group I served as a negative
control with no supplementation of iron or vitamin E. Group II received both iron and vitamin E throughout the study. Group III received vitamin E for 6 weeks followed by iron. Group IV received
iron for 6 weeks followed by vitamin E. The hemoglobin response is
shown in Fig. 2 and 3. At the end of study period, Groups I, III and
IV, which received iron on admission, had higher hemoglobin concentrations than Group I, who had received adequate amounts of protein
and calories but no iron supplementation. This study indicated that
high protein intake alone is not adequate to promote hemopoesis among
children with severe protein calorie malnutrition. Administration
of iron is essential to promote a hemopoietic response.

A second study was conducted to determine the effect of protein
on hemoglobin regeneration in PCM. The study population was 72 children aged 1-5 years with severe protein calorie malnutrition. Patients were classified into 3 different clinical classes: marasmic;
marasmic kwashiorkor; and kwashiorkor, according to the MALAN scoring
system[54]. Children with different types of protein calorie malnutrition were placed in four different dietary groups. All children were
treated similarly during the first week, when all received a constant
amount of protein: 1 mg/kg/d. This regimen, started on day 2 at 25
cal/kg/d, gradually increased the calorie intake to 100 cal/kg/d on
day 7. Fluid and electrolyte imbalance, infection and vitamin dificiency were adequately treated when indicated. Children in all
groups were supplemented with adequate amounts of iron, folic acid,
vitamin E, vitamin B-12, all other vitamins, and trace minerals.

After this one week of stabilization, children with marasmus,
marasmic kwashiorkor, and kwashiorkor were randomly assigned to the
four dietary groups. During the first three week experimental period, Group I was fed 1 mg protein and 100 call/kg/d; Group II received
4 mg protein and 100 cal/kg/d; Group III, 1 mg protein and 175 cal/
/kg/d; and Group IV was fed 4 mg protein and 175 cal/kg/d. After the

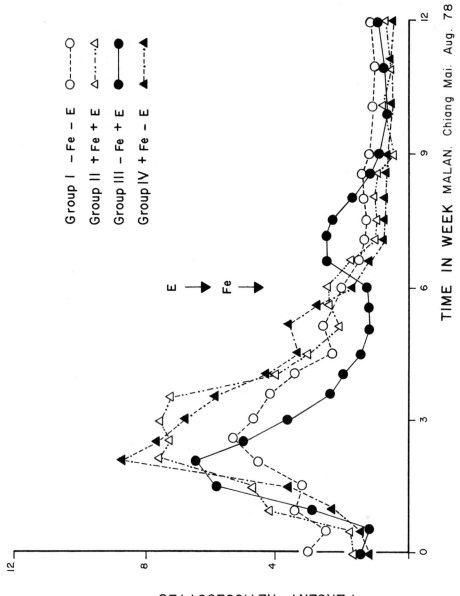

Fig. 2 – Reticulocytes.

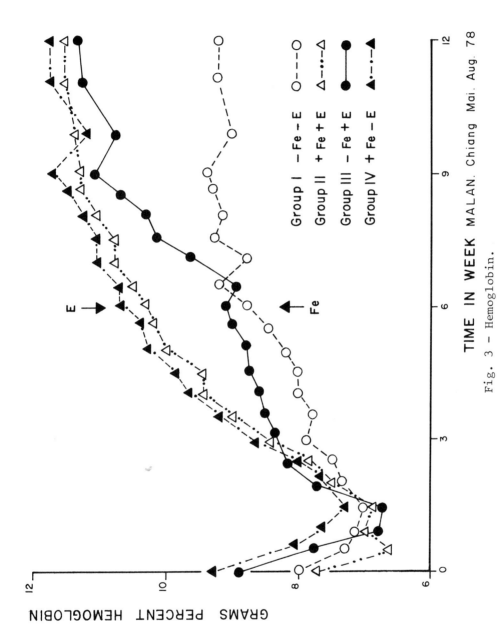

Fig. 3 – Hemoglobin.

TABLE 4

DESIGN OF STUDY INTERACTION OF
IRON AND VITAMIN ON HEMATOLOGICAL RESPONSE

GROUP	THERAPY		
1	NO IRON NO VIT E		
2	IRON PLUS VIT E		
3	FE + E	+ E + FE	
4	+ FE - E	+ E + FE	
	0	6	12

T I M E   I N   W E E K S

three week experimental period, all children were fed an adequate
diet containing 4 mg protein and 175 cal/kg/d for a further 6 weeks.

Reticulocyte response among children on the four dietary regimens
is shown in Fig. 4 and 5. Children with marasmic kwashiorkor and
kwashiorkor independently showed the same response to varying dietary
regimens. It was therefore decided to combine their results together,
and to make a separate analysis of the marasmic group. It was found
that reticulocyte response was much higher among Groups II and IV, who
had received a high amount of protein (4 mg/kg/d). Group IV children,
who were fed high amounts of protein and calorie (4 cm protein and
175 cal/kg/d), showed a far higher reticulocyte response than those
who had received 100 cal and 4 mg protein/kg/d throughout the study
period.

The change in hemoglobin concentration is similar to the reticulocyte changes. Groups II and IV, who received 4 gm protein/kg/d,
showed better responses than those receiveing 1 gm protein/kg/d. The
protein sparing effect of calorie on hemoglobin synthesis was indi-
acted by the fact that the group fed 4 gm protein and 175 cal/kg/d
had the greatest hemoglobin concentration.

Our studies demonstrated that protein plays an important role in
hemoglobin regeneration, and that this role can be significantly
enhanced by the protein sparing effect of adequate calories. There
is a need to give adequate amounts of iron, protein and calories to
promote maximum hemopoeitic response in children with severe protein
calorie malnutrition.

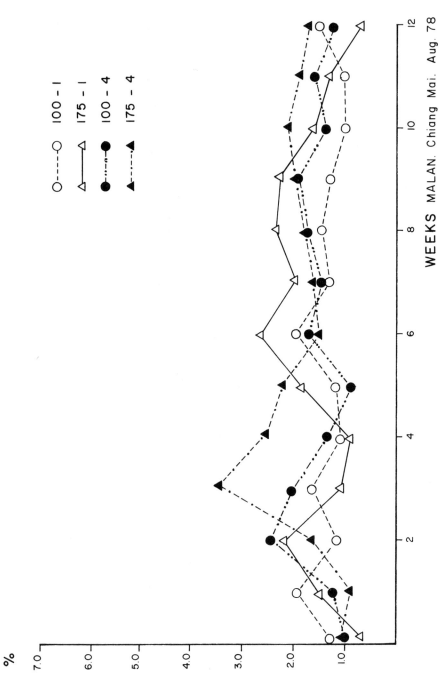

Fig. 4 – Reticulocyte response in marasmus.

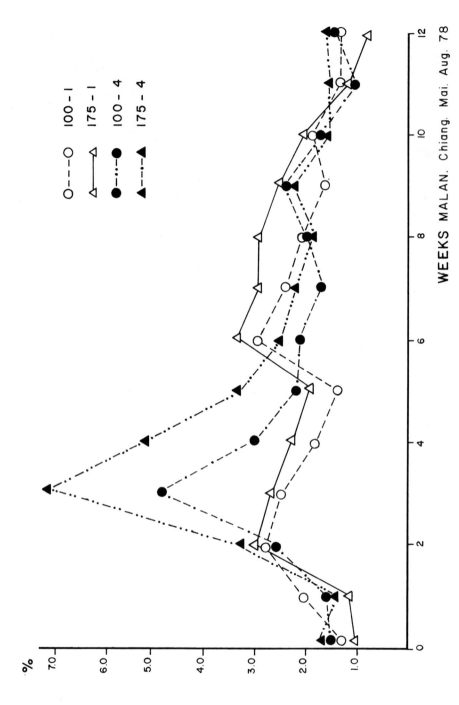

Fig. 5 – Reticulocyte response in MK-K.

In conclusion, iron deficiency is still the primary cause of nutritional anemia among all population groups all over the world. The role of protein in nutritional anemia is crucial, especially among those with protein deficiency. Even though folate deficiency was seen in most countries, it was found only in certain conditions which have a high folate requirement, such as pregnancy, hemolytic anemia, etc. Diagnosis of folate deficiency is difficult. Folate concentration correlates poorly with the presence or severity of megaloblastic anemia. Determination of folate by micro-organisms is complicated; interlaboratory variables were observed. International standardization of folate assays is needed so that the problem can be assessed more accurately.

REFERENCES

[1] Karyadi, Darwin, Soekartijah Martoatmodjo, and Husaini: Studies on nutritional anaemia in Indonesia. Presented at Consultation on Nutritional Anemia, Oct. 1973.

[2] Tantengco, Victor O., Marzen, Anita M. and de Castro, Corazon R.: Nutritional aneaemia in Filipino school children. Southeast Asian J. Trop. Med. Pub. Hlth., 4: 524, 1973.

[3] Than-Batu, Aung, Hla-pe, U. and Nyunt, Khen Kyi: Iron deficiency in Burmese population groups. Am. J. Clin. Nutr., 25: 210, 1973.

[4] Marzen, Anita M., Tantengco, Victor O., CAviles, Alendry P. and Villanueva, L.: Nutritional anaemia in Filipino infants and preschoolers. Southeast Asian J. Trpo. Med. Pub. Hlth., 5: 91, 1974.

[5] Vachananda, R., Pusobha, S., Pobrirksa, R. and Bunvanno, M.: A biochemical survey of blood and urine among the population of Ubolrajtani. I. Hematocrit, hemoglobin and mean corpouscular hemoglobin concentration. J. Med. Assn. Thailand 46: 669, 1963.

[6] Areekul, S., Devakul, K., Smitananda, N., Boonyananta, C., Klongkumnuangarn, K.: Prevalence of anemia in Thai school children. J. Med. Assn. Thailand 55: 475, 1972.

[7] Kulapongs, P.: Prevalence of anemia in school age children in northern Thai villages. J. Paed. Sco. Thailand (to be published).

[8] Turnham, D.I., Migasena, P., Pongpaew, P., Changbumrung, S., Jintakanon, K., and Pavapootanon, N.: A study of some biochemical parameters of nutrition in preschool children in northeast Thailand. First Southeast Asian Reg. Sem. Nutr. Djakarta, 1969.

[9] Na-Nakorn, S.,: Personal communication.

[10] Taguchi, H. et al.: Investigation on nutritional anemia of the farmers in north-eastern part of Thailand, 1974.

[11] Valyasevi, A., Benchakarn V., and Dhanamitta, S.: Anemia in pregnant women, infants and pre-school children in Thailand. J. Med. Assn. Thailand 57: 301, 1974.

[12] Amatayakul, K., Wiersinga, A., Kulapongs P., and Olson, R.E.: A study on anemia of pregnancy in the low income Northern Thai women. J. Med. Assn. Thailand 58 (Suppl. I): 53, 1975.

[13]Rimdusit, S.: Hematocrit values in 39, 915 pregnant women. Siriraj Hosp. Gaz. 27: 1089, 1975

[14]Wasi, P., Pa-Nakorn, S., Piankijagum, A. and Panich, V.; The haematocrit values and the incidence of anemia in the population of Thailand. Siriraj Hosp. Gaz. 25: 584, 1973.

[15]Lourdenadin, S.: Pattern of anaemia and its effects on pregnant women in Malaya. Med. J. Malaya, 19: 87-93, 1964.

[16]Rachmilewitz, M. et al.: Anemia of pregnancy in a rural community of upper Galilee. 1st J. Med. Sci. 2: 472-479, 1966.

[17]Chopra, J.G. er al: Anaemia in pregnancy. Am. J. Public Health, 57: 857-868, 1967.

[18]Elwood, P.C. et al: Evaluation of a screening survey for anaemia in adult non-pregnant women. Br. Med. J. 4: 714-717, 1967.

[19]Natvig, H. et al: Studies on hemoglobin values in Norway. VII Hemoglobin, hematocrit and MCHC values among boys and girls aged 7-20 years in elementary and grammar schools. Acta Med. Scand. 182: 183-191, 1967.

[20]Hallberg, L. et al: Occurence of iron deficiency anemia in Sweden. in: "Occurence, causes and prevention of nutritional anaemias". Blix, G., ed. Symposia of the Swedish Nutrition Foundation VI, Tylösand, 1967. Almqvist & Wiksells, Uppsala, 1968, pp. 19-27.

[21]Sood, S.K. & Ramalingaswani, V. The interaction of multiple dietary deficiencies in the pathogenesis of anaemia in pregnancy. in: "Occurence, causes and prevention of nutritional anaemais". BLIX, G., ed. Symposia of the Swedish Nutrition Foundation VI, Tylösand, 1967. Almqvist & Wiksells, Uppsala, 1968, pp. 135-147.

[22]Scott, D.E. et al: Iron deficiency during pregnancy. in: Hallberg, L., et al., ed. Proceedings of a clinical symposium on iron deficiency, organized by J.R. Geigy S.A. Basle; Arosa, 25-29 March 1969. Academic Press, London 1970, 00. 491-503.

[23]Sanchez-Medal, L. Iron deficiency in pregnancy and infancy. in: Iron metabolism and anaemia. Wahsington, Pan American Health Organization, 1969 (Scientific Publication No. 184), p. 65.

[24]Cook, J.D. et al: Nutritional deficiency and anaemia in Latin America: a collaborative study. Blood, 38: 591, 1971.

[25]Yusufji, D. et al: Iron folate and vitamin B-12 nutrition in pregnancy: a study of 1,000 women from Southern India. Bull. World Healthy Organ. 48: 15-22, 1973.

[26]Senewiratne, B. et al: Some problems in the management of anaemia in tea-estate workers in Sri Lanka. J. Trop. Med. Hyg. 77: 177--181, 1974.

[27]Levy, S. et al: Prevalence and causes of anemia in children in Kiryat Shmoneh, Israel. Am. J. Clin. Nutr. 23: 1364-1370, 1970.

[28]Levy, S. et al: A therapeutic trial in anaemia of pregnancy. 1st J. Med. Sci. 4: 218-222, 1968.

[29]Chanarin, I. The megaloblastic anaemias. oxford & Edinburgh, Blackwells, 1969, p. 275.

[30]Varadi, S. & Lewis. A. Megaloblastic anaemia due to dietary deficiency. Lancet 2: 1162, 1964.

[31]U Hla Pe & Aung Than Batu. Serum, whole blood and red cell folate

levels (L. casei) in some Burmese population groups. Union of
Burma J. Life Sci. 4: 349-353, 1971.

[32] Herbert, V. in; "International Symposium on Vitamin related anaemias". Ed. R.S. harris. I.G. Wool and J.A. Loraine. p. 549.
Acamedic Press, New York and London, 1967.

[33] Vitab, J.J., Velez, H., Bustamante, J., Hellersterin, E.E. and A.
Restrepo. "In Calorie Deficiencies and Protein Deficiencies."
R.A. McCane and E.M. Widdowson, eds. p. 175. J. & A. Churchill
Ltd., 1967.

[34] Klipstein, F.A., and Linderbaun, J. Blood 25: 443, 1965.

[35] Chanarin, I., Rothman, D. and Berry, V. Brit. Med. J. 1: 488, 1965.

[36] Karthigani, S., Gnanasundaram and Baker, S. J. Obstet. Gynec. Brit.
Cwelth LXXI: 115, 1964.

[37] Chatterjear, J.B. Soc. Haematol. p. 120, 1966.

[38] Habbard, B.M. and Habbard, E.D. J. Obstet. Gynec. Brit. Comm 79:
584, 1972.

[39] Tagushi, H., Hare, H., Iwasaki, I,, Kiraki, K., and Wakimoto. Acta,
Haem. Jap. 34: 128, 1971.

[40] Coleman, N., Larson J., Barker, M., Green, R. and Metz, J. : Am. J.
Clin. Nutr. 28: 471, 1975.

[41] Mehta, G., and Gopalan, C. Haematological changes in nutritional
oedema syndrome (kwashiorkor). Indian J. Med. Res. 44: 727, 1956.

[42] Halsted, C.H., Sourial, N., Guindi, S., Mourad, J.A.H., Kattab, A.
K., Carter, J.P., and Patwardham, V.N. Anemia of kwashiorkor
in Cairo: Deficiencies of protein, iron and folic acid. Amer.
J. Clin. Nutr. 22: 1371, 1969.

[43] Kamel, K., Waslien, C.I., El-Ramly, Z., Guindy, S., Mourad, K.A.,
Khattab, A.K., Hashem, N., Patwardm, V.N. and Darby, W.J. Folate
requirements of children. II. Response of children recovering
from protein-calorie malnutrition to graded doses of parenterally administered folic acid. Amer. J. Clin. Nutr. 25: 152, 1972.

[44] Sanstead, H.H., Gabr, M.K., Azzam, S., Shuky, A.S., Weiler, R.J.,
El Din, O.M., Mokhtar, N., Prasard, A.S., El Hifney, A., and
Darby, W.J. Kwashiorkor in Egypt. II. with low serum vitamin
E and a wide range of serum vitamin B-12 levels. Amer. J. Clin.
Nutr. 17: 27, 1965.

[45] Velez, H., Ghitis, J., Pradilla, A., and Vitale, J.J. Cali-Harvard
nutrition project. I. Megaloblastic anemia in kwashiorkor. Amer.
J. Clin. Nutr. 12: 54, 1963.

[46] Allen, D.M. and Whitehead, R.G. Blood 25: 283, 1965.

[47] Vander Weyden, M., Rother, M. and Firkin, B. Megaloblastic maturation masked by iron deficiency: A biochemical basis. Brit. J.
Haematol. 22: 299, 1972.

[48] Arroyave, G. "In Protein Calorie Malnutrition". R.E. Olson, ed.
p. 433, Academic Press, 1974.

[49] Foy, H. and Kondi, A.J. Trop Med. Hyg. 60: 105, 1975.

[50] Sandozai, M.K., Haguani, A.H., Rojeshrari, V. and Kaur, J. Brit.
Med. J. II: 93, 1963.

[51] Latham, M.C. and Stare, F.J. World Review of Nutrition and Dietetics 7: 31, 1967.

[52]Anita M. Marzen, Victor O. Tantengco and Nenita Rapanot. Malnutrition and anemia in young Filipino children. Southeast Asian J.
Trop. Med. Pub. Hlth. 5; 265, 1974.
[53]Untario, S. and S. Pitono: On the incidence of anemia, its nature
and its causes among Indonesian children as surveyed in Surabaya.
Pediatrics Indon. 12: 193, 1972.
[54]Suskind, R., Sirisinha, S., Edelman, R., Vithayasai, V., Kulapongs,
P., Thanangkul, O., Leitzmann, C., and Olson, R.E. : Immune
status of the malnourished child. Proceedings of the 10th International Congress of Nutrition, Kyoto, Japan. August, 1975.

VITAMIN A DEFICIENCY:

THE NATURE AND MAGNITUDE OF THE PROBLEM

Donald S. McLaren, M.D., Ph.D.

Reader in Physiology,

University Medical School - Edinburgh, SCOTLAND

Xerophthalmia due to vitamin A deficiency is one of the major nutritional deficiency disorders affecting mankind at the present time. It is important to consider it at two levels. At the first level, that of vitamin A deficiency itself, entire populations in many developing countries are affected to a varying degree. At the second level, that of severe deficiency resulting in blindness due to xerophthalmia, we are primarily concerned with the vulnerable groups of very young children and to a lesser extent pregnant and lactating women.

Vitamin A deficiency is unique among the major causes of blindness in that it carries a high mortality. Furthermore, it is a tragic and seemingly paradoxical fact that more than half a century after the discovery of the cause and after decades of availability of synthetic vitamin A at a very low cost, xerophthalmia is still the major cause of blindness throughout the world in young children.

Apart from its blinding effects, very little is known about the damage to other organs that may result in man. However, extensive studies in experimental animals have shown that the growth of bone and nervous tissue is impaired; infections, especially of the respiratory and urinary tracts, are very common, and congenital malformations occur in the young deprived of vitamin A in utero.

The economic and social consequences of xerophthalmia in man are enormous. We have no means as yet to estimate the contribution vitamin A deficiency may make to sterility, abortions, malformations at birth and the problems known to be associated with low birth weight, but it is probably considerable. The young child with vitamin A deficiency is usually suffering from other nutritional defi-

ciencies and various infections. Vitamin A deficiency is known to
impair considerably the chances of survival of these already seriously ill children. Survivors from severe xerophthalmia at best have
some impairment of vision, only occasionally capable of correction
by corneal grafting, which will limit their ability to profit from
education and the contribution they can make in work and leisure to
their community and nation. At most, and this is probably true of
the majority, they are totally and permanently blind. The more
fortunate few of these may be fitted for a limited contribution to
society by education in one of the all too few blind schools existing
in the developing countries at the present time. Most are to be
found at home, where their mere survival is continually hazardous,
a constant burden on their families and communities in circumstances
where even the physically fit have difficulty in coping with their
own problems in the struggle for existence.

Children between the ages of 1 and 4 years are most susceptible
and in them the disease usually takes its most severe form. In
highly endemic areas older and younger children are also affected
and occasional cases are seen in pregnant and lactating women and in
other persons at special nutritional risk. Chronically undernourished children with repeated infections and those who suffer from the
severe form of measles that is common in many developing countries
are especially prone to xerophthalmia.

The diet of the mother and her young child is clearly of paramount importance in the pathogenesis of the disease. Animal sources
of vitamin A itself are inexpensive and among lower socio-economic
groups the intake is mainly in the form of carotene precursors from
plant sources such as fruit and vegetables, among which those of the
dark green leafy varieties such as spinach are especially rich.
Red palm oil used for cooking mainly in parts of Africa is a highly
concentrated source. If the dietary intake is low during gestation
the baby will be born with low liver reserves. Normally the liver
contains more than 90 per cent of the vitamin A in the body, sufficient to meet the requirements of adults on a vitamin A-poor diet
for about two years, although the requirements are relatively greater
during the years of rapid growth. Milk is never a rich source of
the vitamin and is poorer still if the child is breast fed by a
vitamin A-deficient mother. Growth, and infections and infestations
frequently precipitated by artificial feeding, increase the nutritional stress. These circumstances are probably the rule among the
poor of developing countries and yet xerophthalmia does not occur in
the majority of malnourished children nor is it universally distributed as we have seen. The crucial factor is undoubtedly the vitamin
A content of the supplementary foods that are given. Rice is so
satisfying to the child, so easy to prepare that it is frequently
regarded as a complete food in itself for babies and other foods are
considered optional. Most cereals have a low carotene content but
the little there, consumed in considerable amounts, is usually suf-

ficient to prevent overt deficiency. However, rice, a good food as it is in itself in many other ways, is devoid of carotene. If a strain of rice containing carotene could be found or bred, most of the problems of xerophthalmia would be overcome. Nevertheless, it is important to realise that carotene is far from being absent from the environments in which xerophthalmia prevails. It is a question of poverty in the midst of plenty, for the tropics and sub-tropics are lush with green leaves in miriad form. The necessary nutrition education would seem to have these three main ingredients: first to convince that a mainly cereal diet for the child is harmful, second to indicate the sources of the missing vitamin locally available, and third to provide assistance in the cultivation and preparation of the green leaves.

Formidable obstacles remain in the path of progress on a world-wide scale towards controlling xerophthalmia. Notable among these are the difficulties in early detection. This is true for the vitamin A deficiency state itself and also for the ocular lesions of xerophthalmia. There is no widely available, simple method of measuring vitamin A concentration in the blood at present, nor is there general agreement on the proportion of values that should be below the normal range in order that there may be considered to be a problem of public health magnitude in a population. Furthermore, the early non-blinding eye lesions of xerophthalmia are notoriously difficult to identify with precision and accuracy. Their sensitivity, specificity and predictive value are also suspect. There is clearly much need for research in these areas.

The eye changes tend to progress to irreversibility without arousing the attention of none too careful parents, who frequently bring the child, often too late, for entirely different complaints. Hospital statistics in no way reflect the prevalence of the disease in the community and from the records, such as they are, no accurate impression can be obtained of the hospital experience of the disease. Only rarely has xerophthalmia been made a notifiable disease. Ophthalmologists are rarely interested in a condition that responds so poorly to treatment (which requires no skill on their part) at the stage when they usually encounter it. Paediatricians are wary about treating an organ with which they are unfamiliar and are preoccupied with systemic disease.

The occurrence of xerophthalmia in a community is a very important public health index of the nutritional status of that community. Because of the accumulation of vitamin A in the body well beyond the day to day requirements, there may normally be considered to exist large reserves of vitamin A. The occurrence of conjunctival and corneal changes characteristic of vitamin deficiency thus betoken exhaustion of these reserves consequent upon severe and usually prolonged malnutrition.

The emergence of xerophthalmia as a new phenomenon in a community thus indicates a serious deterioration in nutritional status. Conversely the disappearance or amelioration of pre-existing xerophthalmia is a token of definite improvement in nutritional status. It is consequently essential that examination of the eyes and if possible plasma retinol levels should be included in all monitoring and surveillance programmes.

In this connection it is important to differentiate between the endemic and epidemic occurrence of vitamin A deficiency and xerophthalmia. The problem tends to be endemic among those communities who subsist on diets in which the staple is deficient in pro-vitamin A carotenoids; the prime example being rice. Public health authorities should be alert to the possibility of epidemics, usually characterized by night blindness, occurring in emergency situations like crop failures and other natural disasters and sometimes associated with the epidemic occurrence of measles.

Our understanding of the magnitude of the problem of xerophthalmia and vitamin A deficiency has not only increased considerably in recent years but the nature of the data upon which it has been based has changed. In this regard one may discern three historic periods. The first extending up to about 1960 was characterized by isolated reports of the occurrence of xerophthalmia based mainly on hospital case reports. This was followed in the early 1960s by the world-wide survey of xerophthalmia sponsored by WHO in which there was a coordinated effort to collect a wide variety of data relating both to vitamin A status and xerophthalmia from many countries. The third period since then has been characterized by a more scientific approach in epidemiologic methodology. A number of country-wide, point-prevalence surveys have been carried out. There is increasing use of the WHO classification of xerophthalmia (Table 1) and its criteria for diagnosis of a public health problem (Table 2) (WHO 1976 Vitamin A Deficiency and Xerophtalmia, Technical Report Series 590, WHO, Geneva) and monitoring and surveillance programmes are being introduced in some countries.

At the Second Session of the Committee on Food Aid Policies and Programmes held in Rome, November 1976, WHO presented a tentative list of 74 countries and territories where vitamin A deficiency is considered to be a public health problem. They are shown in the map (Fig. 1). The highest endemicity and also the vast majority of the total cases of xerophthalmia are found in those countries where most of the classical clinical and epidemiological accounts of the disease come from; what might be termed the 'home' of xerophthalmia. These are the rice-dependent areas of South and East Asia, most notably Central and Southern India, Bangaladesh and Indonesia.

Although the advent of a more scientific and systematic approach to the delineation of the problem of vitamin A deficiency is to be

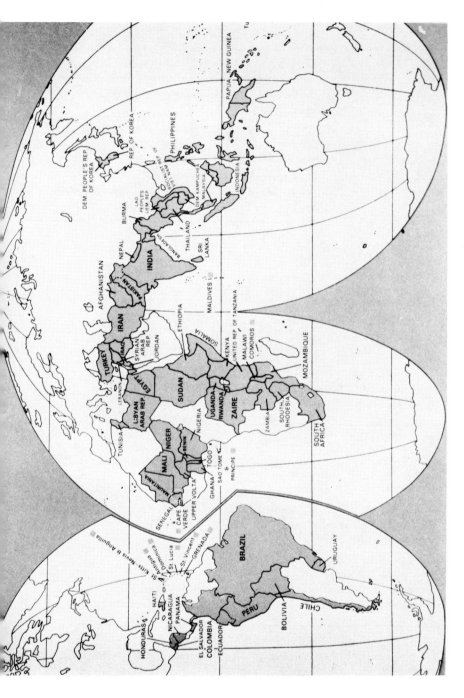

Fig. 1 – Countries and territories where Vitamin A deficiency is considered to be a public health problem.

TABLE 1

WHO XEROPHTHALMIA CLASSIFICATION

CLASSIFICATION	SIGNS – PRIMARY
X1A	Conjunctival xerosis
X1B	Bitot's spots with conjunctival xerosis
X2	Corneal xerosis
X3A	Corneal ulceration with xerosis
X3B	Keratomalacia
	SIGNS – SECONDARY
XN	Night blindness
XF	Xerophthalmia fundus
XS	Corneal scars

TABLE 2

WHO SUGGESTED CRITERIA FOR COMMUNITY DIAGNOSIS
OF XEROPHTHALMIA AND VITAMIN A DEFICIENCY

Clinical

  1. X1B in more than 2.0% of the population at risk

  2. X2 + X3A + X3B in more than 0.01% of the population
      at risk

  3. XS (attributable to vitamin A deficiency) in more than
      0.01% of the population at risk

Biochemical

  4. Plasma vitamin A level of less than 10 µg/100ml
      in more than 5.0 of the population at risk

welcomed and pursued wherever possible, it has to be recognized that
this will not be practicable in many developing countries in the near
future. In these situations there is still great merit in the careful collection and compilation of data on a hospital basis, an exer-

TABLE 3

HOSPITAL DATA ON XEROPHTHALMIA

---

1. Total number of PEM cases

2. Percentage with xerophthalmia

3. Seasonal distribution

4. Sex distribution

5. Age distribution

6. Eye Lesions by WHO Xerophthalmia Classification

7. Mortality

8. Survivors   -   totally blind

              -   partially blind

              -   full vision

---

cise that has been carried out widely in such countries as India and Indonesia. The reporting of data as shown in Table 3 from district hospitals would provide much needed information at both national and international levels.

As a rough guide it may be stated that more than 10 per cent of malnourished hospitalized children with xerophthalmia strongly suggests the existence of a public health problem. Figures as high as 70-75 per cent have been reported in the past, for example, from Indonesia.

WHO has accepted an estimation of somewhere in the region of 100,000 cases of xerophthalmia occurring annually throughout the world. The mortality rate appears to be usually 20-40 per cent and among survivors about 25 per cent remain totally blind, about 50-60 per cent partially blind, and only 15-25 per cent escape with unimpaired sight.

At the World Health Assembly in 1977 WHO passed a resolution in which it set the target of the eradication of the severe clinical forms of malnutrition by the year 2000. The control of xerophthalmia is a major part of this objective and calls for the concerted efforts of all concerned.

GENERAL NUTRITION AND ENDEMIC GOITER

Geraldo A. Medeiros Neto, M.D.

Ass. Prof., Div. of Endcr.; Chief, Thyroid Lab., Hosp.
das Clínicas, Fac. de Med., Univ. de S.P.
São Paulo, BRAZIL

Undernutrition and malnutrition among children and adults in a
number of countries on all continents of the world has been so widespread that the effects of this socioeconomic problem of human euge-
nics cannot be ignored. In the earlier part of this century most
concern was concentrated on vitamins and mineral deficiencies and
resulting diseases (including endemic goiter) but during recent years
investigators have also been concerned about the avaiability of sufficient quantities of high-quality proteins and calories especially
to the vulnerable groups of the human population.

Experimental and clinical studies[1-4] have shown that acute and
chronic starvation and calorie restriction significantly alter the
endocrine system in general and affects the thyroid gland in a number
of ways. Thyroid gland weight, histological structure and glandular
function are known to change depending on nutritional state[5].

1. Experimental malnutrition

It was suggested, as early as in 1940, that suppression of anterior pituitary function occurs in inanition, since the structural
changes observed in the thyroid gland in undernourished guinea pigs
were similar to that of hypophysectomized animals. Weight loss of
20 to 30% of body weight in a period of 2 weeks on a low calorie
diet, resulted in market structural changes in the thyroid characterized by atrophy and flattening of the acinar epithelium and retention of the colloid. Also the thyroid of the undernourished animal
showed a marked increase in sensitivity to thyrotropic hormone.
Later, it was reported a reduced uptake of ^{131}I by the thyroids of
rats maintained on diets containing 3.5% protein, and other investi-

227

gators found that the secretion rate was progressively reduced when
the caloric intake was lowered. It was, also demonstrated that
although the peripheral response to thyroxine was not reduced in male
protein-depleted rats, the response to TSH was low. He interpreted
this to indicated a direct effect of protein depletion on the thyroid
gland.

Histopathological changes in the thyroid gland of protein-calorie
deficients pigs are similar to those of colloid goiter. It was found
reduced glandular activity, flat elongated follicular cells, angular
vesicles and loss of interfollicular tissue in the thyroid glands of
pigs on a low protein calorie diet.

Decreased protein intake by iodine deficient animals results in
an increased thyroid to serum radioactive iodine concentration ratio
(T/S) and histological changes in the thyroid gland. In contrast to
these findings, T/S ratio in protein deficient rats were found to be
lower than those in animal fed a 30% casein diet. It was studied
three groups of female rats on special diets: protein-free, 8% casein
or 60% casein, receiving only 12 μg of iodine daily. Under these
experimental conditions the poor-protein diet impaired the thyroidal
transport of iodine, decreased its concentration in the thyroid, and
was accompanied by an enlargement of the gland. Also the action of
thiourealike substances was enhanced. The administration of protein
reversed these pathological changes. Similar results were also reported by others investigators whom verified that treatment of iodine
deficient, protein depleted rats with TSH failed to increase thyroidal activity over that observed in iodine deficient controls. In
protein depleted animals there was no relative impairment of thyroxine synthesis in the gland. The protein content of the diet does
not become a limiting factor for the thyroxine secretion rate (TSR)
until it is lower than 5%, according to a recent report[3]. These
authors verified that the TSR, body weight and amount of food consumed were significantly depressed in the group of rats fed on a
protein free diet, but these parameters were not decreased significantly in animals fed 5-20% casein diets.

More recently we have studied the TSH response to TRH in rats
given a diet with only 3% casein, and compared the results with a
group of rats fed a 15% casein diet. There was no statistically
significant difference between the two groups with respect to the
basal levels of TSH, but the group of animals fed a low casein diet
exhibit an exaggerated and sustained YSH response to TRH, as compared
with the control group. A diminished peripheral conversion of T_4 to
T_3 could be the main cause of a relatively more active TSH production. The effects of protein deprivation on thyroid function of the
rats were reversible with return to a normal diet.

In summary, experimental data in animals suggest that protein
calorie malnutrition is usually accompained by decrese thyroid func-

tion and histological changes in the gland. Under these experimental conditions both the effects of chronic iodine deficiency and goitrogenic activity are enhanced. At the same time there is a significant decrease of the negative feedback of the thyroid hormones at the pituitary level, with an increase in TSH storage and secretion. These endocrine modifications found in human protein calorie malnutrition, although in the latter condition other aggravating factors are usually present.

2. Infantile malnutrition and thyroid function

As already noted, protein-calorie malnutrition in children has clinical and laboratory differences according to the time of onset, defining two groups of syndorme, maramus and kwashiorkor. It was once thought that infantile malnutrition is followed by hypothyroidism. This concept was based on findings of a low basal metabolic rate a diminished thyroidal uptake of ^{131}I and levels of PBI, T_4 or BEI bellow the normal range for comparable groups of normal children. The low basal metabolic rate in infantile malnutrition was attributed principally to a significant loss of muscle mass. A low thyroid uptake of radioiodine both at 2 and 24 hours was found in marasmus and kwashiorkor, but it was mentioned the difficulty in interpreting the results of the uptake of ^{131}I by the thyroids of small children and suggested that this laboratory parameter should not be used as the sole index of thyroid function in infantile malnutrition.

Measurement of T_4 (either by competitive protein binding or radioimmunoassay) in infantile malnutrition probably reflect in a more precise manner the functional status of the thyroid gland.

However, in chronic malnutrition there are changes in the serum proteins responsible for the transport of the thyroid hormones and this could be a factor in the interpretation of total serum T or T determinations. Thus measurement of free thyroxine levels or indirectly, the free thyroxine index, would be preferable in defining the function in malnourished infants and children. It was observed that total thyroxine (T_4) values in maramus are significantly lower than in control groups, but the free thyroxine levels are either normal or elevated. Similar observations were obtained in children with kwashiorkor. Marasmatic infants at admission had both total thyroxine concentrations and the free thyroxine indices which were at the upper limit of normality (respectively 10.3 ± 1.2 μg/dl and 8.8 ± 1.2 units). At recovery there was a significant decrease in both parameters (respectively 7.1 ± 0.6 μg/dl and 4.8 ± 0.5 units). Thus in this group of infants there was no laboratory evidence for hypothyroidism. On the other hand, children affected with kwashiorkor showed a rise in the total serum thyroxine concentration after recovery (admission: 5.4 ± 0.4 μg/dl and after recovery: 6.7 ± 0.5 μg/dl), with no significant changes in the free thyroxine index.

It was reported a mean serum thyroxine values of 8.65 ± 2.24 and
6.22 ± 3.09 μg/dl in marasmus and kwashiorkor. In the latter group
total thyroxine levels were significantly lower than in normal controls, but the free thyroxine index in both groups was not significantly affected by chronic or acute malnutrition, and it was suggested that the main factor involved in the low total thyroxine concentration in kwashiorkor was probably a decreased serum protein concentration. Others investigators found a diminished concentration of
serum total thyroxine (6.6 ± 0.6 μg/dl) in marasmus, as compared with
control subjects (9.3 ± 0.7 μg/dl). After recovery the mean serum
thyroxine rose to 8.7 ± 0.8 μg/dl, possibly due to an increased thyroid secretion rate of T_4.

Low serum tri-iodothyronine levels have been reported in infantile malnutrition. In studies conducted in Africa it was found a
mean serum T_3 concentration of 55 ± 37 μg/dl in 43 Senegalese children with kwashiorkor that rose to 233 ± 65 μg/dl after three weeks
of treatment. They postulated a partial inhibition of conversion
of T_4 to T_3, starting early in the process of total calorie deficit.
This was confirmed in studies done in Brazil where it was found a
mean value of 164.1 ± 69.3 μg/dl in 8 marasmatic infants as compared
to a mean value of 266.1 ± 71.3 μg/dl in normal subjects. In Chile
it was reported a mean serum T_3 concentration (171.5 ± 24.6 μg/dl)
in marasmatic infants which rose to 205.2 ± 26.4 μg/dl after recovery. These infants were able to increase significantly the serum
T_3 concentration after TRH administration. Diminished peripheral
conversation of T_4 to T_3 could be the main factor leading to the
decreased T_3 concentration in severe infantile malnutrition.

Serum TSH concentration have been normal in kwashiorkor and
reduced in the marasmic infant. Thus it was observed a mean serum
level of 4.6 ± 1.4 μU/ml after three weeks of treatment. They suggested that decrease TSH release from the pituitary could be responsible for the reduced thyroid function in malnourished children.

In Africa it was reported TSH concentration within the normal
range in 43 Senegalese children affected with kwashiorkor. However,
in some of their subjects with protacted protein-calorie malnutrition
there was a significant rise in serum TSH after recovery (mean:
14.7 ± 7.6 μU/ml.

Others investigators found an elevated mean TSH level (6.3 ±
0.7 μU/ml) in 24 malnourished children (presumably with kwashiorkor)
and mentioned that eight subjects had values above the upper level
found in normal controls. Exaggerated TSH responses to TRH were
also observed in this group of children. Those subjects with the
most elevated basal levels showed much exaggerated responses. After
therapy the TSH response to TRH returned to normal.

In South America it was found a mean basal level of TSH in

marasmic infants (0.88 ± 0.10 µU/ml) and a normal TSH response to
TRH, with a mean peak value of 17.3 ± 2.7 µU/ml. After treatment
the serum TSH response in TRH increased to a peak value of 31.5 ±
3.0 µU/ml. This was interpreted as an effect of chronic malnutrition
on the pituitary level of TSH leading to decreased TSH activity which
was promptly corrected by reffeding the infants. Thus one of the
factors in the hypothyroidism of marasmic infants in decreased TSH
secretion rate by the pituitary thyrotrophs.

     In Mexico it was reported normal serum TSH concentrations both
in marasmatic infants and in kwashiorkor. There was no significant
change in thyrotrophin concentration after recovery in either group
of patients.

     In summary, the avaiable data on infantile malnutrition (either
marasmus or kwashiorkor), thyroid function and pituitary thyroid
interrelationship are suggestive of a low serum T_4 concentration with
a normal free thyroxine value, associated with a low serum T_3 concentration. Acute infantile malnutrition may be characterized by a
sharp and significant decrease in serum T_3 concentration that returns
to normal after recovery. This is probably a result of decreased peri
peripheral conversation of T_4 to T_3 rather than a deline in the secretion rate of T_3 or T_4.

     The serum TSH concentration at admission in infantile malnutri-
tion is normal but in marasmic infants a significant rise in serum
TSH concentration follows recovery. The serum TSH concentration
after administration of TRH is exaggerated and sustained. This suggests a relatively diminished negative thyroidal feedback at the pituitary level.

3. Adult man1nutrition and thyroid function

     Evaluation of reports on thyroid function in adults with protein
calorie malnutrition is difficult. The groups of subjects which have
been studied are heterogenous. Also certain of the clinical features, such as lethargy, slow mentation, cool dry skin are compatible
with hypothyroidism. These features might also be systemic effects
of PCM unrelated to the thyroid. Detailed evaluation of thyroid
function in certain other clinical situations, such as hepatic cirrhosis and starvation has suggested the maintenance of a normal thyroid state in most instances. This seems probably to be true also
for adult protein calorie malnutrition.

     In India it was reported 8 subjects (age: 23-38 years) with
clinical and laboratory evidence of PCM (albumin 2.0 µg/dl). Serum
T_4 was within the normal range (4.1 ± 0.5 µg/dl), although the mean
value was considered significantly lower than in normal subjects.
The mean serum T_3 concentration (50.0 ± 10 µg/dl) was significantly

lower than normal (140 ± 10 µg/dl) but the basal TSH values were within the normal range. An exaggerated and sustained TSH response to TRH occured in five subjects, with a tendency for TSH levels to rise for 60 minutes after the TRH injection. This could be interpreted as a relative lack of negative thyroidal feedback at the pituitary level.

In Calcultta (India) ten adults patients with severe PCM (albumin: 1.7 µg/dl) were studied. The serum T4 concentration was normal but the mean serum free T4 concentration (3.8 µg/dl) was significantly greater than the mean post-treatment value (2.2 µg/dl). Both total (21 µg/dl) and free (94 µg/dl) mean T3 concentrations were much lower than the corresponding values for normal subjects. The mean serum TSH concentration was 6.0 µU/ml and was comparable to the mean value of 5.5 µU/ml following treatment. Also, there was no significant change in serum concentration of TBG either before or after treatment. It was suggested that a reversible defect in extra-thyroidal conversation of T4 to T3 might be the major factor determining the changes in the thyroid hormones in PCM.

The same group of investigators reported on the reciprocal changes in serum concentration of reverse T3 and T4 in adult PCM. The mean serum reverse T3 in 10 patients with severe PCM was 53 µg/dl and declined significantly to a mean level of 22 µg/dl after treatment. Serum mean T concentration was clearly sub-normal (22.0 ± 2.9 µg/dl) and rose to 96 ± 14 µg/dl after feeding. The suggestion was made that the diversion of T4 from conversation to the highly potent T3 to the poorly calorigenic reverse T3 could be a defense of the body in protecting protein depleted tissues from excessive metabolic stimulation. The data indicating normalization of high serum rT3 and low serum T3 in PCM after feeding treatment suggest that this abnormality in metabolism of T4 is reversible.

We have studied the responses of prolactin, TSH and thyroid hormones to TRH administration in ten adults with PCM. Basal levels of serum TSH were significantly higher than in normal subjects and an exaggerated and sustained response to TRH was observed in nine subjects. This was not followed by an increase in serum T3 concentration, which remained subnormal during the test. Thus even with maximal TSH stimulation the thyroid was unable to release T3 into the circulation. This was interpreted as a low thyroidal reserve, as commoly observed in other patients either after radioactive treatment for Graves' disease or Hashimoto's thyroiditis.

In Colombia it was reported on the thyroid function of 10 patients with PCM. Two of these patients had clinical and laboratory evidence of primary hypothyroidism. The other eight patients had a normal mean serum T4 concentration (5.1 ± 1.4 µg/dl), a low mean serum T3 concentration (74 ± 17.4 µg/dl) and a normal mean TSH con-

centration (2.9 ± 1.1 µU/ml). Serum T_3 rose significantly after
treatment.

Studies on the kinetics of iodine metabolism demonstrated that
the malnourished individuals consistently had decreased thyroidal
^{131}I uptake and clearance rates, accompanied by decreased total
exchangeable iodine and thyroidal ^{127}Iodine. An impairment of the
transport of iodine through the follicular cell was suggested, and
this appeared to revert after adequate protein calorie intake.

Recently it was reported the thyroid function in 15 goitrous
Indian females with variable degree of malnutrition living in an
endemic goiter area in Chile. The mean urinary excretion of iodine
was 33.4 µg/24 hours. The mean serum T_3 concentration of 116.3 µg/dl
was significantly lower than in a control from Santiago. The serum
T_4 and TSH concentration were normal and the TSH response to TRH was
also normal. It was suggested that a mean low serum T3 concentration
reflects a decreased peripheral conversion of T4 to T3. In another
group of goitrous health and well-nourished individuals studied in
the same area, an elevated mean serum T3 concentration was found.
Thus PCM could be a factor lowering the serum T3 concentration even
in chronic iodine deficiency when preferential thyroidal secretion
of T3 usually takes place.

In summary, thyroid function is affected in adult chronic PCM
at two levels:

1. The peripheral conversion of T4 to T3 is impaired, resulting
       in decreased serum T3 concentration and an increased level
of reverse T3. Secretion of T3 is possibly also diminished as judget
by the impaired elevation of serum T3 following TRH administration.

2. The negative feedback of the thyroid hormones on the pitui-
       tary seems to be impaired and to result in normal or high
basal levels of TSH, but with an exaggerated and sustained TSH response to TRH.

Other factors such as a decrease in thyroid hormone binding globulin, diminished transport of iodine through the follicullar cell,
and a low thyroid hormone production.

Chronic iodine deficiency might add an aggravating factor to
an already stressed thyroid. An argument could be made that iodine
deficiency in PCM could result in thyroid hyperplasia, whereas a
normal subject would be better prepared to adapt to the chronic lack
of iodine. Since goiter is more prevalent in the low socioeconomic
groups of the poulation in endemic goiter areas, and many of these
individuals are undernourished, the low content of protein in the
diet or the low caloric intake or both might play a major role in
the pathogenesis of the goiter.

REFERENCES

[1] Gardner, L.I. & Amacher, P. (Editors) "Endocrine aspects of malnu-
    trition". The Kroc Foundation, Santa Inez, California, 1973
[2] Dunn, J.T. & Medeiros-Neto, G.A. Endemic Goiter and cretinism;
    continuing threats to world health Pan Am Health Oraniz. 292,
    1974.
[3] Medeiros-Neto, G.A. & Ulhoa Cintra, A.B. "Desnutrição humana e
    função tireoidiana" Editamed, São Paulo, 1978.
[4] Manocha, S.L. "Malnutrition and retarded human development",
    Charles C. Thomas, Springfield, 1972.
[5] Endocrinology of adult protein calorie malnutrition. Nutrition
    Reviews 33: 299, 1975.

IODINE DEFICIENCY AND MATERNAL-FETAL

THYROID RELATIONSHIP

Eduardo A. Pretell, M.D.

Chief of the Endcr. Serv., Hosp. Cayetano Heredia & Inst.
de Investg. de la Altura
Univ. Peruana, Cayetano Heredia, Lima, PERU

The deficiency of iodine in the diet still represents a widespread serious problem of human malnutrition. The development of goiter whenever there exists iodine deficiency is the most documented phenomenun and a great number of studies on its pathophysiology have established that the goiter is a compensatory adaptation of the thyroid when the iodine supplementation becomes insufficient. Most important than this, however, is the association of endemic cretinism, an unquestionable geographic and epidemiologic commonplace observation.

Although the relationship between severe iodine deficiency and the more severe manifestations of cretinism have been demonstrated in epidemiological surveys, relatively little is known about the precise mechanisms involved. It is entirely possible that other factors such as protein-calorie malnutrition, in addition to iodine deficiency, may also play a role in the etiology of cretinism. Still less in known about possible critical periods during fetal development and early post natal life when iodine deficiency may most seriously damage the developing child. Theoretically, pregnancy and early infancy should be critical periods of high risk since central nervous system development is extremely rapid during these periods. Moreover, there exists the possibility, as yet unexamined, that a range of mental and neurological disorders results from iodine deficiency such that large numbers of children in iodine deficient areas are at risk of mild to moderate levels of mental retardation. Nonetheless, the World Health Organization has assigned the endemic goiter problem a low priority for urgent attention, mainly because its consequences on the fetal and newborn development have not been well documented.

During the last years extensive studies have demonstrated that thyroid hormones play an important role during the process of fetal development. Their lack during intrauterous and early post-natal life may result in irreversible damage of the central nervous system. It has also been established that the ontogenesis of the hypothalamic-pituitary-thyroid function in the human fetus starts with the embryogenesis during the first 12 weeks of gestation and it gets completed with the maturation of the neuroendocrine control by the 3-4 weeks of extrauterine life. Fetal thyroid hormones secretion starts by 10-12 weeks of gestation and a progressive increase in the serum levels is observed through the end of pregnancy. The fetal pituitary-thyroid system functions autonomously. But what happens if the iodine supplementation during pregnancy becomes inadequate, has not been fully investigated. There is little published information on the maternal and fetal thyroid function in endemic goiter.

With the purpose of contributing to the understanding of such a situation, which might provide some information on the etiopatogenic mechanism of endemic cretinism, I am going to discuss the results of our studies on the effect of chronic dietary iodine deficiency on the maternal and fetal thyroid hormone synthesis, on one hand, and on the intrauterous and postnatal development, on the other hand. I will also present some preliminary observation on the effect of the nutritional status of the pregnant on the maternal and fetal thyroid functions.

During gestation 67% of T4 values among the iodine deficient pregnant women were below the normal range and lower free-T4 and higher TSH levels than control were also demonstrated.

At delivery, the maternal and cord thyroxine values in the iodine deficiency group were significantly lower than normal. Nevertheless, it is important to notice that the differences among the fetal values are less marked than the ones observed with the mothers. Thus, while 73% of the maternal values were below the normal range only 22% met such a condition in the fetal side. It also must be noticed that extremely low fetal values corresponded to very low maternal values as well. This observation is suggestive of a protective role played by the placenta as an iodine pump within certain critical limits of iodine deficiency.

The differences in T3 and TSH values, both between mothers and fetuses were less marked though a tendency for lower T3 and higher TSH values were observed in the mothers with iodine deficiency. A direct relationship between the severity of the deficiency, as demonstrated by the urinary excretion of iodine, and the thyroxine values was established in both mother and fetal sides.

After birth the only source of iodine for the newborn is necessarily the maternal milk. And here again the availability of iodine continues to be a high risk for the developing child since the iodine content of human milk in iodine deficient women is very low or absent, differently from control women, and this is a critical period for completing the central nervous system maturation. Studies on the development follow-up of the newborn children have shown a tendency to lower IQ's in those born to iodine deficient mothers, and some compromise in hearing and language.

Finally, I would like to make a brief comment on our preliminary results about the nutritional status of the pregnant and the maternal-fetal thyroid function. Taking the arm muscle area and the serum albumin concentration as parameters to define nutritional status, we have compared the group of women with a muscle area below the 20th percentile plus an albumin value below 3mg% with those whose values were above the 50th percentile and 3mg%, respectively. Only in the mother side it was found a tendency for lower T4 and T3 values in the less nourished groups than in the control one, but no evidence of fetal compromise was observed within the conditions of the present study. This does not rule out the possibility that in a severe malnutrition, however, a deleterious effect on the fetal thyroid might become apparent as it has been observed in children and adults.

THE PRESENT STATUS OF PREVALENCE

OF ENDEMIC GOITER IN BRAZIL

Amaury de Medeiros Filho, M.D.

Osvaldo Cruz Foundation

Rio de Janeiro, BRAZIL

Brazil covers 8.511.965 square miles, being the fourth country in the world in terms of continuous territorial area.

Situated in a tropical and temperate zone, it presents various climates and different soils.

Its population estimated at 115 million in 1978, is predominantly young with more than 50% of its inhabitants below the 20-years age group.

It presents zones with large human masses along the sea-coast as in the States of São Paulo and Rio de Janeiro, and inland zones of difficult access and with a scant population.

As in other regions of the world, it suffers an intense process of urbanization. It is estimated that the urban population will reach approximately 65% in 1980.

The vital statistics during the last 30 years show a growing life expectancy, at present around 60 years old, which means a distinct improvement in health standard, in spite of the high rates of infant mortality still prevailing in certain areas.

The Decennial Health Plan for the Americas, in accordance with the final information of the III Special Meeting of the Ministries of Health of the Americas held in Santiago, Chile in 1972, published by PAHO (WHO), registers among illnesses caused by lack of nutrients, also lack of iodine and endemic goiter as a Public Health problem. This plan is aimed at lowering the prevalence of endemic goiter to less than 10% in all areas of South America in this decade.

239

Endemic Goiter was unknown to the primitive populations composed of nomad tribes of indians who survived on hunting, fishing, fruit and roots, all practically raw foods.

Only very much later in contact with other civilizations, sedentary life and with the formation of small urban communities in the interior, Endemic Goiter appeared in Brazil.

Many cases of goiter were found by Saint Hilaire (1819) in the region of Goiás and by Von Martius (1820) in Minas Gerais.

Professor Geraldo Medeiros collected a great number of statements in this respect in his well described work "History of Endemic Goiter in Brazil".

Officially, Brazil slowly started the process to fight endemic goiter by iodizing salt in 1955.

The initial legislation, which was introduced by Miguel Couto Filho, anticipated the compulsory use of iodized salt only in areas considered goiter-endemic and delimitated these areas to the western, central and southern states of the country.

Decree-Law nº 39.814 in 1955 gradually extended the use of iodized salt to other states in Brazil. Only 4 years ago in 1974 the law was established to extend the compulsory use of iodized salt for human consumption in the country.

The first inquiry in 1955 which took place before the period of general use of iodized salt showed some areas with a high prevalence of goiter and an overall result of 20.6 of prevalence of Endemic Goiter. To correctly appraise the present problem, the Ministry of Health made a new inquiry in 1975 through SUCAM mostly under our guidance.

The IBGE took into account the Brazilian Municipalities by homogeneous microregions in order to assist the agriculturists, and this division of the land has been useful for various purposes. As endemi goiter is an illness caused by lack of iodine its distribution depend on the nature of the solid which influences directly the compostion of water and food. The inquiry in 1975 followed up the division of Brazil in homogeneous microregions. The inquiry affected 421.734 students aged 7 to 14 spread out in all the local microregions, excep the microregion represented by the Island of Fernando de Noronha, (Table 1).

Comparing the results obtained with those of the first inquiry in 1955, there was an overall decrease in the index of prevalence of approximately 6% in 20 years. The decrease seemed small as the calculation was made considering only the presence or not of the aument

TABLE 1

PREVALENCE OF ENDEMIC GOITER AMONG STUDENTS
ACCORDING TO THE UNITS OF THE FEDERATION
RESULT OF THE SUCAM INQUIRY AMONG 421.734 STUDENTS
AGED 7 to 14 - IN 1975

UNIT OF FEDERATION	PREVALENCE %
Acre	15,4
Amazonas	12,0
Roraima	1,3
Pará	12,3
Amapá	4,9
Maranhão	25
Piauí	3,1
Ceará	6,3
Rio Grande do Norte	0,7
Paraíba	1,0
Pernambuco	10,0
Alagoas	9,6
Sergipe	1,5
Bahia	33,3
Minas Gerais	28,6
Espírito Santo	12,4
Rio de Janeiro	14,5
São Paulo	18,7
Paraná	1,5
Santa Catarina	1,3
Rio Grande do Sul	7,2
Mato Grosso	16,3
Rondônia	31,3
Goiás	13,8
Distrito Federal (Brasília)	3,8
BRASIL	14,1

thyroid, but in actual fact the progress was considerable if we take into account also the size or degree of goiters found.

Contrary to the previous inquiry in 1955, the augmented thyroid cases found among students very rarely applied to grade II, i.e. goiters only visible with the neck in extension.

The criterion used in the inquiry to classify the goiters was as follows:

The thyroid is considered normal size when the lateral lobes are the same size as the distal phalanges of the thumbs of the person under examination.

GRADE 0   : Thyroid not visible to simple examination with the neck in a normal position or extended. Through palpation, grade (zero) can be subdivided into :

        GRADE 0a : Normal sized palpable or impalpable thyroids;

        GRADE 0b : Palpable thyroids increased in volume, but not visible with the neck in normal or extended position.

GRADE I   : Thyroid only visible with head lifted. These thyroids are easily palpable. The presence of a nodule in an apparently normal sized thyroid is reason enough to place it under Grade I.

GRADE II   : Visible goiters with the head in normal position.

GRADE III : Voluminous goiter. Only one case was found.

GRADE IV   : Abnormal voluminous goiters. This type was not found among students.

In case of doubt between these two grades, the lower one should be considered.

The palpably systematic method may allow the difference between grade 0a and 0b, permitting however, the classification of palpable but not visible goiters. However, this refinement would introduce a difficult criterion to be followed uniformly even by specialized doctors; the inquiry would become much more lengthy and it was not requested.

COMMENTS

In the areas in Brazil where iodized salt was established by

Brazilian law was continuously used, was efficient and lowered the prevalence of the illness to residual rates.

However, after 20 years legislation on ionized salt, endemic goiter continues to be a Public Health problem in some areas.

It is presumed that iodized salt is not reaching large needy population contigents which is caused by wrong feeding habits and to the use of non-iodized salt in human food consumption and to important reasons linked to the commercial trade and distribution of the product to rural areas of difficult access.

The inquiry showed by sampling that the results could be much better if the population had uniform access to iodized salt in the proportions recommended by Brazilian law.

It is urged as a correct measure to create a Technical Group to gather government and private entities responsible for the production, iodizing, distribution and trade of iodized salt in order to remove as soon as possible the obstacles of furnishing correct iodine percentage to the population of our country, in general lacking this important element in their diet.

STUDIES IN ENDEMIC GENU VALGUM - A MANIFESTATION

OF FLUORIDE TOXICITY IN INDIA

K.A.V.R. Krishnamachari

Nat. Inst. of Nutr., Indian Council of Medl. Research

Hyderabad-500, INDIA

Fluoride is distributed ubiquitously in nature in several forms and enters the human system by several routes. The most important of these is ingestion of fluoride through drinking water. High levels of fluoride in water have been described from all the continents in the world[1]. However, disease in man due to fluoride toxicity in endemic forms has been identified in only some countries of which India stands distinctly. The disease endemic fluorosis is the result of chronic excessive accumulation of environmental fluoride in human organs; the most important of which is the skeletal system. The basic pathology of fluoride toxicity lies in its affinity for bone wherein the former enters as an integral component of the apatite crystal. Extensive epidemiological studies carried out form time to time in the rural parts of India had indicated that endemic fluorosis is characterized by skeletal changes such as kyphosis, exostoses formation, cervical spondylopathy, sclerosis and also calcification of ligaments, tendons, interosseous membranes and muscular attachments[2,3,4]. In a majority of earlier studies the fluoride content of water in endemic areas had been up to 16 ppm.

Studies carried out by the National Institute of Nutrition, India, since 1973, have conclusively shown that in some parts of the country fluoride toxicity has assumed a new dimension[5,6]. We had termed this newer manifestation of fluorosis as 'Endemic Genu Valgum'. The latter is characterized by the presence of genu valgum, dental fluorosis, osteosclerosis of the spine and simultaneous occurrence of osteoporosis of some other bones such as the lower limb bones. The disease is prevalent in endemic forms among communities which habitually consume high fluoride (2-16ppm) containing sub-soil water obtained from draw wells. While the elderly residents of the endemic areas suffer from chronic crippling skeletal fluorosis, ado-

245

lescents and young adults develop genu valgum deformity. The latter
is an insidious phenomenon and starts at the age of 8-12 years and
is fully manifest by adolescence. Males suffer at least ten times
more often than girls, the reason for which is not understood as yet.

The disease is distinctly different from nutritional rickets in
the age of onset, sex predilection, lack of other evidence of rickets,
exclusive confinement to high fluoride areas and in its recent emergence, and quickly cripples population groups of paediatric age.
Radiologically, the disease is characterized by areas of osteoporosis,
osteosclerosis, cystic expansion of short bones, coarse trabecular
markings of ends of metacarpals, thinning of the cortex and occasional
subcortical areas of bone loss. Horizontal trabecular markings of
lower end of femur and upper end of tibia are prominetly seen in some
instances.

Serum biochemistry is normal as far as calcium, alkaline phosphatase and inorganic phosphorus are concerned. Circulating levels
of immunoreactive parathyroid hormone are markedly elevated in these
subjects[7] while thyrocalcitonin levels are low. Serum growth hormone
levels are also significantly elevated[8]. These observation are in
agreement with radiological evidence of bone resorption. The latter
is further substantiated by the observation that the total 24 hour
urinary hydroxiproline is extremelly elevated in patients of endemic
genu valgum[9]. Most of the observed increase in hydroxiproline is
attributable to the dialysable form i.e., the smaller fragments of
bone collagen breakdown.

Studies using ^{47}calcium had indicated that at the end of the
10th day following I.V. injection of 16-20 μci of ^{47}Ca, the % whole
body retention, the total ^{47}Ca turnover as well as ^{47}Ca accretion rate
into bone are significantly higher in both patients of classical
skeletal fluorosis and endemic genu valgum than in the corresponding
age matched control subjects. However, after correcting for the difference in age, subjects of genu valgum had shown higher values for
these parameters. These studies focussed attention on fluoride induced chages of ^{47}Ca kinetic parameters in endemic genu valgum[10].

Extensive epidemiological studies undertaken at the National
Institute of Nutrition, Hyderabad, had indicated that certain distinctive features are attributable to the population groups among genu
valgum is prevalent. It was observed that the prevalence of this
syndrome varied from none to 17 per cent of the susceptible population groups. It was further realised that while the prevalence on
an average was 1 per cent among those whose staple was rice, it was
around 4 per cent among habitual sorghum eaters. Also, it was observed that the syndrome was not prevalent two decades earlier in
the very same areas. While the actual reason for the sudden emergence of this manifestation in a known endemic area for fluorosis is
not clear, it is interesting to note that some recent changes in the

environment had taken place which may have relevance to the fluorosis problem[11]. There has been a significant elevation in the level of sub-soil water in the neighbourhood of the endemic areas. The latter has been brought forth apparently by the construction of a large dam. In view of the possible role of such changes on trace element content of staples grown in such areas, analysis of samples of foodgrains grown in fluorosis and non-fluorosis areas was carried out for some relevant trace minerals. It may be pertinent to quote that a disease characterised by bone involvement and osteoporosis had been recognised among cattle and sheep reared on pastures naturally containing high molybdenum or low copper. In view of the known metabolic interactions between molybdenum and copper and in view of the latter's role in bone collagen metabolism, these trace minerals were analysed in the food and water samples. It was observed that sorghum had higher content of molybdenum compared to rice[12] and that there existed an inverse correlation between the per cent occurrence of genu valgum and copper concentration in drinking water, of any given village with fluorosis problem.

Studies carried out by Jolly and coworkers (1961-1976) in India among wheat eating population groups exposed to high levels of fluoride ingestion had repeatedly shown that genu valgum is not prevalent among such population groups who however develop classical skeletal fluorosis. Recently, however, Jolly (1978) had confirmed our contention when he had observed genu valgum in another endemic area in India where the staple was either sorghum or bajra. It is clear therefore that the staple may modify or influence the metabolic fate of ingested fluoride. Recent studies carried out at our Institute indeed show that retention of fluoride during a balance study is higher in normal controls when administered with sorghum based diet than when given along with rice based diet[13]. The relevance of this observation with regard to the observed higher prevalence of genu valgum among sorghum eaters is, however, not yet clear.

A systematic analysis of surgical biopsy samples of cortical bone obtained from subjects of genu valgum at the time of corrective osteotomy procedures was carried out for ash content, trace elements such as Zn, Cu, Mn, Cr, Mg, Calcium and Phosphorus and Fluoride contents. These were compared with values observed for the above parameters in normal bone biopsy samples obtained from non-fluorosis subjects. Fluoride content of the bones of endemic genu valgum patients was twice as high as that observed in the normals. While the Ca/P ratio in bone was not different between the two groups, both calcium and phosphorus were observed to be lower in the samples of genu valgum subjects. The ash content was less by 20% in the latter group. A significant finding was that the copper content was only about 35% in genu valgum group as compared to that found in normal bones.

From these studies it is apparent that endemic genu valgum is a
distinct clinical entity attributable to fluoride toxicity. The
recent emergence of this clinical entity in some endemic areas for
fluorosis raises several issues to be answered. How is it that while
endemic skeletal fluorosis has always been linked with osteosclerosis, we found osteoporosis and osteomalacia among our subjects of
fluorosis? Why is endemic genu valgum associated with the staple
sorghum in high fluoride areas? Do trace elements have any role to
play in potentiating or modifying the effects of fluoride? Do different bones respond differently to fluoride induced changes? Apart
from these major issues there are certain facets of research interest
such as the role of sex hormones if any in the genesis of endemic genu
valgum.

There is paucity of information in literature regarding the
occurence of osteoporosis in man in fluorosis. On the contrary it
has been recommended that fluoride can be administered for therapeutic management of senile and postmenopausal osteoporosis with beneficial results. Studies in monkeys have shown that administration
of large doses of fluoride for prolonged periods of 60 months had
resulted in the development of osteosclerosis when high levels of
dietary calcium were made available. On the contrary when administered with low calcium containing diets a picture of osteomalacia
was seen. it is interesting to point out that population who had
endemic genu valgum generally subsisted on diets deficient in several
nutrients including calcium. While the calcium intakes of people of
fluorosis areas of Punjab (wheat belt) was about 1000 mg per day,
the corresponding value for genu valgum areas (sorghum and bajra
belts) was around 300 mg. Whether lower intakes of calcium along
with high fluoride consumption could have been responsible for genu
valgum is not yet clear.

The reasons for the sudden emergence of genu valgum are not
certain. One noticeable feature in the endemic areas for the latter
disease has been a significant elevation of sub-soil water table in
villages of endemic genu valgum. The most reasonable explanation
for this rise of water table may be the construction of a large dam
in the vicinity of these villages[14]. Whether consequent on these
changes the uptake of minerals and trace minerals by the plants
(staple) had changed is a moot point for discussion. Our preliminary
analytical data indicate that further work needs to be carried out
in the direction of interaction between fluoride and molybdenum/cop-
per ratio in the diets. The role played by copper in collagen metabolism (maturation) is well known. A relative deficiency of copper
at a time when bone growth is rapid (children) and significant may
lead to improper bone growth due to inadequacy of collagen formation.
In view of the prepubertal nature of the onset of the disease entity,
such relative inadequacies in bone formation may have to be considered important. A tentative hypothesis to explain the genesis of genu
valgum is presented in Fig. 1.

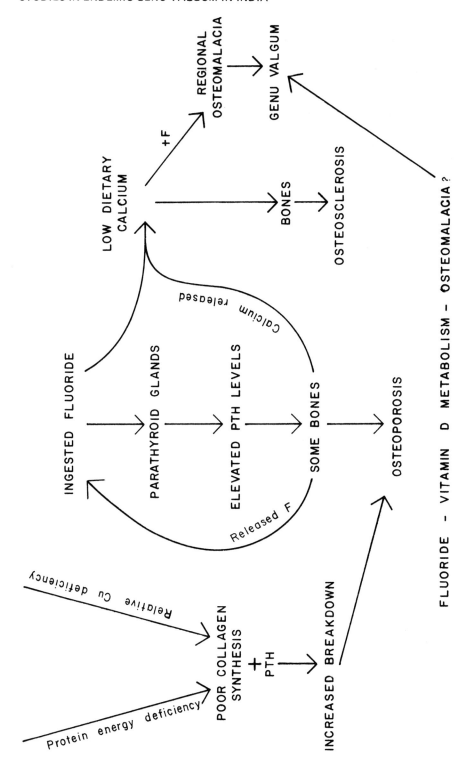

Fig. 1 – A hypothesis for the genesis of genu valgum.

From the presente study it is clear that several factors determine the outcome of chronic fluoride ingestion, the more important
of which are the total quantity of fluoride ingested, duration of
exposure (since birth), nutritional status, dietary status as well
as interaction of fluoride with the other minerals and age and sex
of the individual. The rational approach to control the ill-effects
of fluoride toxicity would therefore be to implement ameliorative
measures to check the total fluoride quantity ingested. While there
have been many attempts to utilise defluoridating agents to reduce
fluoride content of drinking water, these have in general not been
successful under the existing rural set up of endemic areas in India.
In view of the locational advantage, an attempt was made to recommend
the utilization of canal waters (meant of irrigation) for human consumption. A systematic analytical survey conducted by us had
clearly indicated that in these endemic areas surface water of the
canals have lower levels of fluoride compared to those of the drinking water of wells[15].

REFERENCES

[1]WHO. Fluoride and Human Health. WHO Monograph Series no. 59, 1970.
[2]Shortt, H.E., Pandit, C.G., and Raghavachari, T.N.S., Ind. Med.
    Gaz., 72:396 (1937).
[3]Singh, A. and Jolly, S.S., Q.J. Med., 30:357 (1961).
[4]Siddiqui, A.H., Brit. Med. J., 2:1408 (1955).
[5]Krishnamachari, D.A.V.R. and Kamala Krishnaswamy. Lancet. 2:877
[6]Krishnamachari, K.A.V.R. and Kamala Krishnaswamy. Ind. J. Med.
    Res., 62:1415 (1974).
[7]Sivakumar, B. and Krishnamachari, K.A.V.R. Horm. Metab. Res.,
    8:317 (1976).
[8]Sivakumar, B. Horm. Metab. Res., 9:436 (1977).
[9]Krishnamachari, K.A.V.R. and Sivakumar, B. Fluoride., 9:185 (1976).
[10]Narasinga Rao., B.S., Krishnamachari, K.A.V.R. and Vijayasarathy, C.
    Brit. J. Nutr. (In Press), 1978.
[11]Krishnamachari, K.A.V.R. Ind. J. Med. Res., 64:284 (1976).
[12]Deosthale, Y.G., Krishnamachari, K.A.V.R. and Belavady, B. Ind. J.
    Agri. Sci., 47:333 (1977).
[13]Lakshmaiaha, N. and Srikantia, S.G. Ind. J. Med. Res., 65:543
    (1977).
[14]Krishnamachari, K.A.V.R. Proc. Symp. Fluorosis., Hyd., 519: (1974).
[15]Krishnamachari, K.A.V.R. Ind. J. Med. Res., 65:476 (1975).

ORAL USE OF IODIZED OIL IN THE PROPHYLAXIS OF

ENDEMIC GOITER IN ARGENTINA

L. Carneiro, M.D., T. Watanaba, M.D., J.C. Scornavachi,
M.D., H. Niepomniszcze, M.D. & O.J. Degrossi, M.D.
Center of Nuclear Med., Hosp. de Clin. "José de San Martin"
Fac. de Med., Univ. of B.A., & Atomic Energy Commission
Buenos Aires, ARGENTINA

The prophylaxis of endemic goiter is mainly based on the utilization of iodized salt. However, there are many circumstances in which this method becomes impracticable[1]. For these reasons it was suggested the use of iodized oil (lipiodol) as an adequate alternative[2]. Single intramuscular injections of lipiodol have been reported to be effective, in the prophylaxis of endemic goiter, for at least two years[1]. Using oral administration of iodized oil we have shown similar results, in a two years follow up study, to those obtained by using intramuscular injections[1]. In this report we present some relevant findings on the hypothalamic-pituitary-thyroid axis of the patients subjected to oral use of iodized oil in Argentina.

Materials and Methods

The areas under study, located in the west of the Province of Neuquén and the Southwest of the Province of Mendonza, are both Andean foothills zones (Fig. 1). While isolated indian tribes were studied in Neuquén, schoolchildren populations were seen in Mendonza. These geographical areas are isolated during 6 months per year due to weather inclemencies such as snow ans storms. In Fig. 2 it is shown the location of the principal salt mines which provide natural salt to the nearest population, who do not use iodized salt as the rest of the country. Original incidence of goiter is observed in Table 1, and the dosage schedule for administration of iodized oil (Lipiodol Ultrafluide, Laboratories Andre Guerbet) is detailed in Table 2. The effects of iodized oil on thyroidal clearance of iodine, urinary iodine, plasma inorganic iodine, and thyroid absolute iodine uptake, in patients of the area of Neuquen, are shown in Tables 3 and 4, res-

251

Fig. 1 - Areas studied.

Fig. 2 - Location of principal salt mines.

TABLE 1

INCIDENCE OF GOITER

PROVINCE OF NEUQUEN

POPULATION EXAMINED	WITH DIFFUSE GOITER	WITH NODULAR GOITER	WITHOUT GOITER	TOTAL
Chiquillihuin tribe	99 (52.7%)	70 (37.2%)	19 (10.1%)	188
Aucapan tribe	31 (50.0%)	27 (46.6%)	4 (6.4%)	62
Total	130 (52.0%)	97 (38.8%)	23 (9.2%)	250

PROVINCE OF MENDONZA

POPULATION EXAMINED (school)	NUMBER OF CHILDREN	PER CENT OF GOITER
Rio Chico	29	93.0
El Manzano	24	62.5
Huemul	79	56-2
El Alambrado	61	57.3
Bardas Blancas	27	25.9
Total: 220		Mean: 58.6%

pectively. These later studies were previously reported in greater detail[1].

From the total of patients treated orally with lipiodol and from the untreated controls, only those who were subjected to the functional studies with thyroliberin (TRH) will be described.

Eight school-children (4 with diffuse goiter, grade I and II, and 4 without goiter) were medicated with 2 ml lipiodol 6 months

TABLE 2

DOSAGE SCHEDULE FOR ADMINISTRATION OF IODIZED OIL

PROVINCE OF NEUQUEN

	SUBGROUP	ORAL	INTRAMUSCULAR
1	6-12 months old	0.5 ml	0.3 ml
2	1-6 years old	0.7 ml	0.5 ml
3	> 6 years old	1.4 ml	1.0 ml
4	Nodular goiter any age	0.5 ml	0.3 ml

PROVINCE OF MENDONZA

	ORAL
1-6 years old	0.7 ml
> 6 years old	2.0 ml

before; eight school-children (5 with diffuse goiter, grade I and II,
and 3 without goiter) have received 2 ml lipiodol 1 year before, and
four school-children (1 with goiter and 3 without goiter) have only
received 0.7ml lipiodol 2 years before final studies. Twelve school-
-children (8 with goiter, grade I and II, and 4 without goiter) have
not received lipiodol.

The TRH studies were performed by a single i.v. injection of
400 µg TRH[3]. Blood samples were taken at 0, 20, 40 and 120 min.
post-injection. Serum TSH, T_4 and T_3 were investigated in all samples. Similar control studies were also preformed in normal volunteers, whose ages ranged from 11 to 20 years, of a non-endemic area,
Buenos Aires. The ages of the school-children were between 8 and 13
years old.

Patients and controls wer clinically euthyroid and no intestinal,
renal and respiratory illnes were found. Weight and height were in
the normal range for age and sex.

TABLE 3

EFFECTS OF IODIZED OIL ON THYROIDAL CLEARANCE OF IODINE AND ON URINARY IODINE PROVINCE OF NEUQUEN

SUBGROUP	ROUTE OF ADMINISTRATION	RESULTS					
		Basal		One year		Two years	
		CG+ ml/min	E++ µg/d	CG+ ml/min	E++ µg/d	CG+ ml/min	E++ µg/d
1-6 years old	A (oral)		21 ±5		240 ±20		71 ±9
	B (intramuscular)		19 ±4		270 ±30		83 ±17
> 6 years old	A (oral)	103.9 ±27.6	24 ±4	21.7 ±2.5	256 ±22	24.3 ±9.2	82 ±11
	B (intramuscular)	123.6 ±21.3	23 ±8	16.1 ±2.1	406 ±38	18.3 ±6.2	111 ±18
Nodular goiter any age	A (oral)	82.6 ±17.4	23 ±3	28.4 ±9.9	190 ±29	32.1 ±6.4	63 ±8
	B (intramuscular)	102.1 ±22.9	22 ±2	60.4 ±21.9	134 ±34	44.8 ±9.3	51 ±8

+ Thyroid clearance   ++ Urinary iodine

TABLE 4

EFFECTS OF IODIZED OIL ON PLASMA INORGANIC IODINE AND ABSOLUTE UPTAKE OF IODINE BY THYROID PROVINCE OF NEUQUEN

| SUBGROUP | ROUTE OF ADMINISTRATION | RESULTS | | | | | |
| | | Basal | | One year | | Two years | |
		PII+ µg/l	AIU++ µg/d	PII+ µg/l	AIU++ µg/d	PII+ µg/l	AIU++ µg/d
> 6 years old	A (oral)	0.48 ±0.08	72 ±7	4.28 ±0.29	134 ±8	2.05 ±0.19	89 ±9
	B (intramuscular)	0.39 ±0.08	69 ±8	7.63 ±0.70	178 ±15	2.56 ±0.22	93 ±13
Nodular goiter any age	A (oral)	0.48 ±0.06	58 ±5	3.59 ±0.23	147 ±17	1.45 ±0.20	76 ±6
	B (intramuscular)	0.51 ±0.05	75 ±6	2.31 ±0.18	201 ±21	1.26 ±0.21	81 ±7

+ Plasma inorganic iodine   ++ Absolute iodine uptake by thyroid

Results

Thyroid uptake of ^{131}I (RAIU), urinary iodine (E), and serum total thyroxine (TT_4) are shown in Table 5. Group A (school-children treated with 2 ml lipiodol 6 months before) showed the greatest decrease of the RAIU at 24 hs. and the highest increase of urinary iodine, followed by groups B (treated with 2 ml lipiodol 1 year before) and C (treated with 0.7 ml lipiodol), respectively. Group D (non--treated patients of the endemic area) have no modifications of the several parameters under study. The fact that patients of group D have had lower RAIU and higher E than those obtained in groups A, B and C in basal conditions, is due to a distinct dietary intake of iodine, because they belong to different schools. Basal levels of TT_4, although in the normal range, are lower in the four endemic groups than those observed in the controls of the non-endemic area (group E). As a consequence of the prophylaxis with iodized oil there is an increment of TT_4 in groups A and B, being higher in the former. There were no differences after injections of TRH.

The presence of absence of goiter have no influence on the values of RAIU, TT_4, serum THS, serum T_3 (TT_3) and E of the different groups.

Table 6 shows the levels of serum TSH, in each of the five groups, during the TRH test. Since group C has smaller number of patients, and dose and time of administration of lipiodol were markedly distinct than those of groups A and B, these later groups were pooled togehter for comparison with groups D and E (Table 7). Patients from endemic areas have shown higher values of TSH than those from Buenos Aires, mostly in the 20 min. peak. Lipiodol treated children decreased TSH levels at 120 min., nearing the basal values but without reaching them ($p < 0.01$). Group D is definitely far away from group E.

Tables 8 and 9 show TT_3 modifications during TRH test. Basal TT_3 of groups A, B, C and D are significantly higher than those of group E. Maximal TT_3 increment was obtained at 120 min. TT_3 of patients from endemic areas did not return to basal levels at 24 hs post injection of TRH.

Discussion

Although a great number of studies on the mechanism of adaptation to iodine deficiency have been reported in the last few decades[4-7], not much work has been done on this subject by using synthetic TRH as a tool. Despite that people from iodine deficient areas could have a wide spread of serum PBI levels[8-10], it is rather usual to find low normal TT levels contrasting with increased values of TT_3 and TSH[11-13]. This rise in the T_3/T_4 ratio is considered

TABLE 5

EFFECTS OF IODIZED OIL ON THYROID 121-I UPTAKE, URINARY IODINE AND T4 LEVELS IN SERUM

	NO. OF CASES	THYROID UPTAKE OF 131-I (%) 24 hrs.		URINARY IODINE (μg/d)		T4 LEVELS (μg/100ml)	
		PRE	POST	PRE	POST	PRE	POST
Group A 2 ml Lipiodol 5 months before	8	62 ±7	23 ±2	23 ±8	162 ±34	6.9 ±0.9	8.1 ±0.8
Group B 2 ml Lipi 1 year before	8	64 ±6	33 ±3	32 ±9	122 ±15	6.4 ±0.7	7.4 ±0.7
Group C 0.7 ml Lipi 2 years before	4	72 ±4	45 ±2	22 ±11	68 ±10		
Group D Endemic area Controls	12	57 ±6	56 ±2	56 ±5	62 ±4	6.6 ±0.3	6.8 ±0.4
Group E Non endemic area controls	10		25 ±5		189 ±22		8.4 ±0.4

Mean and S.E.

TABLE 6

TRH – TSH TEST

	Nº OF CASES	BASAL	TSH LEVELS (uU/ml) 20 min	40 min	120 min
Group A 2 ml Lipiodol 6 months before	8	7.1 ±0.9	27.5 ±6.2	22.9 ±3.9	12.8 ±2.4
Group B 2 ml Lipiodol 1 year before	8	10.0 ±1.2	31.6 ±4.1	23.3 ±3.4	15.2 ±2.0
Group C 0.7 ml Lipiodol 2 years before	4	13.5 ±3.4	28.8 ±6.6	21.3 ±4.8	15.0 ±5.0
Lipiodol treated pooled groups	20	9.6 ±1.4	27.4 ±3.2	22.7 ±2.2	14.1 ±1.5
Group D Endemic area Controls	12	11.6 ±1.5	33.0 ±4.4	30.5 ±5.0	23.3 ±4.2
Group E Non endemic area Controls	10	3.3 ±0.5	14.3 ±4.0	11.8 ±2.1	4.5 ±1.3

Mean and S.E.

TABLE 7

TRH – TSH TEST

	NO OF CASES	BASAL	TSH LEVELS (uU/ml) 20 min	TSH LEVELS (uU/ml) 40 min	TSH LEVELS (uU/ml) 120 min
Groups A and B 2 ml Lipiodol	16	8.6 ±0.8	29.5 ±3.6	23.1 ±2.5	13.9 ±1.5
Group D Endemic area Controls	12	11.6 ±1.5	33.0 ±4.4	30.5 ±5.0	23.3 ±4.2
Group E Non endemic area Controls	10	3.3 ±0.5	14.3 ±4.0	11.8 ±2.1	4.5 ±1.3
p between AB and D		< 0.05	N.S.	N.S.	< 0.01
p between AB and E		< 0.01	< 0.01	< 0.01	< 0.01
p between D and E		< 0.01	< 0.01	< 0.01	< 0.01

Mean and S.E.

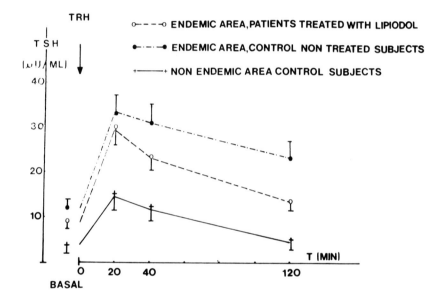

Addendum Table 7 - TSH Serum levels
                    Effect of TRH (400 µG I.V.)

TABLE 8

TRH – T3 TEST

	Nº OF CASES	BASAL	T3 LEVELS (ng/100 ml)		
			60 min	120 min	24 hs
Group A 2 ml Lipiodol 6 months before	8	204 ±34	255 ±25	291 ±36	274 ±26
Group B 2 ml Lipiodol 1 year before	8	246 ±25	292 ±19	393 ±17	288 ±20
Group C 0.7 ml Lipiodol 2 years before	4	250 ±30	302 ±24	318 ±36	254 ±27
Lipiodol treated pooled groups	20	230 ±18	279 ±19	313 ±17	278 ±14
Group D Endemic area Controls	12	259 ±25	319 ±26	257 ±20	290 ±27
Group E Non endemic area Controls	10	133 ±13	125 ±12	198 ±31	125 ±17

Mean and S.E.

TABLE 9

TRH - T3 TEST

	Nº OF CASES	BASAL	T3 LEVELS (ng/100ml) 60 min	120 min	24 hs
Groups A and B 2 ml Lipiodol	16	225 ±25	274 ±20	309 ±36	281 ±16
Group D Endemic area Controls	12	259 ±25	319 ±26	257 ±20	290 ±27
Group E Non endemic area	10	133 ±13	125 ±12	198 ±31	125 ±17
p between AB and D		N.S.	N.S.	N.S.	N.S.
p between AB and E		< 0.01	< 0.01	< 0.01	< 0.01
p between D and E		< 0.01	< 0.01	< 0.01	< 0.01

Mean and S.E.

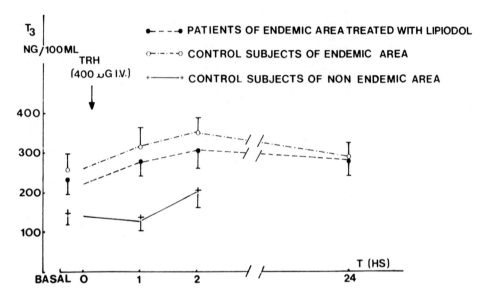

Addendum Table 9 – T3 Serum levels
Effect of TRH

to be part of the adaptation to iodine deficiency and not necessarily parallels the increments of serum TSH. According with our findings (Fig. 3) there is a good inverse correlation between T_3/T_4 ratio and the urinary iodine excretion. As a result of the prophylaxis with iodized oil we have observed a slight decrease of TT_3 and serum TSH levels with a concomitant elevation of TT_4 (Tables 5 to 9).

The salient observation of our studies with TRH are: 1) exagerated and prolonged response of TSH to TRH in lipiodol untreated patients from endemic areas; 2) early rise of TT_3, reaching the maximum at 120 min., and remaining over basal levels after 24 hs., in the same group of patients as above; 3) a diminution of the exagerated and prolonged response of TSH to TRH in patients who have received 2ml of lipiodol; 4) a negative linear correlation between Δ max TSH post TRH (expressed as percentage of increment over basal TSH) and T_3/T_4 ratio (Fig. 4); 5) a positive linear correlation between Δ max TSH post TRH (expressed as percentage of increment over basal TSH) and urinary excretion of iodine (Fig. 5); and 6) a negative linear correlation between Δ max TSH post TRH (expressed as percentage of increment over basal TSH) and basal serum TSH (Fig. 6). These findings are in agreement with the existence of a significant increase of both biosynthesis and secretion of pituitary YSH in individuals from endemic areas, which leads to an absolute increment of the pituitary reserve of TSH to a diminution of the relative hypophyseal TSH reserve; meaning that augmentation of TSH secretion is even higher than augmentation of TSH biosynthesis. This phenomenon would be iodine-intake dependent and its negative correlation with T_3/T_4 ratio could be circunstancial since T_3/T_4 ratio would also be dependent of iodine intake.

The fact that lipiodol treated patients have not the same functional pattern than those form non-endemic areas could be explained for the short period they are under treatment. Most probably, further studies in the future will give no differences with control individuals.

This work was partially supported by the Grant R/00358 of the WHO.

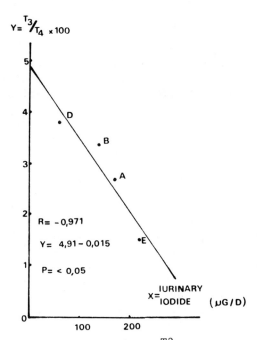

Fig. 3 - Correlation between ^{T3}/T4 serum levels
and urinary iodide.

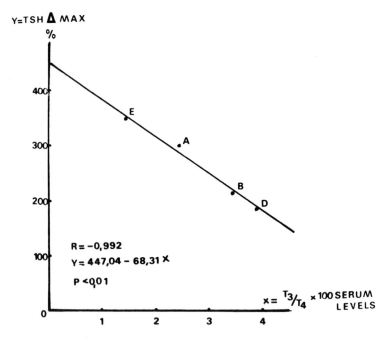

Fig. 4 - Negative linear correlation between
Δ max TSH post TRH and $T3_{/T4}$ ratio

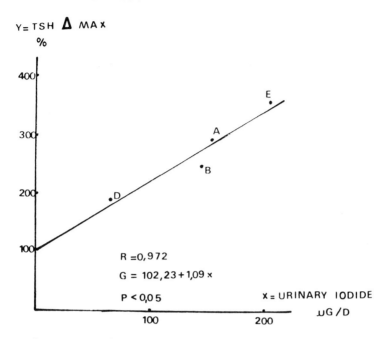

Fig. 5 – Positive linear correlation between
Δ max TSH post TRH and urinary excretion
of iodine

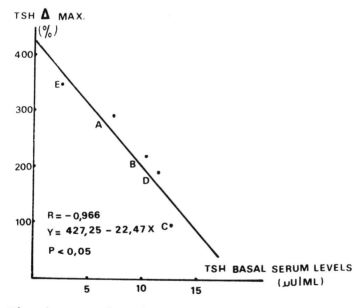

Fig. 6 – Negative linear correlation between
Δ max TSH post TRH and basal serum TSH

REFERENCES

[1] Watanabe, T., Morán, D., El Tamer, E., Staneloni, 1., Salvaneschi, J., Atschuler, N., Degrossi, O.J. and Niepomniszcze. In "Endemic goiter and cretinism.: contuining threats to world health". P.A.H.O., W.H.O. Washington, 1974, pg. 231

[2] Clarke, K.H., McCullagh, S.F. and Winikoff, D. Med. J. Austr. 1:89 1960.

[3] Watanabe, T., Degrossi, O.J., Carneiro, L., El Tamer, E., Scornavachi, J., and Cima, M.E. Rev. Biol. Med. Nucl. 7: 103, 1975.

[4] Degrossi, O.J., Pecorini, V. and Altschuler, N. Bocio Endémico, Com. Nac. Energia Atómica, Buenos Aires, 1970.

[5] Degrossi, O.J., Watabane, T., Altschuler, N., Pecorini, V. and Santillán, C. In "Endemic Goiter", P.A.H.O., W.H.O., Washington, 1969, pg. 159

[6] Stanbury, J.B., Brownell, G.L., Riggs, D.S., Perinetti, H., Itoiz, J. and del Castillo, E.B. In "Endemic Goiter. Adaptation of man to iodine deficiency". Harvard University Press, Cambridge, 1954.

[7] Pretell, R. "El bocio endémico en el Perú". Univ. Peruana Cayetano Heredia, Lima, 1969.

[8] De Visscher, M. Beckers, C. Van der Scneck, H.C., Smet, M. Ermans, A., Galperin, H. and Bastenie, P.A. J. Clin. Endocr. 21:175, 1961.

[9] Tersptra, J. Die Schilklerfunctie bij endemisch krop". Tesis, Leiden, 1956.

[10] Vagenakis, A.G., Koutras, D.A., Burger, A, Malamos, B., Ingbar, S.H. and Braverman, L.E. J. Clin Endroc. 37: 485, 1973

[11] Patel, Y.C. Pharoah, P.O.D., Hotnabrook, R.W. and Hetzel, B.S. J. Clin. Endocr. 37: 783, 1973.

[12] Medeiros Neto, G. de. "Bocio e cretinismo endemicos"., Tesis de Doctors do. Ftad. de Medicina, Universidade de São Paulo, Brasil, 1973.

[13] Greer, M.A. and Rockie, C. Endocrinology, 85: 244, 1969.

INTOLERANCE TO FOODS AND FOOD ADDITIVES

R.K. Chandra

Dept. of Pediatrics, Memorial Unv. of Newfoundland

St. John's, Newfoundland, CANADA

The influence of diet on health has been recognized for centuries. If the ingestion of food results in adverse reactions, this may be the result of hypersensitivity to an antigen or hapten present in the food, or due to a failure of absorption as a result of the deficiency of brush border enzymes such as lactase, or nonspecific release of histamine from mast cells without involvement of reaginic antibodies. Among the sensitizing substances present in diet may be mentioned various proteins, drugs (e.G. penicillin, salicylates), dyes and preservatives (e.g. tartazine, benzoates).

GLUTEN-INDUCED ENTEROPATHY

It is now established that gluten can induce a variety of abnormal reactions and clinical manifestations in sensitive patients. Three distinct clinical entities are recognized. Celiac disease of children and idiopathic steatorrhoea of adults, allergic type of gluten intolerance, and enteropathy associated with dermatitis herpetiformis. Children is first introduced into the diet, but the malabsorptive state may not be detected clinically until later in life and adult cases are being diagnosed with increasing frequency. These patients may have a childhood history of failure to thrive or nonspecific gastrointestinal symptoms. The incidence varies from 1:300 in western Ireland, to 1:2000 in Britain to extreme rarity in Blacks. About 10% of patients give a positive family history. Celiac disease has been reported in uniovular twins. There is recent evidence of linkage with HLA-B1, HLA.B8 and HLA-DW3. There is little information correlation with B cell surface antigens. The mode of inheritance is not established and is often compatible with a dominant gene of low penetrance.

The diagnosis is based on improvement of histological abnormalities (subtotal or total villus atrophy, infiltration with lymphocytes, plasma cells and eosiniphils, and increased number of intraepithelial lymphocytes) on a gluten-free diet and the recurrence of the abnormality on its reintroduction. In addition, objective demonstration of fat malabsorption, faillure of D-xylose absorption and clinical features of malnutrition are supportive evidence. Celiac disease maybe associated with dermatitis herpetiformis, immunoglobulin deficiency, splenic atrophy, diabetes mellitus, fibrosing elveolitis, or external allergic alveolitis.

The basic mechanisms of toxic damage to the small intestial mucosa of susceptible subjects may be either the result of an enzymic deficiency that permits the build up of a toxic polypeptide or via an abnormal immune response. The two theories are not mutually exclusive. A variety of immunological abnormalities have been reported in celiac disease. Circulating antibodies and coproantibodies to a number of dietary proteins, including gluten, have been demonstrated but are nospecific. A majority of patients, especially children, show the presence of reticulin antibodies, their frequency being largely dependent upon whether or not the patients are on a strict gluten-free diet. The reticulin antibody has been proposed as a valuable screening test for celiac disease particularly in family studies. Serum levels of immunoglobulins generally show increased IgA and decreased IgM. Rarely, IgA dificiency may be present. In the small intestinal mucosa, there is an increased number of IgM producing plasma cells and increased immunoglobulin synthesis. Some of this antibody may be anti-gluten in specificity. It has been suggested that tht jejunal tissue releases a soluble mediator of toxicity which in conjunction with gluten produces the tissue damage. Gluten challenge in untreated patients may produce a fall in total hemolytic complement and the appearance of immune complexes. Recently, there are experimental observations suggesting the existence of cell-mediated immune reactions in the genesis of celiac disease, and a specific skin test has been described. Serum of some patients with celiac disease can induce normal lymphocytes to become cytotoxic for gluten-coated target cells. Antibody response as well as PHA-induced lymphocyte stimulation in vitro may be reduced as a nonspecific phenomenon in celiac disease.

Management is based on a strict exclusion of all products containing gluten. Chronic and severe disease may necessitate the use of cortisosteroids. In addition, a diet adequate in calories and proteins, and supplements of hematinics are required.

ALLERGY TO COW'S MILK PROTEINS
AND OTHER FOOD ANTIGENS

I has been estimated that approximately one percent of infants fed cow's milk formulae develop hypersensitivity to milk proteins,

particularly beta-lactoglobulin. The other foods commonly implicated
in gastrointestinal allergy are egg, fish, chocolate, tomatoes,
cereals, nuts and oranges.

The diagnosis can be firmly established by employing strict clinical and laboratory criteria. These include three consecutive periods of relief of symptoms on withdrawal of the offending protein and
relapse on reintroduction. In addition, the presence of specific IgE
antibodies detected by the radioallergosorbent test and complement
activation in vivo following food challenge are supportive evidences.

Infants of atopic parents are at higher risk of developing allergy to foods. There is some evidence to support an increased susceptibility in immunodeficient infants, either primary or secondary
to intrauterine growth retardation. In such groups, the exclusive
feeding of breast milk may have a prophylactic preventive value.

Management depends upon a strict exclusion of the offending
foods. May patients recover from their hypersensitivity after lapse
of a few years. Recently, preliminary trials with the oral use of
disodium cromoglycate have shown beneficial results.

ADVERSE REACTIONS TO FOOD ADDITIVES

Food additives are products of our increasingly industrialized
and urbanized society and are an almost inevitable consequence of
scientific development and technological dependence. A food additive
has been defined as "a substance or a mixture of substances, other
than a basic foodstuff, that is present in a food as a result of any
aspect of production, processing, storage or packaging. The term
does not include chance contaminants". Some common substances included in this definition are aspirin and other salicylates, benzoates, tartrazine, sulphanilic acid, and other dyes. These can produce a variety of clinical manifestations, including gastrointestinal
upset, chronic urticaria, skin rashes, angioedema, rhinitis, asthma,
and purpura. The diagnosis is established by relief of symptoms on
exclusion of the offending agent and by provocation tests. Skin
tests are not reliable. it is also established that there is a frequent cross-reaction between various food additivies, especially
between aspirin, benzoates and tartazine. Management depends upon a
strict exclusion of the offending agents.

INTOLERANCE TO FOODS AND FOOD ADDITIVES

Hugues Gounelle de Pontanel

5 Rue Auguste Maquet

75016, Paris - FRANCE

I - Not all foods are well accepted or tolerated by all consumers. This statement is not surprising, seeing as how each individual has a personality of his own - metabolica, enzymatic, biological - reason why answers to the great questions regarding digestion, anabolism and catabolism are not perfectly comparable.

Such situation is fortunately rare with respect to the basic foods. The need to increase humanity's food supply has led man to utilize productction and preservation processes more conductive to obtaining greater yields; among others, the use of so-called phyto--sanitary substances to protect cultivations from predator attacks, and the incorporation of pharmacotechnics to animal rations in order to protect them from infections which may affect their growth or even cause their death. Whether in the case of pesticides, or in that of chemical complements, no traces should remain, theoritically, in the foods to be consumed, if the utilization of these substances respected the regulations about their utilization; we know, however, that the foods available for consumption are quite capable of containing such residues.

According to the definition by the FAO/WHO Committee, food additives, in the "stricto sensu", are those substances which are intentionally added to foods, usually in small amounts, in order to modify their appearance, taste, or consistency, or even with the purpose of increasing their shelf life.

Food intolerance, then, may be, depending on the case, to the food itself, to the residues of the defensives utilized, or even to the additives employed by food technology. This etiology makes it possible to distinguish intolerance from toxic or infectious reac-

tions occurring from the ingestion of a food which is naturally toxic
due to its chemical composition - and, consequently, non-edible - or
which has become secondarily dangerous in virtue of its microbial or
toxic contamination, or even due to the utilization of an additive,
such as nitrate, capable of being transformed into nitrosamine; such
situations are related to toxicology and not to food intolerance.

II - Let us review, briefly, the clinical symptoms of food intolerance. With regards to the digestive system, various dispeptic disturbances are observed, many times as a consequence of a deficient
digestion and absorption. In the respiratory system we find asthma
and bradipnea with sibilant and bollowing sounds. From the cutaneous
point of view we find itching, urticaria, various types of dermati-
ties, eczema and Quincke's edema. At the level of the nasal mucous
membrane, rhinitis with spasmodical sneezing, swelling and runny
nose; in the ocular globe, conjuntivities.

   The importance of these manifestations is quite variable, since
they range from a simple reaction, without danger, to intense reactions, sometimes disquieting, requiring urgent therapeutic treatment
based on corticoids. Medical supervision should be intensive, seeing
as how the pathology of allergic manifestations is quite broad in our
society of technical progress. We shall not broach the difficult
question of knowing whether such clinical occurrences reveal intolerance due to enzymatic deficiency, whether it is a true allergic reaction, or further still whether it is an anaphylactic shock. What is
important for the doctor, for the nutrologist, and, naturally, for
the consumer, is to be aware of such phenomena, no matter what label.

III - The habitual foods which are poorly tolerated by the organism
       are numerous, beginning with milk. In the case of this food,
it may be due to lactase insufficiency, an enzyme indispensable for
the break-up of lactose; such a fact was largely used to explain the
poor tolerance to milk in the case of individuals in developing countries, where milk was not part of the food habits. In the western
countries, lactase deficiency was estimated at 2% (2 per thousand)
of the population. In some elderly people, in whom lactase production has decreased, intolerance to milk may also be observed. Nutritionally, the protein component of cow's milk most indicated as the
cause for this intolerance is lactalbumin. Recent observations have
also demonstrated intolerance to maternal milk, preventing the occurrence of glucoronic conjugation due to enzymatic disturbances in the
liver and manifesting itself through ictericia; in these cases the
pediatrician should suppress the milk, under penalty of compromissing
the child's growth. More impressive accidents have been described
in particularly sensitive individuals, under the title of anaphylatic
shock.

Allergy to fish is common in countries which are great consumers of this food, such as Scandinavia. It is often due to the consumption of altered fish (not fresh), which leads to in situ liberation of histamine. The mechanism of ictiosism is quite different, and still not very well known. Actually, aside from the species recognized as dangerous, we find many others which, under certain circunstances become noxious; the subject, however, is still quite obscure; in the morays, for example, it is known that the contamination by Gonyaulax Catellan is the factor responsible for toxic reactions, although some severe manifestations of the anaphylactic-type have also been described.

The glycoprotein of chicken egg, ovomucoid, is usually the antigenic agent in the egg. Horse meat is also sometimes poorly tolerated. The syndrome of celiac disease (intolerance to wheat flour) is well known among the victims of digestive malabsorption and manifests itself by severe diarrhea. In this case, the aggressive factor is the alphagliadin of wheat which promotes the appearance of IgE-type antibodies. Seeing as how the immunologic factor does not totally explain the phenomena, it is possible that an enzymatic disturbance may also be involved. Mustard is capable of producing edema in the lips and mouth, as well as vomitting, diarrhea, Quincke's edema, generalized urticaria and asthma. Such anaphylactic-type disturbances sometimes present very severe characteristics of cardio-vascular collapse, comparable to certain reactions which may occur in the therapy with penicillin. Thus it is important to recommend to those individuals who suffer from this symptom that they avoid, in restaurants, cooked preparations, in particular meats, with sauces which may contain traces of mustard; the aggressive factor would be allyl isothiocyanate.

Among the widely-consumed vegetables, celery and carrots are signalled as being responsible for cutaneous and respiratory reactions. Cutaneous exploratory tests by scarring have been positive. Parsley is often blamed, too. Among the fruits we find tomatoes, grapefruit, oranges and, in particular, strawberries. Anaphylactic shocks have also occured after ingestion of so-called tonic drinks containing derivatives of quinquine and quinine.

Considering that no individual contains an identical biological personality, in the presence of intolerance cases it is not possible to discard, a priori, any food, whatever it may be.

IV - The residues of substances utilized in the sense of facilitating the production of foods may also be incriminated. A good example is that of allergic reactions caused by residues of penicillin in poultry and meats and particularly in milk, as consequence of the antibiotic treatment prescribed especially for staphlococcus mamitis in milking cows, with no respect for the time period which should be

observed between the end of the treatment and the killing of the
animals.

The French Academy of Medicine was one of the first to call attention to this important problem (1965). Based on this fact, it
voiced that it should not be permissible to add antibiotics to the
usual rations of stable animals, even in small doses, in virtue that
aside from the allergic phenomena, the possibility of producing anti-
biotic-resistant germs, a well-established fact, now, should also be
considered.

V - Intolerances to intentional additives are becoming better known,
     and in the measure they are researched, the more frequently they
reveal themselves.

Which additives are these? The reason for their employment is
varied. It can be nutritional, when the additive acts as an enricher,
whether of vitamins, minerals, or even amino-acids. In these cases,
it is possible to admit the innocuousness of the procedure, since it
aims at introducing a nutrient which is indispensable to nutritional
equilibrium. The additive may also aim at a technological structural
need, as is the case of the emulsifiers, jelling and thickening
agents, etc. It can be used to facilitate food preservation, as in
the case of anti-oxidants, anti-microbial and anti-enzymatic substances. Finally, it may also provide better taste or visual conditions, as in the case of sweetening and coloring agents, respectively.

One actual example is that of coloring agents, employed with the
purpose of making the food visually more attractive. It has been
proved that some of them are poorly tolerated. The Scientific Food
Committee of the European Economic Community (E.E.C.) had the merit
of signalling tartrazine and erythrosine as possible causes of hyper-
sensibility reactions[1]. Tartrazine is capable of promoting cutaneous, respiratory, ocular and rhinopharingeous manifestations in pre-
disposed individuals, and of provoking a crossed sensibility with
aspirin. Recently we had the opportunity of observing the appearance of cutaneous eruptions and disquieting respiratory disturbances
after the application of a sublingual test carried out with the
amount of tartrazine contained in a strongly-colored chocolate candy[2].

With respect ot the anti-oxidants, there are doubts about the
innocuousness of butylated hydroxytoluene (BHT) and butylated hydro-
xyanisole (BHA) in infant foods. Allergic-type reactions are also
attributed to the gallates, which present the best guarantees from
the toxic point of view. Observations of contact dermatitis among
workers who handled croissants and margarine containing gallates have
also been recorded.

Sulfurous anhydride, a preservation agent, is not protected from similar criticisms, nor is sodium benzoate, responsible for uritica-ria and Quincke's edema. How is it possible not to remember the epidemic of feverish urticaria which affected numerous Dutch as a consequence of an emulsifier in margarine? The flavoring agent L-mono-sodium glutamate is responsible for a syndrome known as that of the "Chinese restaurant".

All these factors demand severe supervision on the part of the nutrologist and the food technologist; this implies the utilization of only those additives which technology considers indispensable - exception made to the nutritional additives.

The authorization for the use of an additive in only granted after it has been tested from the toxicological point of view, and after its innocuousness has been verified with almost total certainty. However, this preliminary condition is not totally satisfactory; whatever the guarantee for innocuousness may be, the food technologist should nor resort to the additive unless it fulfills an evident technological need or substitutes a less guaranteed resource.

It is necessary to conceptualize the expression "technological need"; this means that its presence is imperatively necessary. With LAFONTAINE[3] we took a sanitaristic position contrary to a certain research tendency aimed at facilitating food production. No matter how favorable the toxicological experiments may be in animals, these do not provide precise information about the tolerance in human beings or about the existance of allergic reactions specific to the human species. A typical example òf extreme toxicity is that of thalidomide, which was never expected, based on its chemical formula, to cause such a terrible catastrophe. It is necessary to have present at all times the possibilities of possible intercations with physiological metabolites or with other additives, or even with residues of foreign substances which are frequently utilized in food production and preservation, or which originate from pollution. This concern is valid not only for the so-called additive but also for its degrada-tion products, of which, in the most part, we are still ignorant about their formation inside the organism. Even in small doses, no matter how small, the possibility of an accumulative effect cannot be excluded, either, unless metabolization is complete. The fre-quency of prolonged medical treatments in more and more individuals, linked to therapeutical progress, causes uncertainties about the effects of the medicine/additive performance. Some people object that to simply talk about the possibilities is not the same as demons-trating them, which is true, but is better to prevent than to remedy.

With respect to methods for food preservation, instead of the employment of chemical substances, it would be better to resort to physical processes such as heat, cold, drying and even ionization

radiations. Sanitary experts complain by all means that the number
of additives is progressively increasing each year; that new autho-
rizations should be tied to the suppression of previous authoriza-
tions since, if not so, control will become impossible and, finally,
that in all cases, there be informative labelling, in order to facilitate etiological research by the doctor.

REFERENCES

[1]Scientific Food Committee of the E.E.C., First report series of
      12/31/1975.
[2]Pellegrin A. et Gounelle de Pontanel H. - a food coloring agent
      to be prohibited: tartrazine E 102, in virtue of its allergy
      properties. A demonstrative observation. Bull. Acad. Med.
      162 (1):36-41 (1978).
[3]Gounelle de Pontanel H. et Lafontaine A. - the position of the
      doctor in the presence of new food technologies. Bull. Acad.
      Med. 161(1):29-32 (1977).

APHTHOUS ULCERATION AND MIGRAINE AS MANIFESTATIONS OF FOOD

INTOLERANCE AND ITS RELATIONSHIP TO SUPPLEMENTARY VITAMIN C

C.W.M. Wilson, M.D.

Dept. of Pharmacology, Univ. of Dublin

Dublin - IRELAND

INTRODUCTION

Assessment of the relationship between the appearance of clinical symptoms, and the objective assessment of an antigen-induced irregularity in tissue response, is the purpose of all tests carried out on patients suffering from the manifestations of the allergic syndrome. Evaluation of this irregularity depends, in every test, on challenge of a target tissue with the specific antigen responsible for ultimate production of the clinical symptoms. The resultant physiopathological changes as manifested by reduction in pulmonary volume and/or diminution in nasal free air-way following direct antigen challenge test, change in character of the stools following challenge with lactose in milk protein sensitivity, or frequency and intensity of hay fever symptoms as assessed by the patient, provide objective measurements of the intensity of the antigen antibody reaction in the tissues containing the specific antibodies.

Various types of objective challenge test are used to diagnose antigen sensitivity in reactive tissues (Table 1). In the skin prick test (SPT), specific antigens are injected intradermally. The appearance of a wheal and flare reaction indicates that cell-fixed antibodies in the skin react to challenge with specific antigen by the release of chemical mediators. In the leucocyte ascorbic acid direct antigen challenge test (LAADACT) (Wilson and Loh, 1975), leucocytes sensitive to specific antigens react by reduced uptake of ascorbic acid from the incubation medium in comparison with non-sensitive leucocytes. Elicitation of a positive reaction in both these tissues depends upon deliberate imposition of specific antigen onto cell-fixed antibody in the challenge tissue.

279

TABLE 1

COMPARISON OF RESULTS OF DIRECT ANTIGEN CHALLENGE TESTS IN THE SKIN AND LEUCOCYTES
IN NORMAL AND ATOPIC SUBJECTS USING THE SKIN PRICK TEST AND LAADACT
(LEUCOCYTE ASCORBIC ACID DIRECT ANTIGEN CHALLENGE TEST)
WITH SIMILAR ANTIGENS IN EACH SUBJECT.
SKIN PRICK TEST: mm. LAADACT mcg/10^8 CELLS.

ANTIGEN CHALLENGE TESTS	SUBJECTS		P VALUE
	Normal (11)	Atopic (25)	
Skin Prick Test			
Wheal Size	2.7 $\pm$ 0.7	8.6 $\pm$ 2.9	< 0.01
Flare Size	0	23.6 $\pm$ 4.9	−
LAADACT			
AA uptake (Control)	173 $\pm$ 52	178 $\pm$ 54	> 0.05
AA uptake (in presence of specific antigen)	174 $\pm$ 47	123 $\pm$ 49	> 0.05

As a result of the antigen-antibody reaction chemical mediators are released in the neighbourhood of the cells to which the antibodies are attached including histamine (Chatterjee et al, 1978), prostaglandins (Vane, 1972) and other agents (Piper, 1977). In this paper a method for measuring the effects of challenge with specific food antigens will be described and its implications will be discussed in terms of the tissue responses.

DEVELOPMENT OF PATHOPHYSIOLOGICAL ABNORMALITIES IN FOOD SENSITIVITY

When a patient suffers from rhinits during an allergic cold he begins to mouth-breathe. Polles and other specific antigens are then inhaled into the buccal cavity and may be swallowed. Patients commonly complain of abdominal symptoms during the hay-fever season. Palatal itch and aphthous ulcers are common (Table 2). Patients complain of loss of appetite, and comment that their sense of taste has altered. Oesophageal, gastric, and upper and lower abdominal symptoms are frequently present when the patients are simultaneously suffering from respiratory symptoms. During the period of mouth-breathing associated with high pollen counts, the extrinsic foreign proteins of the specific Pollen antigens come into contact with the mouth mucosa in which antibodies become localized in the same way as occurs in the nasal or conjunctival mucosa. Food is another type of foreign protein with which the buccal mucosa is constantly coming into contact. For this reason buccal sensitivity to food proteins readily occurs.

Any individual can either deliberately ingest or refuse foreign food proteins (Table 3). Many patients deliberately refuse the foods to which they are sensitive as a defence reaction when food sensitivity develops. If the patients do ingest food allergens to which they are sensitive, local antigen-antibody reactions occur in the mouth. Angio-neurotic oedema appears if a severe and wide-spread reaction occurs in the buccal mucosa. Generally however more chronic and selective, but less severe, reactions occur. These affect the normal function of the buccal mucosal cells. Normal individuals detect a slight taste to all foods. When an antigen-antibody reaction occurs in the mouth, the taste receptors are specifically stimulated by the responsible antigen. The normal taste modalities become more intense. Milk, for example, tastes more sweet and creamy. The patient develops an increased craving for the food to which he is becoming sensitive. As the reaction continues, stimulation of particular taste receptors by individual food proteins becomes depresse Taste modalities become differentially and abnormally stimulated. The foods develop a horrible taste. Ultimately the taste receptors normally stimulated by individual foods become uniformly depressed and all sense of taste is lost to them.

TABLE 2

SYMPTOMS ARISING AFTER CHALLENGE WITH EXTRINSIC ALLERGENS SUCH AS POLLENS OR HOUSE DUST,
AND/OR SPECIFIC FOODS FOLLOWING DEVELOPMENT OF CELLULAR SENSITIVITY
AND ASSOCIATED ANTIGENANTIBODY REACTIONS,
GIVING RISE TO INTERMITTENT OR CHRONIC PATHOPHYSIOLOGICAL ABNORMALITIES

RESPIRATORY SENSITIVITY			ALIMENTARY SENSITIVITY	
Nose	Ears, Sinuses		Mouth	Pharynx
Itching	Itching		Changed Taste	Tonsillitis
Sneezing	Pressure		Changed Sensation	Ovulitis
Block	Pain		Palatal Itch	Dysphagia
White Discharge	Buzzing		Aphthous Ulcers	Hiatus Hernia
Mouth Breathing	Loss of Balance		Lingual Ulcers	
	Sinusoidal Pain		Abnormal Salivation	
	Sinusoidal Pressure		Loss of Appetite	
Eyes	Lungs		Upper Abdomen	Lower Abdomen
Itching	Cough		Intercostal Ache	Distension
Pain	Wheezing		Meal related ache	Defecation related ache
Photophobia			Acidosis	Stool Frequency Urgency
Discharge			Eructation	Intermittent Constipation
Difficulty in focussing				Anal Itch
				Fecal Undigested Food
				Fecal Blood Mucus

TABLE 3

BUCCAL DEFENCE REACTIONS AGAINST INGESTED ALLERGES

Buccal Cavity is First Organ Contacting Ingested Allergens.

Buccal Afferent Response produces Efferent Defence Reaction to food or ingested allergen:

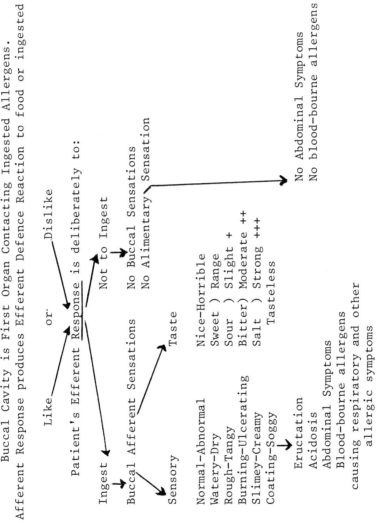

Buccal mucosal sensory receptors undergo a similar process of stimulation followed by depression as allergic sensitivity to individual foods develops. The patients use a standard vocabulary to describe the alteration in buccal sensation. The food is initially described as felling abnormally rough. It then becomes rough and slightly burning, described as tangy. Ultimately it produces a burning feeling. Patients state that buccal ulceration occurs if the food remains in contact with the buccal mucosa in these circumstances. As food sensitivity develops, a creamy or soggy feeling is often described. The feeling of food in the mouth is less acute and is muffled; buccal sensation has become impaired. Another type of buccal sensation described as a feeling of abnormal dryness or wateriness may appear when foods, to which sensitivity has developed, are placed in the mouth. Milk for example is described as being excessively watery, and the patient says that beef is dry and requires excessive chewing before it can be swallowed. In contrast cream and muttom may feel perfectly normal. In such circumstances it appears that salivary secretion becomes stimulated or depressed as part of the allergic response. Transfer of the food down the esophagus may be difficult. An allergic reaction at the gastro-esophageal junction may cause contraction of the sphincter and regurgitation. Such patients frequently give a history of hiatus hernia because an associated inflammatory reaction occurs at the gastro-esophageal junction. If antibodies are localized elsewhere along the alimentary canal, the allergic reactions are associated with smooth muscle contraction, change in vascular supply with consequent alteration in gas exchange across the luminal wall, mucus secretion, and ulceration. Patients complain of symptoms of abdominal pain, flatus, constipation or diarrhoea, and passage of mucus and blood, in the stools. Anal itching is common because of localization of antibodies at the muco-cutaneous junction.

The Diagnosis of Food Sensitivity

The patient's account of the presence of alterations in taste and buccal sensation, and of the occurrence of abdominal symptoms, following the challenge produced by the foods consumed in his everyday diet, can be recorded definitively on the Food Sensitivity Chart. This provides an accurate and sensitive method for determining the development and range of the patient's food sensitivities (Wilson, 1978). Most individuals take a relatively stereotyped diet. The increased consumptiom of a food during the initial stimulatory stage of the food sensitivity reaction, and the subsequent revulsion at the beginning of the defence reaction are generally noted by the patients. The ultimate loss of taste may not be noticed. The present large use of condiments, and addition of food flavourings, indicates the extent of loss of taste to normal foods. The patient may continue to ingest the offending tasteless food for social reasons if it does not also cause abdominal symptoms.

The changes in food taste and buccal sensitivity are shown in
32 patients analyzed by Food Sensitivity Chart during the months of
June-September, 1977 (Table 4). Abdominal symptoms were more common
than nasal or chest symptoms. Of the range of 22 foods analyzed in
the patients, a mean of 9.5 foods were disliked and 3.4 foods were
horrible. A mean of 4.5 foods were said to be tasteless. The patients showed positive skin tests to a mean 4.3 the range of foods
tested. Only 12% of the patients failed to show positive skin tests
to any of these foods. The analysis of the association between the
food consumption, and the alimentary symptoms of which the patients
complained, demonstrated that 52% of their total symptoms could be
attributed to the foods they were eating. 65% of the foods caused
abnormal tastes or feelings in the mouth, and 50% of the foods caused
abdominal symptoms. Of the abdominal symptoms which were produced
after ingestion of particular foods, nausea was most common and was
reported by 82% of the patients. A history of aphthous ulceration
was reported by 64% of the patients. However only 22% could relate
the appearance of the ulcers to the buccal ingestion of specific
foods, presumably because the presence of the ulcer was generally
only appreciated after the offending food had been swallowed.

BUCCAL ANTIGEN CHALLENGE,
APHTHOUS ULCERS AND ASCORBIC ACID

The buccal mucosa has a low positive or a negative charge with
respect to a skin electrode attached to the anterior surface of the
forearm. Similar charges exist in the other gastro-intestinal mucosal surfaces (Edmonds and Godfrey, 1969). This potential difference
is reported to be reduced in ulcerative colitis (Edmonds and Pilcher,
1973) and in aphthous ulcers in the mouth (Huston, 1977).

Six of the subjects in the present investigation presented
during June with hay fever, chronic rhinitis, or asthma. They all
had aphthous ulcers in various stages of formation or healing (Table
5). Their buccal mucosal potentials were measured in front of the
lower incisors, on the surface of the tongue on the right side of the
midline and on the exposed surfaces of the aphthous ulcers in the
buccal mucosa. Antigens were then applied to the buccal surfaces.
It had previously been confirmed that the forearm skin of these patients was sensitive to these antigens by skin prick tests. After
application of the antigens to the buccal mucosal surfaces, it was
found that the negative values were significantly greater in the
ulcers in which pain occurred than in the other ulcers. Application
of one antigen to the ulcers produced pain in each patient, but application of the other antigen caused much less severe, or no, pain.
Generally pain was associated with a higher negative charge on the
ulcer base and the ulcer became more inflamed. A solution of ascorbic acid 100 mg per 1 ml was then retained in the buccal cavity for
one minute after which the antigens were reapplied to the ulcers and

TABLE 4

THE EFFECT OF FOOD ANTIGEN CHALLENGE ON BUCCAL TASTE AND SENSATION,
DEVELOPMENT OF ABDOMINAL AND RESPIRATORY SYMPTOMS, AND ON SKIN REACTIONS.
BUCCAL ABDOMINAL AND RESPIRATORY SYMPTOMS ANALYZED BY FOOD SENSITIVITY CHART.
SKIN REACTIONS ANALYZED BY SKIN PRICK TESTS

Numbers

Patients: 32   Possible Abdominal Symptoms: 23
Foods analyzed per patient: 6-22   Possible Respiratory Symptoms: 19

Patient Symptoms (% Patients)

Total Alimentary	93%
Mouth Ulcer	22%
Nausea	82%
Abdominal Pain	66%
Nasal	81%
Chest	63%

Food Characteristics (Means of Total Foods Tested)

Specific Food Dislikes	9.5
Horrible Food Tastes	3.4
Foods Tasteless	4.5
Positive Food SPT (88% Patients)	4.3

Food Sensitivities (Mean % per patient)

Total Symptoms attributable to Food Sensitivity:	52%
Foods associated with Buccal Sensory changes :	65%
Foods associated with Abdominal Symptoms :	50%

TABLE 5

The effect of antigen challenge on buccal mucosal potential and
induced ulcer pain, after antigen application to lip (L)
tongue (T) and aphthous ulcer surface (U), before and following
mouth wash with ascorbic acid (AA) 100 mg/1 ml for 1 minute,
Potential difference measured between buccal and forearm electrodes.
Tissue values represent differences in MV before and after antigen
application. Pain severity recorded as 0 to 3+

Pat-ient	Diagnosis	Antigens	Before AA Pot. Diff. Pain T L U				After AA Pot. Diff. Pain U	
			T	L	U	Pain	Pot. Diff.	Pain
	Healing Ulcer							
1.	Hay fever	Cabbage	+ 6	- 5	-14	+	+ 2	0
	Abd. Pain	Tomato	- 3	0	-20	0	-14	0
	Haemorrhagic Spot							
2.	Rhinitis	HDM	+ 6	- 5	-14	+	- 2	0
	Asthma	Cheese	- 1	- 2	+10	0	+ 7	0
3.	Hay fever	Tobacco	+ 3	+ 1	- 8	2+	- 9	0
	Abd. Pain	Apple	+15	+ 3	+ 5	0	-18	0
	Active Ulcers							
4.	Asthma	Chocolate	- 9	- 8	-14	+	0	0
		Tomato	- 4	-10	-14	3+	+ 2	0
5.	Rhinitis	Cheese	-12	- 8	-21	3+	- 3	+
	Asthma	Milk	- 2	- 5	-14	2+	- 1	0
6.	Rhinitis	Egg	-21	-21	-14	0	-	-
	Asthma	Pork	-21	-17	-14	+	-	-
	Nausea	Aspergillus	-10	- 3	- 5	0	-	-
		HDM	-15	-14	7	3+	+ 1	0
		Grass	-18	-12	- 9	2+	- 1	0

Mean SD	Before AA	After AA
Ulcers causing Pain (PD)	-11.5 ± 7.4	-0.8 ± 3.5
Ulcers without Pain (PD)	- 4.8±14.5	-8.3±13.4
Pain Value	2	0

the potentials or their bases were measured again. Application of
the antigens to the ulcer bases was associated with such a low ulcer
potential, as was produced before buccal retention of ascorbic acid.
Ulcer pain was not produced on the second occasion by the antigens
which had caused pain on the first occasion.

These results demonstrate that the reaction of the buccal mucosa
to challenge with individual antigens is as variable and specific as
the reaction of the skin to challenge by the prick test. Individual
specific allergens produce a reduction in buccal mucosal potential,
and inflammation and pain in aphthous ulcers. Measurement of buccal
mucosal potential change therefore provides a method for identification of the specific antigen responsible for production of the allergic reaction which precedes the appearance of an aphthous ulcer.
Interference with the uptake of ascorbic acid during the antigen-
-antibody reaction has previously been shown to occur in sensitised
leucocytes (Wilson and Loh, 1975). The present results show that in
high concentration, ascorbic acid prevents the pathophysiological
fall in mucosal tissue electric potential occurring in association
with the antigen-antibody reaction. Through this effect it reduces
the clinical manifestations of the reaction in terms of pain and
inflammation during a period of at least one hour after its onset.

SKIN AND ALIMENTARY FOOD
ANTIGEN CHALLENGE IN MIGRAINE

Carroll (1971) maintains that migraine is a multi-factorial disorder and Dalton (1973) claims that food sensitivity is one of the
major factors which precipitates the migraine attack. An investigation was designed to discover whether challenge of the skin and
alimentary canal by specific food allergens was correlated with the
development of migraine headaches in patients with a clinical history
of migraine. Food sensitivity tests were carried out by means of
skin tests, and analysis with the Food Sensitivity Chart. Twenty
patients were then challenged with specific allergens to which sensitivity had been demonstrated, and were investigated by electro-ence-
phalography.

Wine, cheese, chocolate and spirits were the commonest foods
precipitating a migraine attack. Other individual foods produced
attacks in only a few subjects. The foods have been grouped into
categories in which there is a common antigen, such as milk protein,
which can cause a antigen-antibody reaction (Table 6). Migraine
headaches occurred in 12 patients as a result of taking milk protein
and 12 patients reacted to the common factor of grapes. Eleven
developed headaches as a result of taking chocolate. The patients
who showed positive skin reactions to specific foods developed headaches following ingestion of these foods. In the same way, the foods
which gave rise to abdominal symptoms also caused headaches. There

TABLE 6

The effect of antigen challenge on production of migraine headache,
reaction to skin prick test (SPT) and development of abdominal symptoms
in 20 patients with a clinical diagnosis of migraine.
Positive challenges indicated by subjective description of migraine attack,
positive flare and wheal skin response, and positive Food Sensitivity Chart Analysis (FSC)
indicating alimentary symptoms.
Figures indicate number of patients reacting.
Total Patients, 20.

Antigen Challenges Producing	Antigens							
	Milk Proteins	Currants Wines	Chocolate	Egg	Meats	Vegetables	Fruits	Cereals
Migraine	12	12	11	3	1	0	0	0
Positive SPT	14	13	8	5	5	7	16	5
Positive FSC	17	10	17	7	2	6	4	0

was a significant correlation between the first three types of food
sensitivity and the three types of clinical response(p<0.05). Resting EEG's were abnormal in 7 patients though no focal changes were
noted in them. Ten patients were challenged with specific allergens.
Eight of them developed a headache. In two, increased slow wave activity developed during the resultant migraine.

In this investigation, patient challenge with specific allergens
demonstrated that the same antigen is capable of causing an allergic
reaction in completely different, and separate, target organs. It
appears that the basic abnormality in migraine which gives rise to
the prodromal and associated symptoms is an antigen-antibody reaction
intiated by a food or other extrinsic allergen. This allergen not
only produces the reaction in the brain meninges, but, on account of
the localisation of specific antibodies to the same allergen in the
skin and alimentary canal, associated symptoms also appear in these
organs.

DISCUSSION

The presence of antibodies in the patient's blood is demonstrated by RAST. However as procedure for diagnosis of specific food sensitivity this test is ineffective in comparison with direct tissue
challenge testes. It supplies a laboratory value for specific IgE
concentration but does not provide any indication of food allergen-
icity. It is expensive and time-consuming. The number of foods
which can be tested is limited. Challenge of the skin by skin-prick
or intradermal methods provides highly reproducible data which correlate well with cell sensitivity and clinical illness (Lichtenstein
1972). It is pointed out that with proper methods and well standard-
ised materials, direct skin testing produces a threshold of response
which is at least as accurate as laboratory techniques involving
measurement of histamine release of IgE antibodies (Norma, Lichtenstein and Ishizaka, 1973). When comparing provocative testing of the
responding shock organ with the value of RAST as a diagnostic test,
Berg and Johansson (1974) concluded that the RAST was not as good as
provocation testing of the skin or respiratory organs.

Skin prick testing depends on the presence of antibodies in a
localised area of skin. When these antibodies are challenged with
specific antigen, the antigen-antibody reaction and its associated
release of chemical mediators, take place. These local reactions
are frequently accompanied by other clinical features associated
with the antigen-antibody reaction, such as increase of pulse-rate,
pupillary dilatation or headache. Such symptoms when used in combination with the skin tests are themselves of diagnostic value
(Miller, 1976). In the present results a method has been described
for assessing the response of the patient to food challenge in the
mouth and alimentary canal through his response to the foods which

he normally ingests. It has been shown that the buccal mucosa provides an extremely delicate and discriminatory challenge test organ.
This is because pathophysiological effects can be detected and evaluated not only in the two types of taste and sensory nerve supply in
the buccal mucosa, but also because an inflammatory response in the
form of aphthous ulcers can be observed and measured in the mucosa,
in the same way as the wheal can be observed in the skin.

The fact that a single specific antigen is capable of producing
pathophysiological effects characteristic of the antigenantibody
reaction in a limited number of tissues, indicates that the challeng-
ing allergen is active only in the tissues where the corresponding
antibodies are localised. In the case of pollen or house-dust mite,
effects occur in the skin, nose, eyes and lungs separately or together, and, as demonstrated in the present results, in the buccal
mucosa. Foods produce a positive skin reaction, and also reaction
in the mouth and gastro-intestinal tract, in the brain meninges, and
in the brain itself as shown the appearance of tiredness and EEG
changes. Lehner (1977) also has drawn attention to the wide variety
of clinical features which are associated with aphthous ulceration.
It is clear that any tissue of the body if it contains responsive
antibodies can react to specific allergen challenge by the production of an antigen-antibody reaction and development of tissue-related
symptoms. Coca and Cooke (1922) applied the term atopy to individuals with certain allergic diseases who showed wheal and flare responses on challenge by allergenic materials. Available evidence
now indicates that every tissue of the body is susceptible, and may
respond to challenge by specific allergens. The multiplicity of the
signs and symptoms in the allergic patient creates diagnostic confusion if Coca and Cooke's limited definition of the disease continues
to be accepted.   It is therefore proposed that atopic individuals
responding to antigen challenge in any body tissue should in future
be described as suffering from the allergic syndrome.

Tissue ascorbic acid concentrations are significantly reduced
in the common cold (Hume and Weyers, 1973; Wilson, 1975). They are
also reduced in respiratory allergic conditions (Wilson, 1976), in
alimentary allergic disease (Wilson, 1974a), and in acute leukemia
and other neoplastic conditions (Kakar and Wilson, 1975). Ascorbic
acid uptake into leucocytes is reduced in the presence of specific
antigens to which the leucocytes are sensitive (Wilson, 1974b). Numerous studies have demonstrated that administration of supplementary
Vitamin C is of prophylactic and therapeutic benefit in the common
cold, and in alimentary disease. The present investigation for the
first time has directly demonstrated that locally applied ascorbic
acid has the ability to reduce pathophysiological negative potential
in antigen-induced aphthous ulcers, and simultaneously to remove the
associated inflammatory reaction and pain. This result confirms the
therapeutic benefit of ascorbic acid in antigen-induced, and virus
immune complex, disease, in which evidence of the therapeutic benefit

of ascorbic acid has previously been obtained only indirectly by whole body studies. The mechanism by which ascorbic acid exerts this effect involves an intracellular interaction of ascorbic acid with cyclic AMP and Cyclic GMP through which the release of $PGF_2\alpha$ is inhibited and of PGF_1 is enhanced (Wilson, 1977); Sharma and Wilson, 1978; Sharma, Garg and Wilson, 1976). The synthesis and release of histamine is also inhibited by ascorbic acid (Chatterjee et al, 1975). These results indicate that ascorbic acid plays an important intracellular role through which it modulates cellular release of chemical mediators, and this influences the action of pharmacological anti-inflammatory agents (Wilson and Greene, 1978).

BIBLIOGRAPHY

Berg, T.L.O., Johansson, S.G.O. Allergy diagnosis with radio-allergo sorbent test. J. Allergy Clin. Immunology, 54, 209, 1974.
Carroll, J.D. Migraine - general management. Brit. Med. J. 2, 756-
-7, 1971.
Chatterjee, I. B., Majumder, A.K., Nandi, B.K., Subramanian, N. Ann. N.Y. Acad. Sci. 258, 24, 1975.
Coca, A.F., Cooke, R.A. On the clarification of the phenomena of hypersensitiveness. J. Immunol. 8, 163, 1922.
Dalton, K. Migraine - A personal view. Proc. Roy. Soc. Med. 66, 263-
-6, 1973.
Edmonds, C.J., Godfrey, R.C. Transmucosal electric potentials of human rectum and pelvic colon. Gut, 10, 1044, 1969.
Edmonds, C.F., Pilcher, D. Electrical potential difference and sodium and potassium fluxes across rectal mucosa in ulcerative colitis. Gut, 14, 784, 1973.
Hume, R., Weyers, E. Changes in leucocyte ascorbic acid during the common cold. Scot. Med. J. 18, 3, 1973.
Huston, G.J. Measurement of buccal potential difference in the assessment of mucosal damaging agents. Brit. J. Clin. Pharmacol. 4, 408P, 1977.
Kakar, S.C., Wilson, C.W.M. Ascorbic acid values in malignant disease. Brit. J. Nutr. Suppl. 35, 10A, 1975.
Lehner, T. progress Report. Oral ulceration and Behcet's syndrome. Gut, 18, 491, 1977.
Lichtenstein, L.M. Allergy. "Clinical Immunology", Vol. 1. Bach, F.H., Good, R.A., eds. p. 261, 1972. Academic Press, New York.
Miller, J.B. Food allergy: Technique of intradermal testing and subcutaneous injection therapy. Transactions American Society of Ophthalmologic and Otolaryngologic Allergy, 16, 150, 1976.
Norman, P.S., Lichtenstein, L.M., Ishizaka, K. Diagnostic Tests in ragweed hay fever. J. Allergy Clin. Immunology, 52, 210, 1973.
Piper, P.P. Anaphylaxis and the release of active substances in the lung. Pharmac. Therap. B. 3, 75-98, 1977.
Sharma, S.C., Wilson, C.W.M. The effect of histamine on the in vitro production of prostaglandins E and F by guinea-pig lung tissue.

Prostaglandins, 1978.

Sharma, S.C., Garg, S.K., Wilson, C.W.M. The effect of L-ascorbic
    acid on the In Vitro production of Prostaglandins E and F by
    the guinea-pig lung. ICRS Medical Science, 4, 573, 1976.

Vane, J.R. "Prostaglandins in the inflammatory response. in Inflam-
    mation mechanisms and control", p. 261. I.H. Lepow and P.A.
    Ward, ed. 1972. Academic Press, London.

Wilson, C.W.M. Vitamin C. Tissue Saturation, Metabolism and desa-
    turation. Practitioner, 212, 481, 1974.

Wilson, C.W.M. Colds, ascorbic acid metabolism and Vitamin C. J.
    Clin. Pharmacol. 15, 570-578, 1975.

Wilson, C.W.M. The role of ascorbic acid in allergic reactions.
    Proc. Nutrition Society, 35, 121A, 1976.

Wilson, C.W.M. Pharmacological aspects of Vitamin C in Health and
    disease. Internat. J. Vit. Nutr. Res. 16, 267, 1977.

Wilson, C.W.M. Food sensitivities, taste changes, aphthous ulcers
    and atopic symptoms in the allergic syndrome. Irish. J. Med.
    Sci. In Press 1978.

Wilson, C.W.M., Greene, M. The relationship of aspirin to ascorbic
    acid during the common cold. J. Clin. Pharmacol. 18, 21, 1978.

Wilson, C.W.M., Loh, H.S. Vitamin C metabolism and atopic allergy.
    Clinical allergy, 5, 201-207, 1975.

LACTOSE MALABSORPTION AND MILK CONSUMPTION

IN INFANTS AND CHILDREN

Emanuel Lebenthal, M.D.

Chief, Div. Gastroenterology Children's Hosp. of Buffalo

219 Bryant Street, Buffalo, New York 14222 USA

There is controversy regarding malabsorption of lactose by the infant and child and the desirability of encouraging milk consumption in different age and ethnic groups. In the view of some investigators, research on lactose malabsorption is over-reported and over-interpreted.

The uncertainties arise primarily from the lack of correlation between lactase deficiency and milk intolerance. Clinical symptoms, such as bloating, gaseousness, abdominal pain and cramps, are hard to assess objectively, especially in infants and young children. Other concerns include the quantity of milk or lactose deficient, and, the fact that the large amounts of lactose administered in the lactose tolerance test (2 mg/kg or 50 mg in older children) are rarely encountered in real life. For example, a ten year old child weighing 32 kilograms would have to ingest at least 1000ml of milk, in one sitting, to approximate the amount of lactose administered in the lactose tolerance test. Of additional concern is that lactose in solution can cause a different symptomatic response than whole milk, containing an equivalent amount of lactose. Furthermore, the major diagnostic tool, the lactose tolerance test, if used without information concerning the integrity of the small intestinal mucosa. Thus, a temporary lactase deficiency due to mucosal injury can be interpreted as a primary genetic lactase deficiency. A worldwide problem of special concern, is the role of lactose malabsorption in infants and children who suffer from advanced protein calorie malnutrition. The increased loss of calories in stools may be of little significance to healthy American children with lactose malabsorption, but could be nutritionally important for individuals with borderline caloric intake[1].

Milk is the major food consumed by mammalian infants. The carbohydrate of milk, lactose, is a disaccharide composed of glucose and
galactose in a 1-4 beta linkage. The concentration of lactose in
milk varies from mammal to mammal and bears an indirect relationship
to the concentrations of fat and protein. Lactose is in a higher
concentration in human milk than in cow's milk. The enzyme responsible for the hydroloysis of lactose, "neutral lactase", is localized
in the brush border of the mature absorptive columnar epithelial cell
of the small intestine; the highest specific activity of lactase being in the first part of the jejunum. In the human, two additional
enzymes that hydrolyze betagalactosides are located in the cytoplasm
and lysosomes but do not seem to be involved in the primary digestion
of food[2].

Primary deficiency of lactase can occur on a genetic basis or
as a result of prematurity. Secondary lactase deficiency can occur
as a result of small intestinal mucosal injury.

In the human fetus, intestinal lactase is detectable from the
third month of gestation[3]. Between 26 and 34 weeks of gestation,
lactase activity is approximately 30% of that found in full term
infants. During the same gestational period, sucrase and maltase
activities reach 70% of the activity of full term infants. It is
conceivable that the premature infant between 26 and 34 weeks of
gestation is better equipped to deal with the hydrolysis of the al-
pha-glucosides, sucrose and maltose, than lactose[3]. From 35 to 38
weeks, there is an increase in lactase activity to about 70% of that
found at term. When considering sugars fed to premature infants born
betwwen 26 and 34 weeks of gestation, one has to remember that lactase activity is low. The amounts of lactose which should be fed to
these infants is, thus, very limited, and formulas for the feeding
of prematures should be designed accordingly.

In most mammals, the activity of intestinal lactase is maximal
in the perinatal period, decreases furing weaning and reaches low
levels in the adult[4]. In contrast, human intestinal lactase activity
may be low in adults or may remain high throughout life[4]. There are
distinct ethnic differences and data are being accumulated suggesting
that most adults are lactose intolerant[5].

Prior to the domestication of milk producing animals and the use
of dairy products, lactose was not available to the human infant
after the weaning period. Therefore, adult man, like other adult
mammals, was originally lactase deficient. According to Simoons
hypothesis[5], the frequency of lactase deficiency would be very low
among those societies that have a herding tradition. Cultures without domesticated dairy animals, for example the Chinese, Koreans,
Japanese or American Indians, should have high frequencies of adult
lactase deficiency[5]. On the other hand, northern European countries
with a history of milk consumption should have a low frequency of

lactase deficiency. Prevailing studies of adult lactase deficiency
support this hypothesis. Societies that domesticated and herded
animals had a stable and reliable food source, with milk available
to all age groups, not just to the breast-fed infant. Thus, selective pressures favored the spread of a mutation resulting in a unique
genetic trait in persistent lactase activity, as such individuals or
groups would have gained an advantage in being able to utilize milk
when other dietary staples were less available.

Analysis of the prevalence of lactase deficiency (90% in intolerant populations versus 10% in tolerant groups) yields an assumed
frequency on the gene for adult lactase activity of 0.60 in tolerant
populations and 0.05 in intolerant populations. Calculation of the
selection intensity against homozygotes required to produce a change
in the gene frequency from 0.05 to 0.60, would require the passage
of around 400 generations, following the domestication of milk producing animals[5]. In other words, about 10,000 years of milk consumption are required to change an ethnic group from predominant lactose
intolerance to lactose tolerance, based on a genetic-selective basis.

Only in the last 6,000 years have milk and dairy products served
as dietary staples beyond infancy. The genetic origin of persistent
lactose tolerance is evidence by the ethnic-racial prevalence of the
entity, and the lack of inducibility of lactase by lactose feeding[4].
Studies done in Negeria[6] revealed that, when both parents were lactose intolerant, all of the progeny were intolerant. However, marriage between lactose tolerant and intolerant individuals or low
lactose tolerant individuals usually resulted in a mixed progeny.
The lactose intolerant subjects were presumed to be homozygous.
Therefore, the investigators[6] concluded that the ability of an adult
human to hydrolyze and absorb lactose is inherited as an autosomal
dominant trat.   The mutation, represented by adult lactase persis-
tence, rather than by adult lactase deficiency, appears to afford
a biological advantage to the individual. On the other hand, Sahi[7]
claims that lactase deficiency is inherited as an autosomal recessive trait. No differences in the biochemical properties of the
intestinal lactase of normal infants and adults, with normal or low
lactase activity, have been demonstrated[8, 9].

In ethnic groups with a high prevalence of adult lactase defic-
ency, children after the age of five years who "dislake" milk should
not be encouraged to drink milk. Furthermore, in any child beyond
the age of five who rejects milk, the possibility of adult-genetic
lactase deficiency should be entertained and mill or dairy products
be curtailed. In children with unexplainded recurrent abdominal
pain, the possibility of adult lactase deficiency should be queried.

Since lactase activity decreases in mammalian animals during the
weaning period, one would expect to find decreasing lactase activity
in the human intestine, during a comparable period, between the first

and third year of life. However, in healthy Caucasian children, without morphological abnormalities of the intestinal mucosa, no cases of lactase deficiency were encoutered under the age of five years[10]. After age five, two separate groups emerged, a small group (25%) with low lactase activity, and a second group with lactase activity comparable to that found in the newborn period. This study indicates that the lactase deficiency in the Caucasian population makes its appearance after the age of five years[10]. In India, lactase deficiency could be an important cause of malabsorption in children from three years of age and older[11].

Almost all subjects studied in certain ethnic groups were lactose intolerant as adults despite the consumption of sizable amounts of milk since infancy[10, 12]. This observation supports the concept that intestinal lactase activity is not regulated by dietary lactose and milk consumption. Only one-third of lactose malabsorbers associate abdominal discomfort with milk drinking in their everyday life[10, 12], and many can tolerate up to 200ml. of milk at a time without ill effect.

Primary congenital lactase deficiency is quite rare, if it exists at all. Extensive experience in Boston's Children's Hospital revealed only one probable case in 1500 biopsies performed[13]. It is possible that previous reports on congenital lactase deficiency represented, instead, examples of secondary lactase deficiency due to mucosal injury. To qualify for a diagnosis of congenital lactase deficiency, the infant must have diarrhea starting with the first milk feedings, which ceases after the elimination of milk, and a biopsy, performed after six months without diarrhea, which shows an absence of lactase activity in the presence of normal intestinal mucosa. A critical assessment of all reports of primary congenital lactase deficiency do not show a single case to fit these strict criteria.

In diseases associated with injury to the small intestinal mucosa, the activity of all the disaccharidases may be depressed[14]. Lactase activity appears to be the most sensitive to injury, and is first to decrease and the slowest to recover. For example, celiac disease, tropical sprue, intractable diarrhea of infancy, and acute viral gastroenteritis are all associated with disaccharidase deficiencies, but lactase deficiency is most evident[13]. The transient deficiency of disaccharidases seen with acute viral gastroenteritis may last for four months and is a congent reason for the usual practice of eliminating milk from the diet during such illnesses[13]. Drugs that affect the mucosa, particularly Neomycin and Kanamycin, also lower disaccharidase levels. Relative deficiency of the enzymes may occur in the short bower syndrome, after gastric sugery, in starvation, and with protein-calorie malnutrition. Occasionally, secondary disaccharidase deficiencies, especially low lactase activity, will persist after the clinical remission of celiac disease, intractable diarrhea of infancy, and prolonged gastroenteritis[13].

In early studies of lactase deficiency a number of disease entities, such as duodenal ulcer, "irritable colon syndrome", ulcerative
colitis, viral hepatitis, diabetes mellitus, cystic fibrosis, and
others were reported to be associated with a deficiency of lactase
activity. These associations may be spurious, since there were no
critical evaluations of the racial-ethnic origins and ages of the
patients. In a recent report[15], lactase deficiency was not found
in children with cystic fibrosis under five years of age with normal
intestinal mucosal histology. This supports the view that lactase
deficiency is not related to the disease cystic fibrosis, per se,
but to ethnic genetically determined adult lactase deficiency[15].

For practical purposes, even though the lactose load is relatively large, the traditional lactose tolerance test is the best
screening procedure. It offers a quantitative clinical assessment
of the degree of lactose intolerance. Tolerance tests are relatively
simple to perform, and offer the opportunity to correlate symptoms
with the lactose load.

Several alternatives to the lactose tolerance test have been
proposed. For example, one method measures changes in the breath
hydrogen concentration which is formed by bacterial digestion on
unabsorbed lactose in the colon[16]. A second method mesures $^{14}CO_2$
which is produced from tissue oxidation of absorbed ^{14}C labled lactose[16]. Another technique is small bowel roentogenography performed
after ingestion of a lactose-barium meal. In our opinion, these have
no advantages over the traditional test, and no practical use in
infants and children.

If a definitive diagnosis of lactose intolerance is required,
the most diagnostically reliable and accurate test for lactase deficiency is the determination of lactase activity in a specimen of
small intestinal mucosa, obtained by peroral biopsy[17].

In conclusion, full term infants and children up to the age of
three to five years, irrespective of ethnic background, toletare milk
and lactose unless they have small intestinal mucosal injury or, on
rare occasions, cow's milk protein allergy. Beyond the age of five
years, the ability to tolerate milk and lactose is genetically determined and, in global terms, lactose intolerance prevails. Compelling
evidence indicates that it is unwise to promote milk for adults and
children over three to five years of age in countries and ethnic
groups where tolerance for lactose is low and the nutritional state
of the individual may be still compromised. In the United States,
the majority of children of all ages of northern European background
are able to tolerate milk. Secondary lactase deficiency occurs as a
result of small intestinal mucosal injury in children of all ages.
How much milk can be tolerated by a particular child has to be eval-

uated individually. Even in lactose intolerant subjects, tolerance
of small amounts of milk is not uncommon, and one should not extrapolate from population studies. In secondary lactase deficiency,
due to mucosal injury, if diarrhea reappears with the gradual introduction of milk, ingestion of dairy products should be limited
for at least four months, especially in infants in the first 18
months of life.

The fact that intestinal lactase is not developed in premature
infants, suggests that better nutritional status may be achieved by
formulas which do not contain lactose as the primary sugar.

REFERENCES

[1]Mitchell JD, Brand J, Halbisch J: Weight-gain inhibition by lactose
   in Australian aboriginal children. Lancet 1:500-502, 1977.
[2]Gray GM, Santiago NA: Intestinal β-galactosidases. I. Separation
   and characterization of three enzymes in normal human intestine.
   J Clin Invest 48:716-728, 1969.
[3]Antonowicz I, Lebenthal E: Development pattern of samll intestinal
   enterokinase and disaccharidase activities in the human fetus.
   Gastroenterology 72:1299-1303, 1977.
[4]Kretchemer N: The geography and biology of lactose digestion and
   malabsorption. Postgrad Med J 53:65-72, 1977.
[5]Simoons FJ, Johnson JD, Kretchmer N: Perspective on milkdrinking
   and malabsorption of lactose. Pediatrics 59:98-109, 1977.
[6]Ransome-Kuti O, Kretchmer N, Johnson J, Gribble JT: A genetic study
   of lactose digestion in Nigerian families. Gastroenterology 68:
   431-436, 1977.
[7]Sahi T, Launiala K: More evidence for the recessive inheritance of
   selective adult type lactose malabsorption. Gastroenterology 73:
   231-232, 1977
[8]Lebenthal E, Tsuboi K, Kretchemer N: Characterization of human
   intestinal lactase and hetero-β-galactosidases of infants and
   adults. Gastroenterology 67:1107-1113, 1974.
[9]Freiburghaus AU, Schmitz J, Schindler M, Rotthauwe HW, Kuitunen P,
   Launiala K, Hadorn B: Protein patterns of brushborder fragments
   in congenital lactose malabsorption and in specific hypolactasia
   of the adult. N Engl J Med 294:1030-1032, 1976.
[10]Lebenthal E, Antonowicz I, Shwachman H: Correlation of lactase activity, lactose tolerance and milk consumption in different age
   groups. Amer J Clin Nutr 28:595-600, 1975.
[11]Reddi V, Pershad J: Lactase deficiency in Indians. Amer J Clin
   Nutr 25:114-119, 1972.
[12]Johnson JD, Simoons FJ, Hurwitz R, Grange A, Sinatra FR, Sunshine
   P, Robertson WV, Bennett PH, Kretchmer N: Lactose malabsorption
   among adult Indians of the Great Basin and American Southwest.
   Amer J Clin Nutr 31:381-387, 1978.
[13]Lebenthal E: Small intestinal disaccharidase deficiencies. Pediat

Clin N Amer 20:757-766, 1975

[14]Alpers D, Isselbacher K: Disaccharidase deficiency. Adv Metab Disord 4:75-122, 1970.

[15]Antonowicz I, Lebenthal E, Shwachman H: Disaccharidase activities in small intestinal mucosa in patients with cystic fibrosis. J Pediat 92:214-219, 1978.

[16]Newcomer AD, McGill DB, Thomas RJ, Hofmann AF: Prospective comparison of indirect methods for detecting lactase deficiency. N Engl J Med 293:1232-1236, 1975.

[17]Dahlqvist A: The basic aspects of the chemical background of lactase deficiency. Postgrad Med J 53:57-64, 1977.

PROTEIN PROTEINASE INHIBITOR

FROM EGGPLANT EXOCARPS

M. Kanamori

Dept. Agric. Chem., Kyoto Pref. Univ.

Kyoto, JAPAN

The widespread distribution of a number of proteinase inhibitors have been proved in plants, animals and certain microorganisms. The digestion and absorption of food are influenced negatively by these inhibitors which are considered to be toxic constituents in food. Especially in legumes, a lot of proteinase inhibitors have been studied but the studies on inhibitors contained in vegetables are very few. In our laboratory, a very strong trypsin inhibitor was purified from eggplant exocarp.

Figure 1 shows the preparation method of inhibitor. Heat treatment, salting out, dialysis with the acetylated Visking tube, DEAE cellulose chromatography for the decolorization, DEAE Sephadex A-25 chromatography and finally gel filtration on Sephadex G-25 were carried out. Final preparation was a homogeneous protein electrophoretically.

The most important characteristics of proteinase inhibitor in case of treatment of it as food, are the inhibitory specificity, that is, what kinds of proteinases are inhibited, the stability of it and strength of the inhibition- Figure 2 shows that ETI, namely eggpalnt trypsin inhibitor, inhibited bovine trypsin, chymotrypsin and Nagarse, that is one of bacterial proteinases in case of casein substrate. Pronase which is a mixture of several proteinases produced by Streptomyces griseus, is little inhibited measured with casein. Pepsin is not inhibited at all. However, when we used synthetic substrate, as shown in Figure 3, pronase was inhibited very strongly. This result represents the existence of trypsin like enzyme pronase. No inhibitory activity for papain was observed. Both trypsin and chymotrypsin from bovine pancreas were inhibited by ETI, and also trypsin activity of rat pancreatic extract was inhibited whereas this inhibitor

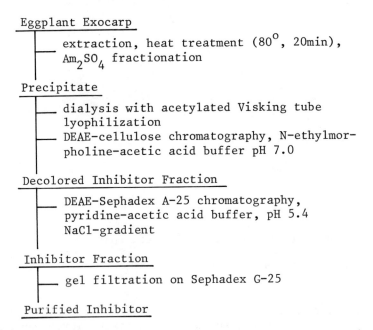

Eggplant Exocarp
└─ extraction, heat treatment (80°, 20min),
   Am_2SO_4 fractionation

Precipitate
├─ dialysis with acetylated Visking tube
│ lyophilization
└─ DEAE-cellulose chromatography, N-ethylmor-
   pholine-acetic acid buffer pH 7.0

Decolored Inhibitor Fraction
├─ DEAE-Sephadex A-25 chromatography,
│ pyridine-acetic acid buffer, pH 5.4
└─ NaCl-gradient

Inhibitor Fraction
└─ gel filtration on Sephadex G-25

Purified Inhibitor

Fig. 1 - Preparation of Eggplant Trypsin Inhibitor (ETI)

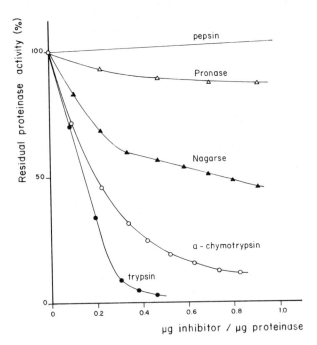

Fig. 2 - Anti-proteolytic activity of the
         eggplant inhibitor.

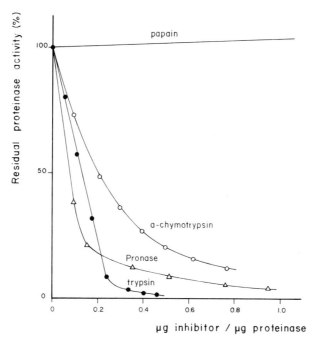

Fig. 3 - Anti-amidase activity of the eggplant inhibitor.

TABLE 1

KINETICS AND THERMODYNAMIC CHARACTERISTICS OF
THE INTERACTION OF TRYPSIN WITH INHIBITOR

25°C, pH 8.0	
k_a (M^{-1} sec^{-1})	1.5×10^{6}
k_d (sec^{-1})	2.3×10^{-4}
K_a (M^{-1})	6.5×10^{9}
K_d (M)	1.5×10^{-10}
ΔG^{o}_{a} (kcal mol^{-1})	-13.2
ΔH^{o}_{a} (kcal mol^{-1})	-2.2
ΔS^{o}_{a} (cal mol^{-1} deg^{-1})	36.9

did not show inhibitory activity for rat chymotrupsin.

About human proteinase, we checked with use of just crude human pancreatic juice. Judging from the results, BANA hydrolytic activity, perhaps trypsin activity was inhibited in fairly large extent but the inhibitory activity for chymotrypsin was a little. Carboxypepetidases A, B and Y did not receive any inhibitory activity.

ETI is very stable in a wide range of pH, from acidic to alkaline, and the inhibitory activity is destroyed quickly in the pH above 10. ETI has strong heat resistability. Below the pH of 7, the loss of inhibitory activity was not observed in the incubation at 90⁰ for 20 min. Also, the inhibitor has the strong resistability for proteinase digestion. No loss of inhibitory activity was observed in the incubation with carboxypeptidase A for long time, namely for 70 hous. For pepsin digestion, ETI was digested gradually and almost 30% loss of activity was noticed after 70 hours' incubation.

Chicken ovomucoid is a representative temporary inhibitor, that is, when the inhibitor-proteinase complex is incubated, the complex dissociates gradually to release free inhibitor. After 50 hours' incubation, the inhibitory activity appeared in reaction mixture in case of ovomucoid. However, in spite of 10 days' incubation, the release of inhibitor from the complex of ETI and bovine trypsin was not recognized at all. This means ETI forms very stable complex with proteinases.

Table 1 shows the various kinetic parameters of ETI. The value indicating the strength of inhibitor is Kd value. This value was calculated from the association and dissociation rate constants which is considered to be the most precise method at present. The order of 10^{-10} M belongs to one of the smallest among inhibitors.

Trypsin inhibitory activity was observed not only in exocarp but also in fruits, leaves, stems and roots, namely in all parts of eggplant, though the data are not shown today.

As a conclusion, considering from the strengh of inhibitory activity, the stability of inhibitor itself and its complex with trypsin, and wide distribution of it, this inhibitor seems to effect negative effect the nutritive value of vegetable.

OBESITY AS A PUBLIC HEALTH PROBLEM

N. Zöllner, M.D.

Medizinische Poliklinik der Universität München
Pettenkoferstrasse 8a
D-8000 München 2, Federal Republic of Germany

Obesity as a public health problem poses two groups of questions: one related to cause and incidence, the other to therapy. In my opinion causes and incidence are closely related, but therapy is the paramount problem which should be tackled irrespective of current pathological theories; it is only the success which counts.

In man, obesity may be a harmless anomaly or even be considered a sign of good health. However, excessive overweight indicates considerable risks for health and life of its bearer. These risks have metabolic, physiological, psychological or plain mechanical reasons.

In so-called affluent societies, overweight is closely correlated with a decreased life expectancy and available evidence indicates that in individuals normalization of body weight may normalize prognosis. However, obesity is not uncommon in nations which do not consider themselves wealthy; and in poor countries (or centuries) it may be a sign of personal wealth, proudly carried. In Germany, during the hungry years of the last war and immediate postwar period it was interesting to observe occasional obese men or women. Today I am often impressed by television reports from very hungry areas showing on the same screen plainly malnourished children and obese men distributing some food. Probably under such condition obesity serves to improve life expectancy.

Obesity as an expression of social behavior has been little studied and one may well question whether such a study will produce results of more importance than the conclusions of plain consideration. However, obesity offers other fascinating problems for research. Certainly, it is agreed that loss and gain of the weight of adipose tissue can be explained by applying the principles of

energy balance. However, the influences of individual differences
in energy expenditure on the one side, biochemical individuality on
the other are far from clear; our own experience indicates that all
differences between individuals must be explained in terms of energy
expenditure above the basal metabolism. We have studied the behavior
of body weight of obese persons under metabolic ward conditions, i.
e., under conditions close to basal metabolism and have found excellent agr ement between actual change in body weight and prediction
based on age, sex, size, weight and food intake of the individual,
as long as sodium and water turnover were considered or controled.
On the metabolic ward the obese behave as predicted from physical
laws, and their eventual weight changes confirm everything which
early investigators like Lusk or DuBois have stated and which has so
amply been confirmed by the more recent research of Pilkington in
England or Bortz in the United States. I dare to voice the opinion
that conclusions to the contrary are based on a methodology which is
too elaborate and therefore involves the changes or experimental
error.

Of course I must add that energy balance is also determined by
energy expenditure for work. The textbook of Human Nutrition and
Dietetics states that the maximum energy expenditure of office workers may be larger than the minimum expenditure of farmers or steel
workers. This indicates very large individual differences. Whether
they are due to different metabolic set ups or different degrees of
industry I would not dare to answer.

Behavioral aspects have recently become a center of interest:
why and how do the obese overeat? Of course, we in Munich in Western
Germany have our opinions based on experiences with a very large patient material. Let me just say that we are convinced that there
are many types of behavior leading to obesity and that most of them
are strongly related to the social situation of the community. If
this is true, results from one social level or from any given country can not be applied to persons in an other. However, these are
just convictions and — as Nietzsche has pointed out — convictions
are not a proof of truth.

Therapy of obesity is based upon the principles of energy balance, sufficient supply of essential nutrients and correction of
behaviour. Obviously there are more ways than one for therapy — and
their efficacy may not only depend on a sound insight into the patients' problems on the side of the therapist but also on local circumstances and even fashions. Yet, any therapy under consideration
must be effective and safe and if possible have a low rate of re-
currences.

I do not want to predict the conclusions of the distinguished
speakers on this panel. But I would like to point out that statistics in the are of therapeutic results may be misleading because

patients entering different studies are a priori not comparable in many respects, most important with respect to motivation. Therefore, let me repeat that in my opinion it is only the success, the lasting success, which counts in therapy or prevention of obesity.

OBESITY: CAUSES, CONSEQUENCES AND TREATMENT

(PATHOGENESIS AND THERAPY)

Martin Richter

Psychiatry Dept.

Univ. Göttingen, FEDERAL REPUBLIC OF GERMANY

Although it is generally acknowledged in the field of obesity research that the pathogenesis of obesity may be considered as a problem of energy balance, this knowledge alone contributes little to a real understanding of the problem itself.

Based on the assumption "hypercaloric nutrition causeso besity", the obvious counter-measure "the patient should eat less" is often the only therapy recommended, a course of action which, in the majority of cases, is condemned to failure.

This pathogenetic relationship appears too simple, because no consideration is made on the real aetiology and a causal relationship is all too readily accepted.

Therapeutic experience has shown that loss of weight and maintenance of weight are different problems. Loss of weight is primarily a question of restricting the caloric intake, i.e., of establishing a negative energy balance, whilst maintenance of weight is based on eating habits. The therapy of obesity therefore involves both dietetic and behavioural aspects.

The elementary pathogenetic model of obesity would thus seem to be in need of revision: the causes of obesity are disturbances of appetite and satiation control by environmental conditions, eg. food surplus.

The first consequence is the hypercaloric nutrition, the second consequence is the overweight. Therefore the first therapy must be a negative energy balance to reduce weight. The second therapy must be a compensation of the disturbances in appetite and satiation con-

311

trol by behaviour modification and self-control techniques to maintain weight.

But the main question is, why do some individuals eat more than
they need?

It had been established by interviews with subjects that stress
led to increases in food intake in many and to decreases in others.
In the hyperphagic reaction, initiated by emotional disturbances, it
is possible to observe an etiological factor in obesity. The question arose whether hyperphagic reactions were reproducible in the
laboratory and, if so, under what conditions and with which subjects.

Stimulated by the work of Schachter, Nisbett & Rodin, we have
conducted a series of investigations on the influence of cognitive,
perceptual factors. Using the food dispenser method, which was
developed by Stunkard, we were able to carry out a number of experiments on human satiation control.

Subjects took their meals on consecutive days from the food dispenser. Stress was induced into the subjects before or during the
last three meals (stressors were noise, flicker-light or task-induced
stress). The result was that up to 30% of all subjects reacted with
an increase in appetite. Subjects who were older, female, and obese
reacted more often with hyperphagia.

Similar studies with children showed that they react to stress
with a decrease in appetite and an increase in the time they take
to eat. No hyperphagic reaction occured in childhood. Subjects over
65 years of age displayed a fixed "automaton-like" food intake that
was not influenced by stress. They did not have significant hypo-
phagic or hyperphagic reactions.

The results suggest that during a lifetime, the detrimental
effects of stress on food intake increases after childhood and decline in old age. The reaction seems to be a conditioned response
— a learnt compensation reaction — and is not primarily biologically
determined.

These findings can give a possible answer to the question, why
some people in certain situations eat more than they need. They have
learnt to eat under the condition stress in a sense of oral compensation. At least for 30% of the population this habit can be an explanation for overweight; mainly in the industrial nations stress
becomes more and more an important problem for the health.

A further possible answer is the concept of externality by
Schachter, which means that external factors have a higher cue sali-
ence for obese than for normal persons. For instance, if an obese
person sees something to eat, he must eat it. If a normal person

sees something to eat, he eats only when he is hungry. But the hypothesis that obese persons are exclusively dependent on external cues,
and normal persons are executively determined by the dependency of
internal cues of hunger and satiation is not right. Normal subjects
were also found to be responsive to external signals in their eating
behaviour. The influence of cognitive factors on food intake is
better defined in terms of the size of the external/internal stimulus
discrepancy. Normal subjects resolved this discrepancy in favour of
the external cues only when the discrepancy was moderate, while obese
subjects oriented themselves to the external cues in their eating behaviour especially when a large discrepancy existed between external
and internal cues.

The results were obtained in an experiment, where subjects estimated their degree of satiation after isocaloric meals in which
the calory content was labeled between 100 to 500 calories during
different meals, in spite the real content always was 100 calories.
Only when a strong discrepancy existed, eg. 100 actual and 400 or
500 putative calories, obese subjects reported their satiation feelings in accordance to the wrong calorie labeling.

To sum up: it seems to be that obese persons oriented themselves
more than normal persons to the external cues in their eating behaviour especially and even when a large discrepancy existed between
external and internal cues, because they have a disturbed internal
satiation regulation.

But how can you describe these disturbances of satiation regulation?

The food dispenser technique allows to register cumulative intake curves. In general we found two characteristic types of curves.
The first one represents a so-called biological satiation process,
which means that the subject at first ingests rapidly and during the
time of meal he slows down his intake. This negatively accelerated
intake resembles a typical biological intake process recorded from
most of the normal weight adults as well as from newborn infants and
from animals.

The second type of curve is more or less linear indicating that
the subjects ingest equal quantities of nutrient at each time inter-
vall.

Studies in newborn infants, children, normal weight and obese
adults, and older subjects revealed that age and weight correlated
with the shape of the intake curves. In older and obese subjects
there was no slowing-down of food intake at the end of the meal.
Another factor influencing the shape is food deprivation, especially
in newborn infants.

The nonbiological, linear intake curves are probably the expression of a disturbance in the regulation of internal satiation mechanism present in obese and older subjects.

One finding important for treatment of obesity was in some eating experiments a proportion of normal weight subjects, reacting in the manner of obese subjects. Psychological examination revealed that these actually normal weight subjects either had been overweight or have "weight problems", which they cope with various control mechanisms. These subjects are referred to as the "latent" obese in opposition to the "manifest" obese with actual overweight.

The therapy of obese persons is to make "manifest" obese to "latent" obese persons. Therefore an exact analysis of the self-control techniques of the "latent" may possibly be the key for the successful therapy of obese with positive long-term effects.

Most of the evaluated dietary treatments of obesity give no final answer to the problem. because the modification of the typical obese eating behaviour is not stressed enough and therefore the stabilization of the reduced weight is not guaranteed for a long time.

Sufficient long-term results were reported by such authors, who worked with behaviour modification techniques. Some components of such behaviour modification strategies are, eg., training in self-observation stimulus control, response control, contract management, etc.

A recent follow-up study in W. Germany shows that about 50% of the patients maintained the reduced weight two years after treatment by behaviour therapy.

Our intention today is to develop methods to slim down the other 50% and to reduce the attrition rate, which is published in the literature to be between 0 to 87%.

OBESITY FACTS, FADS, AND FANTASIES

G.L. Blackburn

New Deaconess Hospital, Harvard Medical School

Boston, Massachusetts 02215, U.S.A.

Diet crazes reflect an epidemic of obesity. Anxiety arising from the threat of obesity is compounded by the public being inundated with the claims of persons who are knowledgeable about nutrition and those who are not. Nutritional faddism remains one of America's favorite indoor sports. Few will argue that dietary goals are needed — e.g. reduction in meal size, and such items as sugar, saturated fat, cholesterol, and meat. Increasing daily exercise as a substitute for eating is also required. Despite the massive industry the probability of success is very limited. Weight loss clubs can claim a short term mean 20 pound weight loss in 1/3 serious participants. Anorectic drugs can produce a 10 pound weight loss, but this is almost approached by the substitution of a placebo. Behaviour modification mean weight loss is also 10 pounds, but it is sustained for prolonged periods (>26 weeks).

Short term success is attained by 8% of the patients making the effort and less than 1% achieve long term effect. The primary problem is not a lack of proper concepts, but the fact that most people are not ready for the required level of discipline. They allow themselves to be exploited by the ever-ready industry. Medically significant weight loss (40 pounds for 2-3 years) appears to require a long term, serious effort involving combined therapies of exercise, behavior modification, assertiveness training, and mental conditioning. Morbid obesity (>100 pounds above ideal body weight) may require the adjunctive therapy of a surgical gastric bypass procedure. Only by such combined programs can a large number of patients lose significant weight. Physicians must take seriously the number one problem of malnutrition in developed countries and become schooled in effective therapy. Often this requires the support of nutritionists, but the physician's input is essential if the patient is to sustain the effort necessary to be rehabilitated.

OBESITY: COMMENTS ON EPIDEMIOLOGY

Frederick H. Epstein, M.D.

Inst. of Social and Preventive Medicine - Univ. of Zürich

Gloriastrasse 32B - 8006 - Zürich, SWITZERLAND

Obesity is said to be a major cause of ill-health, i.e., a public health problem, in developed countries and, if this is true, developing countries would gradually become similarly affected. An attempt will be made to show that obesity presents indeed a serious health hazard but is not — as often suggested nowadays — the predominant and singly most important health problem in industrialized societies.

MORTALITY

Life insurance data indicate that excess weight carries an increased risk of total mortality and death from cardiovascular and a variety of other diseases. By contrast, in prospective epidemiological field studies, such as those in Framingham or in Chicago by Stamler's group, no such simple trend is apparent. On the contrary, the relationship between overweight and total or coronary heart disease mortality is, depending on age and sex, absent, slight, U-shaped and even inverse. Notwithstanding this baffling inconsistency, let it be assumed that the life insurance data are largely valid, as is probably true. Then, the effect of weight reduction in the population as a whole may be calculated, assuming a causal relationship, based on the much-quoted Society of Actuaries data (Table 1). Shifting the whole distribution very markedly to the left would, as shown affect total mortality surprisingly little. It is important here to distinguish between individual and population-attributable risk. An individual with an elevated risk factor level like weight may carry an appreciably increased risk. However, the impact of risk factor elevation in individuals on total disease risk in the population also depends on the frequency of elevated risk factor levels in the pop-

317

TABLE 1

ESTIMATING THE EFFECT OF WEIGHT REDUCTION
ON TOTAL MORTALITY

– Men, Aged 40–69 –

WEIGHT ABOVE DESIRABLE WEIGHT**	PERCENT IN WEIGHT CATEGORY		RELATIVE MORTALITY (b)	NUMBER OF DEATHS*	
	Present** (a_1)	Future (a_2)		Present $(a_1 \times b)$	Future $(a_2 \times b)$
<5%	40	65	100	40	65
5–15%	30	25	130	39	33
15–24%	15	5	140	21	7
25–35%	10	5	150	15	8
>35%	5	–	170	7	–
TOTAL	100	100	–	122	133

*   Arbitrary units.
** Based on Build and Blood Pressure Study Data (Society of Ac-
tuaries, 1959).

ulation. Therefore, even rather drastic weight reduction in the
population as a whole might not, by itself, make a major dent in
total mortality, however much obese individuals might benefit.

These relatively modest, expected reductions in total mortality
are in accord with the findings from epidemiological studies that
the major single cause of death in developed countries, coronary
heart disease, is related to overweight, even if measured in terms
of skinfolds, in an inconstant fashion. As Keys showed over 25 years
ago, geographic regions varying little in the frequency of obesity
may show striking differences in the prevalence of coronary heart
disease. Similarly, prospective investigations have consistently
indicated only weak associations between various indices of obesity
and coronary disease. In Framingham, the relationship is a little
more marked for women than for men, for angina pectoris and at younger ages. In a large Finnish study, the association is confined to
men with high serum cholesterol levels and in the Manitoba Study to
sudden deaths. The trend below age 50 for men is illustrated by the

TABLE 2

RELATIVE WEIGHT AND 10-YEAR RISK
OF MYOCARDIAL INFARCTION AND SUDDEN DEATH - MEN

WEIGHT QUINTILE	RATE/1000	
	Age 40-44 N = 2170	Age 45-49 N = 2121
I.	39.2	75.5
II.	39.2	82.5
III.	50.7	84.9
IV.	53.0	108.5
V.	80.6	120.0
All	52.5	94.3
Ratio Q5/Q1	2.1	1.6
% of Events in O5	30.7	25.5

Source: National Collaborative Pooling Project,
        Am. Health Assoc.

Notes:
  i) Weights are calculated as % of desirable
     weight.
 ii) Trends for ages 50-59 were not statistically
     significant.

Pooling Project data, combining the experience of the largest and
longest investigations in the United States (Table 2).

It is stated rightly that the coronary heart disease findings
take no account of the associations between its precursors and excessive weight. Actually, the association between obesity and serum
cholesterol is weak and confined to younger ages, important though
this is. Triglycerides are strongly related to relative weight but
their significance as an independent risk factor is not established
everywhere. The population attributable risk of diabetes on coronary
disease is relatively small. With smoking, the association is inverse. Hypertension, on the other hand, is strongly and consistently
related to obesity and, for the individual, weight reduction — with
or without salt restriction — reduces blood pressure. It may be

TABLE 3

ESTIMATED EFFECT OF WEIGHT REDUCTION ON THE PREVALENCE OF HYPERTENSION

MEN QUINTILE	N	HYPERTENSIVES %	HYPERTENSIVES N	WEIGHT DISTRIBUTION AFTER WEIGHT REDUCTION Q_1	Q_2	Q_3	Q_4	Q_5	TOTAL
Q_1	200	5	10	180	15	5	–	–	200
Q_2	200	10	20	100	80	15	5	–	200
Q_3	200	15	30	20	100	60	15	5	200
Q_4	200	20	40	10	60	60	60	10	200
Q_5	200	30	60	10	30	60	80	20	200
TOTAL	1000			320	285	200	160	35	1000
HYPERTENSIVES		–	160	16	29	30	32	11	118

Calculated reduction in prevalence $\frac{118}{160}$ x 100 ≈ 75%

calculated, based on informed guesses and assuming again a causal
association, what the effect of weight reduction on the prevalence
of hypertension in the population might possibly be (Table 3).
The population is divided into equal quintiles of weight, each with
200 men. In a less obese population, the weight distribution might
be as indicated, perhaps overoptimistically. The prevalence of
hypertension in this thinner population would now be 12% instead of
16%, by no means a neglibible achievement but not, by itself, an
adequate solution of the hypertension problem in the population at
large.

No mention was yet made, except in relation to cardiovascular
disease, of the morbidity connected with — and presumably partly due
to — obesity, notably osteo-arthritis, gallbladder disease, gout and
others, such as accidents, and — to recall again — diabetes. There
is no reasonable doubt that obesity is a major public health problem
but it would be a mistake to think that the treatment and, above all,
the prevention of obesity will, by itself, reduce the burden of
chronic illness and premature death in the population to tolerable
levels. By contrast, however, the individual has much to gain from
losing weight or better, not putting it on.

How common is obesity as a graded characteristic, measured re-
liably and in defined terms, between countries and by social groups
within the same country? Such prevalence data are regrettably lack-·
ing to a large extent. Therefore, vague and meaningless statements
that 20, 30 or perhaps 50 percent of the population are obese are
frequently heard. In this brief discussion, an attempt was made to
indicate what kind of information would be needed to estimate the
real impact of obesity on ill-health on a national and international
scale.

THERAPY OF OBESITY

E. Deutsch and K. Bauer

Dpt. of Medicine, University of Vienna

A-1090 Vienna, Lazarettg. 14 - AUSTRIA

Weight reduction is only possible if energy input is lower than energy output. Without any doubt the average daily central European diet contains more calories than necessary for basal metabolism plus performance of work. Theoretically, a constant surplus of 75 calories per day (i.e., 28 g white bread or 51 g veal) leads to an increase in weight of 4.5 kg per year or 45 kg in 10 years[1]. Uptake of isocaloric diet in respect of energy need should therefore avoid development of obesity, while moderately hypocaloric food intake should lead to weight reduction. This simple solution of the obesity problem is hindered by two facts: first: life following calorie-tables which thus balances body weight is routinely inpracticable, and second: the body is able to adapt its calorie-balance on lower intake by better utilization of the food[1] Therefore the average weight reduction per day decreases during a prolonged low-calorie-diet.

Surplus calories are stored as fat in the organism. My coworkers K. Irsigler and P. Schmidt[5] constructed a buoyancy scale in order to determine the gas-free body volume of man. They demonstrated, that the normal body contains 20.84% fat, a body with overweight 27.85%, and an obese body 48.38% fat (average of 8 each) (i.e., between 11 and 55 kg fat)[2].

Weight reduction is only possible through a disproportion between energy uptake and energy consumption either by reduction of food intake or by increase of body work. Basically, there is no need in varying the composition of the normal diet. Nevertheless it seems favourable for psychological reasons to gain a faster weight reduction initially. This may be accomplished by excessive reduction of energy intake and especially by ketogenic diets. Ketosis is

created by the strictest withholding of carbohydrates and leads to
faster weight reduction in the first few days via increased waterloss
and reduces the sensation of hunger. It creates, however, some severe
disadvantages. There is no acceleration of fat-utilization[6] but
rather increased protein-loss, since the essential glucose for the
brain is produced by gluconeogenesis from proteins. The brain needs
a period of 3-6 weeks for adapting its metabolism on available ketones
as an energy source. The protein requirement decreases after this
period.

MATERIALS AND METHODS

Our results are based on three groups of patients:

1. 73 patients on complete starvation, 39 of whom could be
   studied for three weeks;

2. 19 patients on a 200 kcal protein diet, 17 of whom could
   be studied for three weeks. The protein was supplied as
   beef or fish, with carbohydrate-free trimmings, containing
   together 300-400 kcal;

3. 5 patients in a not yet completed study with 200 kcal protein formula diet (Bionorm®). Bionorm® is a specially
   prepared durds protein coated with a dragee-mass containing
   very little carbohydrate (9 g), and vitamins 1235 U vitamin
   A, 7.5 mg Vit. E, 0.3 mg Vit. B_1, 0.5 mg Vit. B_2, 0.5 mg
   Vit. B_6, 1.3 mg Vit. B_{12}, 3.3 mg Niacinamid, 2.6 mg Calcium-
   D-pantothenat, 0.1 mg folic acid) per 80 protein calories.

Patients with any other disease, especially cardiac decompensation, any abnormality of the ECG, diabetes, kidney or liver diseases
with exception of fatty liver, and in the last series even latent
diabetics were excluded. The patients were advised to drink at least
2 liters of unsweetened tea or water. The serum level of uric acid
and potassium was determined at least twice weekly, and allopurinol
and potassium given if necessary. A vitamin mixture (Supradyn®)
was supplied to all patients. Urea nitrogen, creatinine, potassium,
sodium and chloride, total protein, albumins and globulins, GOT, GPT,
cholesterol and triglycerides were determined weekly. The largest
deviation of these parameters from the value at entry into the study
independently from the point of time during the treatment period,
and independently of the duration of the change was used for the
judgement of side effects. These changes are summarized in Tables
1 and 2.

The patients performed exercise daily under the supervision of
a physical therapeutist: 30 minutes gymnastic, climbing 4 floor stair

TABLE 1

BIOCHEMICAL PARAMETERS IN 73 PATIENTS WITH COMPLETE FASTING

PARAMETER	BEFORE TREATMENT		DURING TREATMENT		NORMAL LIMIT
	Normal	Patho-logical	Increase	Decrease	
Uric acid	43	28	67	–	<7,5 mg/dl
GOT	63	7	N → N 29	5	<20 U/1
			N ↕ P 21	1	
			P → P 5	1	
GPT	56	9	N → N 33	4	<20 U/1
			N ↕ P 10	2	
			P → P 3	2	
Cholesterol	66	4	27	31	<280 mg/dl
mean			212.16 ± 53.26	195.84 ± 65.07	
Triglycerides	43	17	24	24	<175 mg/dl
mean			162.42 ± 99.91	145.06 ± 76.96	
Total protein			36	22	6.5-8.5 g/dl
mean			7.35 ± 0.44	7.48 ± 0.55	

N → N = Increase or decrease of the parameters within the normal range.
N ↕ P = Increase from normal to abnormal high or decrease form abnormal high to normal.
P → P = Increase or decrease within the abnormal high range.

TABLE 2

BIOCHEMICAL PARAMETERS IN 19 PATIENTS ON 200 KCAL PROTEIN DIET

	BEFORE TREATMENT			DURING TREATMENT	
	Normal	Patho-logical		INCREASE	DECREASE
Uric acid	14	5		16	2
GOT	18	1	N ⟶ N	7	2
			N ↓ P	7	–
			P ⟶ P	–	1
GPT	12	2	N ⟶ N	4	2
			N ↓ P	6	1
			N ⟶ P	–	1
Cholesterol	18	1		7 — 208.75 ± 52.02	180.19 ± 60.93
Triglycerides	16	1		10 — 126.71 ± 32.99	4 — 140.43 ± 40.84
Total protein				8 — 7.50 ± 0.56	7 — 7.41 ± 0.40

For explanation see Table 1.

case twice daily 10 times and bicycle riding twice daily 10 km,
resistance 2.5 kp.

RESULTS

1.   Weight reduction is fastest obtained by complete starvation.
The mean daily weight loss of our 39 patients was 847 g in the first
week, 510 g in the second, and 320 g in the third week. The mean
weight loss over the three weeks was 560 g daily or 11.76 kg totally
(Fig. 1). The weight loss decreased from week to week. This difference is caused by the excess water loss in the first week and by the
adaptation of the organism later on[1, 6]. This adaptation goes so far
that after this period a diet with only 50 g protein may induce a
positive nitrogen balance[2]. The daily protein loss through 14 days
of starvation was calculated by Irsigles and Veitl[2] to be 48±12.07 g
equalling 3.37 kg muscle substance in 14 days.

    The level of uric acid was at entrance into the study increased
in 28 out of our 73 patients and increased further in all patients
during starvation. All patients were treated with allopurinol.

    The level of GOT was elevated in 7 patients at the beginning of
the study and rose further in 55 patients during starvation. It remained in the normal range in 29 and became pathological in 21 patients. In 5 patients with elevated GOT values these increased
further, in 1 patient it normalized (Table 1).

    GPT was normal in 56 cases and high in 9. The values increased
to the high normal range in 33 patients and became abnormal in 10.
Abnormal values rose further in 3 cases and normalized in 2.

    The cholesterol level was elevated only in 4 patients. It in-
creased in 27 and decreased in 31. The mean of the group fell from
212.16±53.26 mg/dl to 195.84±65.07 mg/dl.

    The triglyceride level was normal in 43 and elevated in 17 pa-
tients. During starvation it rose further in 24 cases and was reduced in 24, the mean dropped from 162.42±99,91 mg/dl to 145.06±
76.96 mg/dl.

    The total protein increased during the study slightly in 36,
and decreased in 22 patients; the mean increased from 7.35±0.44 g/dl
to 7.48±0.55 g/dl.

    My coworkers Slany and Mösslacher[3, 4] studied the cardiovascular
function in 14 obese subjects, and found an increase of mean pulmonary artery and pulmonary wedge pressure during exercise in half of
the persons studied indicating depressed left ventricular function.
11 patients could be reinvestigated after weight reduction, revealing a significant fall of pulmonary artery, pulmonary wedge and

right atrial pressures during rest and exercise. A moderate reduction in cardiac output and stroke volume was found in some cases
suggesting loss of intravascular volume, in others unaltered or increased cardiac output suggested improvement of cardiac function.

2.   The 17 patients receiving 200 kcal protein diet lost 560 g daily
in the first week, 425 g in the second and 267 g in the third week.
The mean daily weight loss over the total period was 420 g or 8.77 kg
totally (Fig. 1). Irsigler und Veitl[2] calculated the protein loss
for this type of restricted diet with 22.15 g daily or 1.55 kg muscle
substance in 14 days compared with 3.37 kg during complete fasting.

      The level of uric acid in serum rose in 16 cases during the
study, and was increased in only 5 at the beginning of the study.
The patients received allopurinol like the patients with a complete
fast. The GOT rose in 14 cases, in 7 of these to pathological
values. The GPT was high in 2 patients and increased during the
study in 10, in 6 of these to abnormal values (Table 2).

      The cholesterol increased in 7 and decreased in 8, the mean was
reduced from 208.75±52.02 mg/dl to 189.19±60.93 mg/dl. The triglycerides were high in 1 patient at entry, increased during the observation period in 10, and decreased in 4, the mean rose from
126.71±32.99 to 140.43±40.84 mg/dl. The level of total protein was
slightly reduced in 8 and increased in 7, the mean changed from
7.05±0.56 to 7.41±0.40 g/dl.

3.   The third group in a study now in progress received 200 kcal/
protein as Bionorm® and no additional food, and fluid ad libitum.
The weight loss in the 4 patients who have already completed the
three week period was 660 g daily in the first week, 330 g in the
second, and 360 g in the third week, and the mean weight loss was
daily 450 g or 9.5 kg totally. The other parameters studied behaved
similarly as in the other two groups. (Fig. 1).

      By far most of the obese and overweight patients have a life-
long tendency for gaining weight. Even after reaching the more or
less ideal "Steady State-Weight" a strict control of diet and body
weight for further life is necessary.

      The time period of highly restricted food intake during the
stay in the hospital has to be used for teaching and motivation of
the patient what he has to do after that period. He continues with
a diet less restricted in calories, low in carbohydrates over a long
period of time until the ideal weight is reached. After that he may
eat any convenient type of food, but he has to control his weight
on a scale every day, and he has to restrict food intake at once if
he has gained some weight.

      Finally, I should like to make a few remarks on the special diet

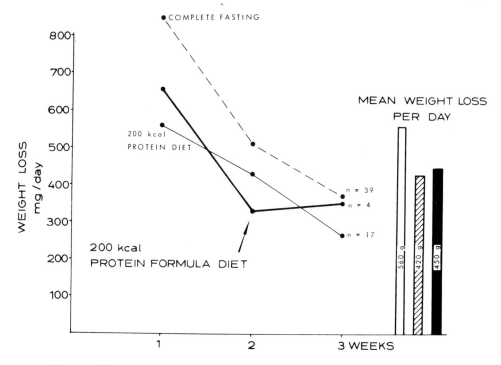

Fig. 1 – Daily weight loss during 3 weeks in the three experimental
         groups. Mean weight loss per day (columns).

    ---- Complete fasting. Mean weight loss: white column.
    ——— 200 kcal protein diet. Mean weight loss: hatched
    column.
    ■■■ 200 kcal protein formula diet. Mean weight loss:
    black column.

which stimulated our interest in these problems. A dermatologist in
Vienna, Dr. Humplik, propagated a diet consisting of meat, vegetables
and fruits with strict exclusion of any other carbohydrates. He
forced his patients to eat as much as possible. His patients were
eating at least 20 eggs and 10 beefsteaks a day and several pounds
of apples, and lost weight. This observation led Dr. Humplik to the
hypothesis that calories ingested with different types of food are
not equal in effect. There is, however, no question that these
patients were not able to eat any more these quantities after a few
days, but they did not confess it to the doctor whose dominating
personality frightened them. We have tried to reproduce this regimen,
but the calculations always demonstrated that the amount of food
really ingested by the patients was hypocaloric and the weight loss
was not a miracle. It is very likely that the same is the case with
all types of one-sided diets where it is propagated that control of
calorie intake is unnecessary.

In conclusion we may state that highly restrictive diets in the beginning of a regimen for weight reduction have the advantage of providing the patient with the feeling of success after a short period of time. A 200 calorie protein diet is superior to a complete fast since the decomposition of structural protein is much less with about the same weight loss. This 200 calorie protein diet is practically danger-free for the patient under the following precautions:

- previous exclusion of any severe disease;
- maximum duration 3 weeks;
- plenty of beverages free of carbohydrates;
- substitution of potassium and vitamins;
- control of uric acid by increase:allopurinol
- hospitalization of the patients.

Lessons in diet theory will enable the patient to compose his personal low calorie diet after discharge from the hospital until his intended ideal weight is reached. Using this mixed reduction diet as a basis the patient is then able to control "steady state weight" by adding or reducing calories by himself.

REFERENCES

[1] Ditschuneit, H. Was ist gesichert in der Therapie der Fettsucht? Internist 17, 622-630, 1976.
[2] Irsigler, K., and V. Veitl. Änderung der Köeperzusammensetzung unter verschiedenen Reduktionskostformen. Infusionstherapie 4, 63-66, 1977.
[3] Slany, J., H. Mösslacher, P. Bodner, K. Irsigler, H. Lageder, P. Schmid, and W. Schlick. Cardiovasculäre Folgen einer Nullkaloriendiet bei Adipösen. Wr. Klin. Wschr. 86, 423-428, 1974.
[4] Slany, J., H. Mösslacher, and K. Irsigler. Beeinflußt Adipositas die Herzfunktion? Z. Kardiol. 64, 851-862, 1975.
[5] Schmid, P., W. Schlick and K. Irsigler. Eine neue Anlage zur Bestimmung der Körperzusammensetzung mit Hilfe der Auftriebswaage und Unterdruckvolumetrie. Wr. Klin. Wschr. 88, 15-19, 1976.
[6] Yang, M.U., and Th. B. Van Itallie. Composition of weight loss during short term weight reduction. J. Clin. Invest. 58, 722-730, 1976.

SPECIAL DIETS IN PREVENTION AND THERAPY OF OBESITY

H.P. Wolf, M.D.

D-6100 Darmstadt 1

FEDERAL REPUBLIC OF GERMANY

There is no doubt that — with regard to the intermediary metabolism — obesity is a balance problem in all parameters involved. Reduction in caloric intake, composition of the food offered and consideration of mobilized and metabolized endogenous substances are thus the main aspects in treating obesity.

Unbalanced intake of nutrients — in the same way as inadequate food intake — may lead to excessive mobilization from endogenous sources. Uncontrolled lipolysis associated with a largely disturbed carbohydrate metabolism by free fatty acids, activation of gluconeogenesis by acetyl-CoA as well as changes in the acid, base and electrolyte metabolism is the most frequent example.

Under these aspects a normal diet reduced in calories, but otherwise not different from the usual eating habits, would be — and is in the long run — the first recommendation. It may also easily be followed in daily working life and at the family table for individual members of such a community. However, loss of weight is in this way achieved only slowly and requires a high degree of consistency — this explains the high failure rate.

Experience shows that a "sense of achievement" at the beginning of the reduction period, that is the loss of several kilograms within a short period of time, makes subsequent keeping to the conditions mentioned considerably easier, especially if simple therapeutic measures on a dietary basis are offered at the same time to make up for lapses which are later on so frequent.

This explains also the many suggestions for special diets since they all claim that with their help this aim is achieved.

Not taking into consideration extreme "outsider" diets (be it the sherry diet, which must be particularly disapproved of, hay's separating diet, which is harmless but lacks any rational basis, or the "point-calculated diet" where one apple may represent five liters of wine, nine pounds of butter or 36 eggs!), some methods should be mentioned which are particularly frequent.

Atkins diet — as an example of those diets which are particularly rich in fat — means a load in addition to the endogenous also by the endogenous fatty acid sources, influencing all the enzymes mentioned before. Where the 4-stage plan over a period of 4 weeks with 0-5-15-30 or 40 g of carbohydrates is strictly followed and assuming that the cerebral requirement alone is about 120 g of carbohydrates, at least 200 g of protein must be made available for gluconeogenesis! Metabolized protein includes essential enzymes of mucosa and liver. If gluconeogenesis from amino acids is in this way activated, there is always an increased amount of acetyl coenzyme A from the mentioned increased lipolysis. At the same time, this compound again is an allosteric effector of an enzymatic reaction of gluconeogenesis. An increased amount of ketone bodies for a prolonged period of time is the inevitable consequence. Competitive inhibition of uric acid excretion by keto-acids and lactate requires in some cases the use of allopurinol in order to restrict uric acid formation a priory.

The Atkins diet may become dangerous and is to be disapproved of, as is the egg diet which combines low caloric intake with an extreme surplus of cholesterol. It already belongs to the "aversion diets" in which the one-sidedness of the food is expected to reduce appetite (often saying at the same time: "Of this you can eat as much as you like"). The potato diet is a less harmful example of this kind.

Extreme conditions as to the adaptation of the intermediary metabolism are present in the case of the much discussed and often spectacularly boosted absolute diet. It is correct that after several weeks a large part of the cerebral glucose metabolism is replaced by the oxidation of keto acids and that treatments have been carried out for up to 200 days. From a physiological point of view, however, it is absolutely deplorable to bring the metabolism out of balance in such an extreme way (this follows alone from the necessity of clinical supervision which is absolutely required in this phase), in order to do then what may be achieved more slowly, but — also in educational sense — more effectively with subcaloric nutrition and adequate protein substitution. Most supporters of the absolute diet have therefore modified it by giving protein right from the the start!

The absolute diet should thus be restricted to cases of extreme overweight.

The use of protein hydrolysates alone under extremely subcaloric conditions ("the last chance") is thus practically nothing else than such a modified absolute diet with all its dangers which can only be brought under control in the clinic. The cases of death reported in the USA have given rise in the lay press to several confusing and unfounded statements on the use of protein-concentrated nutrition or on special diets orientated in this direction.

The original Mayo diet — which is, however, also often too rich in eggs — and its numerous variations (all the suggested diets with which newspapers and TV shower us are hardly anything else) leads back to normal subcaloric nutrition. Here adequate protein supply has to guarantee a steady nitrogen balance after a few days and increased lipolysis should only lead to a brief — usually inevitable — slight increase in ketone bodies. An intermediary increase in uric acid is almost always observed and should, if necessary, be treated with drugs.

In all cases of special diets, as well as in normal diet with caloric reduction, vitamin supply is necessary, since the reduction lasts for a prolonged period of time. This should apply to all vitamins, with the exception of vitamin D.

The same applies to iron — often mentioned but hardly ever paid attention — whereas the question of possible zinc deficiency has not yet been definitely answered, but will probably be affirmed.

We have tried to meet the requirements for a well-balanced nutrition with an optimal protein supply and reduced caloric intake under conditions which are also practicable in everyday life. This is a formula diet, supplementing individual meals with a protein granulate which produces in addition a prolonged sensation of satiety. The recommended diet plan may largely be replaced by the usual food. The investigations on bioNorm available up to now show the required steady balances, no deviations of the parameters examined from the normal range for a prolonged period of time, or in the presence of excessive values return to normal within a few days (e.g., uric acid, ketone bodies) as well as weight reduction, which is initially partic-larly clear due to additional loss of water, and becomes then continuous and is very satisfactory. It is interesting that psychological and radiological investigations have revealed the importance of the parameters feeling of repletion and retention time in the stomach. The product has been used in more than a million cases and no harmful side effects have been reported.

Under the aspect mentioned at the beginning — initial success as stimulating sense of achievement and change to normal food or brief use of the diet when dietary lapses occur, which unfortunately happens again and again — we regard this method as the most promising one for psychological and physiological reasons.

It should however be pointed out that any change from such a subcaloric diet, which is particularly rich in protein, to the usual isocaloric diet often leads to renewed, however limited water retention because of the increased sodium intake (in connection with the carbohydrate intake) which is automatically associated with it. This intermediate weight peak should however be accepted without using diuretics, which unfortunately are taken by the so-called "informed" patient via self-medication occasionally.

I shall try to summarize the subject of my lecture in one sentence:

The therapy of any kind of obesity (of the more than 90% which are due to nutrition, or as part of the therapy in endocrine forms) requires consistent caloric and qualitative balancing; here special diets are an absolutely necessary help at the beginning of treatment and between treatments, but permanent success will in any case only be achieved by changing to a "normal diet" the caloric value and composition of which is controlled.

BEHAVIOR MODIFICATION AND THE CONTROL OF OBESITY

W.H. Sebrell, M.D., F.A.C.P.

800 Community Dr.

Manhasset, New York 11030 - USA

For many years, physicians have been aware of the great difficulty in controlling obesity. Efforts to control the condition by
prescribing very low calorie diets have usually failed as the patients followed the diets only for a limited period of time and then
returned to former habits of overeating, with a resultant return of
the excess fat.

The importance of psychological factors in controlling the obese
patient's desire to eat has been appreciated only in recent years.
The successful application of the techniques of behavior modification to the control of obesity constitutes a great step forward.

Success in controlling obesity should be evaluated not only on
the basis of the loss of excess body fat, but also on the ability
to maintain a normal body weight for a long period of time.

Newly developed programs for the control of obesity rely on
four principal factors, all of which must be integrated and individualized to fit the particular case. These factors are psychological, dietetic, educational, and physical activity.

PSYCHOLOGICAL

The Weight Watchers organization is unique in that for 15 years
it has been using psychological techniques through group interaction
to strengthen and maintain the motivation to lose excess body fat.
Groups of overweight individuals are brought together under a trained
leader for a few hours for one or more sessions a week. Motivation
is strengthened by rewards for achievement, group discussion of in-

335

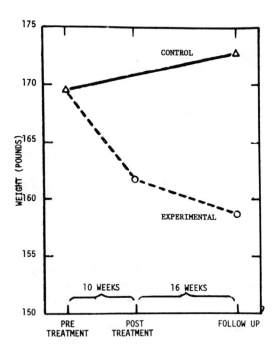

Fig. 1 - Weights for behavior therapy
         and control subjects.

(Harris, 1969)

dividual problems and how to meet them, education in correct eating
habits and related subjects. In 1967, Dr. Richard B. Stuart published his important paper on the behavioral control of overeating[1]. This
study resulted in 8 out of 10 patients losing large amounts of
weight. Although there were no controls, the great loss of weight
over a period of about an year is very impressive. This was accomplished on an outpatient basis with an interview 3 times a week for
4 to 5 weeks, then at 2 week intervals, then monthly sessions, and
finally only as needed. The great contributions that Stuart made
was in the nature of the interviews. These were designed to help
the individual develop positive reinforcing responses. A set of
detailed instructions were given on how to overcome psychological
difficulties and habits that lead to a desire to overeat. For example, how to avoid temptation, how to get help from the family, how
to manage eating at home, and how to manage eating out.

Following Stuart's report, M.B. Harris[2] ran a controlled study
based on behavior modification and also observed a significant weight
loss in a group of overweight female college students. (Fig. 1).

Recent reviews of the results of the behavioral treatment of
overeating conclude that it is the most effective available means of
treating the mildly and moderately obese, Leon[3], Stuart[4]. Stunkard[5] has pointed out that the obese are susceptible to social forces
that affect the effectiveness of behavior modification programs.
He[6] also recently questioned the ability to maintain weight loss, but
Stuart[7] has shown that the results of behavioral treatment are good
when intervention is aimed at both curbing the urge to eat and directly modifying eating behavior when treatment is extended to the
achievement of goal weight, and adequate maintenance services are
provided.

Stuart recently compiled his total approach in a book addressed
to those who are having difficulties with weight control[8]. He has
also published more data on the effectiveness of behavior modification[9] on members of Weight Watchers classes. Data was collected from
721 members of Weight Watchers classes about 15 months after reaching
their goal weights. The group lost an average of 30.1 pounds in an
average of 31 weeks (losing 19% of their body weight). After 15
months, only 28.9% were 11% or more above their goal weights. This
is an extraordinary rate of success.

The failure of medical treatment for obesity before behavior
modification was introduced is illustrated by the report of Christakis[10] who reported the experience of the Bureau of Nutrition in
New York City Health Department in 1967. Over 21% of 2603 patients
dropped out within two weeks and only 9.8% achieved goal weight.

Williams and Duncan[11] compared the results of obese patients
treated in the Royal Adelaide Hospital in Australia with members of

Weight Watchers classes. They found the success rate in the Weight
Watchers classes was clearly superior to the hospital experience,
although the hospital clinic offered individual therapy, anorectic
drugs, diuretics, and inpatient treatment. In the clinic, only 1.6%
of 241 women reached goal weight. In the Weight Watchers classes
21.7% of 5446 women reached goal weight. The clinic had 74% dropouts
and Weight Watchers 43%. The average weight loss in the Weight Watchers classes was about double that in the clinic. They conclude that
it is "highly questionable whether professional resources should be
used at all in the routine treatment of obesity".

The fundamental basis for the improvement in the handling of
obesity by behavior modification techniques is that it focuses on
self-help by the individual in a group. Stuart[12] has defined the
assets of a self-help group as follows:

1. Targets are defined as inappropriate behavior.
2. Developed a faith building morality.
3. Sufficiently informal to attain social objectives.
4. Leader serves as a role model.
5. Members can join and resign with ease.
6. Help members accept their problem as a normal reaction.
7. Payment of fees improves effectiveness of service.

The necessity for the obese individual to develop, strengthen,
and maintain a strong motivation to lose weight cannot be overem-
phasized. Without success in this area, nothing can be accomplished.
However, this is only the beginning. The diet is of equal importance
and must be rigorously controlled by the individual in order to
maintain the negative food energy balance which is essential.

FOOD PROGRAM

There are many factors to be considered in the design of a
successful obesity control food program. The fallacy of the use of
fasting as a method of treating obesity is well illustrated by a
recent study by Johnson and Drenick[13]. They studied 207 morbidly
obese patients. Half were fasted about 2 months, one-fourth less
than 1 month, and one-fourth more than two months. Seventy-nine
(38%) reduced to within 30% of their ideal body weight. The longer
the patient was able to fast, the greater the weight loss. However,
of the entire group, only 7 maintained reduced weight over the
follow-up period and 50% reverted to their original admission weight
within 2 to 3 years. In other words, in terms of maintaining their
weight loss, the fasting program was a failure, and all groups regardless of the length of fast or extent of weight loss returned to
their prefasting weights.

Many reducing diets fail to give attention to nutritional adequacy. They often are grossly distorted diets with high protein, no carbohydrate, or diets with periods of starvation.

If such diets create an energy deficit they, of course, cause weight loss. However, they fail in the important objective of changing the individual's behavior toward food. The weight lost is rapidly regained when the regime is discontinued.

Restricted diets must be followed for long periods of time, if morbidly obese individuals are to attain and maintain normal body weight. Therefore, attention should be given to every aspect of the diet's nutritonal adequacy.

Some of the problems that may be encountered in such programs are well illustrated by the experience with a program which has been called protein-sparing modified fast. This type of program was popularized by a book entitled "The Last Chance Diet", and has been widely used without proper medical supervision. It is based on a very low calorie intake and the use of a variety of liquid protein products of low nutritional quality derived from collagen or gelatin. Although this program results in a large weight loss, the general result has been disastrous. By December 1977, the U.S. Food and Drug Administration had reviewed reports of 40 deaths associated with very low-calorie protein diets[14]. The U.S. Center for Disease Control[15] has found that 15 of these 40 deaths occurred in obese women, aged 25-51, who had lost an average of 83 pounds (40 kg), after being on a liquid digested protein diet for 2 to 8 months. Eleven died suddenly while on the diet and two within two weeks after changing from the liquid diet to regular food. No underlying medical causes that could have caused death were detected and none had a previous history of heart disease or arrhythmias. The remaining deaths are still being investigated. A panel of expert consultants advisory to the U.S. Food and Drug Administration concluded that "very low calorie protein diets are not suitable for use in the absence of careful supervision by medical personnel trained in their use". Twelve of the deaths occurred in patients under medical supervision and 14 had been taking vitamin and mineral supplements.

It is undesirable to use any diet that is severely restricted in energy except under close medical supervision with nutrient and eloctrolyte supplementantion as required. A moderate energy restriction not only makes the regime more acceptable, but also conserves nitrogen. A very restricted diet causes a greater proportionate loss of nitrogen. Since the objective is to lose the maximum amount of adipose tissue, with a minimum loss of nitrogen, the diet should be adequate in protein and only moderately restricted in energy. Furthermore, a severely restricted diet produces the symptoms of semi-starvation including lethargy, inattentiveness, decreased physical and mental activity and a decreased metabolic rate which tends to decrease the loss of adipose tissue.

The proper distribution of energy intake among protein, fat, and carbohydrate is also important. Diets should not be either unduly high or low in any of these nutrients.

In 1975, the Norwegian government was the first to announce a national nutrition food policy in a white paper by the Royal Norwegian Ministry of Agriculture[16, 17, 18]. This comprehensive and important document relates nutritional problems to the food supply and proposes desirable changes, which take into account problems in food production, food marketing, and the economic impact.

Instead of recommending a decrease in energy intake from the present average 2900 Kcals/day, an increase in physical activity is recommended. It is also recommended that the fat intake be reduced to about 35% of the food energy with an increase in the proportion of polyunsaturated fatty acids and a decrease in saturated fatty acids. It is proposed to increase the consumption of high starch foods such as grains and potatoes and reduce the intake of refined sugar.

The Select Committee on Nutrition and Human Needs of the U.S. Senate recently published recommended dietary goals for the United States[19]. This publication has led to a great deal of discussion and differences of opinion among professionals about the proposed changes. Many nutritionists feel that the present U.S. diet of about 42% of the energy from fat, 12% from protein, and 46% from carbohydrate, with 16% of the fat energy from saturated fatty acids, and 24% of the carbohydrate energy from refined sugars is responsible for many of our nutritional problems, with obesity the greatest and most important. The U.S. Senate Committee has proposed goals of 30% of energy from fat (only 10% saturated) and 58% of the energy from carbohydrate (only 10% from refined sugars). It is also recommended that the cholesterol intake be reduced to about 300 mgm/day and salt to about 5 grams per day. The attainment of these goals would make important changes in the food supply. It would mean a considerable reduction in the use of eggs, butter, beef, pork, and sugar with skim milk replacing much of the whole milk.

This general planning is still a long way from an acceptable meal pattern, which must be adjusted to individual habits and preferences. This meal pattern must be analyzed nutritionally and compared with the Recommended Daily Allowances of the National Research Council.

Consideration must also be given to an adequate intake of all vitamins, minerals, and trace nutrients. The accepted basis for nutritional adequacy in the United States is the Recommended Dietary Allowances of the National Research Council's Food and Nutrition Board. These recommended allowances are designed to meet the nutritional needs of practically all normal individuals in the United States[20].

Adequacy is attained through proper food selection and alternates to allow for individual preferences, social and religious customs, cost and availability.

These necessary nutritional considerations form the framework for the eating pattern.

The recommendations of both the Norwegian and U.S. governments are not only that the total fat in the diet should be reduced, but also the proportion of polyunsaturated fatty acids should be increased.

Professional education in nutrition is of the utmost importance, if the obese individual is to be taught how to eat properly. Unless the individual is taught what foods to select and the amounts to be eaten, there is no hope of maintaining good health and a normal body weight. This involves menu planning, portion control, and exact information about the nutrient value of food available in the market.

It is best to confine nutrition educational efforts to teaching what foods to eat, and in what quantities, stressing the fact that ordinary foods in the proper mixture and proper amounts make special dietary or nutritional supplements unnecessary for tthe normal healthy individual.

PHYSICAL ACTIVITY

There are many studies which indicate the great importance of physical activity in obesity and much obesity in young people may be due more to reduced activity than to overeating[21, 22, 23, 24].

Although reduced activity may be a contributory factor to obesity because of decreased energy expenditure, it cannot be the major cause of obesity[25].

The energy expenditure from various forms of physical activity has been carefully measured and interpreted in terms of food and the possible effect on obesity. The Kcal value of 1 kg of adipose tissue is approximately 7488 Kcal (31329 kJ).

Gwinup[26] found that obese women lost a substancial amount of weight on a program of walking without deliberately restricting the diet. However, the walking had to exceed 30 minutes daily over a year or more, and of the 34 subjects who started, only 11 women completed the study.

Stuart[27] has reviewed the use of exercise in the management of obesity. He concludes that the inclusion of an exercise prescription

increases the likelihood that nutritional management will succeed,
and also offers physiological, psychological, and social advantages.

Exercise should be done under medical supervision in order to
avoid exceeding the patient's capacity.

There are several parameters to be considered in measuring the
effectiveness of any obesity control program. The first of these is
whether the program is one which the obese individual will follow
until the body weight approximates the ideal weight. Another, is
whether the individual is emotionally comfortable, stable, and satisfied with the result. Another, is how long ideal weight is maintained. It often happens that the individual gives up the program before
attaining ideal weight for one of many reasons. However, many return
later, and after one or more failures, succeed in controlling their
body weight.

It is almost impossible for the medical practitioner to meet
all of the requirements for success. He usually does not have the
time, knowledge, or facilities to apply the techniques of behavior
modification, either to individuals or groups. He needs dietetic
and nutritional help to develop recipes and food plans that meet nutrient requirements, satisfy the appetite, and avoid diet boredom.
He does not have time to educate the patient in new food habits.

REFERENCES

[1] Stuart, R.B. Behavioral control of overeating. Behav. Res. Ther.
    5: 357-365, 1967.
[2] Harris, M.B. Self-directed program for weight control: a pilot
    study. J. Abnorm. Psychol. 74: 263-270, 1969.
[3] Leon, G.R. Current direction in the treatment of obesity. Psychol.
    Bull. 83: 557-578, 1976.
[4] Stuart, R.B. Behavioral control of overeating: a status report.
    In: "Obesity in Perspective". G.A. Bray, Ed. U.S. Government
    Printing Office, Washington, D.C., 1975.
[5] Stunkard, A.J. Obesity and the social environment: current status
    failure. Prospects Annals, N.Y. Acad. of Sci., 300: 298-320,
    1977.
[6] Stunkard, A.J. Behavioral treatment of obesity: failure to main-
    tain weight loss. In: "Behavioral Self-Management". R. B.
    Stuart, Ed. Brunner Flash Mazel, New York, 1977.
[7] Stuart, R.B. Can they keep it off once they've taken it off.
    In: "Behavior Medicine, Recent Advances in Techniques for Chang-
    ing Lifestyle". P.O. Davidson, Ed. Brunner Flash Mazel, New
    York (in press), 1976.
[8] Stuart, R.B. "Act Thin, Stay Thin". W.W. Norton & Co.., Inc.,
    New York (1978).
[9] Stuart, R.B. and K. Guire. Some correlates of the maintenance of

weight lost through behavior modification. Intl. Jour. of
   Obesity (in press), 1978.

[10] Christakis, G. Community programs for weight reduction: experience
   of the Bureau of Nutrition, New York City, Canad. Jour. Pub.
   Health, 58: 499-504, 1967.

[11] Williams, A.E., and B. Duncan. Comparative results of an Obesity
   Clinic and a Commercial Weight-Reducing Organization. Med. J.
   Aust. 1: 800.802, 1976.

[12] Stuart, R.B. In: Self Help Group Approach so Self-Management.
   Behavioral Self-Management, Strategies, Techniques and Outcome".
   Chapter 12. Brunner/Mazel, New York, 1977.

[13] Johnson, D. and E. Drenick Jr. Therapeutic Fasting in Morbid
   Obesity. Long Term Follow-up. Arch. Int. Med. 137: 1381-1382,
   1977.

[14] FDA Drug Bulletin. U.S. Food and Drug Administration, Rockville,
   Md., Jan.-Feb. 1978.

[15] Center for Disease Control HEW: Deaths Associated with Liquid Protein Diets. Morb. Mort. Wkly. Reports 26(46) 383, Nov. 18,
   1977, 26(52) 443, Dec. 30, 1977.

[16] Royal Norwegian Ministry of Agriculture Report No. 32 to the Storting on Norwegian Nutrition and Food Policy, 1975-76.

[17] Nutrition and Health Appendix to Report No. 32 to the Storting,
   Natl. Nutrition Council, Oslo, 1975-76.

[18] Ringer, Knut. The Norwegian food and nutrition policy. Am. J.
   Pub. Health, 67: 550-551, 1977.

[19] U.S. Senate Select Committee on Nutrition and Human Needs, Dietary
   Goals for the United States, 2nd ed., U.S. Govt. Printing Office,
   Washington, D.C, 1977.

[20] National Academy of Sciences. Recommended Dietary Allowances,
   8th ed., Washington, D.C., 1974.

[21] Mayer, J. and D.W. Thomas. Regulation of food intake and obesity,
   Science, 156: 328, 1967.

[22] Durnin, J.V.Y.A. Activity patterns in the community, Canad. Med.
   Assn. J., 96: 882, 1967.

[23] Johnson, M.L., B.S. Burke and J. Mayer. Relative importance of
   inactivity and overeating in the energy balance of obese High
   School girls, Am. J. Clin. Nut. 4: 37, 1956.

[24] Bullen, B.A., R.B. Reed, and J. Mayer. Physical activity of obese
   and nonobese adolescent girls appraised by motion picture sampling, Am. J. Clin. Nut. 14: 221-233, 1964.

[25] Conzolazio, C.F. and H.L. Johnson. Measurement of energy cost in
   humans, Fed. Proc. 30: 1444-53, 1971.

[26] Gwinup, Grant. Effect of exercise alone on the weight of obese
   women. Arch. Intern. Med. 135, 676-680, 1975.

[27] Stuart, R.B . Exercise prescription in weight management: advantages, techniques and obstacles. Obesity and Bariatric Medicine
   4: 16-24, 1975.

TEN YEARS IN THE TREATMENT OF OBESITY

CARRIED OUT IN SANATORIAL CONDITIONS

T. Tashev

Director, Inst. of Nutr. at the Medical Academy

Boul. Dim. Nestorov 51 Sofia 1431 - BULGARIA

The treatment of obesity is one of the problems in medicine still not completely solved. There are plenty of reasons for that: the different periods during which the treatment is carried out; insufficient taking into consideration of the curative undertakings with the disease pathogenesis and with the individual peculiarities of the case, etc. Besides, the treatment of this most widely spread nutritive-metabolic disease is usually in close relation with the wish of the patient himself to be treated. However, lately the number of obese persons calling for doctor's aid grows considerably. Certainly, it is very difficult to response to the needs, if a general prophylactic and treatment is not undertaken by all medical services of each country.

The present report gives the total results of treatment of about 4000 obese patients, with II, III and IV stage of obesity, carried out in two seaside resorts — Nessebar and Kiten, in Bulgaria, for a period of 10 years (1968-1977).

MATERIALS AND METHODS

The data concerns 842 patients. The number of patients treated in both sanatoria is neraly equal — 1,831 in Nessebar and 2,011 in Kiten, with women being 80 per cent of all patients. The overweight is calculated in percentage according to Sharman's table (for men and women), while the rate of obesity is determined after Egorov and Levitskji.

Out method of treatment is based on: (i) moderately reduced diet — 1,460 cal. daily with composition: 99 g proteins, 47 g fats and 158 g carbohydrates; (ii) regimen of increased motor activity, which results in 4,336 cal. energy consumption in men, and 3,646 cal. in women and energy loss about 2,100 cal. in women and 2,800 in men. We have proceeded following the conception that a decisive etyological role for the ordinary obesity belongs to the positive caloric disbalance related to physical inactivity and a higher nutritive input, characteristic for developed countries.

Our sanatoria are situated at the seaside. The day starts with 15 min. gymnastics, and after breakfast it continues with the complex of fixed curative gymnastics for a period of 30 to 60 min., considered with individual state of the cardio-vascular system and training. After a relax of 1 to 1 1/2 hr., including strolls along the seaside, the so called "chair exercises" are carried out for 30 min. This regimen is observed until lunch. After lunch, follows rest between 1 to 3pm and afterwards 2 hrs sports — basketball, volley-ball, football, handball. After dinner our patients dance in the hall for physical culture and during holidays in some restaurants. Sometimes special dinners for fun and gaiety are organized. Once a week excursions are arranged covering 10 to 15 km first and at the end of the period reaching up to 40-45 km in both directions. Some of the patients make also some water procedures.

A number of investigations are performed before sending the patients to the sanatoria, during the treatment and after their coming back home.

Working according to the described curative complex, we have taken into consideration the fact established by Slabochova et al., that with moderately reduced diet, the physical loading detains the positive nitrogen balance for a long time and determines peculiar "saving" of protein with an effective weight decrease.

RESULTS AND DISCUSSIONS

The mean absolute decrease of body weight for a period of two months is 13,540 kg for women and 16,700 kg for men. Determination of the weight and percentage of teh active body mass and the weight and percentage of body fat, as well as their interrelation is performed after measuring four standard skin-subdermal flaps and calculating after Mohr-Milev formula. It is clear that the body weight decrease is achieved mainly on the account of the body fat, which decreases both relatively and in absolute values. The results received give us ground to agree with Slabochova, that by physical laoding a positive nitrogen balance is obtained and the obese patients lose their weight extremely on the account of the adipose tissus; the patients pass from a higher to lower stage of obesity. A small part of the patients have reached normal weight during their stay in the sanatoria.

An important result of the treatment is the observed change of serum lipid content, parallel to the body weight reduction. There is a significant (p <0,001) decrease of the total cholesterol, the cholesterol in β-lipoproteins, β-lipoproteins, cholesterol/lecitin index, total fats, esterified fatty acids, triglycerides, in conjunction with a significant increase of the nonesterified fatty acids. A study of the 25th day of the stay shows that the decrease is more expressed at the beginning, while at the end of the two month period there is a tendency for turning back to the initial values. However, the difference of the last investigation in comparison to the initial one remains significant. These favorable changes in the lipidogram are more expressed when body weight reduction is greater.

In our century the most conspicuous factor for caloric disbalance facilitating obesity of all age groups is the decreased physical activity. The ergonomists recon that in the developed countries more than 99 per cent of the beneficial work for society is done by machines, and less than 1 per cent by human muscle power. In 1961 Thomson maintained that gain of weight is related to the physical activity decrease. Bullan et al. in 1964 and Harper in 1969 established that obese children eat less than normal weight children, but at the same time are considerably more inactive. However, it is a very complicated problem to distinguish up to what point the physical inactivity is the reason or consequence of the body overweight. Recently, Mayer (1968) underlined again the importance of activity, as a means for preventing the appearance of obesity and its cardiovascular complications. On the other hand, Bloom and Eidex (1967) calculated that obese people are on foot 15 per cent less daily and 17 per cent more in bed in comparison to normal people. The Russian physiologist Pavlov called the activity "joy of the muscles". Modern technical progress takes more and more of this human joy. We believe that physical activity is a reasonable alternative of the unfavorable effects of the technical progress, though we are aceptical that they can be completely compensated.

Bearing in mind the conception for the leading etiological role of the physical inactivity, we put on accent on the strenuous motor regimen. For a period of ten years we have applied seven dietetic regimens. The comparative assessment convinces us that all hypocaloric diets are suitable for obese patients if they are combined with a regimen of increased motor activity. It should be underlined that when such physical regimen is combined with standard caloric food, it leads to body weight reduction, similar to that resulting from more or less low caloric diets. In most of the time we have applied the moderate caloric diet, given in the begining of this report, ensuring sufficient proteins, vitamins and mineral substances, and restricting the calorigenic carbohydrates and fats.

On the basis of the given total data, which are result of a ten year observation, the following conclusions can be made:

1. Rehabilitation of obese persons, based on moderately reduced diets in combination with a regimen of strenuous motor activity leads to significant decrease in body weight.

2. Reduction of overweight is achieved mainly on the account of body fat stocks, whose degradation is stimulated for the needs of endogenous nutrition.

3. The favourable changes in the ratio AMB : BM (active body mass : body mass) in favour of the active body mass is combined with favourable changes in the spectrum of the serum lipids.

GROWTH OF ADIPOSE TISSUE

E.M. Widdowson

Dept. of Medicine, Addenbrooke's Hospital

Cambridge, ENGLAND

Fat is deposited in the body of the foetus in two main forms. First there are the phospholipids which are an essential part of the structure of the cell membranes and nervous tissue, and these are synthesized by the foetus from early in gestation. The fatty acids required reach the foetus through the placenta. Then, at about 6 months gestation triglycerides begin to be deposited in the adipose tissue. Some of the fatty acids required for this undoubtedly come from the mother's circulation, but at about this time the foetus itself begins to synthesize the non-essential fatty acids from glucose and aminoacids, and it has been suggested that it is able to do this because the endocrine pancreas of the foetus then begins to secrete insulin. One of the properties of insulin is to promote lipogenesis, and the foetus cannot begin to store triglyceride in its adipose tissue until its pancreas begins to secrete insulin. There are difficulties about this idea however because although the times coincide in the human foetus, they do not do so in other species.

We do not know how much of the fat deposited in the adipose tissue of the human foetus during the last three months of gestation consists of fatty acids from the maternal circulation and how much from those it synthesizes itself; it used to be believed that the foetus synthesizes at least 80% of its fat from glucose during the last trimester, but recent work by Professor Hull and colleagues suggests that transfer of fatty acids from the mother provides considerably more than 20% of the total.

Adipose tissue in the foetus is of two kinds, generally termed brown and white. Brown adipose cells have a central nucleus and numerous small vacuoles of fat surrounded by large numbers of mito-

chondria. White adipose cells also deposit fat in small vacuoles at
first, but later these fuse into larger droplets, until ultimately
the cell contains only a single fat vacuole which eventually expands
the cell so much that the cytoplasm is reduced to a thin layer and
the nucleus becomes confined to the periphery of the cell.

The distribution of brown and white adipose tissue in the foetus
and newborn differs from one species to another, as does the stage of
development each has reached at birth. In the newborn rabbit there
are large interscapular pads of brown adipose tissue, but very little
white fat. In the rat there is very little fat of either type. In
the human foetus both brown and white adipose tissues are well-de-
veloped. Brown adipose tissue surrounds the muscles and blood vessels of the neck, the great vessels entering the thoracis inlet,
and the aorta, the kidneys and adrenals. White adipose tissue is
almost entirely subcutaneous. Brown adipose tissue produces extra
heat when the body is exposed to cold. First the temperature of the
brown adipose tissue goes up, and this raises the temperature of the
blood passing through it and hence of all the blood and of the body
as a whole. It is in fact a built-in central heating system for the
newborn baby, and some species of animals as well, when they leave
the warm uterus and enter the cooler world outside.

White adipose tissue provides a store of energy to be used as
a standby when expenditure may be exceeding intake. Like all living
tissues it releases heat as a by-product of its metabolic activity,
but it does not release heat specifically in response to cold. A
thick layer of subcutaneous fat provides a certain amount of thermal
insulation, but it is not a very good insulator. Fur is better.

Between the 6th and 9th month of gestation the amount of fat in
the body of the human foetus increases very rapidly, more rapidly
than anything else, so that the percentage of it in the body rises
from about 2 at 6 months to about 16 at term. However, the amount
of fat in babies at birth varies a great deal, just as it does in
adults. It is because fat in the body normally increases so rapidly
during the last part of gestation that preterm babies have less fat
than full term ones. However, the proportion of fat in the body is
very variable even in babies born at full term. We found this in
the original series of bodies of stillborn full term infants that
we analyzed (Widdowson and Spray, 1951), and a more recent study by
Southgate and Hey (1976) has confirmed it. Southgate and Hey also
separated the subcutaneous and deep body fat and found that 80% of
the fat at term was in subcutaneous adipose tissue.

Fat mothers tend to have fat babies (Whitelaw, 1976). Many fat
infants born of fat mothers acquire more normal proportions later,
but an unusually high proportion of them remain fat at least until
7 years.

After birth another influence comes in, the mother's conscious
desire to give her baby the best, the best in this case being as much
food as it will take. Some years ago it was firmly believed that
the overfed fat baby was more likely to become a fat child and a fat
adult than the baby of average weight. Now we are not so sure. Several prospective studies in the United States, Britain and Sweden
have shown that weight during the first six months after birth may
not be an important determinant of shape and size during later childhood. Not unexpectedly some children who were obese and overweight
in infancy were still fat at 7 and 8 years, and many of these had
been fat babies at birth. Many of the overweight infants, however,
had moved into the weight range normal for their age by the time they
were 7 years old, and at the same time others, who were of average
weight during infancy had become overweight.

Recently Durnin and McKillop (1978) have described a follow-up
study of 102 infants whose weights and lengths had been measured at
some time during the first 2 years after birth. Body fat was measured when the children were 13-17 years old. There was no relationship between fatness in infancy and in adolescence either in boys
or girls. On the other hand if fat adolescent girls are selected
these had almost always been fat babies. It would be interesting
to know whether these were the ones that were particularly fat when
they were born.

When I got thus far in preparing this paper I had a strange
feeling of deja-vu. I looked back at the Proceedings of the Second
International Congress of Nutrition in Amsterdam in 1954, when there
was a symposium on Obesity. I gave a paper on Reproduction and
Obesity, and what I said then was:

"The fat mother tends to have a large fat baby. The big baby
of the fat mother is hungry and ready to take more than the average
amount of milk. The child grows fast and is at all ages heavier
than a child with a smaller birth weight. As growth ceases we must
suppose that the big girl, who had a big appetite when she was growing, continues to eat just that little extra each day, and as the
surplus calories can no longer be expended on rapid growth they
will be laid down as fat. The bonny young woman now only has to get
married, (at least she had to in 1954), and the obesity cycle is
complete" (Widdowson, 1955).

Are we much further on now?

BIBLIOGRAPHY

Durnin, J.V.G.A. and McKillop. The relationship between body build
        in infancy and percentage body fat in adolescence - a 14 year
        follow-up on 102 infants. Proc. Nutr. Soc. (in press), 1978.

Southgate, D.A.T. and E.N. Hey. Chemical and biochemical development
    of the human fetus. In: "The Biology of Human Fetal Growth".
    D.F. Roberts and A.M. Thomson, Eds. London, Taylor and Fran-
    cis, 1976.
Whitelaw, A.G.L. Influence of maternal obesity on subcutaneous fat
    in newborn. Brit. Med. J. 1: 985-986, 1976.
Widdowson, E.M. Reproduction and Obesity. Voeding, 16: 94-102,
    1955.
Widdowson, E.M. and C.M. Spray. Chemical development in utero.
    Arch. Dis. Childh. 26: 205-214.

THE RELATION OF OBESITY TO CARDIOVASCULAR DISEASE

F.X. Pi-Sunyer

Dept. of Medicine - St. Luke's Hospital Center

New York, New York 10025 - U.S.A.

With the increasing prevalence of overweight in the technologically advanced countries has come a preoccupation with weight-con-
trol programs, and great amounts of money are being expended by obese
patients to try to attain the magic goal of "desirable" body weight.
In recent years, however, there has been considerable controversy
about the actual effect of obesity on health and therefore about the
necessity of preventive measures.

The 1959 U.S. Build and Blood Pressure Study[1] showed a very considerable proportion of adult Americans to be overweight. It is
disquieting to note that Americans weighed even more at the time of
the Health Examination Survey of 1960-62[2], and that preliminary data
from the HANES (Health and Nutrition Examination Survey) of 1971-74
shows that the trend has continued, with Americans being heavier than
ever before[3].

Reports from a number of other technologically advanced countries show similar trends. For example, data from the Nutrition
Canada Survey published in 1973 show that more than 50 percent of
Canadian adults are overweight[4]. Similar high prevalence, increasing
with age and especially affecting women, were also found in a 1953
survey in Uppsala, Sweden[5]. In 1971, Walker and Richardson reported
that the average weight of young and middle-aged Englishmen were 15
pounds heavier than 30 years previously, with only a slight increase
in height in the same period[6].

In view of this accelerating trend towards heavier weights, the
question as to whether obesity increases morbidity and mortality becomes of particular interest. I am concerned here with the effect
of obesity specifically on cardiovascular disease, and I will confine
myself to this topic.

Basically, as Rimm[7] has suggested, the controversy of whether obesity is harmful or not can be separated into two questions: 1. Does obesity per se "cause" increased morbidity and mortality? or, 2. Is obesity an outward sign of the quantity of food intake and genetic constitution which only co-exists with other conditions, such as hypertension, and has no direct role in causing morbidity and mortality? One model (A) considers obesity as a risk factor along with hypertension, cholesterol level and others which are thought to play a direct role in morbidity and death from cardiovascular disease. If obesity is a risk factor as depicted in Model A, it should operate independently of the other risk factors.

If another model (B) is valid, then obesity is assumed not to create directly any pathological disorder in the vascular system. This implies that whether "excess" lipids are deposited in fat cells per se and cause obesity has no bearing on morbidity. Though obesity might also give rise to excessive deposition of lipid in the vascular system of some people, this would not always occur and thus obesity would not be an independent risk factor. Under these c rcumstances, obesity could serve as a useful marker for the possible presence of other noxious medical conditions, but obesity could just as commonly occur in their absence.

As one examines the published literature, various retrospective studies seem to favor Model A. However, a number of prospective studies make a direct cause and effect relationship unlikely, and suggest that Model B may be a more accurate scheme.

I would like to begin discussion of the epidemiological data by concentrating on life insurance studies. These studies were the first to suggest that obesity is a major risk factor in mortality and particularly mortality from cardiovascular disease. This was first highlighted by the Metropolitan Life Insurance Company's statistical analysis of death rates in different age groups[8]. For this analysis, all of the people in an age group, segregated by sex, were pooled and the number of deaths determined. The mortality rate (deaths per unit population) was represented as 100%. The population was then divided into subgroups, according to deviation in weight from the mean, and compared to the death rate for the entire group. The minimum death rate occurs at below average weight. Thus, life expectancy is increased in individuals who are lighter than average. Conversely, heavier individuals have an excess mortality when compared to lighter members of the peer proup. This suggests that excess weight has a detrimental impact on longevity.

Data depicting percent excess mortality as a function of percent overweight in men in age groups 15-39 and 40-69, shows mortality to be significantly elevated over baseline mean in the obese, and in-

creasing with increasing overweight to reach 42% excess mortality at
30% overweight by Metropolitan Life Insurance median standard. Similar but somewhat lower excess mortality can be seen in women.

   There are few studies that provide a foundation for the proposi-
tion that weight reduction increases longevity. Since it is so difficult to disentangle obesity from other factors which contribute to
mortality and morbidity, it is particularly helpful to examine the
effects of weight loss in formerly obese subjects. The Metropolitan
Life Insurance Company, however, does have data on a group of people
who had initially received sub-standard insurance because they were
overweight but who subsequently were issued policies after they had
reduced their weight and remained reduced[9]. The mortality of those
25% overweight was reduced from 128% to 109% of expected and of those
35-40% overweight from 151% to 96% of expected. Thus, life expectancy in these "reduced" obese improved to that of insured people
with a "standard" risk.

Dublin[10] summarized the mortality experience of 2300 people and
also concluded that excess mortality associated with overweight is
reversible by weight loss.

   Seltzer[11] re-examined the data of the 1959 Build and Blood Pressure study on mortality by transposing weight for height categories
into comparable ponderal indexes or $\dfrac{\text{Height}}{3\sqrt{\text{weight}}}$. The ponderal index
is a frequently used measure of body shape or build. The rationale
of this index is that weight is a measure of volume and that   volume
increases according to the cube of the dimensions. It is a better
index of linearity and laterality that simple broad height-weight
categories. Seltzer's analysis demonstrates that instead of a
straight line relationship of increasing mortality with increasing
weight-for-height, as indicated by the insurance companies, there is
a curvilinear relation of mortality with linearity and laterality of
body form in which there is no significant excess of actual mortality
over expected until a ponderal index of 11.6 or lower is reached (a
level of extreme laterality where men are predominantly frankly
obese). At this level, the mortality ratio rises in a very steep,
almost geometric, progression. Thus, according to Seltzer, there is
no evident risk to health of fat until severe obesity, i.e., in those
who are overweight by more than 30% above desirable weight.

   If one utilizes a different and possibly less controversial
measure of obesity, body mass index, which is the $\dfrac{\text{weight}}{\text{height}^2}$, mortality
does not become prominent until a body mass index of 30 or higher[12],
while an index over 27 is considered overweight, confirming Seltzer's
contention that only at severe overweight of 30% or more above desirable weight is longevity affected.

A number of retrospective studies have attempted to correlate the association of overweight with atherosclerotic changes in the cardiovascular system. I will comment very briefly on five of these. Wilens in 1974[13] carried out a study on 1250 subjects. The prevalance of arteriosclerosis was measured on 1000 consecutive autopsies to which he added 250 additional autopsies in "obese subjects". Presence or absence of obesity was defined by measurement of the abdominal panniculus, with obesity considered present at a thickness greater than 3 centimeters. 395 subjects out of the 1250 were defined as obese. Their prevalence of "severe" atherosclerosis was 2.5 times greater than in lean individuals. This relationship held at all decades of age and even if hypertensives were excluded.

Wilkins[14] in 1959 studied 500 consecutive autopsies in Pittsburgh hospitals. Weight over height was used as a measure of overweight, with 77 of 500 cadavers studied being categorized as obese. Obesity was found to have an effect on the severity of atherosclerosis in the coronary arteries of males, but not females. Overall, the effect was found to be minor.

Studies by Faber and Land[15] in 1949, Spain, Bradess and Huss[16] in 1953, and Kannel and Gordon[17] in 1974 showed no relation between obesity and atherosclerosis. Faber examined 408 aortas of autopsied individuals and could detect no effect on either atherosclerotic lesions or cholesterol content of the aorta. Spain, Bradess and Huss conducted a necropsy study on 111 males under the age of 46 years. Thirty-eight of these died suddenly and unexpectedly from coronary artery disease while the remaining 73 died from sudden, unexpected, violent death. The authors could find no relationship between obesity and the presence of atherosclerotic lesions in the coronary arteries. Finally, Kannel and Gordon in 1974 did a retrospective analysis of subjects from the Framingham Study who died. By using the relative weight a decade before death, they avoided the problem of terminal weight loss which has bedeviled other studies. They found that the effect of overweight on the heart was greater in women than in men and correlated more highly with heart weight and ventricular hypertrophy than with the extent of coronary atherosclerosis. The authors concluded that obesity imposes more of a workload than an atherosclerotic burden on the hearth.

Thus, of these five retrospective studies, only one, that of Wilens, was positive for a definite relationship between coronary atherosclerosis and obesity.

Adequately conducted prospective studies of the effect of obesity on atherosclerosis and cardiovascular disease are more numerous and because of the larger number of subjects and less possibility of bias are probably more valuable in sorting out the effect of obesity on cardiovascular disease. I will briefly review nine such studies.

In 1963, Spain et al.[18] reported on a prospective study on 5000 men studied for 5 years. They found that while overweight individuals who concurrently also were hypertensive and/or diabetic had a higher prevalence of heart disease, the prevalence was not increased in the uncomplicated obese.

Stamler et al.[19] in 1966 published a 10-year report on a prospective study of 756 middle-aged male employees in Chicago. Stamler expressed weight as "relative weight", i.e., weight to height. Thus, those at the median weight for their height were assigned a relative weight of 100. The coronary heart disease incidence rates were lowest in those with relative weights below 90, were 33 percent higher in those with relative weights in the range 90 to 109, 68 percent higher in those with relative weights of 110 to 129, and 180 percent higher in those with relative weights of 130 or above. The data were essentially the same for women.

Dunn et al.[20] in 1970 gave a 14-year report of a prospective study on 13,148 males. They found an association between overweight at greater than 25% and both hypertension and diabetes, but they could find no association between overweight and coronary artery disease.

Rosenman et al.[21] also in 1970 presented a 4-1/2 year followup on a prospective study of 3182 males in California. Studying the risk factors associated with the emergence of coronary artery disease, they could find no statistically significant difference related to overweight as expressed by ponderal index.

Chapman et al.[22] in 1971 published a 15-year report on the incidence of myocardial infarction and angina pectoris in 1859 Los Angeles civil servants. They found that the age-adjusted rates of myocardial infarction showed a trend of increased risk as ponderal index changed from low to high weight in men under 40 years. However the rate was not consistent nor was it statistically significant. Also, while increased weight in men over 40 years seemed to increase angina pectoris, the association was not significant.

Heyden et al.[23], in 1971, reported on 3102 males followed for 7 years in Evans County, Georgia. Using the index $\frac{Weight}{Height^2} \times 100$, they found that overweight was positively correlated with coronary heart disease in smokers but not in nonsmokers.

Carlson and Bottiger[24], in 1972, described their experience with 3168 men after 9 years in the Stockholm prospective study. They found that increasing weight/height ratio could not be considered a risk factor for coronary heart disease.

Keys et al.[25] in the International Cooperative Study of Cardiovascular Epidemiology studied a large group of men in three different
environments, 2442 in the the USA, 2439 in Northern Europe (Holland
and Finland), and 6519 in Southern Europe (Italy, Greece and Yugoslavia). The investigators used body mass index (Weight/Height^2) and
skinfolds as the criterion for presence or absence of obesity. Of
the various indexes of relative weight studied, body mass index is
the most completely independent of height and is possibly the best
in correlating with a direct measure of body fatness[25].

For definition of coronary disease, the various categories of
coronary heart disease were condensed into two: "Hard", meaning death
from coronary heart disease or definite myocardial infarction and
"Other", which included angina pectoris and clinical judgement of
coronary heart disease. Disregarding other variables, an excessive
incidence of coronary heart disease was associated with overweight
and obesity in the USA and S. Europe but not in N. Europe. However,
multivariate analysis of the data showed that no measure of overweight
or obesity made a significant contribution to future coronary artery
disease when the factors of age, blood pressure, serum cholesterol,
and smoking were eliminated. There was, however, a significant relationship between blood pressure and relative weight and fatness in
this study. Since risk of developing coronary heart disease is universally agreed to be related to blood pressure, it is essential to
remove the confounding effect of blood pressure to evaluate obesity
as an independent contributor to risk.

There is evidence to suggest that overweight may have its major
influence on mortality in men age 40 years or younger. If this is
so, then Keys, who examined a cohort of men aged 40 to 59 at the
start, could not be expected to detect much effect of obesity on
coronary heart disease, because he looked at the wrong group, one
which was already too old at the start of the study to show the effect.

The Framingham study, conducted in Framingham, Massachusetts,
over a period of 20 years, is a particularly rich prospective study
of the determinants of cardiovascular disease morbidity and mortality[26]. Biennial exams were done on 5209 men and women. The incidence
of heart disease as manifested by angina pectoris, sudden death,
myocardial infarction, and cerebral vascular disease were followed,
as were various characteristics that might be likely risk factors
for the development of heart disease. These included relative weight
and skin-fold thickness.

For cardiovascular disease in general, there was a highly significant but modest association of relative weight with incidence,
which was greater in women than men and diminished with advancing
age. Comparing multivariate with univariate coefficients suggests
that half the overweight effect in men and two-thirds in women was

mediated through blood pressure, carbohydrate intolerance and cholesterol alterations.

When the data on cardiovascular disease was broken down to clinical manifestations, namely congestive failure, intermittent clau-
dication, atherothrombotic brain infarction and coronary heart disease, only for coronary artery disease in men did there appear to
be a sizeable unique contribution of obesity, not mediated through
its atherogenic accompaniments. In women, there appeared to be some
independent contribution of obesity to congestive failure, but none
for coronary heart disease.

Looking at the coronary artery disease in more detail, the obesity effect was found to be highest for angina pectoris and sudden
death and weakest for myocardial infarction. The coefficients for
angina and sudden death in men show rather substantial gradients of
risk proportional to relative weight.

In summary, although obesity was shown in this study to make a
significant contribution to cardiovascular morbidity, its effect was
modest in comparison to other identified risk factors. Thus, relative weight ranks rather low in men compared to other risk factors.
In women, it ranks somewhat higher.

The impact of obesity on mortality from cardiovascular disease
in the Framingham Study is very small and, in fact, in the 65-74
year category, there are more deaths at the lower relative weights.

I wish finally to review the Manitoba study of Rabkin, Mathewson, and Hsu[27]. They studied a cohort of 3983 men with a mean age
at entry of 30.8 years and compared body mass index at entry with the
26-year incidence of ischemic heart disease. After adjustment for
the effects of age and blood pressure, body mass index was a significant predictor of the 390 cases of ischemic heart disease. The
association with weight was most apparent in men less than 40 years
of age and was not evident until 16 years of followup. Thus, the
authors concluded that overweight is a definite risk factor for
cardiovascular disease, but is so primarily in younger men and after
long periods of observation.

With regard to the length of observation of a cohort, both the
Framingham Study and the Los Angeles Heart Study noted a stronger association between body weight and ischemic heart disease in their
later reports[17, 22] than in their earlier ones[28, 29]. The longer observation period needed to show the effects of body weight may reflect the time required for younger men to reach the age of higher
risk. It is thus possible that body weight could be considered a
"long term" risk factor for ischaemic heart disease, one that perhaps
requires the development of a certain amount of coronary atherosclerosis in order to exert an independent effect. Certainly, this study
reiterates that the obesity risk is more acute in the younger age

group and that 5-10 year prospective studies may be inadequate to
fully assess the risk of obesity on health.

Thus, of the ten prospective studies reviewed, seven are negative for an association of obesity and cardiovascular disease, one
shows a positive correlation in smokers but not in non-smokers, one
shows a modest correlation between obesity and cardiovascular morbidity in men but not in mortality, and one shows a definite association in young men followed over 26 years.

I have tried to briefly summarize the principal data existent
on the relationship of obesity and cardiovascular disease. The data
are conflicting, but the preponderance of evidence seems to suggest
that obesity, as an independent risk factor, is either not important
or minimally important in producing morbidity and mortality from
cardiovascular disease.

This does not mean, however, that obese individuals are not at
risk; it means that obese individuals are at risk when other risk
factors are present. These other risk factors include most impor-
tantly high blood pressure, and then serum low density lipoprotein
cholesterol, diabetes mellitus and serum triglycerides. There is,
I believe, universal aggreement that hypertension, hyperlipidemia,
and impaired glucose tolerance are often associated with obesity.
The data from Framingham[17] suggest that for each ten percent gain in
weight there was noted, on the average, a 6.6 mm HG rise in systolic
blood pressure, a 2 mg/dl increase in blood glucose, and an 11 mg/dl
increment in cholesterol. In women, the changes roughly are half
this order of magnitude. These changes held at all ages and at all
levels of adiposity.

The studies at Tecumseh, Michigan[30] and at Framingham[31] have
both shown well the correlation between obesity and hypertention.
The correlation coefficient was approximately r = 0.3 between blood
pressure and relative weight. In Bjerkedal's Scandinavian study on
67,000 adults[32], there was an increase of 3 mm of systolic and 2 mm
of diastolic pressure for every 10 kg increase in body weight.

Thus, if individuals are overweight and also manifest other
cardiovascular risk factors of a metabolic nature, it seems reasonable that the first line of therapy would be weight reduction in an
attempt to lower or eliminate these risk factors. I have already
quoted the beneficial effects of weight loss in the Framingham study
on blood pressure, cholesterol, and blood glucose. Many other studies
have further documented this beneficial effect. Although not all
patients will improve risk factors by weight loss, enough will that
it is worth trying in all.

Also, the recent report by Gordon et al.[33] suggests that in
women there is an inverse association between relative weight and

high-density lipoprotein cholesterol. Thus, as women become overweight , their HDL cholesterol decreases. Since HDL cholesterol has
been associated with higher coronary risk, this is a relationship
that should be further investigated.

Finally, it may be that a change in body weight after full
linear growth has been achieved may be more important in producing
cardiovascular disease than obesity per se. There is evidence that
life-long obesity is different and has different consequences for
morbidity than obesity that begins after adulthood. The studies of
Heyden[34] and Abraham[35] suggest that cardiovascular and hypertensive
disease incidence is greater in those who have become overweight as
adults than in those who have been overweight since childhood. More
studies are required to elucidate this interesting possibility.

REFERENCES

[1] Society of Actuaries, 1959 Build and Blood Pressure Study, Chicago,
    Ill.
[2] National Center for Health Statistics, Weight by Height and Age of
    Adults, United States, 1960-1962. Vital Health Statistics -
    PHS Publication No. 1000 - Series 11, No. 14, May, 1960.
[3] National Center for Health Statistics. Height and weight of adults
    18-74 years of age in the United States. Advance Data from
    Vital and Health Statistics, No. 3, pp. 1-8, November 19, 1976.
[4] Nutrition Canada National Survey, Nutrition: A National Priority,
    Information Canada Catalogue No. H 58-36, 1973, Ottawa, 1973.
[5] Bjurulf, P. and Lindgren, G., in: Occurrence, Causes and Prevention
    of Overnutrition. Symposium of the Swedish Nutrition Foundation
    II. A Preliminary Study on Overweight in the South of Sweden,
    1963, p. 9.
[6] Walker, A.R.P. and Richardson, B.D., Overnutrition in Children.
    Lancet 2: 1146, 1971.
[7] Rimm, A.A., Controversies in Medicine - Is Obesity Harmful? Obe-
    sity/Bar. Med. 2: 140-144, 1973.
[8] Build and Blood Pressure Study, 1959, Society of Actuaries, as
    derived in Mortality among overweight men. Stat. Bull. Metropol.
    Life Ins. Co. 41: 6 February, 1960 and in Mortality among over-
    weight women. Stat. Bull. Metropl. Life Ins. Co. 41:1, March,
    1960.
[9] Build and Blood Pressure Study, 1959, Society of Actuaries, as
    derived in Overweights benefit from weight reduction. Stat.
    Bull. Metropol. Life Ins. Co. 41: 1, April, 1960.
[10] Dublin, L.I., Relation of obesity to longevity. New England J.
    Med. 248: 971-974, 1953.
[11] Seltzer, C.C., Some re-evaluations of the Build and Blood Pressure
    Study, 1959, as related to ponderal index, somato type and mor-
    tality. New Eng. J. Med. 274: 254-259, 1966.
[12] Bray, J., "The Obese Patient", W.B. Saunders & Co., Philadelphia

1976, p. 219.

[13] Wilens, S.L., Bearing of general nutritional state on atherosclerosis. Arch. Int. Med. 79: 129-147, 1947.

[14] Wilkins, R.H., Roberts, J.C. Jr. and Moses, C., Autopsy studies in atherosclerosis. III. Distribution and severity of atherosclerosis in the presence of obesity, hypertension, nephrosclerosis and rheumatic heart disease. Circul. 220: 527-535, 1959.

[15] Faber, M. and Land, F., Human Aorta: influence of obesity on the development of arteriosclerosis in the human aorta. Arch. Path. 48: 351-361, 1949.

[16] Spain, D.M., Bradess, V.A. and Huss, G.A., Observations on atherosclerosis of the coronary arteries in males under the age of 46: A necropsy study with special reference to somato types. Ann. Intern. Med. 38: 254-277, 1953.

[17] Kannel, W.B. and Gordon, T., "Obesity and cardiovascular disease: The Framingham Study in Obesity", Burland, W., Samuel, P.D. and Yudkin, J., eds., Churchill Livingstone, London, 1974.

[18] Spain, D.M., Nathan, D.J. and Gellis, M., Weight, body types and the prevalence of coronary atherosclerotic heart disease in males. Amer. J. Med. Sci. 245: 63-69, 1963.

[19] Stamler, J., Berkson, D.M. and Lindberg, H.A., Coronary risk factors. Med. Clinic North America, 50: 229-254, 1966.

[20] Dunn, J.P., Ipsen, J., Elsom, K.O., and Ohtani, M., Risk factors in coronary artery disease, hypertension and diabetes. Am. J. Med. Sci. 259: 309-322, 1970.

[21] Rosenman, R.H., Friedman, M., Straus, R., Jenkins, C.D., Zyzanski, S.J. and Wurm, M., Coronary heart disease in the Western collaborative group study. J. Chron. Dis. 23: 173-190, 1970.

[22] Chapman, J.M., Coulson, A.H., Clark, V.A., and Borun, E.R., The differential effect of serum cholesterol, blood pressure, and weight on the incidence of myocardial infarction and angina pectoris. J. Chron. Dis. 23: 631-645, 1971.

[23] Heyden, S., Cassel, J.C., Bartel, A., Tyroler, H.A., Hames, C.G. and Cornomi, J.C., Body weight and cigarette smoking as risk factors. Arch. Int. Med. 128: 915-920, 1971.

[24] Carlson, L.A., and Bottiger, L.E., Ischemic heart disease in relation to fasting values of plasma triglycerides and cholesterol. Stockholm prospective study. Lancet 1: 865-868, 1972.

[25] Keys, A., Aravanis, C., Blackburn, H. et al., Coronary heart disease: Overweight and obesity as risk factors. Ann. Int. Med. 77: 15-27, 1972.

[26] Kannel, W.B. and Gordon, T., Some determinants of obesity and its impact as a cardiovascular risk factor. In: "Recent Advances in Obesity Research: I". A. Howard, ed., Newman Publishing Co., London, 1975, pp. 14-27.

[27] Rabkin, S.W., Mathewson, R.A.L. and Hsu, P.H., Relation of body weight to development of ischemic heart disease in cohort of young North American men after a 26 year observation period: The Manitoba Study. Am. J. Cardiol. 39: 452-258, 1977.

[28] Kannel, W.B., Dawber, T.R., Kagan, A., et al., Factors of risk in

the development of Coronary Heart Disease - 6 year follow-up
experience. The Framingham Study. Ann. Intern. Med. 55: 33-50,
1961.

[29] Chapman, J.M. and Massey, F.J., The interrelationship of serum
cholesterol, hypertension, body weight and risk of coronary
heart disease; results of first 10 year follow-up in Los Angeles
Heart Study. J. Chron. Dis. 17: 933-999, 1964.

[30] Chiang, B.N., Perlman, L.V., Fulton, M., Ostrander, L.D., and
Epstein, F.H., Predisposing factors in sudden cardiac death in
Tecumseh, Michigan. Circul. 41: 31-37, 1970.

[31] Kannel, W.B., Le Bauer, El J., Dawber, T.R. and McNamara, P.M.,
Relation of body weight to development of coronary heart disease. Circul. 35: 734-744, 1967.

[32] Bjerkedal, T., Overweight and hypertension. Acta. Med. Scan.
159: 13-26, 1957.

[33] Gordon, T., Castelli, W.P., Hjortland, M.C., Kannel, W.B. and
Dawber, T.R., Diabetes, blood lipids, and the role of obesity
in coronary heart disease risk for women. Ann. Int. Med.
87: 393-397, 1977.

[34] Heyden, S., Hames, C.G., Bartel, S., et al., Weight and weight
history in relation to cerebrovascular and ischemic heart disease. Arch. Intern. Med. 128: 956-960, 1971.

[35] Abraham, S., Collins, G. and Nordsieck, M., Relationship of childhood weight status to morbidity in adults. H.S.M.H.A. Health
Reports 86: 273-284, 1971.

PRESENT STATUS OF POLYUNSATURATED FATS

IN THE PREVENTION OF CARDIOVASCULAR DISEASE

D. Roger Illingworth, Ph. D., M.D. and
William E. Connor, M.D.
Divs. of Metabolism and Cardiology – Dept. of Medicine
University of Oregon Health Sciences Center
Portland, Oregon 97201 – U.S.A.

INTRODUCTION

The hypocholesterolemic action of polyunsaturated vegetable fats as a replacement for animal fat in the diet of both man and experimental animals has been repeatedly demonstrated over the last 25 years[1-5]. Studies by numerous investigators have shown this effect to be attributable to three main factors: 1) a decrease in saturated fat; 2) a decrease in dietary cholesterol and 3) an increase in polyunsaturated fat particularly linoleic acid. This paper will review briefly the efficacy and mechanisms by which dietary polyunsaturated fats lower blood lipids in man as well as the rationale for current dietary recommendations for the use of polyunsaturated fat for the prevention of coronary heart disease.

THE HYPOLIPIDEMIC EFFECT OF
POLYUNSATURATED FAT ENRICHED DIETS

Since fat typically provides about 40% of the calories in the diets of most industrialized societies, it is informative to initially examine to what extent changes in fatty acid composition of such a diet will affect serum lipids. The basal diet in western populations typically comprises 18-22% of calories from saturated fat, 14-16% from monoenoic fatty acids and 3-5% from polyunsaturated fatty acids[6,7], and contains from 400-800 mg of cholesterol. When a group of normal subjects studied in a metabolic ward were placed on a cholesterol-free eucaloric formula diet in which 40% of calories were derived from cocoa butter (saturated:monounsaturated:polyunsaturated ratio 61:34:3) plasma cholesterol levels fell about 20% as compared to the values on a basal diet[8]. Further isocaloric re-

placement of the cocoa butter by corn oil (saturated:monosaturated:
polyunsaturated ratio 12:31:56) resulted in a further 20% decrease
in plasma cholesterol; with the reinstitution of the cocoa butter
diet, the cholesterol values promptly rose back to their previous
values. Similar changes have been observed by other workers[3,9] and
occur in both normolipidemic individuals and those with hypercholesterolemia[1,2,4]. The predominant polyunsaturated fatty acid in corn
oil and other plant oils is linoleic acid. However, a similar or
even greater hypocholesterolemic effect may be demonstrated upon substitution of marine oils in which C20 to C22 fatty acids possessing
4 to 6 double bonds are present[2,3,10]. By contrast to the opposing
effects of dietary saturated and polyunsaturated fatty acids on
plasma cholesterol trans fatty acids[11] which are produced during
partial hydrogenation of vegetable oils, or monosaturated cis fatty
acids[12], do not appear to have significant hypo- or hyperlipidemic
properties.

The evidence for an independent hypocholesterolemic effect of
polyunsaturated fatty acids that is distinct from that attributable
to removal of saturated fats in the diet derives from a number of
sources. Beveridge et al.[5] showed in separate experiments that isocaloric replacement of 20% and 60% of the calories of a fat free diet
by corn oil resulted in decreases in plasma cholesterol of 7 and 15%
respectively as compared to the fat-free period. Substitution of
butter at similar levels resulted in increases in plasma cholesterol
of 6 and 22%. Bronte-Stewart et al.[2] demonstrated that addition of
sunflower oil or pilchard oil to the low-fat diet of African Bantus,
thereby increasing the proportion of fat derived from calories from
3% to 35-45% resulted in a 25mg% fall in serum cholesterol. More
marked changes were observed with two caucasians on addition of 100
grams safflower oil to the basal diet which contained 50 grams of
animal fat. Hydrogenation of the vegetable oil eliminated this hypocholesterolemic activity. Further evidence for an independent effect
of polyunsaturated fat derives from the work of Keys and colleagues[3]
and of Hegsted et al.[9] in which a wide variety of different fats have
been fed to institutionalized subjects. On the basis of the hypo-
or hypercholesterolemic response to different fats, regression equations have been calculated as follows:

Keys et al.: Δ cholesterol = 2.74 ΔS-1.31ΔP

Hegsted et al.: Δ cholesterol = 2.16 ΔS-1.65ΔP+6.66C-0.53

In these equations, S and P represent the percent of dietary calories
derived from saturated and polyunsaturated fat and C is the dietary
cholesterol content in units of 100mg. Three points are worthy of
emphasis concerning these equations: 1) saturated fatty acids exert
twice as much influence on blood cholesterol levels as the polyunsaturated but each has an independent effect; 2) monounsaturated
fatty acids have no effect, and 3) the hypercholesterolemic action

of saturated fatty acids is greatest with C12 to C16 acids whereas
shorter chain fatty acids appear to have no significant influence.
Both equations provide a good prediction of the response of plasma
cholesterol to changes in dietary fat when the latter provides from
20 to 45% of calories, i.e. values that are within the practical
range for free living subjects to consume.

DIETARY TRIALS AND REVERSAL OF ATHEROSCLEROSIS

     With the exception of patients with genetic hyperlipidemias, the
manifestations of atherosclerosis in man do not usually become clinically evident until after the fourth decade but are felt to have
their genesis in childhood[13]. The rare cases of homozygous familial
hypercholesterolemia in which severe cardiovascular disease develops
within the first 15 years of life provide the strongest evidence for
the causal relationship between hypercholesterolemia and the development of atherosclerosis in childhood. Although this genetic disorder is an extreme example it is well established that children in
the United States have higher levels of plasma cholesterol than their
counterparts in countries where atherosclerotic disease in adults is
less prevalent[14,15] and where lower dietary intakes of animal fats
are the norm. As an example, Table 1 shows a comparison of the
plasma lipid levels and dietary intakes of fat and cholesterol between a group of Tarahumara Indians in Mexico and those of age-match-
ed children in North America. Some 5% of children in North America
have plasma cholesterol levels exceeding 200-220 mg% but in only 1%
of these is the hypercholesterolemia on a genetic basis[19]. Because
of the long time interval involved, it is unlikely that dietary
changes aimed at reducing plasma cholesterol levels in childhood will
ever be shown conclusively to affect cardiovascular mortality in
later life. Nevertheless, this is probably when changes in diet
should begin. Metabolic studies clearly indicate that modification
of the dietary intakes of cholesterol and saturated and polyunsaturated fats will influence plasma cholesterol levels in man. These
studies have provided the basis for a number of trials in which the
feasibility, safety and efficacy of hypolipidemic dietary regimens
in free-living individuals and the role of such therapy in the prevention of cardiovascular disease have been examined. Three basic
but interrelated questions deserve consideration: 1) what has been
the nature of the dietary changes, particularly with respect to polyunsaturated fat; 2) did these dietary modifications reduce serum
lipids; 3) in those studies which looked to this question, was morbidity and mortality from cardiovascular disease decreased?

     The efficacy and safety of a low saturated fat, polyunsaturated
enriched diet (P:S ratio 1:1) containing 200 mg of cholesterol begun
at birth has been shown by Freeman and colleagues[20]. The plasma
cholesterol levels of children on this regimen at three years of age
were 145 mg% compared to 154 mg% in a control group. A group of

TABLE 1

PLASMA LIPIDS AND DIETARY INTAKES OF FAT AND CHOLESTEROL
IN TARAHUMARA INDIAN CHILDREN (5-18) AND SIMILAR
AGED INFANTS IN NORTH AMERICA

	Tarahumara	United States
Plasma cholesterol (mg%)	116	182
Plasma triglyceride (mg%)	110	99
Dietary cholesterol (mg/day)	50	650
% Calories from fat (total)	9	38
Saturated	2	17
Monoenoic	3	17
Polyenoic	4	4
P:S Ratios	2:1	0:23

Data from: Lauer et al.[15], Connor et al.[16], Stein et al.[17] and
McGandy et al.[18].

adolescent boys attending boarding school was studied by Stein and
colleagues in South Africa[17]. In this study, polyunsaturated fat
products were substituted for dairy foods in one phase (phase A) and
polyunsaturated ruminant fats substituted for conventional meat products in the other (phase B). In both of the experimental periods,
the total fat, which provided 38% of calories, and the cholesterol
content of the diets were the same as in the control period, but the
distribution of dietary fats was changed from a saturated:monoun-
saturated:polyunsaturated ratio of 1:8:3 in the control period to a
1:1:1 ratio in the experimental diets. Plasma cholesterol levels
were reduced 14% in both of the experimental groups with a resultant
left shift in the frequency distribution curves. This decrease was
attributable entirely to a reduction in low density lipoprotein. Of
particular interest in this study, is the fact that those children
with higher initial cholesterol values had the greatest response.
Thus, of 34 children with initial plasma cholesterol greater than
230 mg% only four remained above this after consuming the polyunsaturated enriched diets. Reductions of plasma cholesterol of from
10-22% have been reported in other studies on adolescents in which
diets containing from 9-13% of calories as polyunsaturated fats but
with reduced contents of cholesterol and saturated fat have been fed
in place of the normal Western diet[18,21]. The hypocholesterolemic
effects of these dietary modifications appear to be equally efficacious in children heterozygous for familial hypercholesterolemia in
whom reductions of plasma cholesterol of 10-20% may be achieved.
Numerous trials for dietary modification have been undertaken in

adults (reviewed by Leren)[22]. In addition to determining the efficacy and feasibility of lowering blood lipids by dietary changes alone, most of the studies have addressed the question of whether or not cardiovascular morbidity and mortality could be affected. Specific dietary changes varied from one study to another but all involved at least three factors: a decrease in saturated fat and cholesterol and an increase in polyunsaturated fat from 4-5% of calories to 10-15%. In some studies total fat intake was also reduced. These changes were mediated by replacement of butter, cheese and other dairy products by vegetable oil derived substitutes and by reduction in the consumption of eggs and meat fat. Sustained decreases in plasma cholesterol of from 10-18% were observed on the modified diets. For example, in the study of Dayton et al.[23] involving a population of domiciliary Veterans, cholesterol levels were reduced by an average of 12.7%. No escape phenomenon typified by an increase in serum cholesterol after prolonged periods on the diet were observed. Similarly in the Finnish Mental Hospital study[24], a 12-18% reduction in serum cholesterol was observed on the polyunsaturated fat enriched diets. This effect was reduplicated when the diets at the two hospitals were reserved and occured in both men and women. In four of these trials, cardiovascular morbidity and mortality were lower in the treated groups but total mortality was not significantly reduced[23, 27]. As other authors have discussed[28] these trials have been in older subjects in whom the potential for reversal for atherosclerosis is probably less optimal than in younger patients. Indeed, both Leren[25] and Franz[27] found the greatest effects of diet induced hypocholesterolemia in cardiovascular mortality to be in the younger patients. Kinetic studies also indicate that early lesions are more likely to be influenced by lowered blood lipid values than are advanced thick lesions[29]. The former are more apt to predominate in the younger patients. Evidence that atherosclerotic lesions can indeed regress with correction of hyperlipidemia has been documented in both animals and man[30, 31]. For example, Blankenhorn[31] noted regression of femoral lesions in twelve patients studied angiographically 13 months apart in whom blood lipids had been significantly reduced by a combined diet plus drug regimen. A group of 13 other patients with insignificant changes in their lipids showed further progression of the disease during the same time period.

DIETARY POLYUNSATURATED FATS
AND LIPOPORTEIN METABOLISM

The relationships between the concentrations of individual lipoproteins, very low density lipoprotein (VLDL), low density lipoproteins (LDL), and high density lipoproteins (HDL) and the development of cardiovascular disease remains an area of considerable current interest. Epidemiological studies have revealed a positive correlation between the concentrations of LDL and, to a lesser extent, VLDL and development of cardiovascular disease but an inverse cor-

relation between this disorder and HDL levels[32, 33], i.e. higher
levels of HDL are protective. Data from the Framingham study has
also suggested that the ratio LDL cholesterol to HDL cholesterol may
be one of the best predictive indicators for cardiovascular disease.
With this background, it is informative to examine the lipoprotein
changes which occur when diets enriched in polyunsaturated fat are
substituted for the normal Western diet. Studies in the 1950's by
Bronte-Stewart and colleagues[2] showed that decreases in total cholesterol were parallel by decreases in this lipid in the LDL or beta-
lipoprotein fraction of plasma. These results have been confirmed
repeatedly and have been shown to be accompanied by similar reductions[34] in the concentrations of apoprotein B (Table 2), the major
protein in LDL. Some studies have also shown additional effects of
dietary polyunsaturated fats on the concentrations of both VLDL and
HDL. Thus, Chait et al.[35] found significant decreases in the concentrations of both VLDL and LDL but no change in HDL when fat with
a P:S ratio of 2.4 was isocalorically substituted for the basal diet
in which this ratio was 0.2. This effect on both VLDL and LDL has
been observed by other workers and readily explains the hypotriglyceridemic effect of polyunsaturated fat enriched diets noted in many
earlier studies. The hypotriglyceridemic effect of polyunsaturated
fat seems to be most marked in subjects with elevated levels of VLDL
and serum triglycerides above 150mg%[35, 36]. With lower initial VLDL
levels minimal or no changes in the concentration of this lipoprotein have been found[38, 39].

Small but consistent decreases in the concentrations of HDL
cholesterol and apoprotein Al, the predominant apoprotein in HDL,
have also been noted in two recent studies in normolipidemic individuals[38, 39]. Preliminary data from our laboratory confirm these
observations and suggest that the changes in HDL which occur when
normolipidemic individuals consume a cholesterol-free diet containing
40% fat (P:S ratio 4:1) as compared to their basal diet affect apoprotein Al more than total HDL cholesterol (Table 3). The lack of
effect of polyunsaturated fat feeding on HDL observed in some studies, particularly in hypertriglyceridemic patients, may be explained
by the opposing effects of two factors: 1) reduced concentrarions
of VLDL attributable to the increased polyunsaturated fat result in
secondary increases in HDL, and 2) the polyunsaturated fat also has
an independent primary effect in decreasing HDL synthesis. In patients with mild hypertriglyceridemia, these two effects on HDL are
probably about equal and as VLDL levels fall with polyunsaturated
fat feeding, HDL levels remain the same. In contrast, in normo tri-
glyceridemic subjects the influence of decreased synthesis of HDL
predominates and the concentration of this lipoprotein falls. Because of the evidence implicating HDL as an antiatherogenic lipoprotein[32], the observations that polyunsaturated fat enriched diets may
lower HDL level in normolipidemic individuals raises a question as to
their potential detrimental effect. However, the fact that in such
individuals LDL levels fall even more and result in an increased HDL

TABLE 2

THE EFFECTS OF CHOLESTEROL FREE POLYUNSATURATED FAT ENRICHED
DIET ON PLASMA LIPOPROTEIN AND APOPROTEINS IN NORMAL
VOLUNTEERS

	Basal Diet (mg%)	Hypolipidemic Diet (mg%)
Apo AI	129 ± 23	112 ± 26
Apo B	88 ± 14	66 ± 12
Apo CIII	8.8 ± 2.9	7.3 ± 2.2
*Apo E	9.0	5.0 ±
Total Cholesterol	153 ± 30	117 ± 21
VLDL	16 ± 6	6 ± 2
LDL	99 ± 18	68 ± 6
HDL	43 ± 13	41 ± 13

Preliminary data from four subjects.
* Data from two subjects only.

TABLE 3

LIPOPROTEIN CONCENTRATIONS IN TARAHUMARA INDIANS

	n	CHOLESTEROL (mg/dl)			
		Total	VLDL	LDL	HDL
CHILD (5-18)					
Male	10	118	20	75	23
Female	7	143	23	90	30
ADULT					
Male	34	134	21	87	26
Female	6	141	24	89	20

n = number of subjects studied
   Data from Reference 16.

cholesterol to LDL cholesterol ratio suggests that any potentially
detrimental effects of reduced HDL levels are more than offset by
even greater decreases in the concentrations of LDL. This view is
consistent with lipoprotein data from less developed societies where
atherosclerosis in uncommon and in whom levels of LDL and HDL are
lower than western civilizations. Typical values from the Tarahumara
Indians of Mexico are shown in Table 3 and serves to illustrate this
point.

MECHANISMS OF ACTION OF POLYUNSATURATED FATS

        Various mechanisms have been proposed to explain the hypocho-
lesterolemic effects of polyunsaturated fat (reviewed in 37). These
may be categorized into: 1) reduction in the absorption of exogenous
cholesterol from the small intestine; 2) decreased endogenous synthesis of cholesterol; 3) a redistribution of cholesterol between
the plasma and tissue pools; 4) an increase in the fecal excretion
of steroids (cholesterol and/or bile acids) secondary to an increased
secretion of these lipids into bile; and 5) a primary effect on
lipoprotein metabolism with reduced synthesis and/or enhanced catabolism. Todate there is no good evidence that different dietary fats
influence cholesterol absorption. Thus, Grundy and Ahrens[40] and
Nestel et al.[41] both failed to document any change in cholesterol
absorption with substitution of polyunsaturated for saturated fats in
the diet. Studies in our laboratory have also failed to reveal any
differences in the absorption of cholesterol in test meals containing
fats of different degrees of saturation (Table 4). Persuasive evidence that polyunsaturated fats affect cholesterol synthesis in experimental animals[42] or man[43] is singularly lacking as is direct data
to support the tissue redistribution hypothesis[30, 44]. Studies in normolipidemic human subjects including those from this laboratory[8, 43]
have shown increased fecal excretion of neutral steroids and bile
acids when the composition of dietary fat is changed from one rich in
saturated to polyunsaturated fatty acids. A similar increase in fecal
steroid excretion was also observed by Grundy[36] in a series of eleven
hypertriglyceridemic subjects. In contrast, studies in patients with
familial hypercholesterolemia[40] have generally failed to show consistent changes in the fecal excretion of cholesterol or its metabolites after exchanges of saturated and polyunsaturated fat in the
diet. To what extent the latter findings may reflect a general impairment in the catabolism of cholesterol in this disorder is unknown. Considering the central role of the liver in the secretions
of lipids into both bile and plasma changes in the biliary excretion
of cholesterol and bile acids could occur either secondary to or in
association with changes in the rates of lipoprotein synthesis or
catabolism.

        A growing body of evidence suggests that diets rich in polyun-
saturated fats may effect lipoprotein synthesis. Since LDL arises

TABLE 4

EFFECT OF DIETARY FAT COMPOSITION
ON CHOLESTEROL ABSORPTION IN MAN

	Percent Cholesterol Absorption
Iodine #40 (High Saturated)	48.0 ± 12
Iodine #85 (High Monenoic)	49.5 ± 7
Iodine #120 (High Polyunsaturated)	49.0 ± 14

Values are mean ± S.D. (n = 5).

from catabolism of VLDL[45, 46], studies on the rates of synthesis and catabolism of these lipoproteins under different dietary conditions should help clarify the mechanisms by which polyunsaturated fat lowers plasma cholesterol. Two recent preliminary reports[47, 48] have showed that diets rich in polyunsaturated fat, when fed to a heterogenous group of subjects, result in a decreased synthesis of LDL; in one study[49], this was accompanied by an increased fractional catabolic rate of this lipoprotein. In both of these studies, the changes in LDL synthesis could result either from the decrease in saturated fat or the increase in polyunsaturates and it is not possible to determine which effects dominate. Whether these differences result from changes in fatty acid composition of the lipoproteins or represent primary effects of polyunsaturated fat on the synthesis of apoproteins or the assembly of lipoproteins by the liver is not known. Shepherd et al.[37] have also recently documented a 26% reduction in the synthesis of apo-AI, the major apoprotein of HDL upon isocaloric substitution of polyunsaturated fat (P:S ratio 4:1) for saturated fat (P:S ratio 0.25) in the diet of four normal men. The findings of reduced synthesis of LDL with polyunsaturated fats is in accord with the results of Chait et al.[35] who on the basis of comparatively lower incorporation of intravenously injected linoleate as compared to palmitate into VLDL, concluded that VLDL synthesis was also reduced on the polyunsaturated fat enriched diet. Another mechanism by which diets enriched in polysaturated fat may lower the concentration of VLDL is by enhanced removal of the lipoprotein mediated by an increased activity of lipoprotein lipase. Previous studies in man have been conflicting with increased activity of post heparin lipase noted in one study[49] but not in another[35] and this question remains open.

In summary, it appears that in most subjects diets rich in polyunsaturated fat cause an enhanced excretion of biliary and fecal steroids and a reduction in lipoprotein synthesis but the mechanism

by which such diets lower plasma cholesterol in patients with familial hypercholesterolemia remains ill-defined. Reported adverse effects of polyunsaturated fat enriched diets have included an increased requirement for vitamin E and higher incidence of gallstones[50]
noted in the Los Angeles Veterans trial. Although the latter finding
may simply reflect the old age of the patients and has not been confirmed in other trials[52], the potential carcinogenic effects of polyunsaturated fats and their oxidation products remains open and war-
rants further study. The autopsy data from Los Angeles showing an
increased incidence of cholelithiasis is supported by the metabolic
studies of Grundy[36] in which subjects fed polyunsaturated enriched
diets were found to have an increased lithogenicity of bile. This
observation is not totally unexpected in view of previous studies
showing increased fecal steroid excretion with such regimens. Nevertheless, more information on the incidence of gallstones in populations consuming diets comparatively rich in polyunsaturated fats for
long periods of time is clearly needed before any firm conclusions
can be drawn.

SUMMARY AND CONCLUSIONS

     From the foregoing discussions it is clear that diets in which
polyunsaturated fats provide 10-15% of calories, with reduced levels
of saturated fat and cholesterol are efficacious in lowering blood
cholesterol levels in typical Western populations by 10-20%. More
pronounced changes occur under metabolic ward conditions where larger
dietary changes can be effected. Variable reductions in the levels
of plasma triglycerides generally accompany the changes in cholesterol. Such hypolipidemic therapy has been associated with a decreased cardiovascular morbidity and mortality in some but not all
studies. Despite unequivocal proof of benefit, the diet induced reductions in plasma lipids and lipoproteins, conform to a pattern that
is less atherogenic and which, over an extended period of time should
logically lead to lower cardiovascular mortality. Recent surveys in
North America Suggest this is already happening and have documented
a downward trend in the levels of both plasma cholesterol[53] and coronary heart disease[54] over the last decade. These changes have been
paralleled by dietary increases in vegetable fats and small decreases
in animal fat and cholesterol[55]; whether or not they are equitem-
porally related remains speculative. Finally, I hope this brief review will serve as a useful background upon which the dietary recommendations and advice of Dr. Norum can be based.

Supported by the General Clinical Research Centers Program
(RR00334-12) of the DIvision of Research Resources of the
National Institutes of Health and the U.S. Public Health
Service Research Grants HL20910 and HL07295 from the
National Heart, Lung and Blood Institute.

REFERENCES

1 Ahrens, E.H., Jr., H. Hirsch, W. Insull, Jr., T.T. Tsaltas, T.
   Blomstrand and M.L. Peterson, Dietary control of serum lipids
   in relation to atherosclerosis. J. Am. Med. Assoc. 164: 1905-
   1914, 1957.

2 Bronte-Stewart, B., A. Antonis, L. Eales and F. Brock, Effects of
   feeding different fats on serum cholesterol level. Lancet I:521-
   526, 1956.

3 Keys, A., J.T. Anderson and F. Grande, Prediction of serum cholesterol responses of man to changes in fats in the diet. Lancet
   II: 959-966, 1957.

4 Malmros, H., G. Wigand, The effect on serum cholesterol of diets
   containing different fats. Lancet II: 1-8, 1957.

5 Beveridge, J.M.R., W.F. Connell and G.A. Mayer, Dietary factors
   affecting the level of plasma cholesterol in humans: the role
   of fat. Can. J. Biochem. Physiol. 34: 441-455, 1956.

6 Keys, A., Coronary heart disease in seven countries. Circulation
   41, 42, Suppl. 1: I-1-211, 1970.

7 Friend, B., Nutrients in United States food supply. A review of
   trends 1909-1913 to 1965. Am. J. Clin. Nutr. 20: 907-914, 1967.

8 Connor, W.E., D. T. Witiak, D.B. Stone and M.L. Armstrong, Cholesterol balance and fecal neutral steroid and bile acid excretion in normal men fed dietary fats of different fatty acid
   composition. J. Clin. Invest. 48: 1363-1375, 1969.

9 Hegsted, D.M., R.B. McGandy, M.L. Myers and F.J. Stare, Quantitative
   effects of dietary fat on serum cholesterol in man. Am. J.
   Clin. Nutr. 17: 281-295, 1965.

10 Ahrens, E.M., Jr., W. Insull, Jr., J. Hirsch, W. Stoffel, M. L.
   Peterson, J.W. Farquhar, T. Miller and H.J. Thomasson: The effect on human serum lipids of a dietary fat highly unsaturated
   but poor in essential fatty acids. Lancet I: 115-122, 1959.

11 Mattson, F.H., E. J. Hollenbach and A.M. Kligman, Effect of hydrogenated fat on the plasma cholesterol and triglyceride levels
   of man. Am. J. Clin. Nutr. 28: 726-731, 1975.

12 Keys, A., J.T. Anderson and F. Grande, Effect on serum cholesterol
   in man of mono-ene fatty acid (oleic acid) in the diet. Proc.
   Soc. Exp. Biol. Med. 98: 381-387, 1958.

13 McGill, H.C., Jr., Atherosclerosis: problems in pathogenesis. In:
   "Atherosclerosis Reviews", Gotto, A.M. and R. Paoletti (eds.),
   Raven Press, New York, 1977, p. 27-65.

14 Frerichs, R.R., S.R. Srinivasan, L.S. Webber et al., Serum cholesterol and triglyceride levels in 3446 children from a biracial
   community. The Bogalusa Heart Study. Circulation 54: 302-309,
   1976.

15 Lauer, R.M., W.E. Connor, P.E. Leaverton, M.A. Reiter, and W.R.
   Clarke, Coronary heart disease risk factors in school children:
   The Muscatine Study. Pediatr. 36: 697-706, 1975.

16 Connor, W.E., M.T. Cerqueira, R.W. Connor, R.B. Wallace, M.R.
   Malinow, and H.R. Casdorph, The plasma lipids, lipoproteins,

and diet of the Tarahumara Indians of Mexico. Am. J. Clin. Nutr., in press, 1978.

[17] Stein, E.A., D. Mendelsohn, I. Bersohn et al., Lowering of plasma cholesterol levels in free-living adolescent males: use of natural and synthetic polyunsaturated foods to provide balanced fat diets. Am. J. Clin. Nutr. 28: 1204-1216, 1975.

[18] McGandy, R.B. B. Hall, C. Ford and F.J. Stare: Dietary regulation of blood cholesterol in adolescent males. A pilot study. Am. J. Clin. Nutr. 25: 61-70, 1972.

[19] Glueck, C.H. and P. O. Kwiterovich, The lipid hypothesis, genetic basis. Arch. Surg. 11: 35-44, 1978.

[20] Friedman, G., and S.J. Goldberg: An evaluation of the safety of a low saturated diet, low cholesterol diet, beginning in infancy. Pediatrics 58: 655-657, 1976.

[21] Vergroesen, A.J., Dietary fat and cardiovascular disease: possible modes of action of linoleic acid. Proc. Nutr. Soc. 31: 323-329, 1972.

[22] Leren, P., Prevention of coronary heart disease by diet. Postr. Med. J. 51 (Suppl. 8) 44-46, 1975.

[23] Dayton, S., M.L. Pearce, S. Hashimoto et al., A controlled clinical trial of a diet high in unsaturated fat in preventing complications of atherosclerosis. Circulation 40 (Suppl. II): 1, 1969.

[24] Miettinen, M., O. Turpeinen, M.J. Karvonen et al., Effect of cholesterol lowering diet in mortality from coronary heart disease and other causes. Lancet 2: 835-838, 1972.

[25] Leren, P., Effect of plasma cholesterol lowering diet in male survivors of myocardial infarction. Acta Med. Scand. 466: 5-92 (Suppl.), 1966.

[26] National Diet-Heart Study. Final report. American Heart Association Monograph 48, Circulation 37, Suppl. 1, 1968.

[27] Frantz, I.D., E.A. Dawson, K. Kuba et al., The Minnesota coronary survey: Effect of diet on cardiovascular events and deaths. Circulation 41 (Suppl. II) 2-4, 1975.

[28] Glueck, C.J., F. Mattson and E.L. Bierman: Diet and coronary heart disease: another view. N. Eng. J. Med. 298: 1471-1473, 1978.

[29] Jaganathan, S.N., W.E. Connor, W.H. Baker and A.K. Bhattacharyya: The turnover of cholesterol in human atherosclerotic arteries. J. Clin. Invest. 54: 366-377, 1974.

[30] Armstrong, M.L., E.D. Warner and W.E. Connor, Regression of coronary atheromatosis in Rhesus monkeys. Circ. Res. 27: 59-67, 1970.

[31] Blankenhorn, D.H., S.H. Brooks, R.H. Selzer and R. Barndt, Jr., The rate of atherosclerosis change during treatment of hyperlipoproteinemia. Circulation 57: 355-361, 1978.

[32] Castelli, Gordon T., M.C. Hjortland, W.B. Kannel and T.R. Dawber, High Density Lipoprotein as a protective factor against coronary heart disease. The Framingham Study. Am. J. Med. 62:707-714, 1977.

[33] Carlson, L.A. and L.E. Bottiger: Ischemic heart disease in relation to fasting values of plasma triglyceride and cholesterol.

Lancet I: 865-868, 1972.

[34] Durrington, P.N., C.H. Bolton, M. Hartog, R. Angelinetta, P. Emmett and S. Furniss, The effect of a low cholesterol, high polyunsaturated diet on serum lipid levels, apoprotein B levels and triglyceride fatty acid composition. Atherosclerosis 27: 465-475, 1977.

[35] Chait, A., A. Onitiri, A. Nicoll, E. Rabaya, J. Davies and B. Lewis, Reduction of serum triglyceride levels by polyunsaturated fat. Atherosclerosis 20: 347-364, 1974.

[36] Grundy, S.M., Effects of polyunsaturated fats on lipid metabolism in patients with hypertriglyceridemia. J. Clin. Invest. 55: 269-282, 1975.

[37] Shepherd, J., C.J. Packard, J.R. Patsch, A.M. Gotto, Jr., and O.D. Taunton, Effects of dietary polyunsaturated and saturated fat on the properties of high density lipoproteins and the metabolism of Apolipoprotein A-I. J. Clin. Invest. 61: 1582-1592, 1978.

[38] Burslem, J., G. Schoenfeld, M.A. Howald, S.W. Weideman and J.P. Miller, Plasma apoprotein and lipoprotein lipid levels in vegetarians. Metabolism 27: 711-719, 1978.

[39] Jackson, R.L., O.D. Taunton, J.D. Morrisett and A.M. Gotto Jr., The role of dietary polyunsaturated fat in lowering blood cholesterol in man. Circ. Res. 42: 447-453, 1978.

[40] Grundy, S.M. and E.H. Ahrens, The effects of unsaturated dietary fats on absorption excretion systhesis and distribution of cholesterol in man. J. Clin. Invest. 49: 1135-1152, 1970.

[41] Nestel, P.J., N. Havenstein, Y. Homma, T.W. Scott and L.J. Cook, Increased sterol excretion with polyunsaturated fat high cholesterol diets. Metabolism 24: 189-198, 1975.

[42] Bieberdorf, F.A. and J.D. Wilson, Studies on the mechanism of action of unsatuarted fats on cholesterol metabolism in the rabbit. J. Clin. Invest. 44: 1834-1841, 1965.

[43] Nestel, P.J., N. Havenstein, T.W. Scott and L.J. Cook: Polyunsaturated ruminant fats and cholesterol metabolism in man. Aust. N.Z. J. Med. 4: 497-501, 1974.

[44] Frantz, I.D. and J.B. Carey, Cholesterol content of human liver after feeding corn oil and hydrogenated coconut oil. Proc. Soc. Exp. Biol. Med. 106: 800-801, 1961.

[45] Sigurdsson, G., A. Nicholl and B. Lewis, Conversion of very low density lipoprotein to low density lipoprotein. A metabolic study of apoprotein B kinetics in human subjects. J. Clin. Invest. 56: 1481-1490, 1975.

[46] Illingworth, D.R., Metabolism of lipoproteins in nonhuman primates. Studies on the origin of low density lipoprotein apoprotein in the plasma of the squirrel monkey. Biochim. Biophys. Acta 388-38-51, 1975.

[47] Turner, J.D., R. Monell and W.V. Brown, Heterogenous responses to polyunsaturated fat in low density lipoprotein turnover in hyperlipoproteinemia. Circ. 54: II-4, 1976.

[48] Yeshurun, D., A.M. Gotto, Jr., and O.D. Taunton, Effects of poly-

unsaturated versus saturated fat on LDL metabolism in normal subjects. Clin. Res. 24(3) 373A, 1976.

[49] Bangdade, J.D., W.R. Hazzard and J. Carlin, Effect of unsaturated dietary fat on plasma lipoprotein lipase activity in normal and hyperlipidemic states. Metabolism 19: 1020-1025, 1970.

[50] Sturdevant, R.A.L., M. L. Pearce and S. Dayton, Increased prevalence of cholelithiasis in men ingesting a serum cholesterol lowering diet. N. Eng. J. Med. 288: 24-27, 1973.

[51] Pearce, M.L., and S. Dayton, Incidence of cancer in men on a diet high in polyunsaturated fat. Lancet I: 464-467, 1971.

[52] Ederer, F., P. Leren, O. Turpernen and I.D. Frantz, Jr., Cancer among men on cholesterol lowering diets. Lancet II: 203-206, 1971.

[53] United States National Center for Health Statistics. Vital and Health Statistics Ser. 11 No. 202. A comparison of levels of serum cholesterol of adults 18-74 years of age in the United States in 1960-1962 and 1971-1974. Washington DC, Government Printing Office, 1977.

[54] Walker, W.J., Changing United States life-style and declining vascular mortality: cause or coincidence? N. Eng. J. Med. 297: 163-165, 1977.

[55] Pearson, A.M., Some factors that may alter consumption of animal products. J. Am. Diet. Assoc. 69: 522-530, 1976.

DIETARY PROTEIN AND CARDIOVASCULAR DISEASES:
EFFECTS OF DIETARY PROTEIN
ON PLASMA CHOLESTEROL LEVELS AND CHOLESTEROL METABOLISM

K.K. Carroll and M.W. Huff

Dept. of Biochemistry, Univ. of Western Ontario

London, Ontario, CANADA, N6A 5C1

Experiments in our laboratory have shown that the hypercholesterolemia and atherosclerosis observed in rabbits fed semipurified, cholesterol-free diets are due to the use of casein as the protein component of such diets. Both the hypercholesterolemia and atherosclerosis can be prevented by using isolated soy protein rather casein as dietary protein[1-3]. Feeding trials with protein preparations from a variety of different sources indicated that, in general, animal proteins cause a rise in plasma cholesterol while plant proteins give low levels comparable to those in rabbits on commercial feed[1, 2, 4, 5] (Fig. 1).

The differing effects of casein and isolated soy protein appear to be due, at least in part, to differences in their amino acid composition. An enzymatic hydrolysate of casein or a mixture of L-amino acids equivalent to casein caused an elevation of plasma cholesterol comparable to that obtained with the intact protein[2]. The plasma cholesterol remained low in rabbits fed an enzymatic hydrolysate of soy protein, but increased to some extent in animals fed a mixture of L-amino acids equivalent to soy protein[2]. This does not necessarily mean that some non-protein factor in the soy protein preparation is responsible for the low level of plasma cholesterol, since free amino acids in the diet would not be handled in the same way during digestion as amino acids in proteins. Feeding trials have also been conducted with various other combinations of L-amino acids, and with intact proteins supplemented by amino acids[1, 6], but these experiments have not so far resolved the problem.

Most of the semipurified diets used in our experiments have been relatively high in protein (25% w/w) and carbohydrate (60% dextrose) and low in fat (1% corn oil). Increasing the level of polyunsaturated

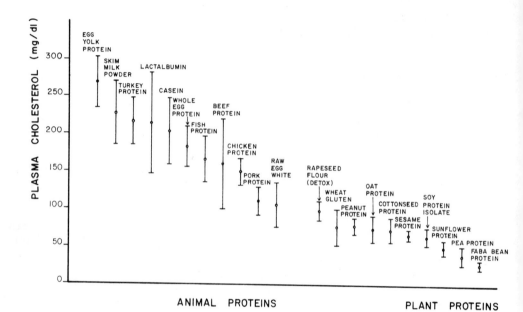

Fig. 1 – Levels of plasma cholesterol in animals fed with animal
and plant proteins.

fat at the expense of carbohydrate in the casein-containing diet
largely prevented the hypercholesterolemic response, in agreement
with earlier studies on this animal model[7, 8]. However, a diet containing 15% butter produced an elevation in plasma cholesterol comparable to that obtained with the low fat diet[1, 4]. The results were
also influenced by the type of carbohydrate used in the low fat,
casein diet[1, 4], but dietary protein appeared to exert the primary
effect on plasma cholesterol levels, since a hypercholesterolemic
response was only obtained with diets containing animal proteins.

The hypercholesterolemia produced by feeding a semipurified diet
containing casein was readily reversed by substituting a similar diet
containing isolated soy protein, as shown by the cross-over experiment illustrated in Fig. 2. Experiments with diets containing mixtures of casein and soy protein isolate showed that a mixture containing more than 50% casein was required to produce a hypercholesterolemic response[2, 9].

Long-term feeding trials with the cholesterol-free, casein,
semipurified diet showed that the hypercholesterolemia produced by
this diet persisted for as long as 10 months, and animals killed
after 6 months or more on the diet showed extensive sudanophilic
lesions in the aorta, in agreement with earlier studies[7, 8]. In animals fed a corresponding diet containing soy protein isolate, the
plasma cholesterol levels remained low and aortic lesions were minimal after 10 months of diet[3, 9]. Analysis of the plasma lipiproteins
midway through these experiments, showed that the excess plasma cholesterol in the casein-fed animals was present mainly in the intermediate density (IDL) fraction, although the levels in both very low
density (VLDL) and low density (LDL) fractions were also elevated[3, 9].

Cholesterol turnover and metabolism have also been investigated
in rabbits on the casein and soy protein diets[10]. These studies
showed that rates of oxidation and turnover of cholesterol were both
faster in animals on the soy protein diet. Analyses of fecal neutral
steroids and bile acids have also shown that both are excreted in
substantially larger amounts by animals on the soy protein diet (Huff
and Carroll, unpublished experiments). These findings suggest that
the higher rate of loss of cholesterol from body pools of animals on
the soy protein diet may help to maintain lower plasma cholesterol
levels in these animals compared to these on the casein diet.

Rabbits have been used in most of our experiments, but studies
in other laboratories have provided evidence that cholesterol-free,
semipurified diets containing casein are capable of producing hypercholesterolemia in other species such as dogs, monkeys and baboons.
This work has been reviewed previously[4, 11].

There is also evidence that dietary protein can influence plasma
cholesterol levels in human subjects[4, 5, 11]. In some of the earlier

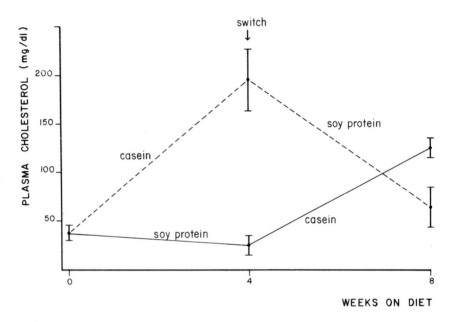

Fig. 2 – Variations of the plasma cholesterol with a diet containing casein and another containing isolated soy protein.

TABLE 1

NUTRIENTS AVAILABLE FOR CONSUMPTION
IN THE UNITED STATES   PER CAPITA PER DAY[a]

YEARS	TOTAL FAT (g)	LINOLEIC ACID/ SATURATED FATTY ACIDS	CHOLES-TEROL (mg)	TOTAL PROTEIN (g)	ANIMAL PROTEIN/ VEGETABLE PROTEIN
1909–1913	125	0.21	509	102	1.07
1947–1949	141	0.27	577	95	1.77
1970	157	0.42	556	100	2.34

[a]Adapted from Gortner[18].

studies, it is difficult to be certain whether the observed effects on plasma cholesterol levels are due to protein or other dietary variables. However, our own recent experiments have shown that replacement of animal protein in the diet of normal human subjects by soybean protein produces a small, but significant decrease in plasma cholesterol, even when other dietary variables, such as fat and sterols, are maintained relatively constant[12]. Sirtori et al.[13] carried out somewhat similar experiments on Type II hypercholesterolemic patients and reported a much larger decrease in plasma cholesterol when animal protein in the diet was replaced by soybean protein. It is not yet certain whether the greater response observed in their experiments is related to the use of hypercholesterolemic patients, or to the different type of diet, compared to that used in our experiments.

In other studies, vegetarians living in the United States were reported to have lower plasma cholesterol levels than the general population[14] and a reducing diet consisting largely of animal protein and fat was found to produce a hypercholesterolemic response in human subjects[15]. Epidemiological data also show a strong positive correlation between animal protein intake and incidence of coronary heart disease in different parts of the world[16, 17].

These appears to have been an increase in the incidence of coronary heart disease in North America since the beginning of this century, and it seems unlikely that any such trend can be attributed to changes in dietary fat and cholesterol. According to information collected by the United States Department of Agriculture (Table 1),

fat available for consumption in the U.S. increased during the period
from 1909 to 1970, but the degree of unsaturation also increased.
There was also a small increase in the amount of dietary cholesterol.
The total protein remained relatively constant during this time span,
but the ratio of animal to vegetable protein increased dramatically
from 1.07 in 1909-1913, to 2.34 in 1970. It seems possible that
this shift from vegetable to animal protein may have had a significant influence on incidence of coronary heart disease.

REFERENCES

[1] Hamilton, R.M.G. and Carroll, K.K., Plasma cholesterol levels in
      rabbits fed low fat, low cholesterol diets. Effects of dietary
      proteins, carbohydrates and fibre from different sources. Athe-
      rosclerosis 24: 47-62 (1976).
[2] Huff, M.W., Hamilton, R.M.G. and Carroll, K.K., Plasma cholesterol
      levels in rabbits fed low fat, cholesterol-free, semipurified
      diets: Effects of dietary proteins, protein hydrolysates and
      amino acid mixtures. Atherosclerosis 28: 187-195 (1977).
[3] Carroll, K.K., Huff, M.W. and Roberts, D.C.K., Dietary protein,
      hypercholesterolemia, and atherosclerosis. In: "Atherosclerosis
      IV, Proceedings of the Fourth International Symposium, Tokyo,
      1976", G. Schettler, Y. Goto, Y. Hata and G. Klose, (eds.),
      pp. 445-448, Springer-Verlag, Berlin, 1977.
[4] Carroll, K.K. and Hamilton, R.M.G., Effects of dietary protein and
      carbohydrate on plasma cholesterol levels in relation to athe-
      rosclerosis. J. Food Sci. 40: 18-23 (1975).
[5] Carroll, K.K., The role of dietary protein in hypercholesterolemia
      and atherosclerosis. Lipids 13: 360-365 (1978).
[6] Huff, M.W., Hamilton, R.M.G. and Carroll, K.K., Effects of dietary
      proteins and amino acids on the plasma cholesterol concentra-
      tions of rabbits fed cholesterol-free diets, in: "Adv. Exp.
      Biol. Med. Vol. 82, Atherosclerosis: Metabolic, Morphologic and
      Clinical Aspects, London, Ontario, 1975", G.W. Manning and M.D.
      Haust, (eds.), pp. 275-277, Plenum Press, New York, 1977.
[7] Lambert, G.F., Miller, J.P., Olsen, R.T. and Frost, D.V., Hyper-
      cholesteremia and atherosclerosis induced in rabbits by purified
      high fat rations devoid of cholesterol. Proc. Soc. Exp. Biol.
      Med. 97: 544-549 (1958).
[8] Malmros, H. and Wigand, G., Atherosclerosis and deficiency of es-
      sential fatty acids. Lancet ii: 749-751 (1959).
[9] Carroll, K.K., Huff, M.W. and Roberts, D.C.K., Vegetable protein
      and lipid metabolism, in: "Soy Protein and Human Nutrition
      Conference at Keystone, Colorado, 1978", D.T. Hopkins, D.H.
      Waggle and H.L. Wilcke, (eds.), Academic Press, New York (in
      press).
[10] Huff, M.W. and Carroll, K.K., Effects of dietary protein on plasma
      cholesterol levels and cholesterol oxidation in rabbits. Fed.
      Proc. 36: 1104 (1977).

[11] Carroll, K.K., Dietary protein in relation to plasma cholesterol levels and atherosclerosis. Nutr. Rev. 36: 1-5 (1978).

[12] Carroll, K.K., Giovannetti, P.M., Huff, M.W., Moase, O., Roberts, D.C.K. and Wolfe, B.M., Hypocholesterolemic effect of substituting soybean protein for animal protein in the diet of healthy young women. Am. J. Clin. Nutr. (in press).

[13] Sirtori, C.R., Agradi, E., Conti, F., Mantero, O. and Gatti, E., Soybean-protein diet in the treatment of Type-II hyperlipoproteinemia. Lancet i: 275-277 (1977).

[14] Sacks, F.M., Castelli, W.P., Donner, A. and Kass, E.H., Plasma lipids and lipoproteins in vegetarians and controls. New Engl. J. Med. 292: 1148-1151 (1975).

[15] Rickman, F., Mitchell, N., Dingman, J. and Dalen, J.E., Changes in serum cholesterol during the Stillman diet. J. Am. Med. Assoc. 228: 54-58 (1974).

[16] Yudkin, J., Diet and coronary thrombosis. Hypothesis and fact. Lancet ii: 155-162 (1957).

[17] Yershalmy, J. and Hilleboe, H.E., Fat in the diet and mortality from heart disease. A methodologic note. N.Y. State J. Med. 57: 2343-2354 (1957).

[18] Gortner, W.A., Nutrition in the United States, 1900 to 1974. Cancer Res. 35: 3246-3253 (1975).

SOME CURRENT CONCEPTS CONCERNING

PATHOGENESIS OF ATHEROSCLEROSIS

Kaare R. Norum

Inst. for Nutrition Research, School of Medicine

University of Oslo, Oslo, NORWAY

The atherosclerosis in the arteries is a slowly developing process which suddenly gives symptoms. These are due to complications of the atheromas caused by thrombosis, partially or complete rupture of the artery, or other mechanism for narrowing or obstructing the lumen of the artery.

The symptoms come usually from disturbed blood flow through the arteries of the heart, i.e. coronary heart disease (CHD), of the central nervous system, and of the legs.

The development of the atherosclerosis and its complications may both have relations to nutrition. In the following a short review will be given on the pathogenesis of atherosclerosis with a discussion of in which way nutrition can have an influence on the cardiovascular diseases.

Atherosclerosis is characterized by deposition of intra- and extra-cellular lipids, mainly cholesteryl esters (CE) and unesterified cholesterol (UC) in the inner part of the arterial wall, proliferation of smooth muscle cells (SMC) in the intima, and accumulation of large amounts of connective tissue matrix in the intima.

Several large epidemiological and clinical studies have revealed certain risk factors for development of cardiovascular diseases[1]. Main risk factors include age, male sex and genetic factors, which we cannot interfere with, and hypercholesterolemia, smoking, high blood pressure and diabetes mellitus, which we more or less can interfere with. Overweight and physical inactivity are also by several considered risk factors. Although the composition of the atheroma is complex, and the different risk factors may have varying impor-

tance in individuals, cholesterol plays a very important role. This
is based upon:

a) persons homozygote for familial hypercholesterolemia develop
fulminant atherosclerosis in childhood without other risk factors as
hypertension, smoking or diabetes mellitus[2];

b) in populations with a mean plasma cholesterol concentration
less than 160 mg/dl coronary heart disease is very seldom, even in~
the presnece of hypertension, smoking and diabetes mellitus[1];

c) the probability of an individual developing CHD increases
with his plasma cholesterol concentration[3];

d) atherosclerosis can be produced in almost every species of
experimental animals in which the plasma cholesterol concentrations
can be raised[4];

e) whereas no atherosclerosis can be induced in primates by
procedures leeding to thickening of the arterial intima unless the
animal has a hypercholesterolemia[5]

Since cholesterol is of crucial importance for the development
of atherosclerosis it is necessary to discuss some basic concepts
concerning cholesterol transport and metabolism.

LIPOPROTEINS AND CHOLESTEROL TRANSPORT

Cholesterol and cholesteryl esters are transported in blood and
lymph plasma as parts of lipoproteins. Lipoproteins generally contain proteins (called apolipoproteins) cogether with polar lipids
like unesterified cholesterol (UC) and phospholipids in a surface
film surrounding a core of apolar lipids like triglycerides (TG) and
cholesteryl esters (CE). The lipoproteins are usually divided into
classes according to their density: high density, low density, and
very low density lipoproteins (HDL, LDL and VLDL, respectively).
Chylomicrons are lipoproteins of very low density made in the intestinal mucosa during absorption of fat. The circulating lipoproteins
are different from the newly made (nascent) lipoproteins. The major
changes occuring in plasma are due to the activity of two enzymes;
lipoprotein lipase (LPL) and lecithin:cholesterol acyltransferase
(LCAT) and exchange and transfer of both apolipoproteins and lipid
between the lipoproteins.

Chylomicrons

On an affluent, fat diet and individual produce 100-150 g chylomicrons per day. About 90% consist of TG. However, the TG transport

needs unesterified cholesterol and apolipoproteins as surface coat
for the neutral lipids. It can be calculated that transport of about
100 g of TG as chylomicrons requires about 1.7 g of cholesterol as
surface coat[6]. Furthermore, 1-2 g of esterified cholesterol formed
in the mucosal cells is transported together with the TG. Thus, a
high fat diet will cause an extra burden on the cholesterol transport
and metabolism. The TG of the chylomicrons are hydrolyzed by lipoprotein lipase, extra surface material (apolipoproteins, phospholipids, cholesterol) are transfered to other lipoproteins and the
rest, the socalled chylomicron remnants are taken up by the liver[7].
Small chylomicrons (intestinal VLDL) produced in the gut may be
metabolized more like VLDL produced in the liver (see below).

VLDL

     VLDL are produced in the liver, the amounts are dependent on
dietary factors. The main function of VLDL is transport of endogenous TG made in the liver from extrahepatic fatty acids or fatty
acids produced in the liver itself. About one third of fatty acids
derived from chylomicron TG may enter the liver. Excess of carbohydrate in the diet may be converted to TG in the liver. In the
post-absorptive state the fatty acids which enter the liver are regulated by the lipolytic activity of the adipose tissue. The liver
utilize only a certain amount of fatty acids, and the excess fatty
acids are esterified to TG and exported as VLDL, from about 15-100 g
daily depending on diet and physiological conditions. The export of
TG from liver requires surface material for the lipoproteins. Thus,
influx of cholesterol and of apolipoproteins in blood plasma from
liver are dependent on dietary fat and excess of carbohydrates. The
main apolipoproteins secreted from the liver on VLDL are apoB, apoE
and some apoC. The two latter leave VLDL as it is metabolized in the
blood plasma; whereas apoB stays with the particle during its intravascular metabolism. The TG are hydrolyzed by lipoprotein lipase,
excess surface lipid material transfered to other lipoproteins or
cellular plasma membranes, and some CE are transfered from HDL. By
this dynamic process the VLDL are converted through particles of
intermediate sizes into particles of low density, i.e. LDL.

LDL

     LDL are formed in blood plasma from VLDL and may be to some ex-
tent from small chylomicrons. The main function of plasma LDL is
probably an easily mobilized store of cholesterol for peripheral
cells. LDL contain more than two third of the cholesterol in normal
plasma. Most of the LDL is metabolized through the socalled LDL-
pathway of peripheral cells, some LDL may also be degraded by other
mechanisms, either in peripheral cells or in the liver. The LDL-
pathway has been clarified mainly through extensive and impressive

studies by Goldstein and Brown[2]. When peripheral cells are in need
of cholesterol they make specific cellular membrane receptors which
bind LDL with high affinity. The bound LDL are invaginated into the
cell in the form of endocytotic vesicles which fuse with lysosomes,
in which the LDL are exposed to several hydrolytic enzymes. The protein part of the lipoproteins is broken down to amino acids and the
CE hydrolyzed to cholesterol and fatty acid. The cholesterol of the
LDL diffuse from the lysosomes to other cellular membranes. This increase in cellular cholesterol triggers at least three different regulatory metabolic events:

a) cholesterol synthesis is reduced through a suppression of the
hydroxymethylglutaryl.CoA (HMG CoA) reductase activity;

b) cholesterol esterification is increased through an increase
in the activity of acyl–CoA:cholesterol acyltransferase (ACAT);

c) the synthesis of LDL receptors is suppressed so that less LDL
are taken up through the LDL pathway.

Patients homozygote for familial hypercholesterolemia lack LDL receptors. However, LDL are also metabolized in these patients, probably through a socalled (scavenger pathway". This operates also in
normal individuals, and probalby is responsible for about one third
of the normal breakdown of LDL. As of today, we have very little
informations concerning the role of liver in the lipoprotein metabolism of man. Indirect studies have suggested that the liver may
have some importance.[8]

HDL

HDL in blood plasma are relatively small, globular particles which
consist of about 50% protein. The main apolipoproteins are apoA I
and apoA II, with small, but important amounts of C apoproteins. The
production sites of HDL are uncertain but are most probably liver and
gut. Based upon studies with perfused rat liver[9] and intestine[10] and
with patients with familial LCAT deficiency[11] it seems likely that
newly formed ("nascent") HDL are disc-shaped, the lipids are mainly
unesterefied cholesterol and phospholipids and the apoprotein composition is different from that of mature plasma HDL, and it depends
on whether the HDL are derived from the gut or from the liver[9, 10].
The main functions of HDL are uncertain. However, apoA I is an activator for LCAT and the HDL/LCAT system most probably plays an important role in the transport of cholesterol from peripheral cells to
the liver[11]. Moreover, some steroid hormon producing glands seem to
have receptor sites for HDL[12]. Rat liver cells bind, take up and
degrade HDL[13]. The HDL/LCAT system has a further role in cholesterol
transport as CE of HDL are transfered to VLDL and LDL, probably by a
specific CE transfer protein plasma[14].

Deposits of cholesterol are mainly found in tissues with a rather poor blood supply, i.e. tendons, skin, cornea and arteries.. For these tissues are the concentrations of lipoproteins in plasma of great significance. The concentrations of the different lipoproteins in blood plasma are results of their production and metabolism. In connection with atherosclerosis it will be preferable to have a low concentration of LDL which mainly supplies extra-hepatic tissue with cholesterol, and a high concentration of HDL which probably has a role in transporting cholesterol to the liver, the only organ which can catabolize and excrete substantial amounts of cholesterol. Both HDL and LDL concentrations are to some extent determined by genetic and hormonal factors. The diet, however, is of great importance. High amounts of dietary fat will increase the production of chylomicrons and VLDL and thereby the LDL. The content of apoB in VLDL (and LDL) is relatively independent of the size of the lipoproteins. Less LDL are therefore produced from VLDL transporting polyunsaturated fat, as this yields larger lipoprotein particles than does saturated fat. Polyunsaturated fat may also have an influence on LDL concentration by an increased turnover[15]. Dietary and drinking patterns may have an effect on the concentration of HDL[16]. Reduction of saturated fat in an ordinary diet led to a significant increase in the plasma HDL in middle-aged men[17], whereas individuals on formula diets got lower plasma HDL cholesterol when isocaloric amounts of saturated fat was changed with polyunsaturated fat[18].

THE CELLS OF THE ARTERY WALL
AND ATHEROSCLEROSIS

The cells lining the arterial lumen are called endothelial cells. Ross and co-workers have based upon studies with cell culture and in vivo experiments with primates set forward an integrated hypothesis for the pathogenesis of atherosclerosis, which they call "the response to injury hypothesis"[5, 19]. The hypothesis suggests that the endothelial cells which line the lumen of the artery are altered by an "injury" so that they desquamate at particular focal sites. Normally the endothelial cells resists the shearing stress of the blood flow. At some particular sites the shearing stress could be too high, specially if the blood pressure is raised, which could lead to focal endothelial desquamation. Chronic hypercholesterolemia seems to injure endothelial cells to that extent that even at normal blood pressure the shearing forces could disrupt the endothelial lining[5]. The focal desquamation expose the underlying basement membrane and collagen fibrils. Platelets will adhere to these sites and aggregate whereby mitogenic factors are released. Furthermore, plasma lipoproteins would more easily enter the arterial wall. The mitogenic platelet factor(s) and LDL stimulate the smooth muscle cells in the media layer of the artery to migrate into the intima and to divide. If the injury is chronic and hypercholesterolemia exists, this normal repair process will go into a vicious circle, in which the arterial wall develops an atheroma. The reaction of the smooth muscle cell in this process is important, but not completely understood. The smooth muscle cells are the repair and scavenger cells of the artery. They are normally located in the media layer, can produce connective tissue material and take up foreign material like excess of plasma proteins in the arterial wall. The cells have LDL receptors like other peripheral cells. However, in the repair process in the artery they seems to take up excess LDL by ꞈ receptor-independent scavenger process that leads to an uncontrolled accumulation of cholesteryl esters[20]. The cholesterol in the cells cannot leave the cell unless it is carried by a particle (HDL?) which can transport it into plasma. The transport of cholesterol from the smooth muscle cells is slow, specially in a thickened artery wall, and the cholesterol accumulates; mostly as cholesteryl esters made through catalysis of the ACAT. The cells are converted into lipid macrophages (foam cells) which may burst or die, resulting in deposits of extracellular cholesteryl ester and cholesterol.

Several aspects in the above outline must be studied in much more details before we understand complete the crucial steps in the multifactorial process of atherogenesis. Based on todays knowledge nutrition may affect several of the steps. A diet high in saturated fat will produce hypercholesterolemia which is harmful for the endothelial cell. The platelets in hypercholesterolemic blood, furthermore, have a greater tendency to aggregate than do platelets in normal blood[21]. Diets leading to a high LDL level will increase the LDL level of the intima[22] and thereby both stimulate the smooth muscle cell growth and accumulation of cholesterol. Polyunsaturated fat in the diet will reduce the aggregatability of the platelets[23]. Especially interesting are the most recent findings that eicosapentaenoic acid, which is an polyunsaturated fatty acid of marin origin is precursor for prostaglandin-like compounds which may lead to an anti-thrombotic state[24].

The last few years have given us much knowledge concerning the pathogenesis of atherosclerosis. We do not understand the whole process. We know for sure that the atherosclerosis is a multifactorial process, which most probably can be somewhat modified by our lifestyle, of which dietary habits play an important role.

REFERENCES

[1] Keys, A., Coronary heart disease. The global picture. Atherosclerosis 22: 149-192, 1975.
[2] Goldstein, J.L. and Brown, M.S., The low density lipoprotein pathway and its relation to atherosclerosis. Ann. Rev. Biochem. 46: 897-930, 1977.
[3] Westlund, K. and Nicolaysen, R., Ten-year mortality and morbidity related to serum cholesterol. Scand. J. Clin. Lab. Invest. 30,

    Suppl. 127, 1972.
[4] Wissler, R.W., Development of the atherosclerotic plaque, in:
    "The Myocardium: Failure and Infarction", Braunwald, E., ed.,
    HP Publishing Co., New York, pp. 155-166, 1974.
[5] Ross, R. and Harker, L., Hyperlipidemia and atherosclerosis.
    Science 193: 1094-1100, 1976.
[6] Kostner, G. and Holasek, A., Characterization and quantitation of
    the apolipoproteins from human chyle chylomicrons. Biochemistry
    11: 1217-1223, 1972.
[7] Fielding, C.J. and Havel, R.J., Lipoprotein lipase. Arch. Pathol.
    Lab. Med. 101: 225-229, 1977.
[8] Sniderman, A., Thomas, D., Marpole, D. and Teng, B., Low density
    lipoprotein. A metabolic pathway for return of cholesterol to
    the splanchnic bed. J. Clin. Invest. 61: 867-873, 1978.
[9] Hamilton, R.L., Williams, M.C., Fielding, C.J. and Havel, R.J.,
    Discoidal bilayer structure of nascent high density lipoproteins
    from perfused rat liver. J. Clin. Invest. 58: 667-680, 1976.
[10] Green, P.H.R., Tall, A.R. and Glickman, R.M., Rat intestine secretes
    discoid high density lipoprotein. J. Clin. Invest. 61: 528-534,
    1978.
[11] Glomset, J.A. and Norum, K.R., The metabolic role of lecithin: cho-
    lesterol acyltransferase: perspectives from pathology. Adv.
    Lipid Res. 11: 1-65, 1973.
[12] Andersen, J.M. and Dietschy, J.M., Regulation of sterol synthesis
    in 15 tissues of the rat. II. Role of rat and human high and
    low density plasma lipoproteins and of rat chylomicron remnants.
    J. Biol. Chem. 252: 3652-3659, 1977.
[13] Drevon, C.A., Berg, T. and Norum, K.R., Uptake and degradation of
    cholesterol ester-labelled rat plasma lipoproteins in purified
    rat hepatocytes and nonparenchymal liver cells. Biochim. Bio-
    phys. Acta 487: 122-136, 1977.
[14] Chajek, T. and Fielding, C.J., Isolation and characterization of a
    human serum cholesteryl ester transfer protein. Proc. Natl.
    Acad. Sci. USA 75: 0000-0000, 1978.
[15] Illingworth, D.R. and Connor, W.E., Present status of polyunsatur-
    ated fats in the prevention of cardi-vascular disease. Pro-
    ceedings from this meeting.
[16] Truswell, A.S., Diet and plasma lipids - a reappraisal. Am. J.
    Clin. Nutr. 31: 977-989, 1978.
[17] Hjermann, I., Enger, S.C., Helgeland, A., Holme, I., Leren, P. and
    Trygg, K., The effect of dietary changes on high density lipo-
    protein cholesterol. The Oslo Study. Am. J. Med. (in press).
[18] Sherpherd, J., Packard, C.J., Patsch, J.R., Gotto, A.M. and Taun-
    ton, O.D., Effects of dietary polyunsaturated and saturated fat
    and the properties of high density lipoproteins and the meta-
    bolism of apolipoprotein A-I. J. Clin. Invest. 61: 1582-1592,
    1978.
[19] Ross, R. and Glomset, J.A., The pathogenesis of atherosclerosis.
    N. Engl. J. Med. 295: 369-377 and 420-425, 1976.
[20] Goldstein, J.L. and Brown, M.S., Atherosclerosis: The low-density

lipoprotein receptor hypothesis. Metabolism 26: 1257-1275,
    1977.
[21]Carvalho, A.C.A., Coleman, R.W. and Lees, R.S., Platelet function
    in hyperlipoproteinemia. New Engl. J. Med. 290: 434-438, 1974.
[22]Smith, E.B., The relationship between plasma and tissue lipids in
    human atherosclerosis. Adv. in Lipid Res. 12: 1-49, 1974.
[23]Hornstra, G., Chait, A., Karvonen, M.J., Lewis, B., Turpeinen, O.
    and Vergroesen, A.J., Influence of dietary fat on platelet
    function in men. Lancet I: 1155-1157, 1973.
[24]Dyerberg, J., Bang, H.O., Stoffersen, E., Moncada, S. and Ane, J.R.,
    Eicosapentaenoic acid and prevention of thrombosis and athero-
    sclerosis? Lancet II: 117-119, 1978.

PREVENTION OF CARDIOVASCULAR DISEASES,

STARTING IN CHILDHOOD

R.J.J. Hermus and F. van der Haar

Institute for Human Nutrition – Agricultural University

Wageningen - THE NETHERLANDS

INTRODUCTION

Knowledge about pathogenesis of atherosclerosis and thrombosis, as well as the nature and epidemic occurrence of cardiovascular complications accounting for about half the deaths in industrialized countries of the West have led to the opinion that prevention will be the only cure[1].

Prevention, as a general term is used in the field of cardiovascular disease in several ways[2]. It embraces all measures which may lead to a reduction or postponement of atherosclerotic complications (see fig. 1). In the context of prevention, starting in childhood we should in fact talk about the prevention of atherosclerosis as a disease process. As the size and extent of atherosclerotic lesions can not be measured in the living intact organism, unless in a very advanced state, the concept of risk factors or, to remain neutral as to a causal relationship, risk indicators, is introduced. A risk factor was defined by Epstein[3] as "an attribute which appears to occur more frequently among persons with coronary heart disease than among control subjects". As risk factors for the development of atherosclerosis Stamler et al.[4] mentioned: age, sex, blood pressure, smoking of cigarettes, diabetes, cholesterolemia and diet in relation to blood lipids. Physical (in)activity and obesity were not considered as independent risk attributes, though very important in determining the risk.

Interest into the pediatric aspects of atherosclerosis arose by two observations. From histopathological investigations it appeared that early atherosclerotic lesions or precursors of the adult fibrous plaque were present in coronary arteries of very young

CURRENT TERMINOLOGY	NONE (= TRUE PREVENTION)	"PRIMARY PREVENTION"	"SECONDARY PREVENTION"
Initiated in:	Childhood	Adults (at Risk)	Survivors of First Attack
Prevention of:	Atherosclerosis	First Episode of Complications	Recurrent Episode

Fig. 1. Usage of the Term "Prevention" in Relation to C.H.D. (Austr. Acad. Sci., 1975:

I RISK DETERMINANTS ⟶	II RISK PROFILE ⟶	III MORBIDITY & MORTALITY
- genetics	- biological	- obesity
- physical environment	- biochemical	- hypertension
- social environment	- behavioral parameters	- diabetes
		- atherosclerosis
		- thrombosis tendency

Fig. 2. Schematic Representation of Pathogenesis of Cardio-Vascular Complications.

people[5,6] second, elevated levels of risk indicators were observed in school-aged children and these levels appeared to occur very frequently in children in industrialized countries. From prospective population studies it is known that the predictive power of risk factors is inversely related to the age of the subjects. Therefore the younger a risk factor is present, the more important the risk is. By consequence, prevention will not only be more effective if the risk is detected and adequately treated at an early age but also more effective if the development of risk in the young is prevented and/or postponed. Consequently prevention of atherosclerosis in a symptom-free population will actually imply prevention of the development of risk indicators or, as the etiology of cardiovascular disease is multifactorial, apparently without thresholds, prevention of development of a risk profile. In a positive way this might be called "health promotion".

Put into a sequence of events we are primarily concerned with the risk determinants (Fig. 2).

These determinants interact in establishing a risk factor of risk profile. Factors from the physical and social environment interact with genetic sensitivity in establishing a risk profile.

RATIONALE FOR PREVENTION

Which preventive approach should be applied ultimately is determined by several factors:

- the tracking phenomenon, which tells about the relative position of a subject over his life span in the distribution curve for the risk indicator under study. Complete tracking thus would provide the possibility to define the high risk adults already in childhood. These might be selected and intensively guided.

- the predictive power of this profile. By selecting the upper quintile of the risk scale only half of future coronary cases can be found. This procedure will leave the remaining half unidentified and without treatment.

- the frequency with which a certain trait is present in a population. If obesity is present in twenty to fourty percent in certain age groups, if increased cholesterol levels are present in the majority of people in industrialized countries, if hypertension has a frequency of 20-40% and if 50% of a population is smoking, then general hygienic measures are indicated, in addition to the generally accepted treatment of known high risk groups, i.e. family members of premature infarction patients, familial hypercholesterolemics and essential hypertensives.

In the following the main risk indicators for children will be discussed shortly with emphasis on both the tracking and the frequency of occurrence aspects. Also aggregation of risk within families will be considered.

OBESITY

There is now abundant evidence that obese babies do not necessarily remain fat[8]. In addition it appeared impossible to predict at the age of 11 how fat a subject was going to be at 16. The heritability of body fatness in childhood is high. Nevertheless there is hardly any information on this heritability for obesity in adults and this may be quite different from that in children[9]. Brook[9] reported longitudinal correlations of skinfold measurements for subjects who were 3-14 year old at first examination and who were reexamined 15 years later (see Table 1). There is an increase in correlation coefficients if the first examination took place at an older age. The older the child, the more it resembles its adult shape. Measurements during puberty of girls are apparently of little predictive value. The data clearly show that body fatness evolves and is not predetermined. Though the correlations are considerable, it should be remembered that the variability in body fatness of adults can be explained only for 30-50% by body fatness variations in childhood.

The prevalence of obesity was recently surveyed in the Netherlands[10] for school children, aged 4-13 years old. The prevalence of borderline plus frank obesity was 15% in boys and 17% in girls (at least 20% of body fat). The prevalence increased from about 6% in the 4-7 year old group to about 25% in 10-13 year old boys and from 9-30% in girls of the same age group. Frank obesity was present in 4% of the boys and 5% of the girls (at least 25% fat), increasing from virtually absent in the youngest to about 7.5% in the oldest groups. Prevalence figures from different countries are hard to compare. However, if we compare[10] (Table 2) the median of several skinfolds it can be demonstrated that the Dutch school children are leaner at the median level than their age mates from Canada, F.R.G., U.K. or U.S.A. This finding should be judged in the light of the presence of 4-5% frankly obese in addition to about 10% borderline obese children in the Dutch school children population. The high prevalence of obesity in industrialized countries, both in children and in adults, added to the already in childhood observed association with blood pressure, increased levels of blood glucose, insulin, triglycerides and possibly cholesterol[10] make it mandatory to shift the whole upperhalf of the distribution of body fatness to lower values.

TABLE 1

LONGITUDINAL CORRELATIONS OF SKINFOLD THICKNESSES IN
CHILDHOOD AND REPEATED 15 YRS LATER (C.G.D. BROOK, 1978)

AGE AT FIRST EXAMINATION	POOLED R FOR TRICEPS AND SUBSCAPULAR MALES / FEMALES
3 – 5	0.43 / 0.42
6 – 8	0.50 / 0.54
9 – 11	0.60 / 0.51
12 – 14	0.70 / 0.24

TABLE 2

90th PERCENTILE VALUES OF SKINFOLD DISTRIBUTIONS
IN SCHOOLBOYS (F.v.d. Haar & D. Kromhout, 1978)

AGE	CANADA	F.R.G.	U.K.	U.S.A.	NETHERLANDS
6	9.5	12.0	12.0	12.0	11.0
8	12.0	13.5	12.0	14.0	10.5
10	14.0	16.0	15.0	16.0	13.0
12	17.0	19.0	17.5	–	15.5

SERUM LIPIDS AND LIPOPROTEINS

Recently very high correlation (r = 0.9) was reported for youngsters between the concentration of plasma low density lipoprotein cholesterol and the rate at which this lipoprotein enters the arterial intima in man. However, cholesterol deposition is the function of not only the lipoprotein concentration but also of the duration of the contact and these constitutes both a strong argument from maintaining low levels of the beta lipoproteins in childhood.

Although there are striking differences in mean serum cholesterol between adult populations, reflecting dietary patterns, these differences are slight or absent at birth[12]. Crawford et al.[12] observed in 0-1 old Europeans and East-Africans cholesterol values of 116 and 120 mg respectively (Table 3). At 6-8 years old they had risen to 165 mg and at 12-16 years to 196 mg in the Europeans. In the East-Africans cholesterol remained stable in this lifespan. As upper limit for normal cholesterol values, Drash[13] suggested a level of less than 200 mg/dl. In industrialized societies between 8-25% of children would be diagnosed as hypercholesterolemic on this basis.

Leonard et al[14] assessed the degree of confidence with which serum cholesterol estimates could be used to identify familial cholesterolemia. In first degree relatives of patients with familial hypercholesterolemia only 4.25% of the children would be misdiagnosed. In the general population in whom the risk is 1:500, the probability of making a correct diagnosis would be only about 10%. Therefore screening should be confined only to the children of patients with premature infarction, in which the genetic occurrence is in the order of 1:8. Hypercholesterolemia may be the most common risk factor for coronary disease identifiable among children. In Wisconsin about one third of schoolchildren (5-14 year) had cholesterol levels over 200 mg/100 ml[17] . Wilmore and McNamara[18] reported a prevalence of 20% and in the Bogalusa study 9% of white and 13% of black children had levels this high. Since only about 2% of this is genetic hyperlipidemia the vast majority is of exogenous origin: the "hypercholesterolemia of affluence"[15]. It may be presumed that children populations with relatively high cholesterol concentrations will retain these high levels to adult life. There is less information about risks for individual children. The Muscatine study[16] in the U.S., a follow-up study of 4 years of 1953 children aged 5-18 years showed a year to year rank correlation for cholesterol of 0.61.

In the Netherlands also a high prevalence for cholesterol levels above 200 mg/100 ml was reported[10]. For boys this figure was 19.5% and for girls 24.8%. Frank hypercholesterolemia (more than 220 mg/ 100 ml) was present in 7% of the boys and 10% of the girls. Comparison with values from other countries revealed that these values were among the highest reported.

TABLE 3

BLOOD CHOLESTEROL LEVELS OF EUROPEAN AND
AFRICAN MALES (M.A. Crawford et al., 1978)

AGE (yrs)	EUROPEAN (mg/100 ml blood ± s.d.)	EAST AFRICAN
0 - 1	116	120 ± 10 (n = 45)
6 - 8	165	109 ± 7 (n = 37)
12 - 16	196	128 ± 13 (n = 26)
25 - 45	210	141 ± 11 (n = 43)

It could be calculated that since 1946 concomitantly with an increased consumption of animal proteins, saturated fats, dietary cholesterol and oligosaccharides the mean serum cholesterol of Dutch school children populations increased from 145-150 to 175-180 mg/ 100 ml[10] (Table 4).

It is evident from these figures that the whole cholesterol distribution curve of the population is shifted to higher values in industrialized countries. This, together with lack of knowledge about the tracking phenomenon or the presence of only a weak correlation between subsequent cholesterol values, added to the low prevalence of genetic abnormalities of lipid metabolism in the young, will lend support to a nation-wide approach to lower cholesterol values of the whole population by general rather than individual means. If the cut-off points proposed by Kannel and Dawber[19] are applied, 160 mg, it would mean that the greater part of children in industrialized countries should be monitored. An immense task if carried out on an individual basis.

SMOKING

Many children experiment with cigarettes[20]. 10% of children in the U.K. had puffed their first cigarette before the age of 9. 7% of boys and 2.5% of girls aged 10-11 smoked 1 or more cigarettes a week. Boys start smoking earlier than girls. The prevalence of smoking increased from 4% at 11.5 years yearly to 30.8% at 15 years. A rapid further increase occurred among those who left school. Factors associated with children smoking have been identified:
- having parents who smoke; mainly sex-linked;

TABLE 4

SERUM CHOLESTEROL AND DIETARY COMPOSITION ON
DUTCH SCHOOL CHILDREN, AGED 6-10, SINCE 1946
(F.v.d. Haar & D. Kromhout, 1978)

YEAR	CHOLESTEROL* (mg/100 ml)	ENERGY (kcal)	FATS	VEG. PROT.	ANIM. PROT.
			(perc. of energy)		
1946	149	1867	28	7	5
1951	146	1862	34	7	6
1974	172	1833	39	4	8
1975	178	1987	42	4	9
1976	182	1947	40	4	9

* All values recalculated to Abell-Kendall values.

- having brothers or sisters who smoke; not sex but number-linked;
- having friends who smoke;
- playing truancy;
- taking part in social activities.

    There seem particularly two locations where smoking patterns
may be influenced: in the school and in the home. The strategy
should aim at making smoking socially unacceptable and to reward non-
smokers and "stoppers". Prevention should probably already be started during foetal life as it was shown that in the placenta a decreased vascularization was found in smokers, perhaps explaining why
children born to heavy smokers are small.

BLOOD PRESSURE

    In early childhood, blood pressure levels are much lower than
those, that are predictive of future disease in adults. Few children
have been followed longitudinally to late adult life to see the relationship of early levels to the adult onset of stroke or coronary
heart disease.

    Normal values in relation to age and sex have been suggested by
the American Task Force on Blood Pressure Control in Children[21]. It
was suggested that all children older than three years of age should
have their blood pressures measured and plotted on grids like height

and weight centile charts as a convenient way to follow blood pressures in a single child over time.

In the Muscatine study[22] 6622 children were measured. Approximately 13% had elevated pressures when first examined but less than 1%, 41 persons, were found to have persistent pressure elevation upon rescreening at school, by a special nurse and finally by a physician. Of these 23 were obese with relative weight in excess of 120%, 5 had secondary hypertension and 13 were considered having essential hypertension. Children therefore should be monitored in the setting of continuing care where repeated pressures can be measured over a period of time to identify those with fixed pressure elevations. In populations where salt intake is high, values for blood pressure increase with age and hypertension is frequently present. There is an epidemiological argument for limiting salt intake in children and also animal experiments support the idea that salt may act as a trigger to provoke hypertension in genetically susceptible individuals.

In general, prevention of obesity and moderation of salt usage to daily intakes below 5 g NaCl, might largely prevent the occurrence of increased blood pressure levels in adulthood. Moreover, an adequate dietary intake of linoleic acid might be required to maintain blood pressure homeostasis in a high sodium environment[23].

Despite the lack of firm data it is clear that 20% or more of the children in industrialized countries are currently destined to become hypertensive adults. Prevention should be focussed on what we know as predisposing factors, i.e. obesity and the high salt intake, as well as on detection after repeated measurements and compliance with continued treatment.

FAMILY AGGREGATION OF RISK

Many of the risk factors aggregate strongly within families. It is already long known that smoking is more prevalent among children of smoking parents than of non-smoking parents. Ayman[24] reported elevated blood pressure in 28% of children in families with one hypertensive parent and 41% in children with two hypertensive parents. This figure rose to 65% of adult siblings of hypertensive parents. Concordance of measurements among parents and their children for several risk factors and diet has also been investigated in Dutch school children[10]. There appeared to be the same picture for standing height, for body fatness, hemoglobin and total cholesterol.

If both parents were in the lower quartile for the parameter under study their children were also for the greater part in the lower quartile. When both parents were high, their children were also likely to be high.

The here and repeatedly observed aggregation of risk between parents and their children perhaps gives the clue to inferring "risk", even in childhood. High degrees of familial resemblance in coronary risk factors will result in a high concordance of the disease[25]. The practical consequence of the familial resemblances in disease and risk factors would be the justification of attempts to prevent or reduce the risk, even in childhood.

The diet of Dutch schoolchildren resembled those of their parents remarkably well[10]. Both were identical to gross consumption in the Netherlands, based on food balance sheets (Table 4). Intra- and inter-individual standard deviations for the main nutrients were the same, indicating a very homogeneous dietary pattern in this population.

CONCLUSION

There seems no doubt that atherosclerosis has its roots in childhood or even foetal life. Many of these roots are related to adult personal habits and lifestyle, which are imposed upon children during the educational process. Preventive lifestyle patterns acquired early in childhood are the preferred alternative. Because members of a family share their genes, live together, eat and drink together, exercise together and smoke together[26] health promotion should also be family oriented.

Risk factor identification as an isolated, individually oriented activity should be considered as useless from a public health point of view. However, continuous monitoring of the risk-profile trends during childhood, would provide the "reality factor" which makes health education personal. It would also allow identification of that small percentage of extremely high risk individuals for who personal medical care is indicated.

The "know your body" program in New York[26] is attempting to achieve the goal of risk reduction and prevention in children by means of screening for risk factors, giving the children their own results in a "health passport" and following up with educational activities integrated to existing school curricula. The final objective in terms of morbidity and mortality may not be demonstrable, for 35 years to come. More work will be needed to develop effective educational programs, evaluation of which by biological and behavioural monitoring should be an integral part.

Part of this study was supported by the Netherlands Heart Foundation, Grant 73.092.

REFERENCES

[1] Holland, W.W., Prevention: the only cure, Prev. Med. 4:387-389, 1975

[2] Australian Academy of Science Diet and Coronary Heart Disease, Report nº 18, March, 1975.

[3] Epstein, F.H., The epidemiology of coronary heart disease. A review. J. Chron. Dis. 18: 735-748, 1965.

[4] Stamler, J., D.M. Berkson and H.A. Lindberg, Risk factors: their role in the eitology and pathogenesis of the atherosclerotic diseases. In: "The pathogenesis of Atherosclerosis", R.W. Wissler and J.C. Geer; Williams & Wilkins Cy, Baltimore, pp. 41-119, 1972.

[5] McGill, H.C., The lesion in children. In: "Atherosclerosis", G.W. Manning and M.D. Haust, Plenun Press, New York, pp. 509-513, 1977.

[6] Woolf, N., The origins of atherosclerosis, Postgrad. Med. J. 54: 156-161, 1978.

[7] Turner, R.W.D., Perspectives in coronary prevention, Postgrad. Med. J. 54: 141-148, 1978.

[8] Food and Nutrition Board, National Research Council, National Academy of Sciences, Fetal and infant nutrition and suceptibility to obesity. Nutr. Rev. 36: 122-126, 1978.

[9] Brook, C.G.D., Influence of nutrition in childhood on the origins of coronary heart disease. Postgrad. Med. J. 54: 171-174, 1978.

[10] Haar, F. van der and D. Kromhout, Food intake, nutritional anthropometry and blood chemical parameters in 3 selected Dutch schoolchildren populations. Thesis, Veenman, Wageningen, Netherlands, 1978.

[11] Niehaus, C.E., A. Nicoll, R. Wootton, B. Williams, J. Lewis, D.J. Coltart and B. Lewis, Influence of lipid concentrations and age on transfer of plasma lipoprotein into human arterial intima. Lancet ii: 469, 1977.

[12] Crawford, M.A., A.G. Hassam and J.P.W. Rivers, Essential fatty acids and the vulnerability of the artery during growth. Postgrad. Med. J. 54: 149-152, 1978.

[13] Drash, A., Atherosclerosis, cholesterol and the paediatrician. J. Paediatr. 80: 693, 1972.

[14] Leonard, J.V., A.G.L. Whitelaw, O. Wolff, J.K. Lloyd and J. Slack, The diagnosis of familial hypercholesterolemia in childhood by serum cholesterol concentration, Br. Med. J. 1: 1566, 1977.

[15] Lloyd, J.K. and R.J. West, The place of the hyperlipidaemias. Postgrad. Med. J. 54: 190-195, 1978.

[16] Clarke, W., R. Woolson, H. Schrott, D. Wiese, and R. Lauer, Tracking of blood pressure, serum cholesterol and obesity in children. The Muscatine Study, Circulation 54, suppl. II: 23, 1976.

[17] Golubjatnikov, R., T. Paskey and S.L. Inhorn, Serum cholesterol levels of Mexican and Wisconsin schoolchildren. Amer. J. Epidemiol. 96: 36-39, 1972.

[18] Frerichs, R.R., S.R. Srinavasan, L.S. Webber and G.S. Berenson

Serum cholesterol and triglycerides in 3446 children from a
biracial community. The Bogalusa Heart Study. Circulation 54:
302-308, 1976.

[19]Kannel, W.B. and T.R. Dawber, Atherosclerosis as a pediatric problem
J. Pediatr. 80: 544-554, 1972.

[20]Bewley, B.R. Smoking in childhood, Postgrad. Med. J. 54: 197-198,
1978.

[21]Task Force on Blood Pressure Control in Children National Heart,
Lung and Blood Institute and National High Blood Pressure
Education Program. Pediatrics 59: 797, 1977.

[22]Rames, L.K., W.R. Clarke, W.E. Connor, M.A. Reiter and R.M. Lauer
Normal blood presure and the evaluation of sustained blood
pressure elevation in childhood. The Muscatine Study. Pediatrics
61: 245-251, 1978.

[23]Vergroesen, A.J. Physiological effects of dietary linoleic acid.
Nutr. Rev. 35: 1-5, 1977.

[24]Ayman, D. Heredity in arteriolar (essential) hypertension: a clinical study of the blood pressure of 1524 members of 277 families.
Arch. Int. Med. 53: 792, 1972.

[25]Rissanen, A.M. and E.A. Nikkilä, Coronary artery disease and its
risk factors in families of young men with angina pectoris and
in controls. Brit. Heart J. 39: 875-883, 1977

[26]Williams, Ch. L., Ch. B. Arnold and E.L. Wynder, Primary prevention
of chronic disease beginning in childhood. The "Know your body"
Program, Prev. Med. 6: 344-357, 1977.

DIABETES: A PUBLIC HEALTH PROBLEM

Virgilio G. Foglia

Instituto de Biologia y Medicina Experimental

Calle Obligado 2490, Buenos Aires - ARGENTINA

Diabetes Mellitus constitues a serious medical-social problem which forces governments to acknowledge it and take the necessary measures to retard its advance. It attacks individuals of all races, ages, sexes and social groups. It constitutes the most common metabolic disease in Latin America.

MEDICAL PROBLEM

The medical problem is characterized by the hereditary and chronic character of the disease and by its tendency to invalidade. There is heredity in the well known case of the identical twins in which the appearance of diabetes in one follows the other, and of the diabetic father and mother who have children suffering from the same affection.

But there are also risk factors which provoke the disease in those who have the genetic potential. Such is the case of obesity, over-feeding or pregnancy. There are also exogenous risk factors, important to know, such as physical or psychic traumas, the use of cigarettes, coffee, etc.

With respect to the complications, they constitute one of the most important problems of the present. The vasculars, including those that affect the nervous system and also those which affect the heart, the kidneys and the eyes, are important causes of invalidity and require, in each case, adequate treatment. To the already-mentioned complications we can add the disturbances caused by diabetes in the pregnant woman or in the fetus, and which causes serious disturbances in the woman or the abortion or malformation

407

of the fetus. We should also consider the dental lesions, with the increased number of cavities and the loss of teeth.

As a consequence of the things already described, the life of the diabetic is permanently threatened by inconveniences, whether temporary or definitive, which shorten his life and demand permanent care.

SOCIAL PROBLEM

The social problems are intimately linked with economical problems. The diabetic must lead an active and normal life and continuously care for his diet and for his medical treatment. This is particularly important in children, who must be treated and educated about their disease.

The treatment is expensive, and it is calculated that for poor families it reflects on 10 to 15% of their income.

Job selection must be compatible with the maintenance of the dietary and medical regime. Diabetics are capable of carrying out any work their health status permits. Their limitation occurs where they might endanger the life of other people who depend on them, as is the case with airplane pilots, bus drivers, etc.

Among the social problems marriage should also be mentioned, due to the hereditary character of this disease.

From the colectivity's point of view, the existence of a great number of diabetics means a serious economic burden. There must be hospitals and medical and paramedical personnel to treat them and to correct their invalidities. Many of them have to miss work. It is estimated that approximately 5,000 million dollars were spent in 1976 in the United States in relation to this disease.

SOLUTIONS

The fight against diabetes must be active and continuous. It is governments' responsibility to understand that diabetes is a public health problem which should be combatted.

The diabetic should be treated by doctors who understand the problem well, and he must, in turn, be educated about his disease. The role of the Medical Diabetes Societies, in this sense, is very important. They serve to inform and to discuss news, and also to organize symposia and congresses related to the subject. In actuality, courses are organized to prepare specialists during the same scientific meetings. This is very important in order to update doctors living in centers distant from the great cities.

Simultaneously, the role played by the <u>Associations</u> <u>for</u> <u>Assis-</u><u>tance</u> <u>to</u> <u>Diabetics</u> is also very important. They <u>fulfill</u> an extraor-dinary social role in the benefit of the patients, helping them with knowledge about the disease, and also financially. The organization of vacation colonies for children are well known and useful.

The present tendency is to help fight diabetes by creating a <u>National</u> <u>Diabetes</u> <u>Council</u>. This already works in many countries, and is used to coordinate the fight. The Council must carry out periodical census in several regions in order to follow the endemic course of the disease; it must undertake the registration of causes of mortality; it must subsidize the interchange of specialists; it must assist its labor with subsidies and assist with the publication of specialized magazines. its function consists of maintaining re-gional hospitals for treatment and research.

However, the fundamental job of all governmental and private entities is to develop scientific research, both basic and clinical. For this it becomes necessary to provide all the necessary financial assistance. This is the most important path to follow. In the study of diabetes successful investigations have changed the course of the disease, in benefit of the patients. But it is necessary to contin-ue, without rest.

The subjects to be investigated are numerous: the study of ani-mals with spontaneous diabetes; the role of genetics; that of virus-es; the function of the peripheral receptors of hormones such as in-sulin and others; pancreas and B-cells transplants; artificial pan-creas; immunological reactions in diabetes; the pathogenesis of vas-cular lesions; etc.

The day these problems, and others which have not been mention-ed, are satisfactorily solved, we can predict that we will have the satisfaction of banishing one of the worst causes of damages and suf-fering by humanity.

DIET IN THE MANAGEMENT OF DIABETES MELLITUS

Seymour L. Halpern, M.D., F.A.C.P.

New York Medical College,

New York, N.Y., U.S.A.

Diet therapy is of critical importance in the management and control of diabetes mellitus. The aims of the diabetic diet are:

1. To supply adequate amounts of all necessary nutrients — carbohydrates, fats, proteins, vitamins and minerals — as well as adequate calories.

2. To help normalize body weight.

3. To help correct the carbohydrate dysmetabolism.

4. To maintain blood lipids within the normal range,

   and

5. To improve the patients feeling of well being and permit normal activity.

Diabetes Mellitus is a heterogenous syndrome with two basic types of lesions. The first involves a disordered beta cell and is characterized by insulin deficiency and thus is insulin dependent.

The second involves a resistence to the effectiveness of insulin by peripheral tissues including muscle and adipose tissue. It is characterized by normal or even elevated amounts of circulating insulin. In some diabetics both defects may be present.

There are marked differences in diet management for these two major groups of diabetic patients. The first type, who are insulin dependent, are usually young diabetics of normal or below normal

weight. The second type, who usually do not require insulin, are
invariably obese when the diagnosis of diabetes is made. Clearly
the priorities of diet therapy for these two groups must be completely different.

In juvenile onset, insulin dependent diabetes, good control cannot be obtained with insulin alone. Diet therapy is essential in its
management. Diabetic children must consume sufficient calories not
only for adequate nutrition, but also for normal weight gain and growth. Meal planning is necessary to help avoid alternating periods of
high or low blood glucose levels. There must be day to day consistency of intake of calories, carbohydrates, proteins, and fat.

Regular spacing of food intake will avert episodes of hypoglycemia, as injected insulin continues to work between meals. Peaks
of hyperglycemia and glycosuria due to excessive ingestion of simple
sugars should be avoided. There must be consistency of timing of
meals, as well as extra food for unusual exercises.

While the non-insulin dependent patient may divide food into the
customary three meals per day, for the insulin dependent individual,
food should be divided routinely into three meals and a bedtime snack.

Occasional mid-afternoon or mid-morning snacks may be indicated
for insulin dependent youngsters. In an active school child, total
calories can be divided such that 2/10 are given for breakfast, 2/10
at lunch, 1/10 as a mid-afternoon snack, 4/10 as the evening meal and
the final 1/10 as a bedtime snack. Clearly, individualization must
be the rule. A source of simple carbohydrate to treat, abort or prevent hypoglycemia should always be available.

The majority of adult onset diabetic patients who do not require
insulin are obese. It is mandatory that obese diabetics lose weight.
In fact all obese diabetics, whether a child or an adult, must reduce
to their ideal body weight if diabetes is to be properly controlled
with a minimum of medication. Most adult onset obese diabetics can
be treated successfully with diet alone.

Studies indicate that those patients who are controlled with
diet alone often have less diabetic complications than those treated
with diet and oral hypoglycemics. The latter must never be used in
order to avoid dietary discipline.

The diet for obese diabetics should be nutritionally adequate
and be restricted in calories. In general an intake of at least
12000 calorie is recommended to minimize catabolism of lean body
tissue while providing adequate vitamins, minerals and other nutrients.

Although at one time it was considered important to restrict

carbohydrate intake, it is no longer considered necessary to dispro-
portionately reduce the intake of carbohydrates in the diet of most
diabetic patients.

Studies utilizing high carbohydrate diets indicate that this type
of diet may actually increase the sensitivity of peripheral tissues
to both endogenous and exogenous insulin. The glucose tolerance also
improves and a lower level of serum insulin has been found. Furthermore, the control of diabetes and insulin secretory capacity appear
to be maintained in spite of the increase in carbohydrate intake.

Since over 1/3 of diabetic patients have significant hyperlipidemia which may contribute to vascular disease, much investigation
has been done in recent years on controlling their blood lipids. It
is now considered that fat should be restricted to a maximum of 30-35%
of the daily calories. The liberalization of carbohydrate intake fa-
cilitates this reduction of total and satured fats and of cholesterol
in the diabetic diet.

In accordance with recent recommendations for the treatment of
hyperlipidemia, it is advisable to increase the proportions of polyunsaturates to saturated fats in the diet.

If a diet higher in carbohydrate and lower in fats is successful
in decreasing blood cholesterol and triglycerides, it is conceivable
that in the future some of the vascular complications of diabetes wil
will be decreased.

In summary, the new concepts of diet therapy provides for a diet
relatively high in carbohydrate — about 45-50% or even more, low in
fat — about 25-30% with emphasis on polyunsaturates, and severe restriction of refined sugar. The protein intake should be about 15% of
total calories.

More recently, high fiber carbohydrates have been emphasized for
diabetic patients. A number of studies have suggested that dietary
fiber influences blood glucose levels and that diabetic patients on
such a diet have both a decrease in blood glucose levels and a decreased requirement for insulin and/or hypoglycemic drugs. In insulin dependent patients there appears to be a significant lowering
of post-prandial blood glucose levels, but not of fasting blood glucose. If the insulin dose is kept constant, there may be an increase
in the number of hypoglycemic episodes. Although the mechanisms by
which a high fiber diet lowers blood glucose is not known, it appears
possible that fibers may slow the absorption of nutrients, thus presenting less of a challenge to the pancreatic beta cells. It has
been suggested that a low fat, high carbohydrate diet, containing
mostly fiber rich foods, may promote a greater decrease in triglycerides and cholesterol than a similar diet of refined carbohydrates.

These new studies on high fiber diets are important because when a patient increases the fiber content of his diet, whether it was recommended by a physician or done without medical advice, a significant fall in blood glucose may result in an increased frequency of insulin reactions in insulin dependent diabetics.

In patients who ingest high fiber diets because of all the publicity being given in magazines, newspaper and books promoting high fiber diets, it is possible that this could be an unrecognized cause of frequent insulin reactions in previously well controlled patients. Furthermore, some current reports indicate that increased dietary fiber may lead to decreased absorption of calcium, zinc, magnesium and phosphorous. On some future date, the traditional diabetic exchange lists which allow for free substitution of foods that are equal in calories, carbohydrates, fats and proteins may have to be revised. For example, apple juice contains 0.1 gram of fiber per 100 grams, while a fresh apple has 1 1/2 grams of fiber. Diabetic diet exchange lists however do not distinguish between them. Thus, difficulty in stabilizing the control of a diabetic patient who rigorously follows the prescribed diet, insulin and other recommende therapy conceivably could be caused by day to day fluctuation in dietary fiber content.

Although some current articles are recommending high fiber content of the diet.

Pregnant diabetics should eat 250 to 300 calories per day in addition to the calorie intake necessary for maintaining body weight. Because pregnant diabetics are more subject to starvation ketosis, the diet should contain a minimum of 200 grams of carbohydrate per day. This will suffice for meeting the energy needs of pregnancy. A total gain during pregnancy of 22 - 26 pounds is optimal, and a reasonable objective which can be achieved. The rate of weight gain should parallel the physiologic pattern of fetal growth which has an S shape. Weight gain should amount to 2 to 4 lbs. during the first trimester (about one pound per month), and approximately 1 pound per week thereafter.

Finally, it is important to realize that more than 50% of diabetic patients evidently fail to follow their prescribed diets. In order to insure patient compliance, diets must satisfy the taste and the ethnic and religious standards of diabetic patients. Furthermore, scales should be abandoned in favor of common household measuring devices such as the cup, tablespoon, ruler and slices. Every diabetic requires instructions in dietary matters. A single session, however, will not achieve behavioral changes that are essential if results are to be expected from educational experiences. Too, some understanding of the patients ability to comprehend and comply with instructions must be assessed in order to best ascertain how to approach the patient.

Motivation is the most essential factor in diet compliance. The personality of the physician and the zeal with which the diet is advocated by him are important in motivating the patient. Family involvement, namely the participation of parents of diabetic children and the spouses of diabetics, can be an important aspect in motivating their adherence to diets. Feedback in which the patient is informed as to the results of his progress helps to maintain a high interest level.

THE PRESENT PROBLEMS OF DIABETES MELLITUS AS

A SOCIAL DISEASE WITH SPECIAL REFERENCE TO ROMANIA

I. Mincu, M.D., Ph.D.

Clinic of Nutrition and Metabolic Diseases

Bucharest, ROMANIA

Investigations on diabetes in Romania have been extended over a period of more than 60 years. In 1921, Paulescu discovered insulin although late, his priority is today recognized by most of the scientific world. In 1938 an Antidiabetic Center was opened in Bucharest, diabetics being treated until then in out-patient units by physicians specialized in this field. Today, more than 38,000 diabetics of Bucharest and the surrounding districts, as well as the problem cases from the entire country, are cared for here. There are similar units in the other county centers of our country. But the unit in Bucharest is the central methodological authority and it works together with the Center of Nutrition within the Institute of Nutrition and Metabolic Diseases.

Epidemiological investigations have been carried out by the Antidiabetic Center of Bucharest already before the World War II on large groups of the population, applying the conventional means — glycosuria. It may be seen, however, that by glycosuria method even postprandial, the proportion of discovered diabetics is much smaller than those of postprandial hyperglycemia method and the latter appears less reliable than the method in two steps recommended by the International Federation of Diabetes or that in three steps (simpler, more economical and just as precise) introduced and used by us in Romania since 1972.

In 1938 the prevalence of diabetes in Romania was estimated on the basis of these investigations at 0.2%. During the same period in the U.S.A. diabetes was considered to have a prevalence of 0.52%, almost identical to that in France, England, Germany, etc.

After World War II the prevalence of diabetes increased in 1954 to 0.07%, in 1962 to 1.2% and in 1969 to 1.9%. Investigations carried out on groups of hundreds of thousands of inhabitants in rural and urban areas throughout the whole country during the 1970-1977 period established the prevalence of diabetes in Romania at 1.4-2.2% in rural areas and 3.7-4.4% in urban districts. The method in two steps of the IFD and that in three steps recomended by us were applied and grave superposed results.

It is noteworthy that all the known cases of diabetes until 1966, that is 17,233 cases, increased twofold by 1976, when there were 36,283 diabetics on the files of the Center, from the district convered by the surveys.

Diabetes was more frequent in males than in females in urban districts (59% as against 41%). In rural areas the incidence was greater in females (53.7% as against 46.3%).

The situation as described in Romania is similar to that of the whole world.

In the U.S.A. it is considered that the morbidity rate from diabetes rose from 0.52% before the second world war to 5% nowadays. The U.S.A. National Committee of Diabetes reported on 10 December 1975 at a Congress held in the country that the prevalence of diabetes had increased fifty-fold during the 1965-1973 period, affecting the health of more than ten million Americans. The incidence of diabetes is still increasing by 6% a year; at this rate it may be envisaged that 1 of 5 individuals born in U.S.A. risks diabetes.

Increase in morbidity from diabetes has also been recorded in other countries: Turkey - 1.58% in rural areas, 3.93% in towns; France - 4%; Egypt - 2.15%; Finland - 1.5% in rural areas, 3.5% in urban districts; Nicaragua - 5%; Costa Rica - 5.4%; Uruguay and Venezuela - 7% (according to West).

The statistics of the Bucharest Center show that the prevalence of diabetes follows an almost uninterrupted ascending slope until now, except for the years 1944-1947, coinciding with the end of the war and the drought that followed, years characterized by severe want of food (Fig. 1). These are not the first of such situations, similar instances were reported in France during the Franco-German war and the Commune of Paris (Butterfield). Germany (Otto) and the Scandinavian countries (Wiegand) during World War LL.

Increase in the frequency of diabetes during the last decades suggests the intervention of certain factors of environment linked on the one hand to the feeding habits of modern man and, on the other, to his way of life. In contrast with the constant incidence of the hereditary factor - 38 to 40% in our statistical material of 20,000 cases (Fig. 2) — one may note a steady increase in diabetes

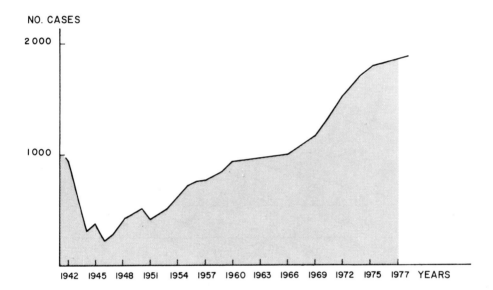

Fig. 1 - New diabetes cases discovered yearly
in the antidiabetic center.

NO. SUBJECTS

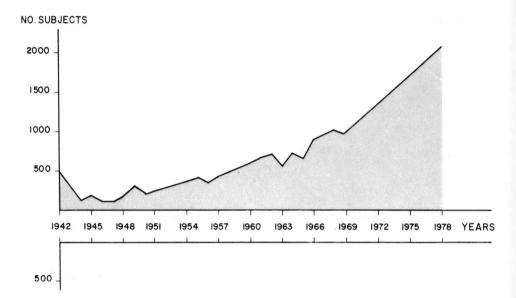

Fig. 2 - New yearly diabetes cases
with and without heredity.

mellitus due to other causes (obesity infectious factors, stress, etc.)

The ascending curve of the prevalence of diabetes also corresponds to our observations concerning the increased incidence of obesity. The relationship between diabetes and obesity can no longer be overlooked, their frequency increasing along parallel lines during the last decades. According to our statistics more than 85% of diabetics are overweight, in good agreement with the statistics of Joslin who found obesity in 70-75% of diabetics. When comparing the number of cases of obesity in terms of age, most of the cases are within the 40 to 65 years age-group, in keeping with the frequency of diabetes (Fig. 3).

THE COMPLICATIONS OF DIABETES

The evolution of diabetes mellitus depends on the subsequent complications and these in their turn are directly correlated with the mode of equilibration of the disease.

It is known that before the discovery of insulin most diabetics died from diabetic coma; since the introduction of insulin the mean survival of these patients has increased and with it, the possible subsequent development of complications chiefly caused by major or lesser vascular disease.

Figure 4, taken from Entmacher, shows the evolution of the causes of mortality before and after the discovery of insulin. After insulin the mortality from diabetic coma fell to less than 2-3%, as it is today. Starting in 1936 with the introduction of antiinfectious chemotherapy and particularly after the introduction of antibiotherapy the mortality rate from infections has decreased. Starting with 1950, which also corresponds with the introduction of oral therapy, the proportion of vascular diseases has gradually increased, exceeding today according to most statistics 78%, whereas the mortality rate from infections has fallen below 16%.

Pell D'Alonzo over a ten years period was following up the mortality from all the forms of diabetes in terms of the general mortality in the case of 90,000 individual between the ages of 17 and 64, reached to the conclusion that the death rate of diabetics is 2.6 times greater than that of the control lot (25.4% as against 9.7%). Recent investigations of the American Association of Diabetes have confirmed D'Alonzo's data, mortality from diabetes being the third on the list of the causes of death, after cardiovascular diseases and after cancer.

In a study we have carried out on 21,116 diabetics within the center of Bucharest over a 30 years period (1942-1972) the proportion

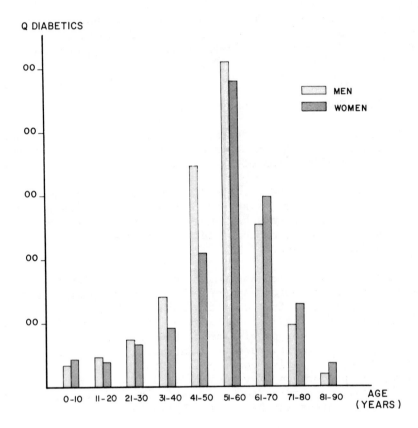

Fig. 3 – Distribution according to sex and age of diabetics
registered during the 1942 – 1977 period.

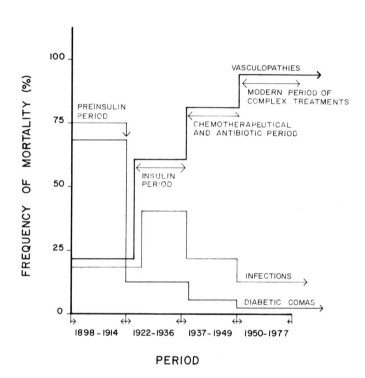

Fig. .4 – Causes of mortality from diabetes
           in the pre – and post insulin era.

and causes of death were calculated in comparison with a control lot.

The mortality rate of diabetics has gradually changed in the course of the thirty years passing from 36.8% in 1946 as compared to 15.5% in the control lot immediately after the war, to 19.9% as against 10.1% in the control lot in 1972, therefore a decrease of almost 1/3 as compared with 1946.

The general mortality among diabetics has slowly and steadily decreased from 88.62% in 1944 to 25.25% in 1972 with a progressive decrease of deaths under the age of 40 years (Fig. 5). Analysis of annual deaths has also shown a gradual decrease from 23% in 1942 to 0.85% in 1972.

CAUSES OF DEATH

The total number of deaths during the three decades were distributed according to the ten more important causes of groups of causes of death, namely: cardiovascular diseases, pulmonary diseases, digestive diseases, hepatic diseases, cancer, infections, etc. (Fig. 6). The total distribution of deaths in terms of the mentioned groups list cardiovascular diseases in the first place, cancer in the second, various causes in the third, diabetic coma in the sixth and infections in the last.

Annual distribution of the mortality indicates cardiovascular diseases increased from 29% in 1942, to 57% in 1972 and 65% in 1976, and deaths from tuberculosis fell from 13% in 1942 to 4.1% in 1972 and 2.9% in 1976. Mortality from infections decreased from 15% in 1942 to 1.4% in 1972 and deaths from diabetic coma from 18% in 1942 to 2.1% in 1972 and 0.3% in 1976.

In our analysis particular attention was given to proportion of deaths in terms of the treatment received by the patient (Fig. 8).

Thus, the proportion of patients receiving only a dietary treatment (the period of hypoglycide dieting) after a wide variations until 1953 had steadily increased until 1963 and with slight variations it remained within constant limits, decreasing nevertheless from 60 to 47% in 1967 and staying at the same level until 1972. Since then a slight decrease occurred to about 31% in 1977 (not shown in Fig. 7).

The proportion of deaths of the patients treated with insulin showed a sharp increase from 25% in 1947 to 70% in 1949 and then a steady decrease to 15.7% in 1972 and 10.2% in 1976.

The proportion of deaths among the diabetics recieving an oral medication and diet progressively increased from 3% in 1962 (when our statistical observations began) to 35% in 1970, 32,6% in 1972 and

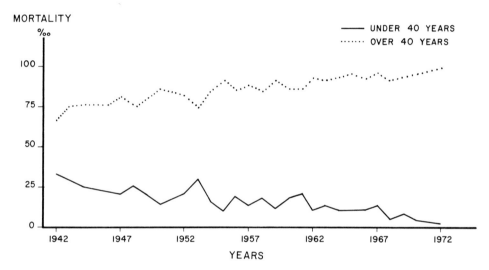

Fig. 5 - Mortality of diabetics registered in terms
         of age at death.

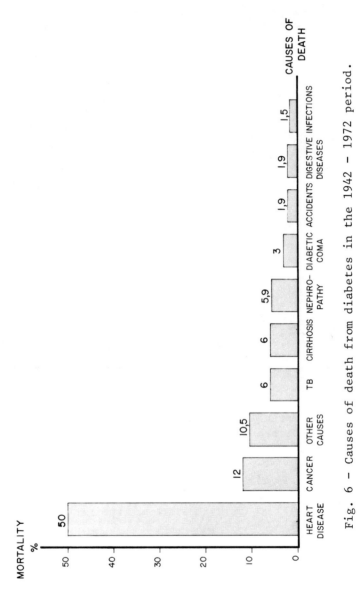

Fig. 6 – Causes of death from diabetes in the 1942 – 1972 period.

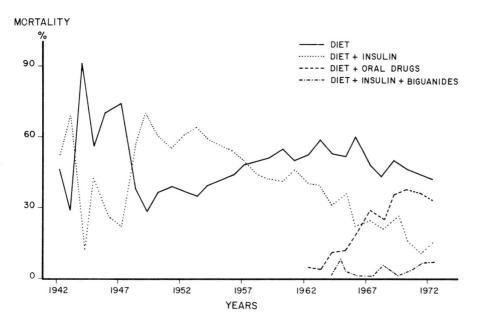

Fig. 7 – Treatment of faral diabetes cases during
the 1942 – 1972 period.

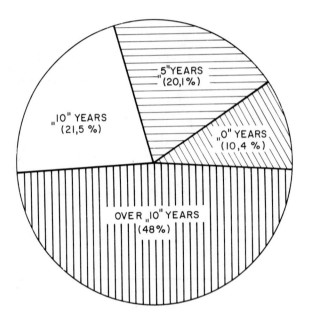

Fig. 8 - Incidence of diabetic angiopathy in terms of the
duration of diabetes mellitus in 3659 subjects.

39.8% in 1976.

These data we have mentioned above seem to show us that a stable balance of diabetes by diet and insulin is the best means to avoid chronic degenerative complications. Although in the question of the oral treatment the results and their interpretation cannot be considered as definitive, the conclusion we have reached is that of the American authors, i.e. the oral treatment is not the best means for balancing the diabetic, as it is suggested by the relationship between the proportion of deaths and the oral therapy.

The evolution of the mortality rate of diabetics runs paralel to our knowledge and the progress made in diagnostic thecniques and management of the cases.

The mortality rate is still very high today and the lifespan of the diabetic is short because of the chronic degenerative complications. Analysing the morbidity by chronic degenerative disease in diabetics we also may explain the exposed type of morbidity.

From our investigations it results that 48% of diabetics have a coronary affection as compared to 24% in the control lot. Stearus found an incidence of 74% among diabetics as against 37% in the controls, and Albrink 56% in diabetics. Similarly, as compared to 2% major vascular disease of the lower extremities in the general population we found an incidence of 20.9% among diabetics. Finally, cerebral atherosclerotic lesions were found in 20% of our patients. Figure 8 shows the incidence of these complications in terms of the duration of diabetes; after more than 10 years of evolution the frequency increases to 48% of cases.

Particular attention was paid to the incidence of microangiopathies (retinopathy, nephropathy). The proportion of 44.9% found in our cases was comparable to the incidence observed in U.S.A., England, Sweden and the countries of south-eastern Europe. Diabetic nephropathy was encoutered in 34.2%

If we refer to the report of the American Association of diabetes in 1975 in which it is stated that "in the diabetic cases, blindness is met with 25 times more often, nephropathies 17 times, arteriopathies 5 times and heart diseases twice as often as in non-diabetics"; this is however sustained by our data, as well, whence the particular severe aspect of diabetes for the public health disease, involving professional disability and retiring 10 to 15 years before time, expenses spend by repeated admissions to hospital and permanent care of these patients, cannot be overlooked.

The main problems stands, in our opinion, in the prevention of diabetes and once the disease occurs in the prevention of complications.

Among the measures taken along national lines in this connection is the prevention and control of obesity — encountered in proportions of 20 to 35% according to the region — by health education, by attentive survey of subjects from a diabetogenic environment, by treatment and surveillance of children during viral disease, by out-patient follow up of mothers with macrosome children or with births followed by a high, unjustified mortinatality. Eight years have passed since these measures were instituted and the results can be seen.

The prevention of the chronic complications of diabetes represents one of the main preocupations of all these centers of medical assistance to diabetics. Out-patient care of all patients with clinically manifest diabets mellitus, their permanent and qualified surveillance, and adequate treatment especially with insulin and diet, a corresponding occupation, survey of the family ambiance of these patients by visiting medical assistants, lead us to hope for beneficial results. It is, however, to soon to draw conclusions.

We consider that diabetes should become a problem of joint responsibility of the medical staff and of the State, as it is a social disease with an extensive character and once it appears will continue a lifetime. Its prevention and treatment is obviously a major social problem.

BIBLIOGRAPHY

Allen, A., L., - "The Kidney". Grune and Stratton, Inc. New York, 1962.

Albrink, I., Labiebs, N., Man, B. - Vascular disease and serum lipids in diabetes mellitus. Amer. J. Intem. Med. 58, 305, 1963.

Butterfield, W., I., H. - "Priorities in medicine". Nuffield Provincial Hospital, Trust, 1968.

Framantin, M., Freyria, I., Brossn, F., Bon, R. - Presse Med, 4, 1272, 1975.

Gronberg, A., Larsson, T., Jung, I. - "Diabetes in Sweden". Stockholm, 1967.

Marks, H., Rakll, P., White, P. - Epidemiology and detection of diabetes, in: "Diabetes Mellitus" (Joslin) Lea and Febiger. Philadelphia, 1969.

Mincu, I. - "Diabetic Micro and Macroangiopathy". ed. De Gruyter, New York, 1975.

Mincu, I. - "Diabetul zaharat". ed. Medicala, Bucuresti, 1977.

Otto, N., Bumek, M., Degenhardt, I. - Über den emfluss des Diabetes auf die Arteriosklerose. in: "Diabetische Angiopathie". ed. by Mohnike G, Akademie Verlag, Berlin, 1964, p. 181-185.

Pell, S., D'Alonzo, C. - J. Amer. Med. Ass 201,10:1832, 1970.

Schlach, V., Gutsche, H. - Epidemiologie des D. Mellitus. Intern--Donau Symposium, Budapesta, 1971.

Stein, G., West, M., Robey, M. - Arch. Intern. Med. 116, 842, 1965.
Stearns, S., Schlessingr, I., Rudy, A. - Arch. Intern. Med. 80, 463, 1947.
West, M., Kalbfleisch, M. - Diabetes, 19, 656, 1970.
West, M., Kalbfleisch, M. - Diabetes, 20, 99, 1971.

THE EDUCATION OF THE DIABETIC PATIENT

Raul Faria, Jr.

Inst. Estadual de Diabetes e Endocrinologia

Rua Moncorvo Filho, 100, Rio de Janeiro, BRAZIL

There is a sentence by Sir William Osler, cited in the preface of one of the books by Prof. Francisco Arduino, which states: if you want to live long, acquire a chronic disease and take good care of it.

The content of this sentence synthesizes the need for educating the diabetic patient. Diabetes mellitus is not a transitional disease, but rather a chronic disturbance which will be sharpened, aggravated and complicated if prophylactic measures are not adopted and an adequate metabolic control maintained. This cannot be done without the active participation of the patient, and will only be possible through his conscientization through education.

Fundamentally, this education should cover five aspects:

1. knowledge about the disease itself and consequences if control is not adequately maintained;

2. the need, advantages and possibilities of maintaining an appropriate diet permanently;

3. the utility of physical exercise and the practice of sports in facilitating the utilization of carbohydrates;

4. the correct application of the prescribed medication and the manner in which to adjust it according to the results of the complementary exams used to evaluate the intensity of the decompensation;

5. the removal of emotional tensions resulting from individual

disadjustments, whether social or environmental.

Oriented by these premises, the educational program for diabetic patients should be structured so as to be able to count with:

- group lectures by specialists, so as to provide basic knowledge about the disease, reasons for its aggravation and complications, as well as ways in which to remove existing problems through the adoption of a life philosophy, hygienic habits and the usase of the necessary therapeutic measures;

- individual or group contacts with nutritionists, to explain the hows and the whys of the diets, adjust them to the individual food habits, teach ways in which to vary them in order to avoid food monotony and attend to the individual preferences and to the environmental or social needs;

- engagement of the patients in physical activity programs or in sports adequate to their physical conditions and age, under the supervision of physical education teachers integrated into the team, who have specific knowledge about the objectives to be reached and who are also capable of recognizing possible occasional episodes of hypoglycemia and how to control them. If this is not possible due to age or some other factor, at least motivate those patients to increase their physical activity within their possibilities;

- participation of nursing personnel specifically trained to teach the patients how to perform the most necessary complementary exams themselves, such as glycosuria and cetonuria analysis, how to correctly interpret the results and utilize them to dose the medication accordingly, adjusting it to the needs of the moment;

- integration of psychologists and social assistants to the treatment team in order to evaluate the individual patient's environmental and social-economic difficulties, acting in such a way so as to obtain the patient's active participation in the removal or minimization of existing problems.

Programs of this type are already implemented in our country. We know about the work of Anita Teixeira and her collaborators, who presented their results in the paper "Program for the Integrated Treatment of the Diabetic Patient", during the II Brazilian Congress of Diabetes, realized in November 1976 in Salvador, Bahia. Also Cristina Conceição Sanches, in the confines of the assistencial service of INPS (*) in Campinas, São Paulo, has been developing similar activities, and in personal communication informed that she would relate her experience in a paper to be presented at the XIII Brazilian Congress of Endocrinology and Metabolism, to be realized in November 1978 in Brasilia, Federal District. Here in Rio, although without publication of the results obtained, Rogerio de Oliveira, at the INPS Hospital da Lagoa, is also working on the development of such type of activity for global assistance to diabetic patients, in particular the younger ones.

At the State Institute of Diabetes and Endocrinology Luiz Capriglione, of the State Secretariat of Health for the State of Rio de Janeiro, which has its Medical Division under our charge, the Diabetes Service, under the orientation of Francisco Arduino, has developed with stronger emphasis this systematic for orienting the patients under its control. Clarification courses, interviews, psychological orientation and interviews, and practical teachings by the Nursing Service are frequently ministered to ambulatory and hospitalized patients.

Our greatest performance in this field, however, is related to the assistance of those patients with infant-juvenile diabetes mellitus.

In order to attend to the education of these young people, in accordance with the objectives we have exposed, IEDE (**) has realized, since 1973, six Vacation Colonies for Diabetic Children, and it is about its planning, implementation and results that we intend to emphasize in this work, relating our experiences with the subject.

The planning of these programs is initiated with the creation of an inter-professional team composed of specialized doctors, nutritionists, nurses, physical education teachers, social assistants, psychologists and recreation specialists.

The selection of the patients is carried out considering preliminarily the age group of the participants who should, in our opinion, be between 12 and 16 years of age, for this is the age group when the teachings ministered are more easily assimilated and the experience acquired permits them to assume, in the future, the control of their disease with greater maturity. General clinical exams, prophylactic vaccination against smallpox, tetanus and typhus, interviews with nutritionists to establish the ideal TCV of each individual's diet and know their food habits; evaluation by psychologists and social assistants of the individual's psychological, socio-economic and environmental or family problems are the next phases of the selection process. The nursing service is responsible for the prevision and requisition of the material, equipment and medication to be utilized.

(*)   Brazil's National Institute for Social Welfare

(**)   State Institute for Diabetes and Endocrinology

The preparation period ends with a meeting of the entire team with the children selected and their parents, in order to inform them about the objectives and norms that have been adopted, and try to obtain cooperation and active participation from all, as well as clarify any doubts.

The implementation of the project includes two weeks during which the children remain in an internship regime, under the permanent supervision of the treatment team. A pre-established daily activity program includes:

- physical activities auch as gymnastics, sports and excursions;

- feeding, programmed in accordance with the diets established based on the stipulated TCV and divided among six meals (breakfast, morning snack, lunch, afternoon snack, dinner and an evening snack);

- control of the basic disease by means of glycosuria tests (and eventually cetonuria) performed by the children themselves before breakfast, lunch, dinner and the evening snack, with annotation of the results in individual forms and the application of the adequate medication (NPH and simple insulin) when necessary;

- recreation through games, reading, theatrical plays, painting, and manual arts.

Throughout the duration of the Colony the children are instructed by the team, in an informal manner and with simple language within, their level of understanding, about the importance of each of these activities in the disease control. Natural curiosity is stimulated, and an environment of freedom and trust created so that all may expose their questions and problems and always receive honest and clear responses towards the solution of their difficulties.

The advantages of this educational method were predictable, considering that in 1929 Dr. Henry John, of Cleveland, installed one of the first enterprises of the kind in the Mita Kod Camp, which functioned for many years and which many authors, such as Marble and Stephens, Etzwiler and Simes, Mc Graw and Trever, have written about it successive articles in medical magazines, revealing the value of such enterprises.

Our experience, based on the six projects already implemented, with approximately 30 participants each, demonstrates that the programmed activities are reached. At short range, at the end of the Colony, at least 60% of the participants have increased, on the average, almost 3% of their corporal weight; close to 50% have decreased, on the average, almost 30% of the NPH insulin dosage previously used,

and more than 90% revealed, by means of questionnaires, to have assimilated the teachings ministered, reduced their anxiety and be more
apt to "assume" the mature form of treatment.

The great value of the Vacation Colonies rests, in our opinion,
in the opportunity it provides for learning about the disease and the
means of controlling it, as well as for favorable modifications of
the psychological position of the young colonists towards the disease
and those responsible for the orientation of their treatment, thus
improving the relationship between the children, their family members
and the members of the treatment team.

THE PSYCHOLOGIST's POSITION IN THE

FACE OF THE DIABETIC PROBLEM

Juan Carlos Kusnetzoff, M.D.

It becomes necessary to clarify the fundaments that orient the action of the professional in psychology, psychiatry, or psychoanalysis when, in diabetic patients, whether young or not.

The framework, we think, is defined in the following sentence: "the psychologist is not, nor can he be, the insulin". What does this mean?

In the field of psychology, there exists a very common tendency on the part of the professional to believe so intensely in the amplitude and efficacy of his theoretical and clinical tool that this becomes evident with the thought of diabetes, like so many other organic disorders, as being produced solely by emotional factors. Here the though defect consists in taking the part for the whole, extrapolating conclusions which are found at different levels of scientific knowledge, and fighting arduously for this knowledge, unknowingly preventing a productive interchange of ideas and experiences.

In the clinical somatic field similar things happen. Here also, the part is taken for the whole. The biologist, the psysiologist, the clinician, the specialist in nutrition, the diabetologician, daily familiarized with concrete elements in the physical-chemical levels of knowledge, also tend to ignore, without knowing it, the emotional and psychosocial factors which, in general, accompany every diabetic patient, who possesses these emotional factors due to the intrinsic quality of being an individual, independently from the diabetes per se. The diabetic of today had emotions of a different nature yesterday, when he did not have diabetes. He had a certain relationship with his family; there were several events in his life

439

which, in several ways, also affected his life, his character, his peculiar way of reacting to life...

The usual attitude of a scientist towards these types of problems is to think with a certain degree of mechanics that the emotional factors are compromised with the diabetes... The logical answer would be: is it possible for it not to be? This does not mean that the classical disturbance of the Langerhans Islets were produced by "an unresolved Oedipus complex", or "an oral frustration from the past", which now manifest itself through the structure of this syndrome. What is even more important to know and to study, in order to intervene with operative efficacy, is that this metabolic disturbance is inscribed in a person who lives in this or that manner, with this or that manner corresponding to the historical pattern of his own historical family development: that the type of reactions or behaviors produced by the diabetic person, produce and are effective, in their turn, in determining certain forms of varying interactions, different in each case, and which the clinician or the nutrition specialist see every day.

The Operative Psychological Action in the Young Diabetic

Thus clarified our position in the face of the problem, it becomes easier to circunscribe now to the "how to proceed" with the adolescent diabetic.

In order to place ourselves, in a very simple way, we must say that, before being a diabetic, an adolescent is an adolescent. And if we know, by definition, that an adolescent does not like being an adolescent, it is easy to imagine the potential complications of this premise, when we are faced with an adolescent, disgusted with nature which forces him to go through this stage: complications which acquire, in certain occasions, tinges of drama when the adolescent does not find someone with whom or against whom to become upset about being a diabetic.

With this we are trying to say that the psychologist will confront, in most cases, a young individual, sometimes pubescent, with a load of anguishing pain, with hardly contained agressiveness and resentment which acquire several forms in their symptomatic manifestation.

On one side of the spectrum we have the adolescents who express their agressive rebellion in a clear and well defined manner. The psychologist here is consulted by clinicians and family members themselves, because the hygienic- dietetic measures, and the specific medication care (insulin, oral antidiabetics, etc.) are not being administered, whether totally or partially.

As a general rule the psychologist or psychoanalyst has little to resolve directly with the patient. It is more valuable and effective to call family reunions, or to work with the clinician in order to investigate with him the best type of action to solve these important obstacles. It is logical that it be so. The derivation to psychologist can only aggravate the case of the rebellious diabetic adolescent. The psychologist or psychoanalyst will also cause strong attacks of rebellious resistance, entering his world of meanings with the false additive that aside from being diabetic they also want to make his crazy.

On the other side of the spectrum is the adolescent with are exterior manifestations of rebellion, these being covered by subtle and complicated behavior mechanisms which lead, in the last instance, to the sabotage of the clinical therapeutic labor. These are the "good" adolescents who rebel through secret violations of the food plan, or through defective or excessive self-administrations of insulin.

Here, the direct operative efficacy of the psychologist is frequent. He may try, with adequate technical administration, to conscientize the series of meanings that the diabetic disease has acquired for the patient, in what way this series of behavior has become inserted in the family environment, and which and how are the secondary benefits which the adolescent has learned to use, taking advantage of being sick. Simultaneously, the psychological action over the family environment, and the permanent group work with the clinician or the nutrition specialist are not excluded, on the contraty, are highly indicated.

In the environment of the psychopathologic spectrum an enormous variety of combinations can be found, presented daily by medical and psychological clinics.

The effective psychologic therapeutic action must part from the previously mentioned premise: the psychology professional can not (nor must he) be the insulin; however — we add — neither can the insulin alone be, metaphorically speaking, the solution to the diabetic.

Profile:

- diabetes resides not only in the adolescent but also, in a prepondenrant manner, in the family environment.

- diabetes prolongs or difficults the individualization or independence of the adolescent subject.

- we do not believe there is a correct and characteristic psychological profile of the diabetic. We believe, to be exact, that it deals with a series of combinations which have become stereotyped with time.

- there is no doubt of the need for psychological attention to the diabetic, as long as this psychological attention is not carried out isolated from the corresponding medical and family context.

DIET THERAPY OF CHRONIC RENAL FAILURE

ON MAINTENANCE HEMODIALYSIS

Prof. C. Delbue, R. de Palma, C. Franco, C. Di Toro
Nutricionista-Dietista Roxana Carreras
Dept. de Med. (Centro de Hemodialisis y Serv. de Nutr.)
Hosp. Nac. "Prof. A. Posadas" - S.E.S.P.
Buenos Aires, ARGENTINA

It is a very known fact the value of diet therapy for the management of patients in Chronic renal failure, either under conservative treatment or during a plan of prolongued dialysis.

Our purposes is the discussion of the follow up to the evaluation of nutritional status and the use of diets.

EXPERIMENTAL

We started our experiences eighteen months ago together in the Hemodialysis Center, the Department of Medicine, "Hospital Posadas".

Subjects

Studies were performed on ten patients from an average of 24, on a Prolongued Dialysis Plan: 5 females and 5 males. They had 3 sessions of 4 to 6 hours each, during an average of six months. The routine parameters were blood urea, creatinine, uric acid, electrolytes, hematocrit, hemoglobin, proteinograms and lipidograms, urine analysis of electrolytes and balances of Nitrogen, creatinine, Na. and K. The weight of the patients was controlled by mechanical scales and one electronic system model for bed use (American Scale, U.S.A.).

Evaluation of nutritional status

First: skinfold thickness was measured. The readings were made

443

with a skinfold thickness caliper, Harpender Model, by V.R. Gersmehl, Cheyenne, Wyoming, USA, specif. 200 G.M. spring 5 cm 100 scale, 3 mm points.

The readings were taken in four places: biceps, triceps, subscapular and suprailiac areas, nearest to 0,5 mm.

Second: The sum of the four skinfold thicknesses gives the equivalent value of the fat contents of the body as the ratio body-weight calculated by Durnim & Womersley[1, 2].

Third : The fat free mass is obtained by substracting from the weight of the body, the fat percentage, and after that by using the regression equation of Hill and others. Total body nitrogen is obtained[3].

Diet Therapy on Chronic Renal Failures

The diet of patients under our care previous to prolongued dialysis plan had been calculated with the ideas and theories that stem from the works of Giordano and Giovanetti[5, 6] Cottini, Gallina y Dominguez[7], and our personal experiences[8]. Selection of proteins with high contents of essential aminoacid, with the idea of minimum nitrogen content, to maintain the balance without uremic symptoms.

The steps are: (Table 1)

F. 1: Basic Protein Requirement
$$BPR = UN + FN + DN$$
BPR = 1.6 x Basal Cal. + 1000 mg + 235 mg x body surface
(UN = urinary nitrogen, FN = fecal nitrogen, DN = dermic nitrogen)

F. 1 should be multiplied by 1.2 correcting factor for digestibility.

F. 2: P.T. = Urea clearance x plasmatic urea x 2.92 (^{0}) + 11.5 (^{00})

(The actual Protein Tolerance - or the real protein that should be given to the patient)

(^{0}) converting factor   (^{00}) non-ureogenic fraction.

If F. 2 is equal to or above F. 1., the patient could be maintained on nitrogen balance. Otherwise if F. 2 is below F. 1, the patient will need dialysis, sooner or later.

Diet during prolongued dialysis plan

Due to the frequency in which patients reach dialysis in very bad conditions, and to the fact of the protein's big loss during

TABLE 1

CALCULATION OF BASAL PROTEIN REQUIREMENTS (F. 1)
AND THE PROTEIN TOLERANCES (F. 2)

<u>EQUATIONS FOR</u> : F. 1) Basal Protein Requirements
 	
              F. 2) Protein Tolerance
 	
F. 1) B.Pr.R. = U.N. + F.N.   + D.N.
 	
       X = 1,6 x B.Cal. + 1000 mg + 235* x B.S.
 	
      B.Pr.R. Corrected for Digestibility = X x 1,2
 	
F. 2) P.T. = (U.Clearance x Pl.U. x 2,92**) + 11,5
 	
  *) Adapted to C.R.I.
 	
 **) Urea Converting Factor

TABLE 2

DIET DURING PROLONGUED DIALYSIS PLAN

CASE: F.P.		N^{cr}: 187.024	
AGE : 43 a.		Weight : 66kg.	
Height: 1,76 m.		IDEAL WEIGHT: 72 kg.	
	CHO: 60,5%	1.464 cal.	366 g.
Cal. 2142	PR : 10,5%	252 cal.	63 g.
	Fat. 29,0%	702 cal.	78 g.

dialysis, a sufficient protein intake has to be provided for. We utilize, based on the work of Heidland and Kult[9], a diet with 1 g/k protein and a high caloric content. On the day of the dialysis we give a suplement of proteins with high essential aminoacids content, divided in two meals (Tables 2, 3a and 3b).

<u>First</u> : Pre-dialysis, a breakfast rich in proteins (milk, ricotta)
<u>Second</u>: During dialysis, a supplement of meat, custard, etc.

C. DELBUE ET AL.

TABLE 3a

DIET ON DIALYSIS - AMINO ACID AND UNSATURATED FATTY ACID COMPOSITION

CASE: F.P. N⁰ 187.024

AMINO ACIDS (mg.)

	ILEV.	LEU.	LYS.	METH.	VAL.	PHE.	THR.	TRYP.	HIS.
DIET	2.950,5	4.074,4	4.430,9	1.772,9	3.591,3	2.768,6	2.666,6	218,8	1.846,8
MINIMUM	700	1.100	800	1.100	800	1.100	500	250	
IDEAL	1.400		1.600		1.600		1.000	500	

UNSATURATED FATTY ACID (g)

	Oleic Acid	Linoleic Acid
DIET	24,7	11,7

SATURATED FATTY ACID NO MORE THAN 10% OF TOTAL CALORIC VALUE (Ideal).

TABLE 3b

DIET ON DIALYSIS – VITAMINS, MINERALS, ASHES AND FIBER COMPOSITION

CASE: F.P. NǓ 187.024

VITAMINS

	A (U.I.)	B1 (mg.)	B2 (mg.)
DIET	12.386	1.581,3	1.184,1
IDEAL	5.000	1.000	1.700

MINERALS

	Ca. (mg.)	P (mg.)	Na (mg.)	K. (mg.)
DIET	585	913,4	420,9	1.959,3
IDEAL				

ASHES (g)	FIBER (g)
8,8	6,06

TABLE 4

IDEAL WEIGHT (*) / ACTUAL WEIGHT

	SEX	
	FEMININE	MASCULINE
Normal	5	3
Undernourished	–	2
	5	5

(*) Jelliffe, D.B. (1968), O.M.S. Mon. NQ 53

TABLE 5

STATE OF NUTRITION AND TOTAL BODY NITROGEN (g)

SEX	NORMALS (*)	UNDERNURISHED (**)
Feminine	1.410,3 (M=5)	---
Masculine	1.895,7 (M=3)	1.544,8 (M=2)

(*)   IDEAL WEIGHT $\pm$ 20%

(**)   < 20% IDEAL WEIGHT

RESULTS

The 5 females were in their ideal weight, 3 of the men were also in their ideal weight and 2 were really in a poor state of nutrition (Table 4) under 20 percent of the ideal weight.

Table 5 shows the correlation between state of nutrition and total body nitrogen.

All the women had normal total body nitrogen. Three of the men had normal values but the two undernourished had low values, the differences were statistically significant. The values obtained from

TABLE 6

SKINFOLDS (mm.)

		BICEPS		TRICEPS		SUBSCAPULAR		SUPRA ILIAC		SUM OF FOUR	
		Pre *	Post *	Pre	Post	Pre	Post	Pre	Post	Pre	Post
FEMININE	Average	8.9	8.4	15.8	14.4	10.6	10.4	13.2	13.0	48.5	46.2
	P (**)	N.S.	(***)	N.S.		N.S.		N.S.		N.S.	
MASCULINE	Average	3.8	2.3	6.4	4.4	5.8	4.9	7.0	6.5	23.0	16.9
	P.	N.S.		N.S.		N.S.		N.S.		N.S.	

(*) PRE = Pre-dialysis, POST = post-dialysis

(**) "T" Test

(***) No significance, P < 0.05

TABLE 7

PRE AND POST - DIALYSIS WEIGHT (kg)

	FEMININE		MASCULINE	
	AVERAGE	P*	AVERAGE	P
PRE-DIALYSIS	58,6	N.S.	61,2	N.S.
POST-DIALYSIS	57,3	**	60,6	

    *   "t" Test
   ** No Significance, p < 0.0. 5.

TABLE 8

PRE AND POST - DIALYSIS TOTAL BODY NITROGEN (g) (*)

	FEMININE		MASCULINE	
	AVERAGE	P**	AVERAGE	P.
PRE-DIALYSIS	1.404,6	N.S.	1.761,8	N.S.
POST-DIALYSIS	1.410,3	***	1.755,4	

   * HILL., G.L. et al., Brit. J. Nutr. (1978), 39,405
  ** "t" Test
 *** No Significance, p < 0.05

the four skinfold thickness are given separatedly in Table 6, for both sexes, before and after dialysis.

As we can see, there is no significant variance in any of them. As Table 7 shows, we can say the same about corporal weight. Table 8 gives the values for body nitrogen and demonstrates no significant difference between pre and post dialysis values.

DISCUSSION

The evaluation of the nutritional status was estimated by many parameters besides the sensation of well being and the general clinical examination, together with the results of the chemical analysis.

The main point is the measurement of the total body fat by skinfolds.

The fat content, no doubt, has physiological and medical importance, it may influence morbidity and mortality.

Most methods for the evaluation of the nutritional status are based on the assumption that the body can be considered divided in two compartments. The body fat and the fat free mass, which includes all the rest of the body, apart from the fat.

Hill et al.'s paper[3] about the close relationship of FFM with total body nitrogen gives us a wide range of body composition of great importance for the evaluation of nutritional status. It can be a useful tool which is not apparently modified by dialysis.

We think it deserves to be incorporated to routine procedures because it is much easier and simpler than other methods like body density[10] total water[11], total body K[12].

Finally, we have to comment on diet therapy. Six out of ten patients complied with 90 percent of the dietary prescriptions. Two of the remaining patients reached a satisfactory standard of 60 to 80 percent.

But the last 2 patients were well below 50 percent on their diet performance. These were the undernourished, with statistically significant low values of total body nitrogen.

There is no doubt that the nephrologist in many cases needs the collaboration of a team composed by an internist, a nutricionist and a psychologist to be able to deal with all the factors belonging to the personality of the patient.

REFERENCES

[1]Durmin J.V., Womersley Br. J. Nutr. 32: 77, 1974.
[2]Durmin J.V., Rahaman M.M. Br. J. Nutr. 21: 681, 1967.
[3]Hill, G.L., Bradley J.A., Collins, J.P. Mc. Carthy I., Oxby, C.B., Burkinshaw L., Br. J. Nutr. 39: 403, 1978.
[4]Best, W.R., USAMRNL Report No. 113, August 31, 1953.
[5]Giordanno, C.J. Lab. Clin. Med. 62: 231, 1963.
[6]Giovanetti, S. Lancet, 1: 100, 1964

[7] Cottini, E.P., Gallina, D.L., Dominguez, J.M. 4⁰ Congreso Arg. Ntr. Dic. 68.

[8] Delbue C., Coordinador grupo de trabajo. Actas 4⁰ Congreso Arg, de Nutr. Mar del Plata 1-7 Dic. 1968.

[9] Heidland A., Kult J., Clinical Nephrology, V⁰ 3, N6, 1975.

[10] Jelliffe, D.B., O.M.S. N⁰ 53, 1968.

[11] Keys, A., Brozek, J., Physiol. Rev. 33: 245, 1953.

[12] Forbes, G.B. Gallup, J., Hursh, J.B., Science, N.Y. 133,101, 1961.

EFFECTS OF DIETARY FIBER ON CARBOHYDRATE METABOLISM

Ruth McPherson Kay, Ph.D.

Dept. of Nutrition and Food Sc. - Univ. of Toronto

Toronto, CANADA M58 1A8

Dietary fiber influences carbohydrate metabolism through a number of direct and indirect mechanisms. The chemical diversity of plant fiber and the paucity of good clinical data preclude a precise estimate of the significance of fiber in the treatment of disordered glucose metabolism. Recent studies, which are discussed below, do suggest that under certain conditions, dietary fiber has an important modulating effect on the glucose absorption rate and attendant hormonal responses.

There is limited evidence that fiber may impede energy intake. The association between excess body weight and impaired glucose tolerance is well-documented[1]. Bulking agents such as guar and cellulose have been used with some success in weight-reducing diets[2]. Titcomb has reported that the incorporation of large amounts of a high fiber bread into a low-calorie diet promoted satiety and facilitated weight loss[3]. These effects may have been due to gastric distention and/or an alteration in the rate of glucose absorption. Fiber may also influence the amount of energy derived from food. Pectin increases fecal fat excretion[4]. Other types of fiber enhance fecal losses of both fat and nitrogen[5, 6]. In a study by Southgate and colleagues[7], dietary fiber supplementation increased fecal energy loss by 20 to 95 kcal per day. The cumulative effects of such a change in net energy balance may be significant.

In a diet-coronary risk factor study involving 200 normal males (35-59 years), we have found a negative correlation between fiber intake and percent ideal body weight. In the same population there was a negative association between cereal fiber consumption and fasting blood glucose concentrations (Table 1) (Kay, Sabry, Csima; in preparation).

TABLE 1

RELATIONSHIP (PEARSON CORRELATION COEFFICIENTS) BETWEEN FIBER
CONSUMPTION, RELATIVE BODY WEIGHT AND FASTING BLOOD GLUCOSE
IN 200 NORMAL MEN. (KAY, SABRY, CSIMA)

	PONDERAL Index	PER CENT IDEAL BODY WEIGHT	FASTING BLOOD GLUCOSE
Total Fiber Intake[b]	.1796[d]	−.1329[c]	NS
Cereal Fiber Intake[b]	.2092[d]	−.1780[d]	−.1233[c]
Ponderal Index	−	.9328[d]	−.1191[c]
Per cent Ideal Body Weight	.9328[d]	−	.1384[c]

[a] (height (in.) $\div 3 \sqrt{\text{weight (lb)}}$ [c] $p < .05$

[b] crude fiber   [d] $p < .01$

By replacing nutrients of high caloric density, promoting satiety and reducing absorption efficiency, fiber may have a role in the prevention or treatment of obesity and thus indirectly influence glucose tolerance.

Dietary fiber also influences carbohydrate metabolism through a number of more direct mechanisms (Figure 1). There may be alteration in gastric emptying[8]. The latter determines the rate at which nutrients are delivered to the absorptive mucosal surface. Fibers also have differing effects on small intestinal transit rate which are discussed below. Since dietary fiber tends to reduce the rate of glucose absorption, there is likely to be alteration in the site of glucose absorption and secondary changes in the release of the gastrointestinal hormones such as gastric inhibitory polypeptide (GIP). GIP acts to potentiate the insulin response to a glucose load and is released in smaller amounts if glucose is absorbed in the distal rather than proximal intestine. It is apparent, then, that by inducing changes in the rate and site of glucose absorption, dietary fiber may influence postprandial blood levels of glucose and glucose-regulatory hormones.

The viscous or mucilaginous fibers appear to have the greatest effect on postprandial glycemia. Jenkins and coworkers fed seven normal subjects a meal containing 106 grams of carbohydrate with or without the addition of a 25 gram mixture of guar and pectin. Following the fiber-enriched meal, blood glucose levels were significantly lower; blood insulin levels were also decreased[9]. In the same study, the authors noted a reduction in the insulin to glucose ratio during the 45 to 90 minute postprandial period. Such an effect may have been due to alteration in GIP release. Goulder and cowork-

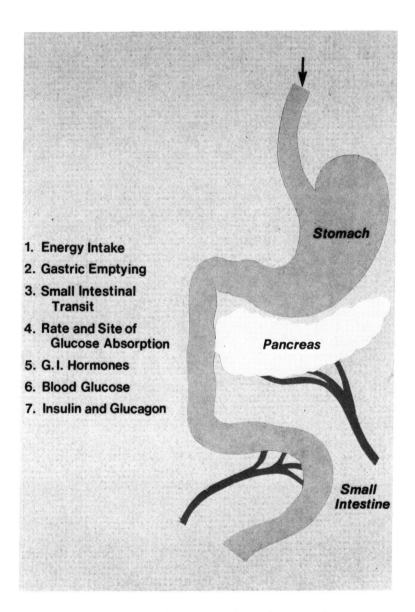

1. Energy Intake
2. Gastric Emptying
3. Small Intestinal
   Transit
4. Rate and Site of
   Glucose Absorption
5. G.I. Hormones
6. Blood Glucose
7. Insulin and Glucagon

Fig. 1 - Mechanisms by which dietary fiber
may influence c.h. metabolism.

have reported that guar induced a marked fall in GIP release in both normal and diabetic subjects[10].

More recently Jenkins and associates have investigated the mechanisms by which fiber may influence glucose tolerance[11]. Normal volunteers ingested a test meal containing glucose, xylose and lactulose with and without 12 grams of various fibers. The latter included viscous fibers (guar, pectin, gum tragacanth), nonviscous fiber (hydrolysed guar), particulate fiber (wheat bran), synthetic fiber (cellulose) and a fiber-analogue (cholestyramine). The rate of glucose absorption (as assessed by the urinary appearance of xylose, an absorbable but non-metabolisable sugar) was reduced by both the viscous fibers and wheat bran. Small intestinal transit time was determined as the peak rise in breath hydrogen induced by the cecal appearance and subsequent fermentation of lactulose, a nonabsorbable carbohydrate. Mouth to cecum transit time was prolonged by the viscous fibers, seemingly in relation to the degree of viscosity. Particulate fiber (bran) by contrast, reduced small intestinal transit time. All of the fibers studied, with the exception of hydrolysed guar, tended to reduce postprandial glucose and insulin; the viscous fibers had the greatest effect (Table 2).

In summary, these experiments indicate that viscous fiber isolates reduce the rate of glucose absorption and prolong mouth to cecum transit time. These effects appear to be a function of the degree of viscosity since if a mucilaginous fiber such as guar is hydrolysed, it loses its viscosity and no longer influences glucose absorption, intestinal transit or postprandial glycemia. It has been suggested that mucilaginous fibers impose a barrier to glucose diffusion by increasing the viscosity of the unstirred layer.

Particulate fibers such as bran also have an immediate effect on glucose absorption. Although not viscous, wheat bran has a high water holding capacity and acts as an intraluminal bulking agent. It may thus reduce the rate at which water soluble nutrients diffuse towards the absorptive mucosal surface.

The experiments of Jenkins, Leeds and others have opened up new therapeutic possibilities for mucilaginous fiber isolates. Guar and pectin have been used successfully in the treatment of post-gastrectomy dumping syndrome and diabetes mellitus[12-14]. Under these conditions, the fibers in question have been administered at much higher levels than could be obtained from a normal diet. We, and others, have been interested in the effect of fiber present in foods commonly eaten on the postprandial rise in blood glucose.

In a preliminary experiment, normal subjects were fed 30 grams of glucose in a fiber-rich (whole oranges) or fiber-depleted (orange juice) form. The former meal resulted in a slight decrease in the postprandial glycemic peak and a significant improvement in late

TABLE 2

GASTROINTESTINAL AND METABOLIC EFFECTS OF FIBERS: RELATIONSHIP
TO PHYSICAL CHARACTERISTICS (Adapted from Jenkins et al. (11))

	PHYSICAL CHARACTERISTICS	GLUCOSE ABSORPTION RATE	MOUTH TO CECUM TRANSIT TIME	POSTPRANDIAL GLUCOSE (60 min)	POSTPRANDIAL INSULIN (60 min)
Guar	viscous	↓	→	↓↓	↓↓
Hydrolyzed guar	non-viscous	0	0	0	0
Wheat Bran	particulate	↓↓	←	→	→

TABLE 3

EFFECT OF A HIGH CARBOHYDRATE-HIGH FIBER DIET ON BLOOD GLUCOSE
AND LIPIDS IN DIABETIC MEN (Adapted from Kiehm, Anderson and Ward (19))

	DRUGS	F. BLD. GLU.	PL. CHOL	PL. TG
Sulfonylurea (5)	↓↓	↓33%	↓24%	↓15%
15-28 u. insulin (5)	→			
40-55 u. insulin (5)	0	0		

postprandial satiety as measured by the 120 to 180 minute blood
glucose levels[15]. Similarly, Haber, Heaton and Murphy have reported
a flattening of the post-meal blood glucose and insulin-response
curves in normal subjects fed a carbohydrate load in the form of
whole apples as compared to apple juice[16]. These results indicate
that in normal individuals, physiological amounts of food-derived
fiber have significant effects on postprandial glycemia and satiety.

Perhaps of greater interest is the effect of food-derived fiber
on glycemic control in diabetic subjects. We have been examining
this question in elderly non-insulin dependent diabetics. Patients
consumed a high or low fiber diet, each for a 14 day period. On the
last day of each diet, a standard meal was given with individually-
fixed energy, protein, fat, available polysaccharide and sugar content. Dietary fiber averaged 1.1 grams (low fiber) or 14.3 grams
(high fiber). Blood samples were taken at 15 to 30 minute intervals
after the start of the meal. In all subjects tested, the high fiber
meal resulted in a demonstrable flattening of the glucose response
curve (Kay, Grobin & Track, in preparation).

Miranda and Horwitz have also recently reported that fiber has
a beneficial effect on diabetic control[17]. Eight insulin-dependent
diabetic patients were placed for ten days, on diets containing 3
or 10 grams of crude fiber. The added fiber was predominantly in
the form of cellulose fed as a high fiber bread. Diurnal fluctuations in glucose, insulin and glucagon were measured. Fasting blood
glucose was not altered during the high fiber diet, however the mean
blood glucose level throughout the day was reduced. Insulin levels
did not change; this was not unexpected since insulin was administered exogenously. Plasma glucagon concentrations were however
lowered during the period of fiber ingestion.

These short-term studies indicate that dietary fiber has an
acute effect on glycemic control. Other studies suggest that fiber
may also have a prolonged effect on glucose absorption or metabolism.

Brodribb fed 24 grams per day of wheat bran to 37 subjects for
a 12 month period. Body weight did not change but the glucose tolerance curve was significantly lowered following one year of bran
ingestion. This effect was observed even though the subjects had
not taken any bran for at least 12 hours before the glucose tolerance
test[18]. These results suggest that fiber may induce morphological
alterations in the gut affecting the rate and/or site of glucose
absorption.

Fiber is now being included in diabetic diets, the composition
of which would have seemed radical ten years ago. Anderson has
treated diabetics with a diet containing 75% of the calories as
carbohydrate and large amounts of dietary fiber. He has compared
the effects of this diet with that of a standard American Diabetic

Association — type diet, low in carbohydrate and fiber. Subjects tested included patients managed by oral hypoglycemic agents and patients requiring low or high doses of exogenous insulin. The high carbohydrate-high fiber diet resulted in total cessation of sulfonylureas in the first group. Insulin was decreased or discontinued in subjects requiring low amounts of insulin. In these two groups, fasting blood glucose fell by 33%. A similar response was not observed in those patients managed on high doses of exogenous insulin, that is patients with little residual pancreatic function[19] (Table 3. Serum lipids decreased in all groups, an observation of some importance since diabetics are known to be inordinately susceptible to atherogenesis. Since carbohydrate per se improves glucose tolerance, the beneficial effects of this diet cannot be solely attributed to fiber. However it does suggest that a high carbohydrate, high fiber diet may be the therapy of choice for mild diabetes.

In conclusion, dietary fiber modulates the glycemic and hormonal responses to glucose ingestion. Mucilaginous fibers appear to have the greatest effect on glucose absorption rate; more information is required on the effect of fibers present in foods as eaten and on the mechanism of their action. Fiber may be of clinical relevance in the treatment of obesity, diabetes mellitus, postgastrectomy dumping syndrome and reactive hypoglycemia.

REFERENCES

[1] Kay, R.M. Nutritional aspects of diabetes mellitus. "C.R.C. Handbook of Nutrition and Food", M. Rechcigl, edit., in press.

[2] Evans, E. and Miller, D.S. Bulking agents in the treatment of obesity. Nutr. Metabol. 18: 199-203, 1975.

[3] Titcomb, S.T. Use of reduced calorie, high fiber bread in special diets and weight loss experiments. Cereal Foods World 21· 421, 1976.

[4] Kay, R.M. and Truswell, A.S. Effect of citrus pectin on blood lipids and fecal steroid excretion in man. Amer. J. Clin. Nutr. 30: 171-175, 1977.

[5] Southgate, D.A.T. and Durnin. J.V.G.A. Calorie conversion factors. An experimental reassessment of the factors used in the calculation of energy value of human diets. Brit. J. Nutr. 24: 517-535, 1970.

[6] Southgate, D.A.T. Fibre and other unavailable carbohydrates and their effects on the energy value of the diet. Proc. Nutr. Soc. 32: 131-136, 1973.

[7] Southgate, D.A.T., Branch, W.J., Hill, M.J., Drasar, B.S., Walters, R.L., Davies, P.S., McLean-Baird, J. Metabolic responses to dietary supplements of bran. Metabolism 25: 1129-1135, 1976.

[8] Grimes, D.S. and Goddard, J. Gastric emptying of wholemeal and white bread. Gut 18: 725, 1977.

9 Jenkins, D.J.A., Leeds, A.R., Gassull, M.A., Cochet, B. and Alberti, K.G. Decrease in postprandial insulin and glucose concentrations by guar and pectin. Ann. Int. Med. 86: 20-23, 1977.

10 Goulder, T.J., Morgan, L.M., Marks, V., Smythe, T. and Hinks, L. Effects of guar on metabolic and hormonal responses to meals in normal and diabetic subjects. Diabetalogia 14: 235, 1978.

11 Jenkins, D.J.A., Wolever, T.M.S., Leeds, A.R., Gassull, M.A., Haisman, P., Dilawari, J., Goff, D.V., Metz, G.L. and Alberti, K.G. Dietary fibres, fibre analogues and glucose tolerance: importance of viscosity. Brit. Med. J. 1: 1392-1394, 1978.

12 Jenkins, D.J.A., Gassull, M.A., Leeds, A.R., Metz, G., Dilawari, J.B., Slavin, B. and Blendis, L.M. Effect of dietary fiber on complications of gastric surgery: prevention of postprandial hypoglycemia by pectin. Gastroent. 72: 215-217, 1977.

13 Jenkins, D.J.A., Wolever, T.M.S., Hockaday, T.D., Leeds, A.R., Howarth, R., Bacon, S., Apling, E.C., and Dilawari, J. Treatment of diabetes with guar gum. Lancet 2: 779-780, 1977.

14 Jenkins, D.J.A., Wolever, T.M.S., Nineham, R., Taylor, R., Metz, G.L., Bacon, S. and Hockaday, T.D. Guar crispbread in the diabetic diet. Brit. Med. J. 2: 1744-1746, 1978.

15 Kay, R.M. and Stitt, S. Food form, postprandial glycemia and satiety. Am. J. Clin. Nutr. 31: 738-739, 1978.

16 Haber, G.B., Heaton, K.W. and Murphy, D. Depletion and disruption of dietary fibre. Lancet 2: 679-682, 1977.

17 Miranda, P.L. and Horwitz, D.L. High-fiber diets in the treatment of diabetes mellitus. Ann. Int. Med. 88: 482-486, 1978.

18 Brodribb, A.J.M. and Humphreys, D. Diverticular disease: three studies. Part III - Metabolic effects of bran in patients with diverticular disease. Brit. Med. J. 1: 428-430, 1976.

19 Kiehm, T.G., Anderson, J.W. and Ward, K. Beneficial effects of a high-carbohydrate, high-fiber diet on hyperglycemic diabetic men. Am. J. Clin. Nutr. 29: 895-899, 1976.

INFLUENCE OF FIBER ON LIPID METABOLISM*

David Kritchevsky[1], Jon A. Story[2] & George V. Vahouny[3]

[1]The Wistar Inst. of Anatomy and Biol.
 36th Street at Spruce - Philadelphia, Penn. 19104
[2]Dept. of Foods and Nutrition - Purdue Univ.
 W. Lafayette, Indiana 47907
[3]Dept. of Bioch. - George Washington Univ. Med. Center
 Washington, D.C. 20037   USA

Long-term studies correlating various metabolic and behavioral parameters have shown that elevated levels of serum cholesterol represent a risk factor vis-a-vis the development of coronary dissease[1,2]. Implication of serum cholesterol in the development of coronary disease has precipitated many studies aimed at controlling serum lipid levels.

Blood cholesterol is derived from the diet and endogenous synthesis. The major catabolic pathway of cholesterol is conversion to bile acids, and the main excretory route of both cholesterol and bile acids is via the feces. However, because of the many molecular changes that take place in the gut, the final fecal products are more aptly designated as neutral and acidic steroids. Substances that inhibit synthesis or facilitate excretion of steroids may also affect serum cholesterol levels.

Burkitt et al.[3] and Trowell[4-6] have equated the absence of coronary disease in the African black with the large amount of fiber in his diet. Their publications have stimulated new research concerning the effects of fiber on lipid levels and cholesterol metabolism.

---

* Supported, in part, by grants HL 02033, HL 03299 and AG 00076, and a Research Career Award (HL 0734) from the National Institutes of Health.

Work on the metabolic effects of dietary fiber dates from the 1930s[7-11], but most early work was concerned with the laxative effects of fiber. The newer literature on fiber and lipid metabolism is no more than 15 to 20 years old. The development of methods for fiber analysis[12-14] and a better understanding of fiber chemistry[15-16] have expanded the area to be investigated.

LIPID METABOLISM IN ANIMALS

Cellulose has little effect on lipid metabolism in experimental animals. Wells and Ershoff[17] added 5% cellulose to a rat diet containing sucrose, cottonseed oil, casein and 1% cholesterol. Serum cholesterol levels subsequently rose by 30% and liver cholesterol levels by 25%. A similar effect was observed by Kiriyama et al.[18] and by Tsai et al.[19]. We[20] have also seen a slight hypercholesteremic effect in rats fed 1% cholesterol and 5% cellulose. When rats were fed cholesterol-free, semipurified diets containing cellulose, their serum cholesterol levels were higher than those of controls fed laboratory ration[21-22].

In contrast, pectin significantly reduces liver cholesterol levels in rats fed semipurified, cholesterol-augmented (1%) diets[17-20, 23-25]. One of the earliest studies on the dietary effects of pectin was carried out in 1957 by Lin and his associates[26]. They added 500 mg/day pectin to a rat diet that contained 50 mg/day cholesterol and found a 16% increase in the saponifiable matter of the feces. Pectin (5%) has been shown to increase excretion of bile acids but not of sterols in rats fed 1% cholesterol[26].

Among the other complex saccharides that have been tested, vegetable gums such as guar gum or carageenan have been found to lower serum and liver cholesterol levels in cholesterol-fed rats. Agar, however, causes a sharp rise in liver cholesterol[17-19], as does alginic acid[18].

Portman and Murphy[27] showed that, when a semipurified diet was substituted for laboratory ration in rats, the result was a marked decrease in fecal excretion of bile acids and neutral steroids. In studies with special diets, we[22] have found that, compared to cellulose, alfalfa increases fecal excretion of acidic steroids by 49% and of neutral steroids by 76%. Beher and Casazza[28] fed rats 4% of a hydrophilic colloid (Metamucil) and observed change in the cholic acid pool but an expansion of 123% in the chenodeoxycholic acid pool.

The addition of 5% lignin to a rat diet containing 0.5% cholesterol resulted in a 66% decrease in liver cholesterol; in contrast, the same level of lignin in a 1% cholesterol diet lowered liver cholesterol by only 19%. Serum cholesterol was not affected

in either case, and liver triglycerides were reduced by about 20%
in both experiments[20]. Judd et al.[29] fed rats 30 g/day lignin for
3 weeks and observed a 15% drop in serum lipid levels.

LIPID METABOLISM IN MAN

In the early 1950s, Hardinge and Stare[30] showed that vegetarian
diets result in lower cholesterol levels in man than do diets that
include meat. Compared to populations subsisting on standard diets,
lacto-ovo vegetarians ingest 50% more fiber and exhibit cholesterol
levels 11% lower. True vegetarians, whose cholesterol levels are
about 28% below those of the standard population, eat 125% more
fiber[4, 31]. In 1954, Walker and Arvidson[32] suggested that heart disease was almost unknown among the South African Bantu because of
their massive daily intake of crude fiber. Several years elapsed
before the research community, thanks to prodding of Burkitt and
Trowell, began a systematic inquiry into the effect of dietary fiber
on lipid levels in man.

Cellulose fed in quantities of 15 g/day has no effect on serum
cholesterol levels in man[33]. Studies of bran, the substance most
widely tested for hypocholesteremia, have had uniformly negative
results. Truswell and Kay[34] summarized the literature in early
1976. They cited 10 studies using an average of 14 subjects fed an
average of 37 g of bran daily (14-100 g) for periods of 3-19 weeks.
In only one instance was serum cholesterol lowered (by 7%); in one
other, triglyceride values dropped (by 18%). Bran does seem to
increase fecal steroids. Jenkins et al.[35] fed 30 g/day bran to six
men and found that the concentrations of neutral and acidic steroids
in feces (milligrams per gram) fell by 44% and 43%, respectively,
but that absolute excretion (milligrams per 24 hr) rose by 40%.
Cummings et al.[36] fed 45 g/day bran and found a similar increase in
neutral steroid excretion. Walters et al.[37] found bran to have essentially no effect.

Pectin (15 g/day) was found by Keys et al.[33] to be hypocholesteremic in man. Palmer and Dixon[38] found a similar effect with
6-10g of pectin daily. Durrington[39] and Jenkins[40] and their coworkers
fed 12 and 36 g of pectin, respectively, and observed cholesterol
level decreases of 9% and 12%. Kay and Truswell[41] administered 15
g/day pectin to nine men and observed a 15% drop in cholesterol
levels. Excretion of neutral steroids rose 16% during the feeding
period, and excretion of bile acids rose 40%. Upon cessation of
pectin feeding, all values returned to pretreatment levels.

Guar gum has been reported to lower cholesterol levels in man.
Fahrenbach et al.[42] fed 23 subjects 6 g/day for 66 days and achieved
a 5% reduction in cholesterol levels.

When they fed 12 g/day for 45 days (to 25 subjects), the reduction was 11%. Jenkins et al.[40] fed seven patients 36 g/day guar gum for two weeks and observed an average decrease in serum cholesterol of 16%. Bagasse (10.5 g/day) did not affect serum lipids but increased fecal bile acids by 50%, while reducing fecal neutral steroids by 10%[37].

It has been recognized that complex carbohydrates are less lipidemic than simple sugars. Cleave[43] has atrributed the increase in disorders such as heart disease and diabetes to increased ingestion of refined flour and sugar. Such complex carbohydrates contain various levels of fiber, which may also play a role in the observed effects.

Antonis and Bershon[44] fed white and black prisoners in South Africa diets in which two factors were varied: the men received either 15% or 40% calories in the form of fat and either 4 g or 15 g fiber. In general, subsistence on the high-fiber regimen resulted in increased excretion of fecal steroids. In the white subjects on the 40% fat diet, acidic and neutral steroid excretion of high-fiber was 3% and 48% higher, respectively, than in those on the low-fiber diet. In the Bantu, on the high-fiber diet, no effect on acidic steroid excretion was seen, but there was a 90% increase in neutral steroid excretion. For the low-fat/high-fiber diets, acidic and neutral steroid excretion percentages were increased by 43% and 81% in the white subjects and by 24% and 96% in the Bantu, respectively. The lipid patterns were similar for the two groups and depended upon the fat level of the diet. For the diets containing 15% fat, the serum lipid patterns of both groups resembled those usually observed in Bantu; for the 40% fat diets, the serum lipid levels of both groups resembled those usually seen in South African whites[45]. The two factors varied in the diets of the two groups must be important, since analysis of serum lipoproteins of control populations in South Africa has shown that the difference in these figures is far less marked than the gross racial difference in coronary heart disease mortality[46].

Grande[47] has summarized results from 12 studies in which starch was substituted for sucrose in the diets of normolipemic patients. The sources of starch were fruit, vegetables, legumes, cereals, leguminous seeds, potatoes and bread, and the average amount of calories exchanged was 23%. The exchange obviously resulted in the addition of fiber to the diets of the subjects. Every study showed a reduction in serum cholesterol levels (average reduction 12.6%) with seven of the reductions statistically significant.

Most recently, Kiehm et al.[48] altered the 2200-calorie diet of 13 diabetic men so that the carbohydrate level was increased by 77% but the oligosaccharides were unchanged. The ratio of complex to simple carbohydrate was increased from 1:1.15 to 1:2.63 and the level

of fiber from 4.7 g to 14.2 g. Serum cholesterol and triglyceride
levels fell by 24% and 15%, respectively, and blood glucose levels
were reduced by 26%. Therefore, the data derived from human subjects, like those from animal studies, suggest that fiber exerts a
wide range of effects on serum lipid levels, depending on its composition and possibly on the other dietary components as well.

FIBER AND EXPERIMENTAL ATHEROSCLEROSIS

About 15 years ago, research literature contained a number of
studies of atherosclerosis in rabbits fed saturated fat but no cholesterol. The data were confusing, since some of the studies pur-
ported to find saturated fat atherogenic per se and others did not.
Collation of the literature[49] showed that saturated fat had no atherogenic potential when added to a diet based on laboratory ration.
In seven experiments in which the average fat content was 19% in a
diet fed for an average of 5.3 months, no atheromata were observed.
In contrast, Lambert et al.[50] and Malmros and Wigand[51] found that a
semipurified diet containing as little as 8% saturated fat led to
atherosclerosis. In seven experiments utilizing semipurified diets,
an average of 18% saturated fat fed for 7.3 months led to an average
atheromata of 1.7, compared to a rate of 0.2 in rabbits fed unsaturated fat. When 5% of either coconut or corn oil was added to rabbit
laboratory ration and fed for 12 months, no lesions were observed
in either group, and the serum cholesterol levels found were 24 and
37 mg/dl, respectively[52].

Since the dietary fat was the same in diets based on a semipurified mix and in laboratory ration, it was clear that some other
component was responsible for the atherogenic properties of the semipurified diet. Speculation[49] led us to test the type of fiber included in each diet. It was found that the fat normally present in
rabbit feed prevented neither cholesteremia nor atherosclerosis.
The residue from the extraction of the ration fat limited cholesteremia and atherosclerosis.

Moore[55] fed rabbits a semipurified diet containing 20% butter
oil and 19% fiber. The fiber used was cellulose, cellophane, wheat
straw or cellophane-peat (14:5). Cholesteremia (milligrams per deca-
liter) and degree of atherosis were respectively: cellophane, 216
and 37.5; cellulose, 133 and 20.8; wheat straw, 114 and 12.7; and
cellophane-peat, 141 and 10.7.

Cookson et al.[56] fed rabbits a diet of either standard calf
meal or calf meal:alfalfa at a ratio of 1:9, augmenting each diet
with a daily oral dose of 600 mg cholesterol. After 10 weeks, cholesterol levels were 1345 mg/dl in the control group and 45 mg/dl
in the calf meal-alfalfa group. The former group exhibited atherosclerotic lesions; the latter showed none.

Howard et al.[57] fed rabbits an atherogenic, cholesteremic diet
containing 20% beef fat and 15% cellophane. Mixing the diet with
an equal weight of stock diet reduced plasma cholesterol to normal
levels and inhibited atherosclerosis.

Baboons fed a semipurified diet containing 40% carbohydrate,
25% casein, 14% hydrogenated coconut oil and 15% cellulose showed
more severe aortic sudanophilia (8.4%) than did controls fed bread,
fruit, and vegetables (0.02%)[58]. Vervet monkeys fed similar diets[59]
also showed much more severe aortic sudanophilia than did controls
(9.1% and 5.0%, respectively).

The influence of fiber on other atherogenic dietary variables
has been tested. Studies have indicated that casein is more atherogenic for rabbits than is soy protein[60, 61]. We fed rabbits a
semipurified diet containing soy protein or casein plus one of three
types of fiber: wheat straw, cellulose or alfalfa[62]. Wheat straw
had no effect on serum cholesterol levels, but alfalfa was hypocholesteremic. With every type of fiber, soya protein was less cholesteremic than casein. Wheat straw, despite its lack of effect on
cholesteremia, reduced the severity of lesions by 32% in the casein
group and by 28% in the soy group. Alfalfa negated the difference
between casein and soya protein.

It has been suggested that the mechanism by which the fiber
inhibits atherogenesis involves the binding and excretion of bile
acids. This increases turnover of cholesterol into bile acids and
provides less of this sterol to the circulation. Indirect evidence
supporting the hypothesis is available from studies involving baboons[58], vervets[59], and rabbits[60].

Eastwood and Boyd[63] examined bile salts in terms of their distribution between the soluble and insoluble material in the small
intestine of the rat. They found that appreciable quantities of
bile salts were bound to the insoluble material and surmised that
this accounted for the differences in amount of bile salts found in
different parts of the intestine. Eastwood and Hamilton[64] subsequently carried out an examination in vitro of these binding properties using a "dry grain" preparation and measuring the amount of
various bile salts bound to the fiber. They established many of
the characteristics of the binding process and concluded that the
binding of bile salts is mainly hydrophobic.

Kritchevsky and Story[65] compared the binding of taurocholate
and glycocholate by several types of fiber commonly used in animal
diets. Alfalfa bound an appreciable quantity of taurocholate (21%
more than cholestyramine), and wheat straw, bran and oat hulls bound
small quantities of taurocholate. Cellulose and cellophane did not
actively adsorb taurocholate. Balmer and Zilversmit[66] found that
alfalfa and several other grains, as well as ground stock diets,

bound taurocholate and cholesterol from a micellar solution. Birkner and Kern[67] investigated the binding of bile salts to various non-digestible food residues prepared from celery, corn, lettuce, potatoes and string beans and found that these residues adsorbed large quantities of bile salts. By comparing several bile salts, they also suggested that the binding was hydrophobic in nature. Condiments, vegetables and spices also bind appreciable quantities of taurocholic acid[68].

Story and Kritchevsky[69] examined the binding of most of the bile acids and bile salts in man to several types of fiber and cholestyramine. The adsorption affinities of the different types of fiber varied greatly. These data brought into question the earlier assumptions that the level of adsorption could be measured from data concerning only one bile salt. It is also apparent that the adsorption process is not totally hydrophobic and is most likely a combination of several types of interactions.

Eastwood et al.[70] have reported a standardized procedure for measuring the amount of bile salts adsorbed by different types of fiber with different water-holding capacities. They have also measured the reversibility of bile salt adsorption and have found, for example, that carrot fiber binds 80% deoxycholate and that 40% remains adsorbed to the fiber after 2- to 16-hr washings with buffer. This indicates that the fiber actively adsorbs bile salts but that previous testing methods also measured an appreciable quantity of bile salt that had been reversibly held by the fiber.

Further work in the area of fiber, lipids and atherosclerosis must delineate the mechanism(s) of action of each chemical component of fiber and must clarify the influence of dietary fiber and other nutrients, especially on trace metals. Once the various roles are defined, it is conceivable that combinations of fiber will be administered to achieve desired effects, i.e., one for hyperlipidemia, one for bulk, etc.

REFERENCES

[1] Kannel, W.B., T.R. Dawber, A. Kagan, N. Revotskie and J. Stokes, III. Factors of risk in the development of coronary heart disease — six year follow up experience: The Framingham Study. Ann. Internal Med. 53: 33, 1961.

[2] Kagan, A., W.B. Kannel, T.R. Dawbar and N. Revotskie. The coronary profile. Ann. N.Y. Acad. Sci. 97: 883, 1962.

[3] Burkitt, D.P., A.R.P. Walker and N.S. Painter. Dietary fiber and disease. J. Am. Med. Assoc. 229: 1068, 1974.

[4] Trowell, H. Ischemic heart disease and dietary fiber. Ann. J. Clin. Nutr. 25: 926, 1972.

[5] Trowell, H. Dietary fiber and coronary heart disease. Eur. J.

Clin. Biol. Res. 17:345, 1972.

[6] Trowell, H. Dietary fibre, ischaemic heart disease and diabetes mellitus. Proc. Nutr. Soc. 32:151, 1973.

[7] Cowgill, G.R., and W.E. Anderson. Laxative effects of wheat bran and "washed bran" in healthy man. A comparative study. J. Am. Med. Assoc. 98:1866, 1932.

[8] Cowgill, G.R., and A.J. Sullivan. Further studies on the use of wheat bran as a laxative. J. Am. Med. Assoc. 100:795, 1933.

[9] Olmsted, W.H., G. Curtis and O.K. Timm. Cause of laxative effect of feeding bran pentosan and cellulose to man. Proc. Soc. Exptl. Biol. Med. 32:141, 1934.

[10] Williams, R.D., and W.H. Olmsted. The effect of cellulose, hemicellulose and lignin on the weight of the stool: A contribution to the stydy of laxation in man. J. Nutr. 11:433, 1936.

[11] Williams, R.D., and W.H. Olmsted. The manner in which food controls the bulk of the feces. Ann. Internal Med. 10: 717, 1936.

[12] Van Soest, P.J., and R.W. McQueen. The chemistry and estimation of fiber. Proc. Nutr. Soc. 32:123, 1973.

[13] Southgate, D.A.T. Fibre and other unavailable carbohydrates and their effects of the energy value of the diet. Proc. Nutr. Soc. 32:131, 1973.

[14] Southgate, D.A.T. The analysis of dietary fiber. In: "Fiber in Human Nutrition", G.A. Spiller and R.J. Amen, eds. Plenum Press, New York, 1976, p. 73.

[15] Cummings, J.H. What is fiber? In: "Fiber in Human Nutrition", G.A. Spiller and R.J. Amen, eds. Plenum Press, New York, 1976, p. 1.

[16] Eastwood, M.A., and W.D. Mitchell. Physical properties of fiber: A biological evaluation. In: "Fiber in Human Nutrition", G.A. Spiller and R.J. Amen, eds. Plenum Press, New York, 1976, p. 109.

[17] Wells, A.F., and B.H. Ershoff. Beneficial effects of pectin in prevention of hypercholesterolemia and increase in liver cholesterol in cholesterol-fed rats. J. Nutr., 74:87, 1961.

[18] Kiriyama, S., Y. Okozaki and A. Yoshida. Hypocholesterolemic effect of polysaccharides and polysaccharide-rich foodstuffs in cholesterol-fed rats. J. Nutr. 97:382, 1969.

[19] Tsai, A.C., J. Elias, J.J. Kelley, R.S.C. Lin and J.R.K. Robson. Influence of certain dietary fibers on serum and tissue cholesterol levels in rats. J. Nutr. 106:188, 1976.

[20] Story, J.A., S.K. Czarnecki, A. Baldino and D. Kritchevsky. Effect of components of fiber on dietary cholesterol in the rat. Federation Proc. 36:1134, 1977.

[21] Kritchevsky, D., R.P. Casey and S.A. Tepper. Isocaloric, isogravic diets in rats. II. Effect on cholesterol absorption and excretion. Nutr. Rept. Internat. 7:61, 1973.

[22] Kritchevsky, D., S.A. Tepper and J.A. Story. Isocaloric, isogravit diets in rats. III. Effects of non-nutritive fiber (cellulose and alfalfa) on cholesterol metabolism. Nutr. Rept. Internat. 9:301, 1974.

[23] Ershoff, B.H., and A.F. Wells. Effects of gum guar, locust bean gum and carrageenan on liver cholesterol of cholesterol-fed rats. Proc. Soc. Exptl. Biol. Med. 110:580, 1962.

[24] Riccardi, B.A., and M.J. Fahrenbach. Effect of guar gum and pectin N.F. on serum and liver lipids of cholesterol-fed rats. Proc. Soc. Exptl. Biol. Med. 124:749, 1967.

[25] Chang, M.L.W., and M.A. Johnson. Influence of fat level and type of carbohydrate on the capacity of pectin in lowering serum and liver lipids of young rats. J. Nutr. 106:1562, 1976.

[26] Lin, T.M., K.S. Kim, E. Karvinen and A.C. Ivy. Effect of dietary pectin, "protopectin" and gum arabic in cholesterol excretion in rats. Am. J. Physiol. 188:66, 1957.

[27] Portman, O.W., and P. Murphy. Excretion of bile acids and hydro-xysterols by rats. Arch. Biochem. Biophys. 76:367, 1958.

[28] Beher, W.T., and K.K. Casazza. Effects of psyllium hydrocolloid on bile acid metabolism in normal and hypohysectomized rats. Proc. Soc. Exptl. Biol. Med. 136:253, 1971.

[29] Judd, P.A., R.M. Kay and A.S. Truswell. Cholesterol lowering effect of lignin in rats. Proc. Nutr. Soc. 35:71A, 1976.

[30] Hardinge, M.G., and F.J. Stare. Nutritional studies of vegetarians. II. Dietary and serum levels of cholesterol. Am. J. Clin. Nutr. 2:83, 1954.

[31] Hardinge, M.G., A.C. Chambers, H. Crooks and F.J. Stare. Nutritional studies of vegetarians. III. Dietary levels of fiber, Am. J. Clin. Nutr. 6:523, 1958.

[32] Walker, A.R.P., and U.B. Arvidson. Fat intake, serum cholesterol concentration and atherosclerosis in the South African Bantu. I. Low fat intake and age trend of serum cholesterol concentration in the South African Bantu. J. Clin. Invest. 33:1366, 1954.

[33] Keys, A., F. Grande and J.T. Anderson. Fiber and pectin in the diet and serum cholesterol concentration in man. Proc. Soc. Exptl. Biol. Med. 106:555, 1961.

[34] Truswell, A.S., and R.M. Kay. Bran and blood-lipids. Lancet 1:367, 1976.

[35] Jenkins, D.J.A., M.J. Hill, and J.H. Cummings. Effect of wheat fiber on blood lipids, fecal steroid excretion and serum iron. Am. J. Clin. Nutr. 28:1408, 1975.

[36] Cummings, J.H., M.J. Hill, D.H.A. Jenkins, J.R. Pearson and H. S. Wiggins. Changes in fecal composition and colonic function due to cereal fiber. Am. J. Clin. Nutr. 29:1468, 1976.

[37] Walters, R.L., I. McLean Baird, P.S. Davies, M.J. Hill, B.S. Drasar, D.A.T. Southgate, J. Green and B. Morgan. Effects of two types of dietary fibre on faecal steroid and lip excretion. Brit. Med. J. 2:536, 1975.

[38] Palmer, G.H., and D.G. Dixon. Effect of pectin dose on serum cholesterol levels. Am. J. Clin. Nutr. 18:437, 1966.

[39] Durrington, P.N., A.P. Manning, C.H. Bolton and M. Hartog. Effect of pectin on serum lipids and lipoproteins, whole gut transit

time and stool weight. Lancet 2:394, 1976.

[40] Jenkins, D.J.A., A.R. Leeds, C. Newton and J.H. Cummings. Effect of pectin, guar gum and wheat fibre on serum cholesterol. Lancet 1:1116, 1975.

[41] Kay, R.M., and A.S. Truswell. Effect of citrus pectin on blood lipids and fecal steroid excretion in man. Am. J. Clin. Nutr. 30:171, 1977.

[42] Fahrenbach, M.J., B.A. Riccardi, J.C. Saunders, I.N. Lourie and J.G. Heider. Comparative effects of guar gum and pectin on human serum cholesterol levels. Circulation 32:11, 1965.

[43] Cleave, T.L. "The saccharine Disease". Bristal: John Wright and Sons Ltd., 1974.

[44] Antonis, A., and I. Bersohn. The influence of diet on fecal lipids in South African white and Bantu prisoners. Am. J. Clin. Nutr. 11:142, 1962.

[45] Antonis, A., and I. Bersohn. The influence of long-term high-fat diets on the serum lipid patterns of white and Bantu South African subjects in a prison research study. South African Med. J. 37:440, 1963.

[46] Joubert, F.J., A. van Bergen, I. Bersohn, A.R.P. Walker and W. Lutz. Serum lipoprotein concentrations (S_f values) in South African Bantu and white subjects. South African J. Lab. Clin. Med. 8:10, 1962.

[47] Grande, F. Sugars in cardiovascular disease. In: "Sugars in Nutrition", H.L. Sipple and K.W. McNutt, eds. Academic Press, New York, 1974, p. 401.

[48] Kiehm, T.G., J.W. Anderson and K. Ward. Beneficial effects of a high carbohydrate, high fiber diet on hyperglycemic diabetic man. Am. J. Clin. Nutr. 29:895, 1976.

[49] Kritchevsky, D. Experimental atherosclerosis in rabbits fed cholesterol-free diets. J. Atheroscler. Res. 4:103, 1964.

[50] Lambert, G.F., J.P. Miller, R.T. Olsen and D.V. Frost. Hyper-cholesteremia and atherosclerosis induced in rabbits by puri-fied high fat rations devoid of cholesterol. Proc. Soc. Exptl. Biol. Med. 97:544, 1958.

[51] Malmros, H., and G. Wigand. Atherosclerosis and deficiency of essential fatty acids. Lancet 2:749, 1959.

[52] Kritchevsky, D., and S.A. Tepper. Cholesterol vehicle in ex-perimental atherosclerosis. VI. Long-term effect of fats and fatty acids in a cholesterol-free diet. J. Atheroscler. Res. 4:113, 1964.

[53] Kritchevsky, D., and S.A. Tepper. Factors affecting atheroscle-rosis in rabbits fed cholesterol-free diets. Life Sci. 4:1467, 1965.

[54] Kritchevsky, D., and S.A. Tepper. Experimental atherosclerosis in rabbits fed cholesterol-free diets: influence of chow components. J. Atheroscler. Res. 8:357, 1968.

[55] Moore, J.H. The effect of the type of roughage in the diet on plasma cholesterol levels and aortic atherosis in rabbits. Brit. J. Nutr. 21:207, 1967.

[56] Cookson, F.B., R. Altschul and S. Fedoroff. The effects of alfalfa on serum cholesterol and in modifying or preventing
cholesterol-induced atherosclerosis in rabbits. J. Atheroscler. Res. 7:69, 1967.

[57] Howard, A.N., G.A. Gresham, I.W. Jennings and D. Jones. The
effect of drugs on hypercholesterolaemia and atherosclerosis
induced by semi-synthetic low cholesterol diets. Prog. Biochem. Pharmacol. 2:117, 1967.

[58] Kritchevsky, D., L.M. Davidson, I.L. Shapiro, H.K. Kim, M. Kita-
gawa, S. Malhotra, P.P. Nair, T.B. Clarkson, I. Bersohn and
P.A.D. Winter. Lipid metabolism and experimental atherosclerosis in baboons: Influence of cholesterol-free, semi-
synthetic diets. Am. J. Clin. Nutr. 27:29, 1974.

[59] Kritchevsky, D., L.M. Davidson, H.K. Kim, D.A. Krendel, S. Malhotra, J.J. Vander Watt, J.P. Du Plessis, P.A.D. Winter, T.
Ipp, D. Mendelsohn and I. Bersohn. Influence of semi-purified
diets on atherosclerosis in African green monkeys. Exptl.
Mole. Pathol. 26:28, 1977.

[60] Meeker, D.R., and H.D. Kesten. Experimental atherosclerosis and
high protein diets. Proc. Soc. Exptl. Biol. Med. 45:543, 1940.

[61] Howard, A.N., G.A. Gresham, D. Jones and I.W. Jennings. The prevention of rabbit atherosclerosis by soya bean meal. J. Atheroscler. Res. 5:330, 1965.

[62] Kritchevsky, D., S.A. Tepper, D.E. Williams and J.A. Story. Experimental atherosclerosis in rabbits fed cholesterol-free
diets. 7. Interaction of animal or vegetable protein with
fiber. Atherosclerosis, in press.

[63] Eastwood, M.A., and G.S. Boyd. The distribution of bile salts
along the small intestine of rats. Biochim. Biophys. Acta
137:393, 1967.

[64] Eastwood, M.A., and D. Hamilton. Studies on the adsorption of
bile salts to non-absorbed components of diet. Biochim.
Biophys. Acta 152:165, 1968.

[65] Kritchevsky, D., and J.A. Story. Binding of bile salts in vitro
by non-nutritive fiber. J. Nutr. 104:458, 1974.

[66] Balmer, J., and D.B. Zilversmit. Effects of dietary roughage on
cholesterol absorption, cholesterol turnover and steroid excretion in the rat. J. Nutr. 104:1319, 1974.

[67] Birkner, H.J., and F. Kern, Jr. In vitro adsorption of bile salts
to food residues, salicylazosulfapyridine and hemicellulose.
Gastroenterology 67:237, 1974.

[68] Story, J.A., and D. Kritchevsky. Binding of sodium taurocholate
by various foodstuffs. Nutr. Rept. Internat. 11:161, 1975.

[69] Story, J.A., and D. Kritchevsky. Comparison of the binding of
various bile acids and bile salts in vitro by several types
of fiber. J. Nutr. 106:1292, 1976.

[70] Eastwood, M.A., R. Anderson, W.D. Mitchell, J. Robertson and S.
Pocock. A method to measure the adsorption of bile salts to
vegetable fiber of differing water holding capacity. J. Nutr.
106:1429, 1976.

DIETARY FIBER, BILE ACIDS AND

CHOLESTEROL METABOLISM

G.V. Vahouny

Dept. of Biochemistry - The George Washington University

2300 Eye Street, N.W. - Washington, D.C. 20037 USA

There is considerable evidence that high dietary fiber intake is associated with a reduced incidence of ischemic heart disease in humans[1,4] and a reduced rate of atheroma development in laboratory animals[5,7]. Experimental approaches to elucidate the underlying mechanisms of these effects have yielded inconsistent results both with natural fiber, such as wheat bran, and with purified fiber components like cellulose[4,7]. This has been due, in part, to the criteria employed to determine the physiological response to changing fiber content of the diet, and in part, to the composition of the diets themselves. In some cases, the determination of serum of liver cholesterol levels, for example, has not permitted definitive conclusions regarding the overall effects of dietary fiber, since in the human, and in experimental animals like the rat, there are efficient homeostatic mechanisms for regulating circulating sterol levels.

An alternative approach to determining the effect of dietary fiber on cholesterol metabolism has been the measurement of fecal excretion of neutral sterols and acidic steroids. Based on this approach, and on in vitro studies of bile acid binding by various full fiber and fiber constituents[8,10], it has been suggested that one major mechanism responsible for the hypocholesteremic or antitherogenic actions of dietary fibers is the binding of bile acids in the intestinal lumen and the subsequent increased excretion of these metabolites as fecal acid steroids[4,7]. This effect has been easier to demonstrate with fiber sources containing high concentrations of the fiber constituents, pectin and lignin, such as alfalfa meal[11,12], or with these commercially available fiber constituents, per se[8,13,14].

473

Among the individual fiber constituents, cellulose has been the most commonly employed bulking agent in diets, largely because it has been considered inert with respect to alterations in lipid metabolism. This fiber has little or no capacity to bind bile acids[11, 15] and has given equivocal results in humans and experimental animals with respect to effects on changes in serum and liver lipids, transit times and fecal acidic and neutral steroids[4, 16]. Pectins and lignins have generally been reported to have hypocholesteremic effects[4, 16], and available evidence suggests that this effect is likely related to bile acid binding and a subsequent increase in fecal steroid excretion[13].

Among the full fiber sources, white wheat bran has received the greatest attention, and has given uniformly negative results with respect to serum lipid lowering[4, 16]. This is presumably due to the inability of this fiber to effectively sequester bile acids[11, 15]. However, there is some evidence that bran intake results in increased fecal steroids[4, 16], which may be related to decreased intestinal transit times induced by this material. Alfalfa meal, in contrast, exhibits significant bile acid binding characteristics in vitro[11, 15], and this fiber source has been reported to be hypocholesteremic and antiatherogenic in rabbits[17], and to increase fecal acidic steroids in rats[12].

From a variety of studies in humans and experimental animals[4, 16], several mechanisms have been proposed for the overall long-term effects of dietary fibers on cholesterol absorption and metabolism. These include: reduced intestinal transit time; bile acid binding and increased fecal excretion of acidic and/or neutral steroids; altered bile composition; redistribution of serum and tissue cholesterol pools, and sterol binding due to the saponin content of crude fiber sources.

Despite the significant increase in interest in the mechnism of action of dietary fibers, the results to date have been equivocal and lead to certain unavoidable impressions: The first is that although the bile acid binding capacity of a dietary fiber may represent an important aspect of fiber action, other mechanisms, either singly or in combination may modify the overall observed effects of the fiber source. In addition, data obtained with full fiber materials is not entirely consistent with observations obtained with "purified" fiber constituents, such as pectins and lignins. This is likely related to the alterations in chemical and physical properties of these individual fibers resulting from isolation procedures.

Secondly, it has become apparent that the largely indirect approaches to elucidating fiber effects on cholesterol metabolism, such as measurement of changes in serum lipid levels, has not allowed unequivocal conclusions on the mechanisms of actions of dietary fibers.

Our own interest in the physiological and therapeutic effects of the dietary fiber have been largely based on elucidating direct and indirect effects of dietary fiber on intestinal absorption of endogenous and exogenous cholesterol. To data, the effects of dietary fiber on cholesterol absorption have been studied only by the conventional balance technique employing radioactive tracers[19]. A more direct approach to this question involves measurement of lipid transport directly into mesenteric lymph in rats with indwelling catheters in the left thoracic lymphatic channel[15,20].

In the present studies we have compared the effects of chronic (6-week) feeding of purified diets containing various types of fiber and fiber constituents on the lymphatic transport of administered cholesterol and triolein in the rat. In addition data are presented on intestinal transit times, serum and hepatic lipid levels, and on bile acid and cholesterol "binding" of these fiber inclusions in vitro.

The isocaloric, isogravic diets[12,15] contained the following components in g/100 g diet: dextrose, 40; casein, 25; corn oil, 14; USP XIV salt mix, 5, vitamin mix, 1; test fiber, 15. The fibers included white wheat bran, alfalfa, cellulose, pectin and baker's yeast cell wall glycan. In addition, control groups were given diets with no fiber inclusion or with cholestyramine as 2% of the diet.

Cannulation of the left thoracic lymphatic duct[15] was performed on all animals after 6 weeks on diet, and at the time of surgery, an indwelling infusion catheter was placed into the duodenum for saline infusions and administration of the lipid test emulsion. The animals were maintained in restraining cages and on the following day, were administered 0.8 ml of an aqueous emulsion[21] containing 0.07 mg glyceryl tri 1-^{14}C oleate 1,2-^{3}H cholesterol and 6.8% nonfat dry milk. Techniques for lymph collection, extraction and analyses have been described earlier[15,22].

In vitro studies on binding of bile acids by test fibers were carried out essentially as described earlier[11,15,23]. Binding of cholesterol was determined in a similar manner except that the test substances were incubated with 10 ml of micellar solutions[24] containing 50 μmoles sodium taurocholate, 2.5 μmoles monolein, 5.0 μmoles oleic acid, and 2.5 μmoles 4-^{14}C cholesterol.

RESULTS

In Vitro Binding of Bile Salts
and Cholesterol

The extent of binding of representative unconjugated or conjugated bile salts and of cholesterol to the dietary test materials, excepting pectin, is shown in Figure 1. Pectin forms a gel in aqueous systems, and studies on bile acid adsorption cannot be performed in the same manner as that which is possible with insoluble fibers or resins. The results of the present study were largely confirmatory of those reported earlier[11]. Thus, except for binding of cholate, which was approximately 60% bound, cholestyramine bound deoxycholate, chenodeoxycholate and taurocholate almost completely. As we reported earlier[15], this anion-exchange resin bound cholesterol quantitatively from micellar solution, probably by hydrophobic interaction with the apolar core of the resin.

With the exception of alfalfa meal, the remaining test materials had little or no bile acid sequestering activity, as had been shown earlier[11]. Largely based on its pectin and lignin content[11], alfalfa sequestered approximately 20-30% of the amount of the various bile acids bound by cholestyramine. However, this fiber source only bound 10% of the available cholesterol from micellar solution. It was of interest that wheat bran which has little bile acid-sequestering ability, removed almost 40% of the cholesterol from micellar solution.

TRANSIT TIMES

As shown in Figure 2, significant reductions in intestinal transit times (carmine red) were evident only in the groups fed diets containing wheat bran or celluse ($p < 0.01$). Also in these groups, the variance between animals was significantly reduced compared to controls ($p < 0.05$ for bran-fed rats; $p < 0.001$ for cellulose-fed rats). These fiber sources also, and perhaps coincidentally, had a significant effect on cholesterol absorption in short-term feeding studies reported earlier[15].

LYMPHATIC ABSORPTION OF CHOLESTEROL

The absorption of intraduodenally-administered cholesterol into thoracic duct lymph of rats on various dietary fibers is shown in Figure 3. These data are presented to indicate that with all dietary regimes, the cumulative absorption curves were comparable, i.e., there was no marked delay in absorption of cholesterol in any of the dietary groups. Thus, as shown in Figure 3, the observed effects of fiber on cholesterol absorption were clearly evident during the period of maximal cholesterol absorption, 0-8 hr[25], and the extent of cholesterol absorption from 12-24 hours after dosing was comparable in all groups.

The collective data for all studies on cumulative absorption of cholesterol are summarized in Figure 4. Despite an apparent effect of the wheat bran diet on lymphatic absorption of cholesterol, these

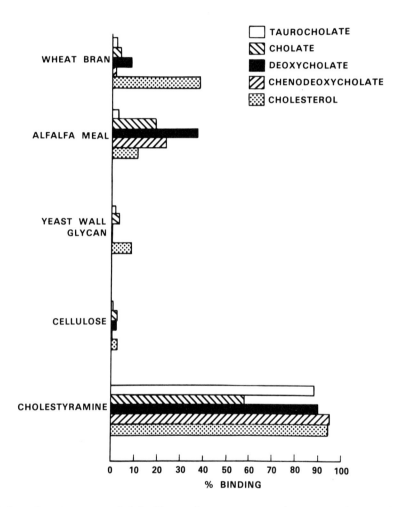

Fig. 1 - Extent of binding of representative unconjugated
or conjugated bile salts and of cholesterol to
the dietary test materials, excepting pection.

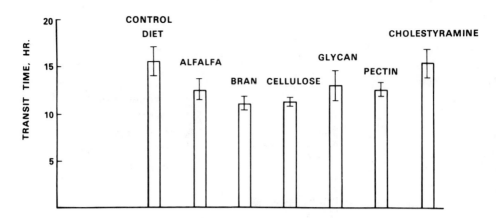

Fig. 2 - Intestinal transit times of different diets

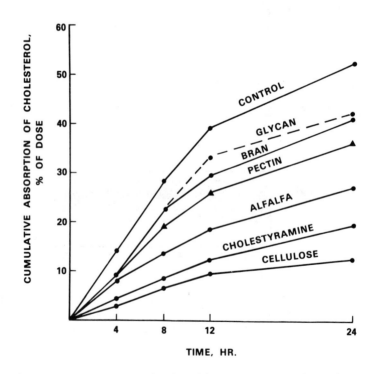

Fig. 3 - Absorption of intra duodenally-administered cholesterol into thoracic duct lymph of rats on various dietary fibers.

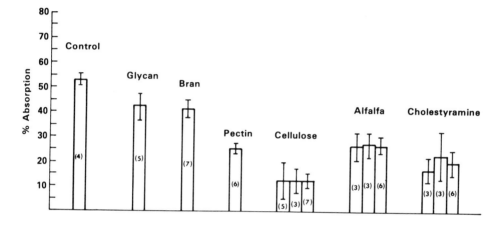

Fig. 4 - Collective data for all studies on
cumulative absorption of cholesterol

differences were not statistically significant at a 5% level. There was also no effect of this diet on total lymph volume or lymph flow rates. While it has been reported[26] that baker's yeast wall poly-glycan has a hypocholesteremic effect in rats, it was neither possible to confirm these results (see Table 1) nor to show a significant effect of the glycan-containing diet on lymphatic absorption of cholesterol. There was a diminished lymph flow in these animals, but this was not accompanied by any change in cholesterol absorption and transport in lymph.

Chronic intake of diets containing alfalfa meal, cellulose, pectin and cholestyramine all resulted in significant inhibition of cholesterol absorption. The least dramatic of these was seen with the pectin-containing diet, which was also the only group of the four in which there was a significant reduction in lymph flow compared to controls. The results obtained with pectin-containing diets are comparable to those reported by us earlier[20].

The dramatic reductions in cholesterol absorption in animals fed either the cellulose or alfalfa-containing diets for six weeks were comparable to that observed in the group fed the 2% cholestyramine diet. As shown in Figure 4, these results were reproducible.

CORRELATION OF CHOLESTEROL AND
TRIGLYCERIDE ABSORPTION

Data on the relative effects of dietary fiber on both cholesterol and triglyceride absorption into lymph are shown in Figure 5. These results indicate a direct correlation between the dietary fiber effects on the absorption of both lipids from the intestine. Thus, when effects on cholesterol absorption were significant, similar effects on triglyceride digestion and absorption were noted. The line was statistically fitted (least mean squares) and the correlation coefficient between effects on absorption of both lipids was 0,962.

SERUM AND LIVER LIPIDS AND HEPATIC
CHOLESTEROL 7α-HYDROSYLASE

Data on serum and liver lipids, and hepatic 7α-hydroxylase activities[27] for the various dietary groups are summarized in Table 1. There were no significant differences in mean values for animals in any of the dietary groups. Also, serum cholesterol and triglyceride levels in all groups fed their respective diets for six weeks were statistically similar.

The chronic feeding of various fiber sources in the isocaloric isogravic diets used earlier[12, 15] had no significant effect on hepa-

TABLE 1

ANIMAL WEIGHTS, SERUM AND HEPATIC LIPIDS AND
HEPATIC 7α-HYDROXYLASE ACTIVITY

GROUP[a]	ANIMAL WEIGHT	SERUM LIPIDS[e]		HEPATIC LIPIDS[e]			HEPATIC CHOLESTEROL 7α-HYDROXYLASE
		CHOL	TG	CHOL	TG	PL	
	g	mg/dl		mg/100 g wet weight			p moles/min/mg
Control	340 ±32	53 ± 1	350 ±10	427 ±23	428 ±43	2150 ±160	58 ±11
Wheat Bran	342 ± 7	73 ±11	298 ±27	400 ±19	396 ±36	2530 ±120	52 ± 6
Alfalfa Meal	348 ±24	44 ±12	312 ±43	459 ±18	310[b] ±26	2750[b] ± 90	50 ± 3
Yeast Wall Glycan	302 ± 6	45 ± 3	299 ±38	464 ±34	460 ±44	2200 ±350	N.d.[d]
Cellulose	338 ±16	54 ± 4	317 ±15	386 ± 7	280[b] ±10	2660[b] ± 90	45 ± 5
Cholestyramine	336 ±13	46 ± 5	333 ±27	398 ±15	300[b] ±28	2490[b] ± 50	94[c]

[a] all animals received the purified diet consisting of the following in g/100g: Dextrose, 40; casein, 25; corn oil, 14; Salt mix, 5; vitamin mix, 1; test fiber, 15. Cholestyramine was added as 2% of the diet and dextrose was increased to 53 g/100 g diet.

[b] $p < 0.05$ from control.

[c] average from two animals.

[d] not determined.

[e] Chol, cholesterol; TG, triglycerides; PL, phospholipids.

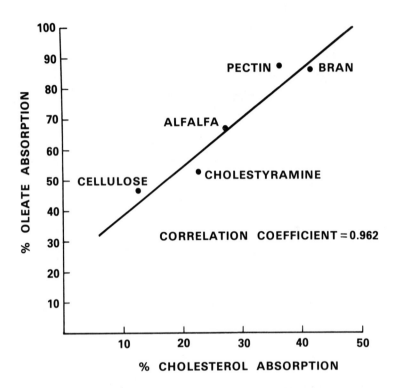

Fig. 5 - Relative effects of dietary fiber on both cholesterol
and triglyceride absorption into lymph.

tic cholesterol levels or on hepatic 7α-hydroxilase levels. As suggested earlier[12], this is likely due to the interaction and effects of all components of this diet rather then to the unique effects of fibers or fiber constituents per se. Even with 2% cholestyramine feeding there was no significant reduction in hepatic cholesterol concentrations although there was some indication of increased bile acid production based on limited analysis of hepatic 7α-hidroxylase activities.

While wheat bran and the yeast cell wall glycan had no effect on hepatic triglycerides and phospholipids, animals on diets containign alfalfa, cellulose or cholestyramine had significantly lowered triglycerides and elevated phospholipides compared to controls. As indicated below, these are the same groups in which lymphatic cholesterol absorption was depressed. Although the data is not yet definitive, the results suggest an effect of these dietary inclusions, on hepatic lipid metabolism. This in turn may be a reflection of altered lipoprotein metabolism or turnover in these particular conditions.

The present study represents an initial attempt to define the direct effects of certain fiber sources and individual fiber constituents on specific aspects of cholesterol and bile acid metabolism. It is well recognized that the complex interactions between all components of the diet can influence cholesterol metabolism[12, 18]. The present study emphasizes that the influence of diet on cholesterol absorption need not represent a direct interaction of dietary components, including fiber, on dietary cholesterol. It also reemphasizes the suggestion[7, 28] that no single mechanism of action can be applied to the action of individual dietary fiber constituents on the absorption and metabolism of cholesterol. This is even more apparent in our recent studies in which we have demonstrated that cholestyramine and certain dietary fibers, particularly those which bind bile acids in vitro, have dramatic effects on the three dimensional topography of the jejunum and colon[29]. Whether these changes alter the functional capabilities of the intestinal epithelial cells or modifies their replication and rate of maturation remains to be determined.

This research was supported in part by U.S.P.H.S. Grants HL-02033, HL-03299, Career Research award (D.K.) HL-00734 and the Research Fund of the Biochemistry Department, George Washington University

REFERENCES

[1] Trowell, H., Ischemic heart disease and dietary fiber. Am. J.
     Clin. Nut. 25: 926, 1972.
[2] Trowell, H., Dietary fiber and coronary heart disease. Eur. J.
     Clin. Biol. Res. 17: 345, 1972.
[3] Burkitt, D.P., A.R.P. Walker and N.S. Painter. Dietary fiber and
     disease. J. Am. Med. Assn. 229: 1068, 1974.
[4] Kay, R.M. and S.M. Strasberg. Origin, chemistry, physiological
     effects and clinical importance of dietary fibers. Clin. Invest.
     Med. 1: 9, 1978.
[5] Kritchevsky, D., S.A. Tepper, D.E. Williams and J.A. Story. Ex-
     perimental atherosclerosis in rabbits fed cholesterol-free
     diets. Part 7. Interaction of animal or vegetable protein with
     fiber. Atherosclerosis. 26: 397, 1977.
[6] Moore, J.H., The effect of the type of roughage in the diet on
     plasma cholesterol level and aortic atherosclerosis in rabbits.
     Brit. J. Nutr. 21: 207, 1967.
[7] Kritchevsky, D., Effect of dietary fiber on lipid metabolism and
     atherosclerosis. In: "International Conference on Atheroscle-
     rosis". L. Carlson, R. Paoletti and G. Weber, eds. Raven Press,
     New York, 1978, p. 169.
[8] Eastwoo, M.A. and D. Hamilton. Studies on the adsorption of bile
     salts to non-absorbed components of the diet. Biochim. Bio-
     phys. Acta 152: 165, 1968.
[9] Kritchevsky, D. and J.A. Story. Binding of bile salts in vitro
     to nonnutritive fibers. J. Nutr. 104: 458, 1974.
[10] Story, J.A. and D. Kritchevsky. Binding of sodium taurocholate by
     various foodstuffs. Nutr. Rep. Int. 11: 161, 1975.
[11] Story, J.A. and D. Kritchevsky. Comparison of the binding of
     various bile acids and bile salts in vitro by several types of
     fiber. J. Nutr. 106: 1292, 1976.
[12] Kritchevsky, D., S.A. Tepper and J.A. Story. Isocaloric, iso-
     gravic diets in rats. III. Effects of non-nutritive fiber
     (alfalfa or cellulose) on cholesterol metabolism. Nutr. Rep.
     Int. 9: 301, 1974.
[13] Leveille, G.A. and H.E. Sauberlich. Mechanism of the cholesterol
     depressing effect of pectin in the cholesterol fed rat. J.
     Nutr. 88: 207, 1966.
[14] Story, J.A., S.K. Czarnecki, A. Baldino and K. Kritchevsky. Effect
     of components of fiber on dietary cholesterol in the rat. Fed.
     Proc. 36: 1134, 1977.
[15] Vahouny, G.V., T. Roy, L.L. Gallo, J.A. Story, D. Kritchevsky,
     M. Cassidy, B. Grund and C.R. Treadwell. Dietary fiber and
     lymphatic absorption of cholesterol in the rat. Am. J. Clin.
     Nutr. 31: 208, 1978.
[16] Kritchevsky, D., Fiber, lipids and atherosclerosis. Am. J. Clin.
     Nut. 31: 65, 1978.
[17] Cookson, F.B., R. Altschul and S. Fedoroff. The effects of alfalfa
     on serum cholesterol and in modifying or preventing cholesterol-

induced atherosclerosis in rabbits. J. Athero. Res. 7: 69,
1967.

[18] Kritchevsky, D., R.P. Casey and S.A. Tepper. Isocaloric, isogravic diets in rats. II. Effect on cholesterol absorption
and excretion. Nutr. Rep. Int. 7: 61, 1973.

[19] Kelly, J.J. and A.C. Tsai. Effect of pectin, gum arabic and agar
on cholesterol absorption, synthesis and turnover in rats. J.
Nutr. 108: 630, 1978.

[20] Hyun, S.A., G.V. Vahouny and C.R. Treadwell. Effect of hypocholesteremic agent on intestinal cholesterol absorption. Proc.
Soc. Exp. Biol. Med. 112: 496, 1963.

[21] Zilversmit, D.B., A single blood sample dual isotope method for the
measurement of cholesterol absorption in rats. Proc. Soc. Exp.
Biol. Med. 140: 862, 1972.

[22] Vahouny, G.V., M. Ito, E.B. Blendermann and C.R. Treadwell. Puro-
mycin inhibition of cholesterol absorption in the rat. J.
Lipid Res. 18: 745, 1977.

[23] Hagerman, L.M., D.A. Julow, and D.L. Schneider. In vitro binding
of mixed micellar solutions of fatty acids and bile salts by
cholestyramine. Proc. Soc. Exp. Biol. Med. 143: 89, 1973.

[24] Thornton, A., G.V. Vahouny and C.R. Treadwell. Absorption of
lipids from mixed micellar bile salt solutions. Proc. Soc. Exp.
Biol. Med. 127: 629, 1968.

[25] Treadwell, C.R. and G.V. Vahouny. Cholesterol absorption. In:
"Handbook of Physiology, Alimentary Canal". American Physiological Society, Washington, D.C., 1968, p. 1407.

[26] Robbins, E.A. and R.D. Seeley. Cholesterol lowering effect of
dietary yeast and yeast fractions. J. Food Sci. 42: 694, 1977.

[27] Nicolau, G., S. Shefer, G. Salen and E.H. Mosbach. Determination
of hepatic cholestegol 7α-hydroxylase activity in man. J.
Lipid Res. 15: 146, 1974.

[28] Kritchevsky, D., S.A. Tepper and J.A. Story. Nonnutritive fiber
and lipid metabolism. J. Food Sci. 40: 8, 1975.

[29] Cassidy, M.M., B. Grund, F. Lightfoot, G.V. Vahouny, L.L. Gallo,
D. Kritchevsky, J. Story and C.R. Treadwell. Alterations in
topographical ultrastructure of rat jejunum and colon induced
by feeding with alfalfa and cholestyramine. Fed. Proc. 37:
543, 1978.

PROTECTIVE ROLE OF FIBER IN CARCINOGENESIS

Baruch Modan, M.D.

Dept. of Clin. Epidem. - Chaim Sheba Medical Center
Tel Hashomer - Tel Aviv University Medical School
Tel Aviv, ISRAEL

The notion that specific food components may be implicated in cancer etiology has first been raised at least in the beginning of the century[1]. By 1940 there were already hundreds of laboratory experiments supporting this concept, usually referring to such substances as fats, carbohydrates or total caloric intake. All along, fiber has been left in a dormant stage, being considered as an inert substance, with no after-affects following its travel along the gastro-intestinal tract.

The recent resurgence of fiber as a potentially protective agent against cancer into the focus of the medical community is usually ascribed to Burkitt. However, although Burkitt's 1971[2] contribution to the popularization of this — once considered odd — idea is invaluable, one cannot forego that this hypothesis was brought forward in this way or another by Malhorta in 1967[3], Oettle in 1964[4], Cleave in 1962[5] and Walker as far back as 1954[6]. Supportive animal data, though with no attempt to extrapolate to humans, can be dug out of the comprehensive report of Rusch in 1944[7], and sporadic reports are found primarily in the Japanese medical literature of the late nineteen thirties[8-10].

The following effort will be directed at a short summary of the current thinking, with regard to the mechanism by which fiber may exert its protective effect, report of supporting human and animal data, discussion of the problems involved in confirming or disproving the hypothesis, and finally, some thoughts towards the future.

MECHANISM

The possible mechanisms by which fiber may exert its protective effect have been repeatedly reported and one may feel that any additional survey would seem redundant. Nevertheless, such a survey is unavoidable for the completeness of this presentation. To date, the following properties of fiber have been implicated:

1) Shortening of the intestinal transit time[11, 12]
2) Bulkier stool[13]
3) Lowering the absorption of potential carcinogens[14]
4) Reduction of the conversion of bile salts to potentially carcinogenic sterols[15]
5) Interference with the cholesterol pool[16]
6) Leading to lower anerobic/aerobic ratio in the intestinal flora which in turn affects bile salt degradation[17, 18]

All but the last of these characteristics have been confirmed by experimental evidence in humans and/or laboratory animals. The effect on intestinal flora has been only indirectly inferred[19].

The potential pathogenetic pathway implies either a shorter or lesser contact with the gut mucosa and/or a smaller availability of carcinogenic metabolites. While this hypothesis seems very reasonable, questions have been raised to what extent it is supported by real data. The epidemiological inference is based mainly on three types of observations:

1) Indirect correlation between disease incidence and fiber consumption,
2) Case control dietary studies,
3) Animal experimentation.

INDIRECT CORRELATIONS

These refer practically exclusively to cancer of the colon, and are based on the fact that this neoplasm is extremely rare in developing countries where fiber consumption is abundant. In addition there is evidence of a shorter transit time, bulkier stools, lesser amount of bile degradation products and a lower anerobic/aerobic ration in people of such societies, as well as among individuals on experimentally high fiber diet[15, 20].

There is also a weak inverse correlation between consumption of cereals in various countries and their colon cancer incidence[21]. More recently, an international collaborative group, led by McLennan[22] has demonstrated a higher fiber intake in rural Finland where colon cancer is relatively rare, than in Copenhagen, where its incidence is quite high. More specifically 30.9 grams per day versus

17.2 in the presence of a four fold differential in colon cancer incidence.

The above observations are however hampered by two facts:

a) A low fiber diet is usually accompanied by a higher fat consumption, which fat ingestion may affect some of the alternative pathways listed above as possible mechanisms for fiber action[19, 23].

b) There is no good index for high fiber diet and therefore international disease correlations are limited to a crude analysis of food items.

CASE CONTROL DIETARY STUDIES

Dietary studies have been criticized as an inaccurate tool for providing information on the actual consumption of various food substances by human individuals. The main limitations of this approach are as follows:

a) inaccurate and inconsistent recall;
b) difficulties in quantification of food consumption;
c) the fact that high consumption of one food item is not independent of either a higher or a lower consumption of other items;
d) the problem of food interaction;
e) changes in diet over time before the disease started, combined with an inability to determine th e point in time when the disease process started, especially in view of the possibly long latent period;
f) possible changes in food habits once the disease becomes symptomatic;
g) complexity of delineating definite food groups by major component.

The above limitations are, however, not insurmountable, since the probability for consistent bias in repeated studies conducted with different methodologies in distinct population groups is quite low. Thus far, fiber containing items have emerged as protective agents for colon cancer in two and perhaps three case-control studies conducted in four entirely different settings, i.e. by our group in Israel[24], Bjelke in Norway and Minnesota[25, 26], and to a certain extent Dales et al. in San Francisco[27].

Our study, which has been reported before, was based on 198 histopathologically confirmed colon cancer patients, and two matched control groups — surgical and neighborhood. Each individual was interviewed with regard to the consumption of 243 food items, up to one year prior to diagnosis. Indirect quantification was attempted by referring to frequencies ranging from at least once daily to several times a week, several times a month, rarely and never. If any major changes in frequencies occurred during the past 10 years they were also recorded.

Analysis was made for various food groups by comparing intrapair frequencies between the case and each control. Fiber turned out to be the only food group where significant differences were noted in comparison to both controls.

These observations are strengthened by two additional findings. First, no such difference was found in other sites of gastro-intestinal cancer similarly studied by us — i.e., stomach and rectum where such an association seems less plausible; second, the higher frequency of fiber consumption among controls was not limited to a few food items, but was present in most of them. Thus, out of 73 items, originally labelled by us as having a relatively high fiber content, 61 and 57 respectively were consumed less frequently by the cancer patients than by the neighborhood and surgical controls. Again, no such effect was noted for other sites, and the respective numbers for rectal cancer are 37 and 38.

It should be pointed out, however, that this finding has not been found in at least one well-designed dietary study[28] where, in fact, an opposite effect was noted. Moreover, in both our and Bjelke's studies the term fiber is not exactly the one usually referred to by Burkitt and Walter, i.e., our evidence points primarily to fruits and vegetables rather than to bran. This topic will be discussed in more detail later.

ANIMAL EXPERIMENTATION

This is not an attempt to cover the wide range of experiments in this field. Fiber enriched diet has been repeatedly proven to lead to a lower rate of tumor induction, when animals were exposed to various carcinogens such as cyclamate, irradiation or nitrosourea[29-32]. The protective effect was achieved with various fiber components, such as bran, cellulose, alfalfa and pectin[29] among others, and was observed primarily, but not exclusively in the colon. This may be the right context to mention that various fiber components have proved consistently to lead to the reduction of serum cholesterol level[32]. The implication of this finding will also be discussed later.

On the other hand, these laboratory results have to be viewed very cautiously, due to the difficulties in extrapolating from animals to human beings, especially with regard to dosage.

PROSPECTS

Where does all this absorbed bulk pool of degraded information take us in transit? Nowhere is one possible answer, somewhere is another, everywhere is the third.

"The fiber theory in cancer etiology is too good to be true to lead us anywhere". This is a quite frequent claim. In other words — "the hypothesis is consistent with logic, but not with real data". We dare to say — the fiber theory is too promising to be abandoned, since, if true, it implies a possibility of protection by adding a substance to the diet, which is far more acceptable to the public than deletion of a favorite item.

One of the most intriguing recent epidemiologic observations has been the international correlation between breast and colonic cancer, and between the latter and arteriosclerotic heart disease[19, 34]. A correlation of this kind, i.e. a higher incidence of one disease in a specific country or population group accompanied by a high incidence in other disease categories is usually construed as implying common causality. However, this is not necessarily so since the correlation may be indirectly related to a third common factor. The reason for the association of this triad is unclear, but the fact that in a 20 year follow-up just reported by Morris et al.[35], low consumption of fiber from cereals was the only food group of influence on the development of coronary heart disease. Fiber not fats, fiber not meat, fiber not carbohydrates.

Does this mean that we are dealing with one common cause to three modern epidemics? Possibly. We have tried to further explore this hypothesis by repeating our dietary study in a group of breast cancer patients. Using the same definition of fiber, the study is still in progress, but preliminary results do not indicate any difference in fiber consumption among the cases, as compared to two similarly matched control groups.

Does this then mean that the etiologies of breast and colon cancer differ? Not necessarily. The crux of the issue is the qualification "using the same definition of fiber". What do we really mean by fiber[36, 37]? How do we define high fiber items, and — most important — do we, or can we assume that fiber is one homogeneous group, as far as disease etiology is concerned? By now we know amply well that the major components of "fiber" differ from each other in numerous characteristics, even though they have common properties. Thus, one cannot rule out the possibility that one component, e.g. cellulose or hemicellulose, is the determinant one in protecting against colon cancer through its effect on stool composition and its sequelae, while the other, say, pectin or lignin will be most influential in protecting against arteriosclerotic heart disease, and perhaps breast cancer, through its cholesterol binding capacities.

Needless to say, by making this statement we move from scientific evidence to speculative theory, but, then, once a theory is developed, even a speculative one, scientific evidence is bound to be collected.

In conclusion, fiber has been shown to have a protective capability at least against colon cancer, by case control studies, indirect disease correlations and animal experimentation. The effect is exerted by a combination of physical and metabolic effects such as contact shortening, dilution and/or absorption of a carcinogen or limiting its internal production. The common denominator for all these pathways is a lesser availability of the carcinogen for the population at risk. Further prospective studies, using new, more applicable, and more defined criteria, of distinct fiber components are urgently needed.

REFERENCES

[1] Russel, F.A.R. "Notes on the causation of cancer". Longmans, Green & Co. London, New York, Bombay, Calcutta, Madras, 1916.
[2] Burkitt, D.P. Epidemiology of cancer of the colon and rectum. Cancer 28:3-13, 1971.
[3] Malthortra, S.L. Geographical distribution of gastro-intestinal cancers in India with special reference to causation. Gut 8:361-371, 1967.
[4] Oettle, G.G. Cancer in Africa, especially in region south of the Sahara. J. Natl. Cancer Inst. 33:385-439, 1964.
[5] Cleave, T.L. Diseases of Western civilization. Br. Med. J. 1:678-679, 1973.
[6] Walker, A.R.P. & Arvidsson, U.B. Fat intake, serum cholesterol concentration and atherosclerosis in the South African Bantu. J. Clin. Invest. 33:1358-1365, 1954.
[7] Rusch, H.P. Extrinsic factors that influence carcinogenesis. Physiol. Rev. 24:177-204, 1944.
[8] Kinosita, R. Gann 33:225, 1939.
[9] Ando, T. Gann 33:371, 1940.
[10] Fischer-Wasels, B. Zentralb f allg Path. u Path. Anat. 66:359, 1936.
[11] Harvey, R.F., Pomare, E.W. and Heaton, K.W. Effects of increased dietary fibre on intestinal transit. Lancet, 1278-1280, 1973.
[12] Payler, D.K., Pomare, E.W, Heaton, K.W. & Harvey, R.F. The effect wheat bran on intestinal transit. Gut 16:209-213, 1975.
[13] Walker, A.R.P. Diet, bowel, motility, faeces composition and colonic cancer. S. Afr. Med. J. 45:377-379, 1971.
[14] Dwyer, J.T., Goldin, B., Gorbach, S. and Patterson, J. Drug therapy reviews: Dietary fiber and fiber supplements in the therapy of gastrointestinal disorders. Am. J. Hosp. Pharm. 35:278-287, 1978.
[15] Walker, A.R.P. & Burkitt, D.P. Colon cancer: Epidemiology. Sem-

inars in Oncol 3:341-350, 1976.

[16] Mathe, D., Lutton, C., Rautureau, J., Coste, T., Gouffier, E., Sulpice, J.C. & Chevallier, F. Effects of dietary fiber and salt mixtures of the cholesterol metabolism of rats. J. Nutr. 107:466-474, 1977.

[17] Burkitt, D.P. Large-bowel cancer: An epidemiologic jigsaw puzzle. J. Natl. Cancer Inst. 54:3-6, 1975.

[18] Renwick, A.G. & Drasar, B.S. Environmental carcinogens and large bowel cancer. Nature 263:234-235, 1976

[19] Hill, M.J. & Wynder, E.L. The etiology of colon cancer. CRC Critical Reviews in Toxicology. 31-82, 1975.

[20] Walker, A.R.P. Colon cancer and diet, with special reference to intakes of fat and fiber. Amer. J. Clin. Nutr. 29:1417-1426, 1976.

[21] Irving, D. & Drasar, B.S. Fibre and cancer of the colon. Br. J. Cancer 28:462, 1973.

[22] McLellan, R. Dietary fibre, transit time, faecal bacteria, steroids and colon cancer in two Scandinavian populations. Lancet:207-211, 1977.

[23] Hill, M.J., Crowther, J.S., Drasar, B.S., Hawksworth, G., Aries, V. & Williams, R.E.O. Bacteria and aetiology of cancer of large bowel. Lancet:96-99, 1971.

[24] Modan, B., Barell, V., Lubin, F., Modan, M., Greenberg, R.A. & Graham, S. Low-fiber intake as an etiologic factor in cancer of the colon. J. Natl. Cancer Inst. 55:15-17, 1975.

[25] Bjelke, E. Epidemiologic studies of cancer of the stomach, colon and rectum, with special emphasis on the role of diet. Scand. J. Gastr. 9:1-235, 1974.

[26] Bjelke, E. Case control study of cancer of the stomach, colon and rectum. Oncology 5:320-333, 1970.

[27] Dales, L.G., Freidman, G.D., Ury, H.K., Williams, S.R. & Grossman, S. Colorectal cancer and diet in blacks. Am. J. Epid. 106:230-231, 1977.

[28] Haenszel, W., Berg, J.W., Segi, M., Kurihara, M. & Locke, F.B. Large-bowel cancer in Hawaiian Japanese. J. Natl. Cancer Inst. 51:1765-1777, 1973.

[29] Kritchevsky, D. Modification by fiber of toxic dietary effects. Fed Proceed 36:1692-1695, 1977.

[30] Ershoff, B.H. Synergistic toxicity of food additives in rats fed a diet low in dietary fiber. J. Food Sci. 41:949-951, 1976.

[31] Watanabe, K., Bandaru, Reddy, S. & Kritchevsky, D. Effect of various dietary fibers and food additives on azoxymethane (AOM) or methylnitrosourea (MNU)-induced colon carcinogenesis in rats. Fed Proceed 37:262, 1978.

[32] Wilson, R.B., Hutcheson, D.P. and Wideman. Dimethylhydrazine-induced colon tumours in rats fed diets containing beef fat or corn oil with and without wheat bran. Am. J. Clin. Nutr. 30:176-181, 1977.

[33] Kritchevsky, D., Tepper, S.A. & Story, A.J.A. Symposium: Nutri-

tional perspectives and atherosclerosis; Nonnutritive fiber and lipid metabolism. J. Food Sci. 40:8-11, 1975.

Rose, G., Blackburn, H., Keys, A., Taylor, H.L., Kannel, W.B., Paul, O., Reid, D.D. & Stamler, J. Colon cancer and blood-cholesterol. Lancet:180-183, 1974.

Morris, J.N., Marr, J.W. & Clayton, D.G. Diet and heart: A post-script. Br. Med. J. 2:1307-1314, 1977.

Eastwood, M.A., Fisher, N., Greenwood, C.T. and Hutchinson, J.B. Perspectives on the bran hypothesis. Lancet:1029-1033, 1974.

Spiller, G.A. and Amen, R.J. Plant fibers in nutrition: Need for better nomenclature. Am. J. Clin. Nutr. 28:675-676, 1965.

NUTRITION AND CANCER

J. Masek

Head, Metabolism & Nutrition Research Center
Institute for Clinical & Experimental Medicine
Budějovická 800, Praha 4, CZECHOSLOVAKIA

Views on the role of nutrition in the genesis and development of neoplasms are not uniform so far, as apparent from the extensive discussions at the recent symposium[1] and reviews of literature (e.g.[2,3,4]).

Essentially the experience is accepted that (1) neoplastic growth may be iniciated and stimulated by various nutritional carcinogens of different types which are found in food, water and ever in the atmosphere. Evidence of this is experience that the prevalence of tumourous diseases may change as a result of migration (Alcantara, Speckmann[5]), whereby (2) the role of nutrition is often rather a modifying than causal factor of proliferation.

Carcinogens of the atmosphere – air pollution – play most probably a role in the development of pulmonary tumours. The role of water – a part of the diet, discussed a long time ago – appears to be mainly a carrier of radionuclides. This pertains in particular to deep waters, for instance those which flow through old galleries in mines (uranium) (Wynder[6]).

As to foods, contamination with mycotoxins is considered as well as nitrates, nitrites, nitrosamides, in gastric cancer in particular aflatoxin[7].

THE DIRECT EFFECT OF FOODS AND THEIR CONSTITUENTS RESPECTIVE IS SUGGESTED BY VARIOUS EPIDEMIOLOGICAL AND EXPERIMENTAL INVESTIGATIONS

Epidemiological data are not always reliable; it is probably an important finding that the incidence of some carcinomas may change

rapidly with migration[3, 5, 6]. Already in the second generation of
Japanese in the USA the number of cancers of the colon increases, as
compared with the incidence in Japan; the prevalence of gastric cancer declines somewhat[6].

Differences in the diet pertain not only to its basic composition
but also its technology. Mechanisms which act in this respect are
mainly hypothetical. Among various factors nowadays mainly the role
of fats is emphasised as they can alter not only the consistence of
the diet, as the known change from cereals to meat (where the predominance of fibre disappears) but there are also changes of the intestinal microflora which influences the constellation of enzymes. Thus
in experiments a marked increase of beta-glucuronidase was observed
(Goldin and Gorbach[8]); these changes can be influenced only to a
minor extent by addition of Lactobacillus casei cultures. Administration of dimethylhydrazine potentiated the disbalance of enzymes,
but did not alter markedly nitroreductases and the microsomal mixed-
-function oxidase system (Rogers et al.[9]). After intubation of 1,2-
-dimenthylhydrazine to rats the number of tumours of the colon per
rat declined when large doses of vitamin A were administered. A
preventive role of 13-cis-retinoic acid was described also by others
(Newberne[10]).

In human pathology also the importance of polyenic fatty acids
was emphasized as well as their possible effects, incl. that of
peroxides. For instance in discussions of the experiments of Pearce
and Dayton on the reduced coronary lethality of old men after administration of essential fatty acids. The authors do not admit a
negative effect of technology (Pearce, Dayton[11, 12]), nevertheless
suspicion of the effect of polyenic acids persists as a threat if
the antioxidative amount of alpha-tocopherol is not sufficient. It
is therefore possible that the vitamin E content in some natural
foods has a protective action. however, it cannot interfere with the
action of chemical carcinogens (nitrites and nitrosamines) and it
would be necessary to study also other anticarcinogens - perhaps
homologues of vitamin A (Poirier, Boutwall[13]) or other similarly
acting substances of plant origin.

The influence of fats on membranes is also considered, as well
as immunocompetence and perhaps also DNA (Hopkins et al.[14], Miller[15]).

Evidence from epidemiological surveys in humans points in particular to a high consumption of fats and oils in relation to carci-
noma of the mammary gland, but also of the ovaries, prostate and
rectum; there are few data in relation to leukaemia (Lea[16], Basu et
al.[17]).

In endocrine - dependent tumours we must take into consider-
                                                    ation the comprehensive ac-
tion of intestinal bacteria on lipid substrates and cholesterol.

Hill et al[18, 19, 20] assume that intestinal bacteria may on a fat-
-rich diet produce carcinogens and steroids of an oestrogenic nature,
as proved by some experiments in vitro with cholesterol and bile
acids as a substrate.

In populations the different incidence of cancer of the breast
in Asian and western populations is emphasized (McSheehy[21, 22]. He
mentions differences in mouse adenocarcinoma in conjunction with nutritional factors and reaches the conclusion, that in humans we must
consider a generally richer diet with large amounts of protein and
fat and that in particular polyenic fatty acids may be high in populations suffering frequently from adenocarcinoma of the breast (Mc-
Sheehy[21], Carroll et al.[23, 24].

So far the role of mineral substances in carcinogenesis remains
obscure. It is obvious that the pathogenetic influence of food on
the genesis and development of neoplasms is not exhausted by what
has been said. The effects of external carcinogens could according
to the symposium held in Florida in 1975 be classified as follows:

1. Nutrients, food additives and food concomitents
   supposed acting as carcinogens / iniciators of
   neoplastic growth:

   in gastrointestinal tract   – esp. chemical carcinogens plus
                                          co-carcinogens, food aditives etc.

   in gastrointestinal tract
   + kidnays and urinary bladder – Ra-nuclides

   oesophagus   – some food additives, whiskey etc.

   stomach   – Ra-nuclides, polycyclic hydro-
                                          -carbons, ethanol (whiskey etc.)

   hepatomas   – aspergillus flavus

   colon   – fiber deficiencies?

                                        – fat (polyenoic F.A. ?)

                                        – food excess (+ polycyclic hydro-
                                          carbons)

   lungs (bronchi)   – smoking, air pollution

2. "Endocrine" dependent neoplasmas – fat excess (polyenic f.A.)
                                        (stimulating stereoneogene-
                                        sis? or others)

<u>Anticarcinogens</u> (veget. origin): vitamin A (carotens) vit. E?
                                    anticarcinogens of vegetable
                                    origin (?)

Obviously a patient with a tumour develops malnutrition and there
is competence between the tumour and its host (Munro[25]). The tumour
cumulates at the expense of the other tissue the necessary amino
acids and some other factors and thus cause e.g. pyridoxine deficiency (Potera et al.[26]), it cumulates L-ascorbic acid (Kakar[27]) and
others. During treatment of patients we must therefore consider
corrections due to the developing nutrition deficiencies.

It seems therefore that in the diet of patients with cancer only
few dietetic interferences play a part. We must not restrict protein but give a normal protein allowance, possibly by the parenteral
route nutrition. The system of p.n. has not been fully developed for
cancer patient, as after the traditional glucose with insulin there
is a rise of lactate in carcinomatous patients rise of the percentage
of CO_2 (from ^{14}C-glucose) and we often encounter a significant rise
of the RQ above 1.0; endogenous glucose and glycogen synthesis is
supposed to be suppressed (Holroyde et al.[28]). No doubt, the malnutrition of carcinomatous patients must be controlled by a differ-
entiated diet with the suitable substitution of deficits.

Although the Symposium or the American Cancer Society in Florida
1975 aroused some critical comments (e.g. Aurswald[29]) it is essential
to deal with this problem and to analyze in greater detail nutrition, in particular in the function of a permissive factor in neoplastic start and growth.

The fundamental problems of research will therefore probably be
in addition to classical and non-classical chemical and biological
carcinogens:

(1) the nutritional factors, acting as (primary or secondary) co-
     carcinogens

(2) in endocrine dependent neoplasmas, the role of polyenoic acids,
     cholesterol and bacteria

(3) naturally occuring anticancerogens?
     (retinoic acid, vitamin E, L-ascorbic acid?, etc.)

(4) biological substances acting as stimulants (aflatoxin, etc.)

(5) metabolic changes in patients with neoplastic growth

(6) dietetic treatment, respecting metabolic changes in such pati-
     ents

Let us hope the seminar will contribute towards the elucidation
of at least some of these points.

REFERENCES

[1] Nutrition in the Causation of Cancer. Proc. of a Symposium, Key
     Biscayne, Florida, May 1975. Cancer Res. 35, (11): 3231-3550,
     1975.
[2] Gori, G.B.: Diet and Cancer (an overview for perspective). J. Am.
     Diet. Ass. 71, (4): 375-379, 1977
[3] Nutrition, Diet and Cancer. Dairy Council Digest, 46 (5): 25-30,
     1975.
[4] Proceedings of the 1977 Workshop on Large Bowel Cancer (Nat. Large
     Bowel Cancer Project), Houston, Texas (USA), 1977.
[5] Alcantara, E.N., Speckmann, E.W.: Diet, Nutrition and cancer. Am.
     J. Clin. Nutr. 29 (9): 1035-1047, 1976.
[6] Wynder, E.L.: The dietary environment and cancer. J. Am. Diet.
     Ass. 71: 385-392, 1977.
[7] Anonymous: Diet and aflatoxin toxicity. Nutr. Rev. 29, (8): 181-
     -182, 1971.
[8] Goldin, B., Gorbach, S.L.: Alterations in fecal microflora enzymes
     related to diet, age Lactobacillus supplements and dimenthylhy-
     drazine. Cancer 40, (5): 2421-2426, 1977.
[9] Rogers, A.E., Herndon, B.J., Newberne, P.M.: Induction by dimethy-
     lhydrazine of intestinal carcinoma in normal rats fed high or
     low levels of vitamins A. Cancer Res. 33: 1003-1009, 1973.
[10] Newberne, P.M., Suphakarn, V.: Preventive role of vitamine A in
     colon carcinogenesis in rat. Cancer (Suppl.) 40, (5): 2553-
     -2556, 1977.
[11] Pearce, M.L., Dayton, S.: Incidence of cancer in men on a diet
     high in polyunsaturated fat. Lancet. I., 464-467, 1971.
[12] Pearce, M.L., Dayton, S.: Unsaturated fat and cancer. Lancet II.,
     610, 1971.
[13] Poirier, L.A., Boutwall, R.K.: Nutrition and cancer. Feder. Proc.
     35: 1307-1338, 1976.
[14] Hopkins, G., West, C.: Possible roles of dietary fats in carcino-
     genesis. Life Sci. 19: 1103-1116, 1976.
[15] Miller, A.B.: Role of nutrition in the etiology of breast cancer.
     Cancer (Suppl.) 39: 2704-2708, 1977.
[16] Lea, A.J.: Dietary factors associated with death-rates from certain
     neoplasms in man. Lancet II., 332-333, 1966.
[17] Basu, T.K., Dickerson, J.W.T., Williams, D.C.: Inter-relationships
     of nutrition and cancer. Ecol. Food and Nutr. 2, (3): 193-199,
     1973.
[18] Hill, M.J., Crowthers, J.S., Drasar, B.S., Hawksworth, G., Aries,
     V., Williams, R.E.D.: bacteria and etiology of cancer of the
     large bowel. Lancet I., 95-97, 1971.
[19] Hill, M.J., Drasar, B.S., Williams, R.E.D., Meade, T.W., Cox, A.G.,
     Simpson, J.E.P., Morson, B.C.: Fecal bile-acids and clostridia
     in patients with cancer of the large bowel. Lancet I., 535-536
     1975.
[20] Hill, M.J., Goddard, P., Williams, R.E.O.: Gut bacteria and aetio-
     logy of cancer of the Brest. Lancet II., (7722): 472-473, 1971

[21]McSheehy, T.W.: Nutrition and brest cancer. Ecol. Food, Nutr. 3, (2): 141-146, 1974

[22]McSheehy, T.W.: The onset of mammary adenocarcinoma in mice. A possible correlation with nutrition. Ecol. Food, Nutr. 3, (2): 147-156, 1974.

[23]Carroll, K.K.: Experimental evidence of dietary factors and hormone-dependent cancers. Cancer Res. 35: 3374, 1975.

[24]Carroll, K.K., Gammal, E.B., Plunkett, E.R.: Dietary fat and mammary cancer. Canad. Med. Ass. J. 98, (12): 590-594, 1968.

[25]Munro, H.N.: Tumor-host competition for nutrients in the cancer patient. J. Amer. Diet. Ass. 71, (4): 380-384, 1977.

[26]Potera, C., Rose, D.P. Brown, R.R.: Vitamin B-6 deficiency in cancer patients. Am. J. Clin. Nutr. 30, (10): 1677-1679, 1977.

[27]Kakar, S.C., Wilson, C.W.M.: Ascorbic acid metabolism in human cancer. Proc. Nutr. Soc. 33: 110A, 1974.

[28]Holroyde, Ch. P., Myers, R.N., Smink, R.D., Putman, R.C., Paul, P., Reichard, G.A.; Metabolic response to total parenteral nutrition in cancer patients. Cancer Res. 37, (9): 3109, 1977.

[29]Auerswald, W.: Zur Frage von Ernährungs einflüssen auf die Krebsentstehung. Wien. med. Wschr. 126: 281-283, 1976.

CANCER PREVENTION THROUGH PRUDENT DIET

Gio B. Gori, Ph.D.

Deputy Dir., Div. of Cancer Cause & Prevention
Nat. Cancer Inst., Nat. Inst. of Health
Bethesda, Maryland USA

CANCER AND THE PRUDENT DIET

Diet and nutrition as an issue of public interest has been gathering strength and momentum during the past decade. Expressions of concern regarding the healthfulness of dietary practices and requests for guidance in developing nutritionally-sound and health-promoting diets are heard with increasing frequency. Considerable attention has been given to research findings linking diet and nutrition to many chronic diseases, including cancer as well as coronary artery disease, diabetes, and hypertension.

One response to this growing concern with diet and nutrition and their relationship to health has been the promulgation of dietary guidelines and recommendations by a variety of groups worldwide. In the United States perhaps the best known pronouncements to date are those formulated by the Senate Select Committee on Nutrition and Human Needs, chaired by Senator George McGovern. The Committee's Dietary Goals were originally issued early in 1977 and, predictably, stirred a controversy of emotions and reactions.

Attempts by the Committee to address early criticisms led to some modifications and revisions in a later edition of the Goals. Nonetheless, they remain controversial.

Unquestionably, nutrition is not the sole determinant of currently-observed disease patterns. Furthermore, adjustment of eating habits to conform to the recommended dietary goals may not be necessary or even appropriate for all segments of the population; more sophisticated techniques and procedures are required to adequately assess the particular nutritional requirements of each individual.

501

Yet, it is apparent that the lack of such individualized nutritional assessments should not deter from offering general, prudent dietary guidance to the public based on accumulated evidence. That evidence is sufficient to warrant at least cautionary advice regarding the ingestion of excessive amounts of fats, cholesterol, alcohol and calories. In essence, this is the message of the Dietary Goals. One may quarrel with the numerical values assigned to suggested levels of sodium, cholesterol or fat intake, but given what we know about the correlation between these substances and certain diseases, some form of guidance is clearly indicated.

Much reserach remains to be done before we can fully appreciate or understand exactly how these dietary components contribute to carcinogenesis or other disease processes. Still, the accumulated epidemiologic and laboratory findings which correlate imbalances of these and other dietary components with incidence of diseases such as cancer or cardiovascular disease are convincing. It is likely that, while man has slowly adapted to changing environmental conditions throughout his evolution, technological changes in the last few millenia have been so rapid that there has not been enough time for genetic selection and adaptation to occur. This is most evident for changes in lifestyle and food consumption, and variations in food use within a wide range of environments. The resulting imbalances and altered metabolic processes are affecting the expression of human disease.

EPIDEMIOLOGIC STUDIES

Disease patterns and resulting mortality have changed considerably since the beginning of the 20th century. Infectious diseases are no longer a major threat; rather, chronic problems such as cancer and cardiovascular diseases are the major factors in disease and m mortality[1].

It has been estimated that 80-90% of the cancer incidence rate in the United States is attributable to environmental factors[2]. It has also been estimated that the portion of total cancer incidence related to diet and nutrition is 60% for women and greater than 40% for men[3]. Thus, diet and nutrition appear related to the largest number of human cancers, with a specificity that is second only to tobacco smoking the next largest contributing etiologic factor. These data are not meant to imply that cancer is caused directly by diet, but they do reflect the observed relationships between increased cancer incidence for particular sites and certain nutritional practices.

Migrant Studies

Some of the best evidence of nutrition's role in the causation
of cancer comes from studies of migrant populations[4]. Generally,
cancer incidence patterns of migrants change from that of their native
country to that common to the population of the host country. These
shifts require a few generations to occur because dietary habits
learned in the country of origin are slowly changed in the process
of acculturation to the host country's way of life.

This can be demonstrated by the mortality of Japanese migrants
to the U.S.[5]. In Japan, colon and breast cancer are low, while stomach cancer is high[6]. The reverse is true in the United States.
Within two of three generations, Japanese migrants to the United
States show a shift of cancer incidence patterns from those common
in Japan to those prevalent in the United States[7]. This observation
correlates with a shift of dietary habits, and particularly with increased caloric and fat intake.

One could argue that the environment of lifestyle of different
countries could explain the observed differences. However, the
history of general pollution and food contamination has been similar
in Japan and the Untied States; therefore, these factors are not
likely to be responsible for observed incidence differences in colon
and breast cancer between the two countries, nor for the shifting
patterns of incidence rates observed in migrant populations. This
is particularly true when considering that some epidemiological
studies leading to these conclusions were performed on Japanese populations migrating to Hawaii, an area notably low in man made environmental contamination.

One could argue that the environment or lifestyle of different
countries could explain the observed differences. However, general
pollution and food contamination are similar in Japan and the United
States; therefore, these are not likely to be responsible for observed incidence differences in colon and breast cancer between the two
countries, nor for the shifting patterns of incidence rates observed
in migrant populations.

Also, within a given population group smokers are exposed to
enormous quantities of carcinogenic substances similar to, and in
far larger quantities than, those normally found in air, water, or
food pollution[8]. If these substances were important etiologic factors for those cancers associated with diet, such as colon and
breast, then high incidence rates could be expected among tobacco
smokers. In fact, smokers do not have an excess of these cancers,
and it is reasonable to surmise that carcinogens present in smoke
— and, by inference, in many environmental conditions — are not
likely causative factors for these particular diseases[9]. Such evidence suggests that environmental pollutants are not significant

factors in the observed relationship of nutrition and some cancer
site incidence and mortality. Also, breast and colon cancer are not
known to be more prevalent in occupational groups where exposure to
environmental carcinogens is much higher than in the general population.

Special Populations

Other major studies reinforce the evidence that diet, rather
than other environmental or genetic factors, is involved in the
causation of certain types of cancer. Particularly significant evidence comes from studies of homogeneous population groups which live
and work in the same environment as their cohorts and are exposed
to the same environmental pollutants[10], [11].

Comparison of the cancer incidence and mortality rates in Seventh Day Adventists living and working in Los Angeles show significant
differences in cancer incidences for almost tumor types[12]. Some of
these differences, such as for lung cancer, are clearly linked to
their abstinence from smoking, even though they are exposed to the
same environmental pollutants. This epidemiologic evidence is consistent with the hypothesis that the Adventist lifestyle can account
for a major portion of the reduced risk of non-smoking-related cancer.
Phillips contends that the most distinctive feature in this lifestyle
is the unique vegetarian diet, substantially lower in protein and
higher in dietary fiber and unrefined carbohydrates than that of
their counter-parts[13].

Comparison of the cancer incidences of Mormon and Non-Mormon
populations living in Utah show similar significant differences in
cancer incidences[14]. The diet of the Mormons is different than that
of their cohorts, but it is not the vegetarian diet prescribed by
the Seventh Day Adventists.

Worldwide High/Low Correlations

Breast, colon, esophageal and pancreatic cancers, which have
high correlation with diet, have great variation in incidence around
the world[15]. Many epidemiologic studies have correlated specific
dietary intakes of different populations with high and low incidences
for certain form of cancers, and have found worldwide correlation
between bowel and breast cancer and fat intake. Leveille has correlated the incidence of cancer of the large intestine with foods of
animal origin[16]. Colon cancer has also been shown to correlate highly with the consumption of meat, even though it is not clear whether
the protein, fat content, or an associated factor is the real correlating factor. Seventh Day Adventists have a restricted fat and meat
intake when compared to other populations living in the same district

and, as indicated, they suffer considerably less from some forms of cancer, notable breast and colon[17]. Other observations have postulated that a low fiber intake in western countries may be responsible for a high incidence of colon cancer. Indeed the Japanese diet is high in fiber content, and several observations indicate that African populations with high fiber intake in their diet, like the Japanese, experience a low incidence of colon cancer. The same appears true for the Seventh Day Adventists.

U.S. Trends

Other studies that reinforce the link of nutrition and certain forms of cancer have noted changes in cancer rates over time. The sharp decrease in stomach cancer incidence for the United States in the last 20 years suggests the likely introduction of protective factors in the diet. Gortner's recent study on nutrition in the U,S. indicates changes in the consumption of specific dietary components during this same time period that may explain the decrease in stomach cancer incidence.

It is interesting to note that trends in certain dietary habits in the U.S. during recent decades correlate with colon and breast cancer trends to reinforce the evidence for the etiologic role of certain dietary components, as already discussed.

ANIMAL STUDIES

It seems possible to design and conduct animal studies in dietary carcinogenesis to produce most any effect the experimenter wishes, and conflicting findings are not unusual. In general, however, animal studies in nutrition and cancer etiology tend to corroborate and support epidemiologic evidence. For example, increased amounts of fat in animal diets have resulted in an increased incidence of certain tumors, notably breast tumors, and these findings are consistent with epidemiologic evidence on the incidence of breast cancer in women[18]. Of all dietary modifications in animals, caloric restriction has had the most regular influence on tumor formation[19]. With few exceptions calorie restriction inhibits tumor formation and increases life expectancy. While not conclusive, investigations using animal models to study the role of vitamins, minerals, antioxidants, and non-nutritive components such as fiber in experimentally-induced cancers have all reinforced evidence of the relationship between diet and the occurrence of cancer in humans.

THEORIES ON THE ROLE OF DIET AND
NUTRITION IN THE CAUSATION OF CANCER

Despite the epidemiologic evidence and animal studies, the specific role of diet and nutrition in cancer etiology is still unclear. Theories on the relationships between diet, nutrition, and cancer are not lacking, and are occasionally advanced as the sole causative link. It is unlikely, however, that the interrelationships are as simple and unconfounded as any one of these theories suggests. At the time, it is likely that they all reflect occurences in a far more complex chain of causative events. The brief summary of some of these theories provided here perhaps serve to indicate the extraordinary diversity of relationship which have already been identified.

Bile Acids

The metabolic derivatives of bile acid degradation by intestinal microorganisms are thought to act as direct carcinogens in the etiology of large bowel cancer[20]. In addition, the levels of specific bile derivatives may be affected by the functioning metabolic pathways of endogenous microorganisms, which in turn are affected by psychological stress, diet, and other factors[21]. Bile production is directly affected by diet. A diet high in fat content stimulates increased bile production, thus increasing the risk of these carcinogenic processes occurring.

Protease Inhibitors

These compounds inhibit enzymes which catalyze the hydrolysis of peptide bonds[22]. Protease enzymes act at the extracellular level in promotion of malignant transformations and may do so by inhibiting an anti-inflammatory response such as that found for glucocorticoids[23], interfering with mechanisms for repressing abnormal cell growth[24], or by directly promoting tumor cell growth[25, 26]. At the extracellular level, protease inhibitors oppose these processes. It has also been suggested that the inhibitors may be absorbed intact through pinocytosis and exert their anti-tumor effects intracellularly through inhibition of gene activation[27, 28]. Experimental studies indicate that leupeptin and other protease inhibitors do have anti-tumor activity[29, 30, 31]. Because seed foods such as soybeans or lima beans are rich natural sources of protease inhibitors, they have been suggested as a possible factor in the lowered cancer incidence among vegetarian populations.

Antioxidant Deficiency

A variety of natural and artificial dietary antioxidants —

— Vitamin E, selenium, BHT, BHA, and ethoxyquin — are known to negatively affect chemically-induced tumor incidence[32], [33], [34]. They
may affect the mixed-function oxidase system and hence have an indirect mode of action, or they act directly by combining with free radical carcinogens, thereby preventing tumorigenic cellular reactions.
Recent studies of ultraviolet light-induced tumor formation suggest
that antioxidants can also affect this process[35], [36]. In contrast
to earlier studies, these findings imply a common etiologic mechanism
for both chemical and ultraviolet-induced carcinogenesis.

Fiber

Fiber is thought to exert its effect on the carcinogenic process
in several ways. First, dietary fiber increases fecal bulk, thereby
reducing the concentration of carcinogens in the intestinal tract.
This then would reduce carcinogenic exposure of the intestinal epithelium and could also reduce carcinogenic and precarcinogenic absorption. It may also affect the metabolic processes of endogenous microflora, as well as their population per se. Increased fecal bulk also
decreases intestinal transit time, thereby shortening potential contact time with epithelial cells and absorptive surfaces[37-40].

Hormone Production

Nutrient intake and dietary components, along with nutritional
status, play a major role in determining an individual's hormone profile. Significant alteration of hormone profiles is known to increase the likelihood of cancer development in laboratory animals[41].
This theory appears to be potentially viable in the development of
hormone tissue-specific tumors such as breast, uterine, testicular,
or prostate. Altered hormone profiles may also repress tumors whose
growth is affected by the anti-inflammatory hormones[42].

Excess Nutrient and Caloric Consumption

Excessive consumption of a variety of nutrients and excessive
caloric intake have been linked to cancer incidence in both animal
and epidemiologic studies. Animals on calorie-restricted diets are
known to have a reduced cancer incidence for both spontaneous- and
chemically-induced tumors[43], [44].

Excessive protein may exert tumorigenic actions through metabolic by products which may be mutagenic, such as ammonia[45]. Excessive lipid consumption may affect hormone profiles[46-49]. In addition,
fat or cholesterol metabolites may be either carcinogenic or precarcinogenic, e.g., epoxides of unsaturated fatty acids[50]. Several of
the minor nutrients, including selenium, may be carcinogenic per se

at high enough dosage levels[51, 52, 53]. Actually, most minerals appear to have an optimum range for dietary consumptions, and intake above or below this range may increase susceptibility to tumorigenesis.

Deficient Nutrient Consumption

Nutrient deficiencies, especially of the trace elements, have been linked to cancer etiology. Both animal and epidemiologic studies have linked a selenium deficiency with an increased cancer incidence. Chronic deficiencies of vitamins A, C, and E, as well as most of the B-complex vitamins, have been shown to increase susceptibility of animals to chemically-induced tumors[54-64]. Overt deficiencies in animal studies are usually associated with reduced caloric intake so that starvation or other complications of malnutrition claim the animal before alterations in spontaneous tumor incidence can be observed.

Carcinogenic Absorption

Certain dietary components may enhance or inhibit absorption of carcinogens. Among the strongest evidence for this specific role of diet is the increased incidence of upper alimentary tract cancers among persons who both smoke and drink. Alcohol and cigarette smoke appear to have a synergistic effort on the development of cancers of the mouth, pharynx, and laryns. Only a relatively small risk increase for these forms of cancer is experienced among nonsmokers who drink alcohol, suggesting that the alcohol acts as a cocarcinogen rather than the primary causative agent[65-69]. Other dietary constituents or imbalances may also affect absorption; the possible role of fiber in this respect has already been mentioned.

Food-borne Carcinogens

The theories presented thus far have dealt primarily with metabolic and hormonal processes, and nutrient excesses and deficiencies which can have cancer-promoting or inhibitory effects. Directly carcinogenic or precarcinogenic substances are present in foods as well. These substances may be classified in the following way:

- natural inherent (e.g., cycasin)

- natural contaminant (aflatoxins, molds)

- condiments (saffrole, chili powder)

- food additives (aniline dyes)

- metabolites of additives (nitrosamines)

- metabolites of nutrients (epoxides)

- compounds used in food production (herbicides, pesticides, fertilizers)

- residues of food processing and preparation (bene(a) pyrenes, the mutagens in meats identified by Commoner and Sugimura)

Amazingly little research has been conducted in this area, although food additives have been tested for toxicity and carcinogenicity for some time. More recently, potential food contaminants introduced in food production, processing, and preparation have been investigated — in some cases by accident. Comparatively few studies have looked at the safety of endogenous but exotic chemical constituents of foods, although food technologists have often exploited these constituents in food processing and chemical production procedures[70-81].

With the exception of food additives, little effort has been made to identify the levels of risk associated with any one of these categories of substances. Appropriate risk-benefit analysis procedures will also need to be developed in order to guide recommendations and decision-making regarding the use of some of the substances. For example, how does one weigh the benefits in increased food production against the possible dangers posed by certain pesticides and fertilizers?

Dietary Inducers of the Mixed-function Oxidase System

Chemical compounds occurring naturally in some foods — but with no known nutritive role — have been shown to inhibit carcinogenesis. They appear to achieve this effect by increasing activity of the mixed-function oxidase system, which metabolizes many known chemical carcinogens. Plants of the Cruciferae genus, such as brussels sprouts, cabbage, cauliflower, and broccoli, contain specific indoles which have been associated with increased aryl hydrocarbon hydroxylase (AHH) activity. In turn, this increased AHH activity from exogenous food sources has been demonstrated to protect against chemical induction of mammary tumors and forestomach neoplasia[82]. The inclusion in the diet of foods in which these compounds are present may therefore modulate the effect of environmental carcinogens[83].

It is probable that all of these theories, and more, contain elements of truth which can contribute to our understanding of the overall process of human carcinogenesis. That process is complicated by the interaction of nutrients and dietary compounds with each other, the many factors affecting metabolic processes, and both dietary and non-dietary carcinogens. For example, studies of radiation-induced tumors in animals consuming either a natural food stock diet or a purified diet showed that mice consuming the purified diet had a significantly increased tumor incidence when compared to those maintained on a natural food stock diet. Non-irradiated mice fed either diet were virtually free of tumors. Caution in releasing results of this type of research is well-advised, proper, and ethical. But there would be little justification for extending such caution to the recommendation of prudent dietary practices such as those contained in the Dietary Goals.

Dietary Guidelines

While it is true that we cannot guarantee the efficacy of these dietary guidelines in disease prevention, neither can we guarantee the results of most modern medical and public health practices. The line of demarcation between research findings which should responsibly be withheld, and those which should be released in the public interest, is not always as sharp and well-defined as one might wish. Yet to abdicate responsibility in providing guidance on important health-related issues may well lead to public reliance on less scientifically sound and frequently less scrupulous sources of information. In nutrition, as in other areas where research and experiment constantly produce new insights and information, one must rely on one's best judgment in applying the knowledge available, and do so in a manner which maximizes health-promoting effects while minimizing associated risk factors.

We must also maintain an awareness of the dynamic context within which dietary goals evolve; age, genetics, physical activity, environment, and other demographic and lifestyle factors influence each individual's nutritional status and requirements. A given individual could safely consume a diet that would be unwise for another based on a number of genetic, somatic, behavioral, or environmental variables. Furthermore, current prudent dietary recommendations are based in large part on epidemiologic data amassed over the past fifty or sixty years, and consequently reflect conditions that may be unique to the twentieth century rather than fundamental to man's physiology. If environmental, lifestyle or dietary changes occurred, nutritional requirements and dietary goals undoubtedly would be affected.

For example, changes in consumption levels of a specific nutrient or food stuff might well influence metabolic processes and other nutritional requirements in a way that is not altogether predictable at this time. Increased fiber intake is a case in point. Such an increase would likely affect the current prudent diet message regarding fat, vitamins, minerals, and perhaps even cholesterol[84].

Reduction in stress levels, smoking cessation or increased physical exercise may all have influences on nutritional status. For instance, the anorectic effects of smoking are popularly recognized, and gaining weight is much feared by smokers who consider quitting, women in particular[85]. At the same time, smoking may also influence the utilization of some nutrients.

The beneficial effects of exercise would likely go beyond caloric consumption and affect dietary requirements in other ways. The modification of blood lipids is but one of the more likely consequences[86, 87, 88].

In addition, dietary patterns are not formed on the basis of nutritional needs alone. Preferences in food, methods of food preparation, and the ritual surrounding mealtime are all part of our individual social and cultural heritage. The alteration of eating habits — even when such alterations are considered desirable — is not always easy to achieve, as evidenced by the constant struggle against overweight among a large number of Americans. And, as conditions in some Third World countries make all too abundantly clear, dietary patterns may also be economically determined. Even were we able to accurately predict the nutritional consequences of a prescribed diet, under controlled conditions, the influence of culture, habit and economic necessity could not be readily or easily overcome.

It is obvious that dietary requirements — and therefore dietary goals — are as plastic as the many factors that affect them. Equally obvious is the fact that dietary habits do not lend themselves to rapid change. But while such changes may occur slowly, they do take place. Indeed, American dietary patterns are now changing. Evidenced by the increasing popularity of restaurants and stores promoting "natural" and "health" foods, and subtle shifts in marketing by the food industry, much of this change reflects an interest in developing healthier eating habits.

Those of us concerned with health maintenance and disease prevention have both an opportunity and a responsibility to see that this change is informed and guided by our knowledge of the risk and benefits of certain dietary practices. It also presents us with the challenge to intensify our efforts to expand and refine that knowledge. Recent research suggests certain relationships between diet, nutrition, and health which require more investigation before their implications for dietary practices can be fully assessed. New relationships will undoubtedly emerge as a result of studies now underway or planned.

At the same time we continue to search for new knowledge on the role of diet and nutrition in health and disease, we must utilize the information currently available to us in a wise and judicious manner. The prudent application of this knowledge by an informed public is not likely to produce radical and abrupt shifts in eating habits, with undersirable nutritional and economic consequences. Rather, important steps toward the implementation of prudent dietary practices should take place, and a favorable environment created for continued efforts to monitor nutritional processes and develop more precise and individualized recommendations for a healthy diet.

REFERENCES

[1] Perry, T.M. The new and old diseases. A study of mortality trends in the United States, 1900-1969. Am. J. Clin. Pathol. 63:453. 1975.

[2] Doll, R. Strategy for detection of cancer hazards to man. Nature 265:589-596. 1977.

[3] Wynder, E.L. and G.B. Gori. Contribution of the environment to cancer incidence: an epidemiologic exercise. J. Nat'l. Cancer Inst., 58:825-832. 1977.

[4] Weisburger, J.H. and R. Raineri. Dietary factors and the etiology of gastric cancer. Can. Res. 35:3469-3474. 1975.

[5] Haenszel, W. and M. Kurihara. Studies of Japanese Migrants. I. Mortality from cancer and other diseases among Japanese in the United States. J. Nat'l. Cancer Inst. 40:43-68. 1968.

[6] International Union Against Cancer (UICC). "Cancer Incidence in Five Continents, Vol. II". R. Doll, C. Muir, and J. Waterhouse (eds). New York, Springer-Verlag. 1970

[7] Haenszel, W., M. Kurihara, M. Segi, and R.K.C. Lee. Stomach cancer among Japanese in Hawaii. J. Nat'l. Cancer Inst. 49:969-988. 1972.

[8] U.S. Department of Health, Education, and Welfare. Toward less hazardous cigarettes. Publ. No. (NIH) 76-905.

[9] National Cancer Institute. Monograph 19, 1966.

[10] Lemon, F.R. and R.T. Walden. Death from respiratory system disease among Seventh Day Adventist men. J. Amer. Med. Assoc. 198:117--126. 1966.

[11] Phillips, R.L. Cancer and Adventists. Science 183:471. 1974.

[12] Lemon, F.R., R.T. Walden, and R.W. Woods. Cancer of the lung and mouth in Seventh Day Adventists. Cancer 17(4):486-497. 1964.

[13] Phillips, R.L. Role of lifestyle and dietary habits in risk of cancer among Seventh Day Adventists. Can. Res. 35:3513-3522. 1975.

[14] Enstrom, J.E. Cancer mortality among Mormons. Cancer 36:825-841. 1974.

[15] UICC, Cancer Incidence.

[16] Leveille, G.A. Issue in human nutrition and their problable impact on foods of animal origin. J. Animal Science 41(2):723-731. 1975.

[17] Phillips, R.L., Cancer among Seventh Day Adventists.

[18] Tannenbaum, A. The genesis and growth of tumors. III. Effects of a high-fat diet. Can. Res. 2:468-475. 1942

[19]Tannenbaum, A. Nutrition and cancer. in: "Physiopathology of Cancer", 2nd Ed. F. Homburger (ed), Hoeber-Harper, pp. 517-562. 1959.

[20]Reddy, B.A., A. Mastromarino, and E.L. Wynder. Further leads on metabolic epidemiology of large bowel cancer. Can. Res. 35: 3404-3406. 1975.

[21]Moore, W.E.C. and L.V. Holdeman. Discussion of current bacterio-logical investigations of the relationship between intestinal flora, diet, and colon cancer. Can. Res. 35:3326-3331. 1975.

[22]Troll, W. and T.G. Rossman. Protease inhibitors in carcinogenesis: Possible sites of action. in: "Modifiers of Chemical Carcino-genesis: An Approach to the Biochemical Mecahnism and Cancer prevention", T.J. Slaza, ed. Raven Press. 1978 (in press).

[23]Belman, S. and W. Troll. The inhibition of croton oil-promoted mouse skin tumorigenesis by steroid hormones. Can. Res. 32: 450-454. 1972.

[24]Giraldi, T., M. Kopitar, and G. Sava. Anti-metastic effects of a leukocyte intrecellular inhibitor of neutral proteases. Can. Res. 37: 3834-3835. 1977.

[25]Schnebli, H.P. and M.M. Burger. Selective inhibition of growth of transformed cells by protease inhibitors. Proc. Nat'l. Acad. Sci. USA 69:3825-3829. 1972.

[26]Burger, M.M. Proteolytic enzymes initiating cell division and escape from contact inhibition of growth. Nature 227:170. 1970.

[27]Hodges, C.M., D.C. Livingston, and L. M. Franks. The localization of trypsin in cultured mammalian cells. J. Cell Sci. 12:887--902. 1973.

[28]Troll, W., T. Rossman, J. Katz, M. Levitz, and T. Sugimura. Pro-teinases in tumor promotion and hormone action. in: "Proteases and Biological Control". E. Reich, D. Rifkin, and E. Shaw (eds), Cold Spring Harbor Laboratory, Cold Spring Harbor, New York, pp. 977-987. 1975.

[29]Matsushima, T., T. Kaziko, T. Kawachi, K. Hara, T. Sugimura, T. Takeuchi, and H. Umezawa. Effects of protease-inhibitors of microbial origin on experimental carcinogenesis. in: "Funda-mentals in Cancer Prevention", P.N. Magee et al. (ed), pp. 57--69. 1975.

[30]Troll, W. Blocking tumor promotion by protease inhibitors. in: "Fundamentals of Cancer Prevention". P.N. Magee et al. (ed), University of Tokyo Press, Tokyo/University Park Press, Balti-more, pp. 41-53. 1976.

[31]Troll, W., T. Rossman, J. Katz, M. Levitz, and T. Sugimura. Pro-teinases in tumor promotion and hormone action. in: "Proteases and Biological Control". E. Reich, D. Rifkin, and E. Shaw (eds), 1975.

[32]Griffin, A.C. and M.M. Jacobs. Effects of selenium on azo dye hepatocarcinogenesis. Cancer Letters 3:177-181. 1977.

[33]Harber, S.L. and R.W. Wissler. Effect of vitamin E on carcinogeni-city of methylcholanthrene. Proc. Soc. Exptl. Biol. Med. 111: 774-775. 1962.

[34]Harman, D. Prolongation of the normal lifespan and inhibition of spontaneous cancer by antioxidants. J. Gerontol. 16:247-254. 1961.

[35]Black, H.S. and J.T. Chan. Suppression of ultraviolet light-induced tumor formation by dietary antioxidants. J. Invest. Dermatol. 65:412-414. 1975.

[36]Black, H.S., J.T. Chan, and G.E. Brown. Effects of dietary constituents on ultraviolet light-mediated carcinogeneses. Can. Res. 38: 1384-1387. 1978

[37]Ward, J.M., R.S. Yamamoto, and J. H. Weisburger. Brief communication: cellulose dietary bulk and azoxymethane-induced intestinal cancer. J. Nat'l. Cancer Inst. 51:713, 1973.

[38]Leveille, G.A. Importance of dietary fiber in food. Chic. Section IFT and Mid-West Section Am. Assoc. Cereal Chem. 1974

[39]Burkitt, D.P., A.R. Walker, and N.S. Painter. Effect of dietary fibre on stools and transit-times and its role in the causation of disease. Lancet 2:1408. 1972.

[40]Harvey, R.F., E.W. Pomare, and K.W. Heaton. Effects of increasing dietary fibre on intestinal transit. Lancet 1:1278. 1973.

[41]Belman, S. and W. Troll. 1972

[42]Viaje, A., T.J. Slaga, M. Wigler, and I.B. Weinstein. Effects of antinflammatory agents on mouse skin tumor promotion, epidermal DNA synthesis, phorbol ester-induced cellular proliferation and production of plasminogen activator. Can. Res. 37:1530-1536. 1977.

[43]Tannenbaum, A. 1959.

[44]Ross, M.H. and G. Bras. Lasting influence of early caloric restriction on prevalence of neoplasms in the rat. J. Nat'l. Cancer Inst. 47:1095. 1971.

[45]Topping, D.C. and W. J. Visek. Nitrogen intake and tumorigenesis in rats injected with 1,2 -dimethylhydrazine. J. Nutr. 106(11): 1583-1588. 1976.

[46]Benson, J., M. Lev, and C. G. Grand. Enhancement of mammary fibroadenomas in the female rat by a high fat diet. Can. Res. 16:135 135-137. 1975.

[47]Carroll, K.K. and H. T. Khor. Effects of dietary fat and dose level of 7,12-dimethylbenz(a)anthracene on mammary tumor incidence in rats. Can. Res. 30:2260. 1970.

[48]Carroll, K.K. and H.T. Khor. Effects of level and type of dietary fat on incidence of mammary tumors induced in female Sprague--Dawley rats by 7,12-dimethylbenz(a)anthracene. Lipids 6:416. 1971

[49]Tannenbaum, A. 1942.

[50]Kotin, P. and H.L. Falk. Organic peroxides, hydrogen peroxide, epoxides, and neoplasia. Radiation Res. Suppl. 3:193-211. 1963.

[51]Shapiro, J.R. Selenium and carcinogenesis: a review. Ann. N.Y. Acad. Sci. 192:215. 1972.

[52]The selenium paradox. Food Cosmet. Toxicol. 10:867. 1972

[53]Sunderman, F.W. Carcinogenic effects of metals. Fed. Proc. 37(1). 1978.

[54] Hancock, R.L. and M.M. Dickie. Biochemical, pathological, and genetic aspects of a spontaneous mouse hepatoma. J. Nat'l. Cancer Inst. 43(2):407-415. 1969

[55] Sporn, M.B., N.M. Dunlop, D.L. Newton, and J.M. Smith. Prevention of chemical carcinogenesis by vitamin A and its synthetic analogs (retinoids). Fed. Proc. 35:1332-1338. 1976.

[56] Chu, E.W. and R.A. Malmgrem. An inhibitory effect of vitamin A on the induction of tumors of forestomach and cervix in the Syrian hamster by carcinogenic polycyclic hydrocarbons. Can. Res. 25: 884. 1965

[57] Maugh, T.H. Vitamin A: potential protection from carcinogens. Science 186:1198. 1974.

[58] Raineir, R. and J.H. Weisburger. Reduction of gastric carcinogens with ascorbic acid. Ann. N.Y. Acad. Sci. 258:181. 1975.

[59] Mirvish, S.S. Blocking the formation of N-nitroso compounds with ascorbic acid in vitro and in vivo. Ann. N.Y. Acad. Sci. 258: 169. 1975.

[60] Kamm, J.J., T. Dashman, A.H. Conney, and J.J. Burns. Effect of ascorbic acid in amine-nitrate toxicity. Ann. N.Y. Acad. Sci. 258: 169. 1975.

[61] Cameron, E. and L. Pauling. Ascorbic acid and the glycosaminoglycans: an orthomolecular approach to cancer and other diseases. Oncology 27: 181. 1973.

[62] Tannenbaum, A. and H. Silverstone. The genesis and growth of tumors. V. Effects of varying the level of B vitamins in the diet. Can. Res. 12:744-749. 1952.

[63] The Nutrition Foundation. Riboflavin metabolism in cancer. Nutr. Rev. 32:308. 1974.

[64] Rivlin, R.S. Riboflavin and cancer: a review. Can. Res. 33:1977. 1973.

[65] Wynder, E.L., I.J. Bross, and R. Feldman. A study of etiological factors in cancer of the mouth. Cancer 10:1300-1323. 1957.

[66] Vincent, R.G. and F. Marchetta. The relationship of the use of tobacco and alcohol to cancer of the oral cavity, pharynx or larynx. Am. J. Surg. 106:501-505. 1963.

[67] Keller, A.Z. and M. Terris. The association of alcohol and tobacco with cancer of the mouth and pharynx. Am. J. Public Health 55:1578-1585. 1965.

[68] Wynder, E.L., I.J. Bross, and E. Day. Epidemiological approach to the etiology of cancer of the larynx. J.A.M.A. 160:1384-1391. 1956.

[69] Rothman, K.J. and A.Z. Keller. The effect of joint exposure to alcohol and tobacco on risk of cancer of the mouth and pharynx. J. Chron. Dis. 25:711-716. 1972.

[70] Gross, R.L. and P.M. Newberne. Naturally occurring toxic substances in foods. Clin. Pharm. and Therap. 22(5) Part II:680-698. 1977.

[71] Miller, J.A. Naturally occurring substances that can produce tumors. in: "Committee on Food Protection, Toxicants Occuring Naturally in Foods", 2nd Ed. National Academy of Sciencies,

Washington, D.C. 1973.

[72]Shank, R.C., J.E. Gordon, G.N. Wogan, A Nondasuta, and B. Subhamani. Dietary aflatoxins and human liver cancer. III. Field survey of Rural Thai families for ingested aflatoxins. Food Cosmet. Toxicol. 10:71-84. 1972.

[73]Shank, R.C., N. Bhamarapravati, J.E. Gordon, and G.N. Wogan. Dietary aflatoxins and human liver cancer. IV. Incidences of primary liver cancer in two municipal populations of Thailand. Food Cosmet. Toxicol. 10:171-179. 1972.

[74]Boffey, P.M. Color additives: botched experiment leads to banning of red dye no. 2 Science 191:450-451. 1976.

[75]Shubik, P. Potential carcinogenicity of food additives and contaminants. Can. Res. 35:3475-3480. 1975.

[76]Miller, J.A. and E.C. Miller. Carcinogens occurring naturally in foods. Fed. Proc. 35:1316-1321. 1976.

[77]Issenberg, P. Nitrite, nitrosamines and cancer. Fed. Proc. 35:1322-1326. 1976.

[78]Grover, P. How polycyclic hydrocarbons cause cancer. New Sci. 58:685-687. 1973.

[79]Stout, V.F. Pesticide levels in fish of the northeast Pacific. Bull. Environ. Contam. Toxicol. 3:240-246. 1968.

[80]Rice, J.M. Environmental factors: chemicals. in: "Prevention of Embryonic, Fetal and Perinatal Disease".R.L. Brent and M.I. Harris (eds), DHEW Publ. No. (NIH)76-853, National Institutes of Health, Behesda, Maryland, pp. 163-177. 1976.

[81]Marx, J.L. Drinking water - getting rid of the carbontetrachloride. Science 196:632-636. 1977

[82]Wattenberg, L.W. and W.D. Loub. Inhibition of polycyclic aromatic hydrocarbon-induced neoplasia by naturally occurring indoles. Can. Res. 38:1410-1413. 1978

[83]McLean, A.E.M. and H.E. Drives. Combined effects of low doses of DDT and phenobarbital on cytochrome P450 and amidopyrine demethylatin. Biochem. Pharm. 26:1299-1302. 1977

[84]Eastwoo-, M.A. Fiber and enterohepatics circulation. Nutr. Reviews 35:42-44. 1977.

[85]Blitzer, P.H., A.A. Rimm, and E.E. Giefer. The effect of cessation of smoking on body weight in 57,032 women: cross-sectional and longitudinal analyses. Great Britain, J. Chron. Dis. 30:415--429. 1977

[86]Holloszy, J.O., J.S. Skinner, and G. Toro et al. Effects of a six month program of endurance exercise on the serum lipids on middle-aged men. Amer. J. Cardiol. 14:753-760. 1964.

[87]Kahn, H.A. The relationship of reported coronary heart disease mortality to physical activity of work. Amer. J. Pub. Health 53: 1058-1067. 1963.

[88]Werko, L. Can we prevent heart disease? Ann. Intern. Med. 74:278--288. 1971.

LONG-TERM DEVELOPMENT IN AN AREA WITH

A HIGH INCIDENCE OF GASTRIC CANCER

Stanislav Hejda, M.D., Dr. Sc.

Institute of Hygiene and Epidemiology

Prague, CZECHOSLOVAKIA

In collaboration with Křikava, Ošancová and other members of
the Institute of Human Nutrition in Prague we investigated at the
beginning of the fifties the diet, nutritional and health status of
the population in an area with a high incidence of gastric disease,
in particular gastric cancer. Research on this subject was stimulated by a report of a health community doctor / general practitioner
from Southern Bohemia who drew attention to the strikingly high incidence of gastric disease in his health community, in particular
peptic ulcer and gastric cancer. The views of this physician were
based on his subjective impressions and not on comparison with other
areas. Our preliminary findings were from the very onset so interesting that we decided to investigate the above in detail and asked
about ten other, mostly non-medical departments to cooperate.

The suspected region formed a defined area of about 90 square
kilometres in Southern Bohemia near the Austrian frontier inhabited
by a fairly stable population of more than 8000 people without much
migration. The majority of the population worked in agriculture,
forestry, fisheries and crafts. As compared with the whole region,
the investigated area can be characterized as slightly backward, with
little industry. In the investigated area were large moors extending
to the close vicinity of some villages. In this area up to the end
of the last century iron, silver and gold-containing ores were mined
which were processed on the spot. From the geological aspect the
territory belongs mostly to tertiary lake sediments. The soil is
very acidic and of little value for agriculture.

In the investigated area we selected a representative population
sample in 24 communities and investigated the food consumption, made
a clinical examination focused in particular on the nutritional stat-

us and some biochemical and haematological tests. Our collaborators
from other research departments examined some indicators of the external environment which could be related in some way to the pathogenesis of malignant growth. In addition to the representative population sample we examined in the course of several years about 800
patients with gastric disease and found that about 38% suffered from
peptic ulceration, 44% from chronic gastritis and 18% from malignant
gastric tumours. Among the patients with peptic ulceration gastric
ulcers predominated slightly. The ratio of gastric and duodenal
ulcers was 5 : 3. According to statistical data pertaining to that
period the incidence of peptic ulceration and gastric cancer was
several times higher than in the region, as well as in the whole of
Bohemia and Moravia, that is an area with some 10 million population.
The incidence of gastric cancer was eight times higher.

A more detailed analysis of the assembled data revealed that
peptic ulceration of the stomach and duodenum as well as malignant
gastric tumours are not distributed in a diffuse manner in the entire
area but are concentrated in "nests" formed by several or only one
house. We found that in some houses as much as three generations
died from gastric cancer. This "nest-like" incidence reminded very
much of Delbet's "maisons de cancer", although we were unable to
confirm or prove his theory of magnesium deficiency in patients with
malignant growths. A total of 65 houses in 11 communities were thus
detected. We investigated in addition to the living population also
the causes of deaths in all subjects who died after 1850. I must add
that the incidence of malignant tumours was obviously higher than
would appear from the death certificates and registers because in
this century the corpses were still examined by laymen without any
special medical training. We were surprised to find e.g. in death
certificates of people only just over 50 "senile decay" as the cause
of death. Marasmus senilis was given as the cause in almost 12% of
all deaths. We assume that this and other diagnoses such as cachexy,
marasmus, consumption were made in subjects who actually died from
malignant tumours. A malignant tumour was given as cause in more
than 16% of deaths. In several localities - six from 24 - the mortality from malignant tumours was much higher and varied between 20
and 31%. In addition to the diseases mentioned, in the investigated
area a higher incidence of anaemia and leukaemia was found. In more
than 100 families genetic relations and the familial incidence was
investigated. In many of the deceased it was striking that for some
time - and sometimes for many years - they were treated for gastric
ulcers and then gastric cancer was detected - frequently already
inoperable. In the whole region where the investigated area is
situated there was the highest mortality from cancer of all 14 regions and amounted to 0.42 pro mille.

In the diet of the population in the affected and investigated
area we did not detect any special features, as compared with the
population living in the close vicinity. The dietary intake was

fairly satisfactory, although at the beginning of the fifties protein deficiency as well as deficiencies of some vitamins, iron and calcium were not uncommon. At that time our country had not yet overcome all sequelae of the Second World War. I shall not deal with the findings which gave negative results, though there were by no means few. We devoted considerable attention also to potable water and found that water from many sources contained elevated amounts of humic acids and some heavy metals and that the radioactivity of most sources is usually within a permissible range but the radiocativity of the general environment is raised. A higher radioactivity was assessed in the water of some deep wells. Later in the investigated region more than 1300 specimens of potable water from wells were examined for radioactivity and the concentration of Ra^{226} and radon Rn^{222} was assessed. Only in three wells a radium activity higher than the maximum permissible concentration / 3 pCi/l /, about double the amount, was found. Certain relations were revealed between the hydrogeological formation and the activity of underground waters. In the area with the higher incidence of tumours of the digestive tract no significant differences in the activity of potable waters, as compared with other areas in the region, were found.

At the time of our investigation, at the beginning of the fifties, the majority of the population used water from wells. Most wells were shallow and the water was soft and often unwholesome. The composition of the water promptly reflected the position in the nearby river. The mineral content of drinking water was negligible, incl. a low to zero fluorine content. Some of the water contained traces of arsenic and a relatively large amount of chromium. It is worth reminding that the investigated area is surrounded by moors 13 - 17 m deep with many so-called "dead lakes", i.e. lakes without any animals and hardly any plants.

Interesting results were obtained also by analysis of rain-fall and snow-fall which, as longitudinal observations revealed, had a five-six-year cycle. In our historical and recent mortality records there was a similar cycle, however, shifted by one year, as compared with the peak of rain-fall and snow-fall.

From what has been said it is obvious that we did not find out exactly which factors are responsible for the high incidence of gastric ulcers and malignant gastric tumours. We summarized our investigation to the effect that the following should be considered: genetic load, noxious substances in potable water, humic acids and heavy metals, elevated radiocativity of the environment. We recommended to ensure wholesome tap water in the investigated area and to provide dispensary care for all patients with gastric disease and complaints which may imply neoplastic growth in the stomach.

Within several years in the affected area an adequate system of water mains was built, the number of doctors increased, the area

became industrialized and there were shifts of the population. Many
subjects from the original population sample died and patients were
examined not only by their health community doctor but also by an
oncologist. Food supplies of local origin represent a much smaller
proportion of the total intake than at the time of our investigation.
This applies in particular to staple foods. According to recent statistics the incidence of gastric ulcers and gastric cancer is roughly
the same as in other parts of the country.

Our observation is consistent with a report by Roumanian authors
who found at the beginning of the seventies high radioactivity of
water in some mountainous regions where tumours of the gastrointestinal tract are about three times more frequent as compared with the
general population. Similarly as in our population sample the disease begins as gastroduodenal ulceration and within 3 - 5 years develops into a tumour which unfortunately is often already inoperable.

Our investigation is an example showing that sometimes a useful
provision can be made although the cause of the high incidence of
malignant tumours was not known. When trying to re-evaluate the
findings assembled then, we realize the shortcomings of our work and
the opportunities which were missed. Perhaps it may serve as an explanation and apology that this work was completed more than a quarter of a century ago.

DRUG ADDICTION, CHRONIC ALCOHOLISM AND LIVER FUNCTION WITH

REFERENCE TO PROTEIN METABOLISM

H.A. von Köhn

Department of Medicine, Luitpoldhospital,

Würzburg, FEDERAL REPUBLIC OF GERMANY

Disorders of liver function with subsequent impairment of protein metabolism are frequent in drug addicts and alcoholics. In both groups of subjets the liver is in danger to become damaged either by infection with hepatitis virus type B or Non A- Non B or by the toxic effect of alcohol, respectively. Nutritional influences such as under - or malnutrition may act in addition. Since drug abuse and alcoholism often are combined, a combination of Virus-hepatitis, alcohol toxic liver damage and malnutrition may be expected. In such cases both liver damage and disorders of protein metabolism will be more severe.

The question, whether the drugs used are per se noxious to the liver and to protein metabolism is matter of debate. Most of the authors don't see any connection. Brooks et al (1963) did not find any kind of altered liver function in morphin dependent rhesus monkeys. In human drug addicts disorders of the subcellular structure of hepatocytes, i.e. of the endoplasmatic reticulum are occasionaly observed, but this finding was not interpreted as sign of toxic liver damage due to the drug. The main target organ of these drugs is the central nervous system, but not the liver. On the other hand in drug addicts both mal- and undernutrition are frequent. They alter protein metabolism and thus liver damage may be an indirect consequence of drug addiction. Tartakow (1971) stressed that in drug addicts the liver mainly is in danger to become affected by virus hepatitis which often has a severe and prolonged course (Cherubin et al, 1972), especially in case of protein deficiency. Fulminant hepatic failure as consequence of marked liver cell necrosis may be a further consequence of malnutrition. In this concern, it is of interest, that after the end of the second world war the number of cases with fulminant

521

hepatitis increased in Vienna which happened collaterally with a
period of low protein and calories intake (Davidson, cit. by Thaler,
1967) Similar observations were made by Kallei (1967) in Yugoslavia.
In Bosnia the course of hepatitis during pregnancy was more severe
in islamic patients with low protein intake and the death rates
were significantly above the related number observed in pregnant
women who had a higher protein intake. These observations justify
the conclusion that hepatitis in drug addicts with protein
deficiency has to be expected to be severe.

In alcoholics damage of intestinal mucosa by alcohol is a
further factor which has to be discussed if related disorders of
protein metabolism are debated. The mucosa of the small gut
decreases in height in animal experiments, if alcohol is administered
during a period between three and four weeks. This happens together
with an increase of both mitatic rates and thymidinkinase activity.
The activity of some vileous enzymes of enterocytes decreases
simultaneously (Baraona et al, 1974)- By electron microscopy some
other disorders were observed and interpreted as sign of cell damage
(Rubin et al, 1972). Decreased absorption of vitamins, minerals and
amino acids will be the consequence (Leevy a. Smith, 1974).Otherwise,
increased absorption of l-methione was found during intestinal
alcohol concentrations already present after a medium alcohol intake
(Israel et al, 1969). This was supposed to be due to blocked
synthesis of nucleic acids subsequent to a direct toxic action of
alcohol on the lipid and protein components of the mucosa. Low
protein intake in alcoholics with subsequent damage of intestinal
mucosa and disorders in absorption may be of additional significance
(Tandon et al 1968). Halstadt et al (1971) demonstrated that in
alcoholics absorption of folic acid is more marked when combined
with malnutrition.

Thus metabolic disorders of protein, vitamin, lipids and
minerals may occur in alcoholics without additional liver damage.
Malabsorption will be more severe in cases in which the toxic effect
of alcohol and the consequences of lack of protein act together.
Finally, the different influences in combination potentiate each
other thus causing a more severe disorder of protein metabolism.

In concern of the influence of alcohol and nutrition and
protein synthesis of the liver, one has to differentiate between
the consequences of acute and chronic alcohol action. (Lieber,
1977). Alcohol in a single dose blocks the protein synthesis within
the liver parenchymal cells as shown by decreased synthetic rates
of albumin, fibrinogen and transferrin (Jeejeebhoy et al. 1975),
and a number of enzymes. Rothschild et al (1971, 1974) used the
isolated perfused rat liver model and demonstrated a decrease of
albumin synthetic rates from a normal value of $33mg/12.5^h/100g$
liver weight down to values between 8 and $10mg/12.5^h/100g$ liver

weight, if alcohol in a concentration of 220mg/100ml was added to the perfusate medium. This effect could be abolished by the administration of tryptophan, arginin, ornithin and lysine. The cause of alcohol-induced disorders in protein synthesis was supposed to be due to a blocked synthesis of DNA and ribosomes as shown by related experiments.

In chronic alcoholism an increased protein synthesis was described by Kuriyama et al (1971) and Renis et al (1975), but chronic alcoholism in combination with malnutrition leads to decreased protein synthetic rates (Banks et al., 1970 Morland, 1975). Since these results were evaluated by animal experiments, their relation to clinical medicine has to be interpreted with caution.

The secretion of protein by liver parenchymal cells obviously is blocked by alcohol too. Baraona et al (1975) demonstrated in rats to which alcohol was administered over a long period an accumulation of protein, especially albumin and transferin within liver cells, thought to be indicative of disturbed secretory function.

Malnutrition, especially lack of protein as seen frequently in alcoholics, may lead - as mentioned already above - to decrease of DNA synthesis. Subsequently, other organ functions may be involved, such as decrease of immuno reactivity.

Hsu and Leevy (1972) observed a decreased lymphocyte transforma tion after phythämagglutinin in malnutritioned alcoholics, especially when liver cirrhosis had developed. Decrease of DNA-synthesis in severe malnourished subjects is accompanied by a decrease of the activity of plasmatic and microsomed enzymes, and activities of serum transaminases will be the result. This latter event is accentuated in case of vit. B 6-deficiency. (Ning et al, 1966). The activity of alcoholdehydrogenase decreases in malnourished states preferentially when due to protein deficiency. (Leevy et al, 1967, Bode and Goebell, 1970). In consequence, degradation of alcohol will be prolonged, and the thus increased alcohol toxicity will initiate a vicious circle.

An other point of interest is the possibility of induction of drug metabolising enzymes by alcohol. It is well known, that the tolerance against different drugs is increased in alcoholics. More recently this fact was thought as due to an adaption of the central nervous system, but Misra et al (1971) demonstrated that a more rapid clearance of drugs (e.g. pentobarbital and meprobamate) is causative.

One may speculate therefore that the increased tolerance of chronic alcoholics against tranquilizers is due to the induction of drug metabolising systems.

The situation in acute alcohol intoxication is different, especially in the absence of previous chronic alcoholism, since the activity of drug metabolising enzymes transitorily is decreased. This fact may cause an increased susceptibility of the pharmacological properties of sedatives and tranquilizers in acute alcohol intoxication.

In conclusion: disorders of protein metabolism in drug addicts may be caused by different mechanisms. In most of the cases the situation is complex, since protein deficiency, disorders in intestinal absorption and liver damage act together. Thus, an individual analysis of each drug addict is necessary including the history of viral hepatitis, malnutrition and alcoholism, which latter both may be causative for a more severe course of the disease.

Clinical action includes the avoidance of all mentioned factors which potently lead to disorders in protein metabolism.

THE RELATION BETWEEN CHRONIC ALCOHOLISM, DRUG ADDICTION AND

NUTRITION WITH SPECIAL REFERENCE TO THE THIAMINE STATUS

J.C. Somogyi

Director, Institute for Nutrition Research

Rüschlikon-Zürich, SWITZERLAND

In this paper I shall deal with the effect of chronic alcoholism and drug addiction respectively on the thiamine status of humans. While the relationship between chronic alcoholism and thiamine status has been thoroughly investigated, much less is known about the effects of drug addiction on the thiamine metabolism.

Chronic alcoholism is frequently associated with deficiencies of vitamins, minerals and proteins, questions which the subsequent presentations will discuss. There are several reasons for the vitamin deficiencies and therefore also for that of thiamine. They are generally due to a decreased food intake, caused by reduced appetite, to a decreased intestinal absorption or reduced storage of ingested vitamins and to a partial loss of the ability to convert vitamins into their metabolically active forms.

The clinical symptoms of thiamine deficiency in alcoholism are dependent upon the severity of the vitamin deprivation. This deficiency extends from the mildest form, for example from peripheral neuropathy to the most severe one: Wernicke's syndrome. Several methods were proposed for the assessment of the thiamine status, among others: determination of blood pyruvate levels, estimation of the thiamine excretion in urine, the microbiological assay of thiamine in blood and the determination of transketolase activity in red blood cells.

Several authors[1,2,3] (Ritzel, 1968; de Wijn, 1972 and Dirige et al., 1973) and we also are of the opinion that the transketolase activity itself is not a good measure for the thiamine status (Somogyi and Kopp, 1975)[4]. The enhancement in enzyme activity resulting from the in vitro addition of thiamine pyrophosphate (TPP) to the hemolyzed blood, the so called TPP effect as was proposed e.g. by Säuberlich (1961)[5] (1967)[6], gives in this respect the better results. In many cases this might prove right, but according to our own investigations the determination of the TPP effect before and after oral thiamine administration represents the more specific test (Somogyi, 1973 and Somogyi and Köpp, 1976)[7, 8].

According to Brin and co-worker (1965)[9] the thiamine status is normal if the TPP effect varies between 6-15%, marginally deficient between 15-24% and severely deficient at 25% and more. This classification should in my opinion be considered only as a guideline, since a certain degree of overlapping can occur.

The determination of the TPP effect was generally accepted for the assessment of thiamine status with exception of certain cases of alcoholics. Fennely, Frank, Baker and Leevy (1967)[10] as well as Konttinen et al. (1970)[11] reported that the in vitro addition of TPP to hemolyzed blood from alcoholics with liver cirrhosis did not increase the transketolase activity, while in the hemolysates from thiamine deficient patients without liver disease a marked rise of enzyme activity was observed.

As a consequence of these investigations the use of the TPP effect for the assessment of thiamine status of chronic alcoholics would in my opinion be rather limited since many of these patients suffer from liver cirrhosis.

A reinvestigation of this question seemed to us rather important. Therefore new experiments have been carried out on these lines by my co-worker, Dr. Kopp, and myself in cooperation with Dr. Filippini and Dr. Monnat from the Medical Department of the General Hospital of Canton Luzern (Somogyi, Filippini, Kopp and Monnat (1977)[12]. The TPP effect was investigated in 36 alcoholics 1) without significant hepatic abnormalities and 2) with liver cirrhosis of various degrees, i.e. a) compensated, b) slightly decompensated and c) severely decompensated liver cirrhosis cases.

The difference of transketolase activity before and after addition of TPP to hemolyzed blood is expressed in units — as it was done by Fennelly et al. (1.c.) and the increase of transketolase activity is given in %, which is the more usual way. So the transketolase activity showed an increase after TPP addition also in patients with severe liver cirrhosis.

May I summarize now our investigations on the thiamine status of 52 chronic alcoholics, 30 former alcoholics and 50 healthy volunteers. The thiamine status was established by determination of the TPP effect before and after oral thiamine administration. The dose of vitamin B_1 was varied and the smallest quantity causing a decrease of the TPP effect was used.

For the determination of the transketolase activity, a modified version of the method by Schouten et al. (1964)[13] was used. In a part of our experiments, the thiamine content of blood was determined microbiologically by the method of Sarett and Cheldelin (1944)[14].

In a group of chronic alcoholics consisting of non-hospitalized persons[14] the oral administration of 10 mg thiamine during 7 days did not cause any decrease of TPP effect and 20 mg did so only in a few cases. A reduction of the TPP effect was observed only after increasing the dose to 50 mg or more thiamine per day and person.

The mean value of the TPP effect is high. After oral administration of 50 mg thiamine during 14 days a marked decrease of the TPP effect occurred. After further 7 days, the initial values were nearly reached again.

Principally the same results were obtained in the other 38 investigated chronic alcoholics, inhabitants of a so called open home and patients of the Psychiatric Clinic of the University of Zürich respectively.

As mentioned, the thiamine content of the blood was determined microbiologically and the results of these assays have been compared with the TPP effect of the same person.

We were interested to find out whether there are any changes in these parameters among former alcoholics, who had abstained from drinking alcohol fro at least 4-10 months. The average TPP effect in this group shows an increase of 15%, a borderline case of a biochemical vitamin B_1 deficiency state — a value, as you will hear later, we also found in healthy persons. However similarly to healthy volunteers, the TPP effect could be reduced by the oral administration of 10 mg thiamine   daily during 14 days.

Also in this case a good correlation between thiamine content of blood and the TPP effect has been observed.

From this part of our investigation follows that the thiamine status of former alcoholics improved in a short time. This result is surprising since it cannot be excluded that among the investigated volunteers were persons who might have certain damages, e.g. of the intestinal mucosa or of the liver, interfering with the thiamine utilization.

The increase of the transketolase activity after in vitro addition of TPP in healthy persons is lower than in alcoholics. Doses of 10 mg thiamine per person and day given during and 14 days respectively caused a marked reduction.

In the last part of this paper the effect of drug addiction on the thiamine status should be discussed. As I mentioned in the introduction of my talk our knowledge in this field is limited. It is known that a great number of various drugs used in pharmaceutical preparations can affect appetite, food intake and the absorption of nutrients. On the other hand vitamin inadequacy influences in some cases the drug utilization by increasing the residence time of the drug in the organism, which might cause a potentiation of the drug action. I do not enter into the interesting interrelation between active agents and nutrients, since this is - due to the shortage of time - out of the scope of such a paper. I would like to refer, however, to recently published very good review articles by Sullivan and Cheng (1978)[15], Brin (1978)[16] and Roe (1978)[17].

On the effect of drug addiction on the nutritional status and especially on thiamine status, with exception of chronic alcoholism, only a small number of papers have been published. Therefore several authors, e.g. Halsted (1978)[18] and Sullivan and Cheng (1.c.) emphasize the needs of further research in this interesting and important field.

We have been interested whether various drugs exert similar effects on the thiamine status as chronic alcoholism and therefore we started a few years ago with this kind of investigations (Somogyi and Kopp (1978)[19]. The methods used were the same as I mentioned above and are described in several of our publications.

56 drug addicts have been investigated all hospitalized in special homes and clinics. We began with the investigation immediately after their admission to avoid interferences caused by a better food intake.

Due to the diversity of substances to which these volunteers were addicted we divided them into three groups. Group A consisted of patients addicted to so called "hard" drugs such as morphine, heroin, LSD etc. Into the group B belonged patients who consumed excessive amounts of various pharmaceutical preparations containing, e.g. hypnotics, mainly different barbiturates, analgetics on basis of phenacetin, propyphenazon, codein etc., then amphetamins or psychopharmaca mainly the type of benzodiazepins. Group C was formed of patients who consumed both drugs and alcoholic beverages. The thiamine status of these three groups of addicts were determined twice before and four times after oral thiamine administration.

The initial TPP values were about the same or slightly higher than we observed in healthy people. Already by small oral doses of thiamine, that is 10-20 mg daily during two weeks, the TPP effects were lowered significantly. This in contrast to chronic alcoholics who had to receive doses of 50 mg thiamine per head and day to achieve such an effect.

From this investigation follows that the thiamine status of the majority of the investigated drug addicts show no (Group A and B) or slight/medium thiamine deficiency, as do the patients of Group C, who were addicts to alcohol and to drug simultaneously.

These first results indicate that according to this investigation drug addiction seems to have much smaller effect on the thiamine status than chronic alcoholism. These findings are surprising however also present investigations of heroin addicts by Baker (1978)[20] show that only 25-50% of the subjects tested had a reduced thiamine content in blood.

An interpretation of our findings is not simple but maybe explained by the observation that drug addiction in general does not cause strong pathological-anatomical damage to the organism. Therefore the intestinal absorption of thiamine and its conversion to the metabolically active form in the liver seems — in contrast to chronic alcoholics - to be only slightly affected.

More research and a close cooperation among nutritionists, clinicians and psychologists is needed to obtain a better insight in this important and interesting problem.

REFERENCES

[1] Ritzel, G., Untersuchungen über die Vitamin B_1-Versorgung bei einem städtischen Kollektiv mit Hilfe des Transketolase-Tests. Int. Z. Vitam. Forsch. 38: 508 (1968).

[2] Wijn, J.F. de : personal communication.

[3] Dirige, O.V., Jacob, M., Wang, M., Swenseid, M.E., and Kopple, J.D., Transketoale activity in tissues of uremic rats. J. Nutr. 103: 1723 (1973).

[4] Somogyi, J.C. and Kopp, P.M., Alcohol and nutritional status. Proc. 9th Int. Congr. Nutrition, Mexico 1972, 1: 212 (1975).

[5] Bunce, G.E., and Sauberlich, H.E., Burma nutrition survey. Report of ICNND (1961).

[6] Saurberlich, H.E., Biochemical alterations in thiamine deficiency. Their interpretation. Am. J. Clin. Nutr. 20: 528 (1967).

[7] Somogyi, J.C., Nutrition and alcoholism: connection between chronic alcoholism and thiamine status. Symposium "Nutrition and psyche", Grimaldi di Ventimiglia 1973.

[8] Somogyi, J.C. and Kopp, P.M., Alkoholismus und Ernährungsstatus. Bibl. Nutr. Dieta 24: 17 (Karger Basel 1976).

[9] Brin, M., Dibble, M.V., Peel, A., McMullen, E., Bourquin, A., and Chen, N., Some preliminary findings on the nutritional status of the aged in Onondaga County, New York. Am. J. Clin. Nutr. 17: 240 (1965).

[10] Fennelly, J., Frank, O., Baker, H., and Leevy, C.M., Red blood cell-transketolase activity in malnourished alcoholics with

cirrhosis. Am. J. Clin. Nutr. 20: 946 (1967).

[11]Konttinen, A., Louhija, A., and Härtel, G., Blood transketolase in
     assessment of thiamine deficiency in alcoholics. Annls Med.
     Exp. Biol. Fenn. 48: 172 (1970).

[12]Somogyi, J.C., Kopp, P.M., Filippini, L., and Monnat, A., TPP ef-
     fect in chronic alcoholics with different degrees of liver
     cirrhosis. Paper in preparation, 1977.

[13]Schouten, H., Eps. L.W. van, and Boudier, A.M., Transketolase in
     blood. Clinica chim. Acta 10: 474 (1964).

[14]Sarett, H.P. and Cheldelin, V.H., The use of lactobacillus fer-
     mentum 36 for thiamine assay. J. Biol. Chem. 155: 153 (1944).

[15]Sullivan, A.C., and Cheng, L., Appetite regulation and its modula-
     tion by drugs, in: "Nutrition and drug interrelations", Hath-
     cock and Coon, p. 21, Academic Press, New York, London, 1978.

[16]Brin, M., Drugs and environmental chemicals in relation to vitamin
     needs, in: "Nutrition and drug interrelations", Hathcock and
     Coon, p. 131, Academic Press, New York, London, 1978.

[17]Roe, D.A., Drugs, diet and nutrition. Contemporary Nutrition
     3 (6), 1978.

[18]Halsted, C.H., Drugs and water-soluble vitamin absorption. In:
     "Nutrition and drug interrelations", Hathcock and Coon, p. 83,
     Academic Press, New York, London, 1978.

[19]Somogyi, J.C. and Kopp, P.M., Drug addiction and thiamine status,
     (in press), 1978.

[20]Baker, H.: personal communication, 1978.

RELATIONSHIP BETWEEN THIAMINE AND ETHANOL INTAKE

Motonori Fujiwara, M.D.

Professor, Faculty of Medicine, Kyoto University

Yoshida-Konoe-Cho. Sakyoku, Kyoto, JAPAN

Recently, the relationship between thiamine and ethanol intake has drawn public attention, because alcoholism is often accompanied by thiamine deficiency. The possible mechanisms involved in thiamine deficiency may be summarized as follows:

1. Low intake of thiamine in alcoholic patients.
2. Absorption of thiamine from gut is impaired by ethanol administration.
3. Impairment of thiamine retention in the body by ethanol administration.

To clarify these possibilities, many studies are now being carried out in our laboratory.

A few years ago, we, in Japan, had a recurrence of beriberi sickness which we believed had been completely eradicated after the war. About 400 typical beriberi patients have been detected by a retrospective epidemiological research of the Committee organized by the Government. The incidence was markedly high in adolescents, in the south west area in Japan, and in summer.

The cause for this disease was, in general, low intake of thiamine from foods and high intake of carbohydrate, due to poor dietary habits such as unbalanced diet and fondness of quick lunch such as instant noodles and canned juice which contains a high percentage of sugar.

But it cannot be denied that cases of heavy drinkers have been found in these patients. At least, some of them were found in the center of Tokyo. Two of them had such severe symptoms that they had a narrow escape from the jaws of death. Of course, all patients were cured by the treatment of allithiamine.

Fig.1 shows our experimental results using volunteers and animals. It is worthy to note that thiamine concentration in blood decreased rapidly after 200-300 ml of whisky was given over a period of 2 hours to the human volunteers. Transketolase activity also decreased in corresponding with thiamine concentration of blood and TPP effect increased in consistent with these phenomena.

Fig. 2 and 3 show the experimental results with rabbits. Similar results were found in human volunteers. Namely, thiamine concentration of blood and also transketolase activity decreased rapidly and TPP effect increased. Such a same relationship between thiamine concentration, transketolase activity and ethanol administration was found in the experiments of rats.

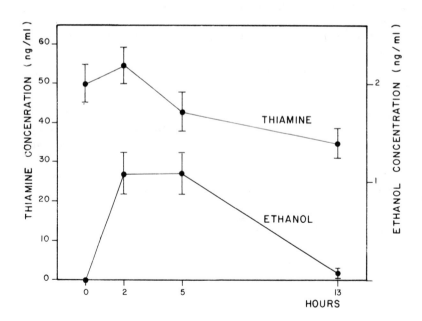

Fig. 1 - Thiamine and ethanol concentration in blood of human sub-
         jects after ethanol ingestion (means ± of 7 subjects)
         (Dose given; 200-300 ml of whisky/person)

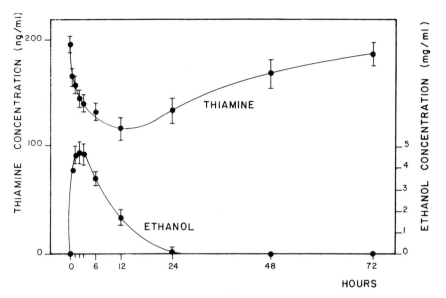

Fig. 2 - Thiamine and ethanol concentration in blood after admini-
         stration of ethanol to rabbits (means ± of 6 rabbits)
         (Dose given; 30% ethanol, 13 ml/kg).

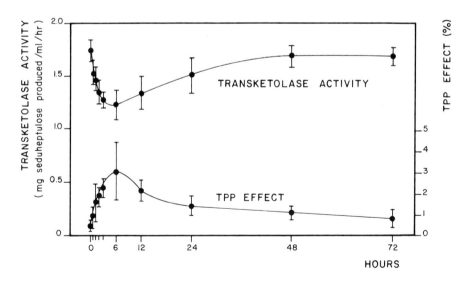

Fig. 3 - Transketolase activity and TTP effect to erythrocytes
         after administration of ethanol to rabbits (means ± of
         6 rabbits) (Dose given; 30% ethanol, 13 ml/kg)

It is a perfectly established fact that long term administration of ethanol damages liver and there comes degeneration and
fatty liver. But, it is an interesting fact that our experiment
showed that coexistence of thiamine deficiency clearly reinforced
the damage of liver.

In conclusion, it might be surmised that continuous and
longtime ingestion of alcohol results in thiamine deficiency.

BIBLIOGRAPHY

1. Abe,T. & Itokawa,Y.; Effect of ethanol administration on thiamine
   metabolism and transketolase activity in rats. Intn. J. Vit.
   Nutr. Res. 47, 307-314 (1977)
2. Balaghi,M. & Neal,R.A. ;Effect of chronic ethanol administration
   on thiamine metabolism in the rat. J.N. 107, 2144-2152 (1977)
3. Lieber,C.S. & DeCarli,L.M.; Quantative relationship between
   amount of dietary fat and severity of alcoholic fatty liver.
   A. J. C. N. 23, 474-478 (1970)
4. Lieber,C.S. Jones,D.P. & DeCarli,L.M.; Effects of prolonged
   ethanol intake; Production of fatty liver despite adequate diets.
   J. Li. Mv. 44, 1009-1021 (1965)
5. Nakazawa,T & Itokawa,Y.; Alcoholic and thiamine. Vitamin 52,
   287 (1978)
6. Thomson,A. Baker,H. & Leevy,C.M.; Pattern of ^{35}S-thiamine
   Hydrochloride absorption in the malnourished alcoholic patient.
   J. Lab. Clin. Med. 76, 34-45 (1970)

EFFECT OF DRUG ADDICTION AND CHRONIC ALCOHOLISM ON

MINERAL AND TRACE ELEMENTS ABSORPTION AND METABOLISM

E. Grassmann

Institut für Ernährungsphysiologie
Technischen Universität
München, FEDERAL REPUBLIC OF GERMANY

The acceleration of drug abuse in the last two decades has lead to an intensive work concerning the problems that are connected with its incidence. This means in the psychological field, investigations about the motives and ways how to overcome the addiction, as well as in the physiological and biochemical field work on the metabolic effects and their treatment by therapeutics. This fact induced a rapid increase of literature. It is therefore not possible to give a complete review in this presentation not even concerning the small field of mineral and trace element absorption and metabolism. Because alcohol is the most common and widespread drug, more is known about the effects of chronic alcoholism. They in many instances are better understood. Consequently, much more attention will be given to these questions. Additionally, on the basis of today's knowledge about mineral and trace element metabolism, some common problems of interest from the standpoint of nutrition will be considered.

The two major groups of drug abusers, the addicts of hard drugs and alcoholics, commonly suffer from hepatic disfunction (Orkin and Chen, 1977). The deviations from normal metabolism and the alteration in the nutrition due to the reduced appetite may lead to deficiencies of protein, vitamins, and essential mineral elements (see Somogyi and Kopp, 1976). From this one can deduce that the interrelationship between drug addiction and nutritional status have to be considered under two aspects: first, the nutritional status affects the metabolism of drugs, second, chronic consumption of these compounds results in radical changes in metabolism. This, in turn, affects the nutritional status. With regards to mineral elements: Some of the major elements and trace elements are involved directly in the catabolism of the different drugs or indirectly by maintaining essential structures. Marginal or deficient supply with these elements,

535

as is well established for calcium, magnesium, iron, zinc, and copper may therefore limit the capacity for metabolism of drugs and alcohol. Chronic consumption of these compounds, in many cases, simultaneously results in an alteration of the mineral supply, e.g. by decreasing ingestion, by impairment of the absorptive and metabolic efficiency or by disturbing homeostatic regulation. Consequently, the supply status turns to deficiency, and hence the catabolism of drugs is reduced. Although basic experiments concerning many of these questions are still lacking, some of the effects on the utilization of mineral and trace elements can often be derived from known interactions with other dietary and bodily constituents.

EFFECTS ON THE METABOLISM OF MINERAL ELEMENTS

Major Elements: Sodium, Potassium, Calcium, and Magnesium

Alterations in metabolism of mineral elements often are due to an elevated renal excretion of the elements. So an increased urinary loss was observed for calcium, magnesium, potassium, in some cases also for sodium following chronic consumption of alcohol or other drugs, e.g. morphine derivatives (see Walser, 1969; Martin et al., 1971). On the other hand Van Thiel and coworkers (1977) found significantly elevated Na concentrations in plasma of alcohol-fed rats because of reduced renal function. These authors compare these results with prior clinical studies in man, that patients with cirrhosis (presumable due to alcohol abuse) had both reduced osmolar clearance and increased Na retention. In addition, absorption of sodium from the intestine seems to be impaired (Mekhjian et al., 1975).

This may also be true for potassium. Besides an elevated urinary excretion, poor absorption is said to be responsible for deficiencies in this element, following chronic alcohol abuse. The deficit can induce metabolic alkaloses.

In alcoholics, even without liver cirrhosis, a reduced Ca concentration in plasma could be demonstrated. This response was, in part, attributed to an impaired absorption (Vodoz et al., 1977; Luisier et al., 1977). It has not yet been unequivocally established whether the reduced capacity of the liver and kidneys to synthesize 25-hydroxy-calciferol and 1.25-di-hydroxy-calciferol, respectively, is responsible for this effect (Avioli and Haddad, 1973; Lund et al., 1977). Chronic applications of morphine derivatives leads to alterations in Ca metabolism of the brain. With experimental animals, it has been demonstrated that morphine inhibits the Ca-activated ATPase in isolated synaptosomes due to an impaired uptake of calcium (Lin, 1978). In addition, the Ca-binding capacity of synaptic plasma membranes seemed to be diminished. No such effects were observed for sodium, potassium, and magnesium (Ross et al., 1976).

Deficiency in magnesium was shown in 1954 by Flink and coworkers to result from alcoholism. Since then, many experiments have been done with this element. The Mg deficiency comes about by diuretic loss of magnesium (Kalbfleisch et al., 1963; Heaton et al., 1962); McCollister et al., 1963). Hypomagnesemia is also induced by drugs, especially by barbiturates (Hellman et al., 1962). An additional, and probably more important factor may be malnutrition resulting from the ingestion of nearly Mg-free foods. Therefore, in patients with hypomagnesemia, there often is no increase in renal excretion of magnesium (Sullivan et al., 1966; Dunn and Walser, 1966; Lim and Jacob, 1972).

Low Mg concentrations in muscle of alcoholic addicts were some-times found at the beginning of withdrawal, in degree about as low as in kwashiorkor and celiac disease (see Flink, 1976). Furthermore, along with Mg deficiency neurological disorders were observed such as delirium tremens and cardiac arhythmias (Chadda et al., 1973; see Somogyi and Kopp, 1976). There seems to be a better correlation of the clinical symptoms with intracellular than with extracellular magnesium. Thus, in all patients with delirium tremens, the Mg concentration in erythrocytes was reduced, whereas its concentration in plasma was diminished in only about 50 percent of the cases (Smith and Hammarsten, 1959).

Effects on Utilization of Trace Elements

Trace elements predominantly function as integral constituents or activators of enzymes. Therefore, trace element deficiency, as in the case with vitamins leads to reduced enzyme activities and to metabolic disorders. Effects of drug addiction and chronic alcohol-ism on the metabolism of trace elements are well known for zinc, iron and, to a lesser degree, also for copper. Deviations in the metab-olism of other essential trace elements have been stated in a few reports. These and other studies suggest interactions with utiliza-tion of iodine, chromium, selenium, and molybdenum. Pathologically elevated urinary excretion observed in patients with liver cirrhosis includes most of the essential trace elements (Culebras-Poza et al., 1974).

Zinc - Zinc status is strongly affected by addiction. Vallee and coworkers (1956; 1957) had already observed an abnormal Zn metab-olism in patients with alcoholic liver cirrhosis about 20 years ago. They suggested a conditional Zn deficiency. In their experiments as well as in later ones by other authors the Zn content was signifi-cantly reduced in serum, liver, and erythrocytes, and the renal ex-cretion of zinc was elevated (see Sullivan and Burch, 1976). In addition, variations in plasma distribution have been described.

Zinc bound to albumines decreased from 81 to 54%, the portion associated with α_2-macroblobuline increased from 19 to 46% (Schechter et
al., 1978). Similar effects were observed in response to drug application. Spencer and coworkers (1976) suggest catabolic reactions
leading Zn loss and hence depletion of the organism as well as direct
effects on Zn metabolism itself.

Zinc is known to be a constituent of numerous enzymes, with
examples in all main classes of enzymes. Diverse biochemical changes
are the consequence of dietary depletion or impaired utilization of
this trace elements (See Kirchgessner et al., 1976a; 1976b). Regarding metabolism of drugs, special attention has to be paid to the
alcohol dehydrogenase (Fig. 1), ornithin transcarbamylase, δ-amino-
levulinic acid dehydratase, and possibly the glutamate dehydrogenase.
As a result of an impaired function of these enzymes, one has to expect problems in catabolism of alcohol, amino acids, and porphyrine
synthesis. According to the function in δ-aminolevulinic acid dehydratase there is close correlation between plasma Zn contents and
the cytochrom P-450 in liver biopsies of alcoholics (Hartoma et al.,
1977). This may be an explanation for the high sensitivity of chronic alcoholics with advanced liver damage to drug administration as
has been pointed out by Sullivan and Burch (1976).

Decreased activities of alcohol dehydrogenase not only affects
catabolism of ethanol. Moreover, utilization of vitamin A is disturbed, especially noticeable in an impaired dark adaptation (Morrison et al., 1973). As to the function of zinc in glutamic acid
dehydrogenase as well as in enzymes of nucleic acid metabolism,
negative effects can be expected in relation to protein synthesis
(see Reinhold, 1975; Kirchgessner et al., 1976a).

Furthermore, zinc seems to be of importance for stabilizing
membranes of cells and organelles by inhibiting fatty acid peroxida-
tion (Chvapil et al., 1972). Possibly, an interaction with selenium
may be involved. Selenium, an integral component of glutathione
peroxidase, protects fatty acids such as arachidonic acid from per-
oxidation and, in this way, favours prostaglandin synthesis (see
Ganther et al., 1976). Since prostaglandins, according to Evans
and Johnson (1978), presumably participate in Zn absorption, the
membrane-stabilizing effect could also be due to selenium.

Iron - An increased absorption and an enhanced storage of iron
in liver of alcoholics has been reported several times (Anguissola,
1970; Simon et al., 1977; Somogyi and Kopp, 1976). In some other
cases there was no increase in the Fe content of the liver, based on
dry matter, however, there was a significant shift to the debit of
nonheme iron (Lundvall et al., 1969; Lundin et al., 1971). The observation that absorption of iron is enhanced in the presence of
alcohol can probably be explained by an induction of secretion of
gastric juice (Charlton et al., 1969). As has been shown by in vitro

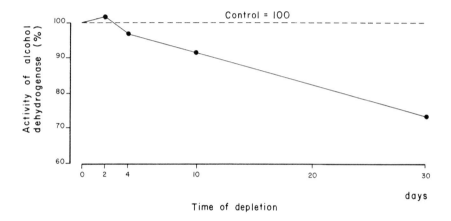

Fig. 1 - Effect of dietary Zn depletion on the activity of alcohol
dehydrogenase (Roth and Kirchgessner, 1974).

experiments, acute and chronic alcohol application affects iron absorption only in an indirect way (Forth and Rummel, 1973). From experiments with rats, it is known that depletion of easily mobilizable
iron stores like ferritin-Fe precedes the decrease in heme-Fe (Grassmann, 1977). Therefore, latent iron deficiency induced by drugs
could promote absorption of this metal. This concept agrees well
with the finding that chronic alcoholism had no effect on iron absorption rate nor on the proportion of nonheme-Fe in the liver (Loh
and Juggi, 1975). In addition, there is strong evidence that the
binding form of iron and also the interaction with dietary components
must be taken in consideration (Grassmann and Kirchgessner, 1974;
Grassmann, 1976; Celada, 1977).

Intermediary metabolism of iron is distinctly affected by many
drugs, primarily due to an enhanced porphyrin synthesis. The activity of enzymes involved in this pathway is very susceptible to feedback regulation and to a series of trace elements including iron itself. Thus, synthesis of cytochrome P-450 increases in iron deficiency. Becking (1976) suggests that the ferric/ferrous level in
tissues is important. Ferric iron, e.g. ferritin-Fe, seems to
inhibit, ferrous iron to activate the synthesis of the heme body.
Strongly negative effects were observed with other divalent ions,
such as Co^{2+}, Mn^{2+}, Cd^{2+}, and Pb^{2+} (Tephly et al., 1978). According
to Held (1977), alcohol exhibits its stimulatory effect solely on the
activity of δ-aminolevulinic acid synthetase without influencing the
concentration of cytochrome P-450 and the microsomal oxygenase system.

Anemia, frequently observed in chronic alcoholics and drug addicts (see Boettiger, 1973), may be attributed not so much to impaired Fe utilization as to other factors like deficiency of folacin
or pyridoxine (Eichner and Hillman, 1971). Certainly, reduced intake
of nutrients, especially of protein, is of importance. Erythrocyte
survival is reduced in addicts but this defect is compensated by a
greater erythropoetic rate (Loh and Juggi, 1975).

Copper - Only a few studies have been reported on the significance
of drug addiction and chronic alcoholism in respect of Cu supply
and metabolism. Chronic alcoholism is found to slightly reduce the
activity of cytochrome oxidase in skeletal muscle (Kiessling et al.,
1975). Hartoma and coworkers (1977) recently reported significantly
higher Cu concentrations in serum of chronic alcoholics, independently of the degree of parenchymal liver damage. Contrary to other results with experimental animals, they were not able to find any correlation between serum-Cu and metabolism of drugs, such as antipyrine.
On the other hand, copper seems to stimulate ferro-chelatase and
therefore, promotes the synthesis of cytochrome P-450 to about 22%
higher values (Tephly et al., 1978). Results with Cu-deficient rats
show the same trend (Grassmann and Kirchgessner, 1973). A single
dose of copper leads to a 24% higher activity of the blood catalase
than in controls with adequate Cu supply. These values agree well

the observation of Tephly and coworkers (1978). There are indications that the catalase in liver and blood respond similarly (Grassmann et al., 1978). There is evidence pointing to an involvement of
this enzyme in drug metabolism, at least in respect to alcohol
(Thurman, 1977). Cu depletion of the organism might bring about
radical alterations in the drug-metabolizing system by this mode of
action, especially by influencing Fe metabolism.

Secondary effect of protein supply

Cu depletion of tissues on the one hand can be an consequence
of an enhanced renal excretion as stated above. On the other hand,
however, copper is presumably absorbed mostly in the form of amino
acid complexes (Kirchgessner and Grassmann, 1970). Reduced dietary
protein intake following addiction should therefore impair Cu absorption. In addition, metabolic efficiency of trace element utilization is influenced by alimentary protein deficiency (Fig. 2). In
experiments with rats given adequate trace element supply, activity
of blood catalase, a critereon of Fe utilization correlated directly
with the dietary protein level; activities of ceruloplasmin and alkaline phosphatase in blood plasma, well established criteria for the
metabolic availability of copper and zinc, respectively, showed
quadratic responses (Grassmann et al., 1978). The high values for
ceruloplasmin activity at zero and low dietary protein may be explained by the wellknown stimulation of the synthesis of this enzyme
under stress conditions (Evans et al., 1969).

In summary, drug addiction and chronic alcoholism may exert a
great number of effects on mineral and trace element status. These
effects may be either direct or indirect, e.g. by increasing urinary
excretion, by decreasing absorption or the metabolic efficiency, by
altering the distribution in tissues, by inducing drug metabolizing
enzymes, or by affecting the nutrition habits. Even if the action
of drugs is a direct one, an altered nutritional status is the
ultimate result, and hence a change in the intake and requirements
for mineral and trace elements. This, in turn, may have consequences
for the metabolism of the drugs including alcohol. This aspect has
not been much emphasis in this paper. (Fig. 3).

Similarly, the great number of possible interactions between
minerals and trace elements (see Kirchgessner et al., 1978) has not
been considered. Nevertheless, it must be borne in mind that they
may pose and additional burden on the health of the addicts. This
particularly applies to the ecologically important elements cadmium
and lead, which affect the metabolism of many drugs, in part because
of their interacting with essential trace elements, and in part because of their inhibiting the participating enzyme systems (Tephly
et al., 1978.

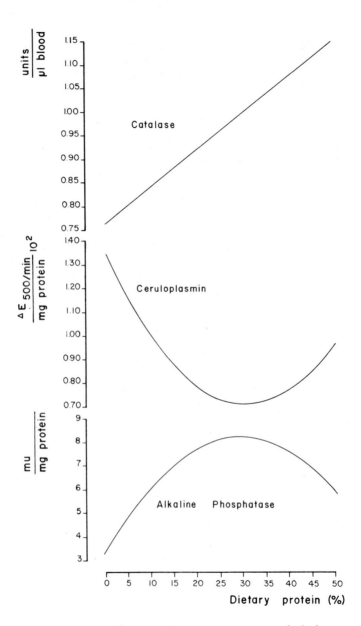

Fig. 2 - Protein supply and enzyme activities
(Grassmann et al., 1978).

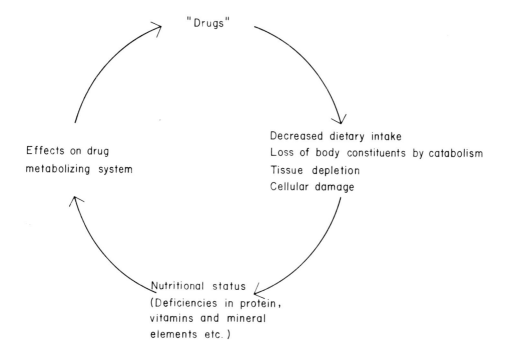

Fig. 3 - "Circulus vitiosus" of addiction.

BIBLIOGRAPHY

Anguissola, A.B., 1970, The nutritional value of wine as regards its
    iron content, in: "Iron Deficiency", Hallberg, L., Harwerth,
    H.-G., and Vanotti, A. (eds.), p. 71-74, Academic Press,
    London and New York.
Avioly, L.V., and Haddad, J.G., 1973, Vitamin D: current concepts.
    Metabolism 22: 507-531.
Becking, G.C., 1976, Hepatic drug metabolism in iron-, magnesium-
    and potassium-deficient rats. Fed. Proc. 35: 2480-2485.
Böttiger, L.E., 1973, Alcohol and the blood. Scand. J. Haematol.
    10: 321.
Celada, A., Rudolf, H., and Donath, A., 1977, Effect of alcohol on
    the absorption of iron. Schweizerische Medizinische Wochen-
    schrift 107: 1471.
Chadda, H.D., Lichstein, E., and Gupta, P., 1973, Hypomagnesemia and
    refractory cardiac arrhythmias in a nondigitalized patient.
    Amer. J. Cardiol. 31: 98-100.
Charlton, R.W., Jacobs, P., Seftel, H., and Bothwell, T.H., 1969,
    Effect of alcohol on iron absorption. Brit. Med. J. 2:1427-
    1429.
Chvapil, M., Ryan, J.N., and Zuroski, C.F., 1972, Effects of zinc on
    lipid peroxidation in liver microsomes and mitochondria.
    Proc. Soc. Exp. Biol. Med. 141: 150-153.
Culebras-Poza, J.M., Dean-Guelbenzu, M., Santiago-Corchado, M., and
    Santos-Ruiz, A., 1974, Normal and pathological excretion of
    trace elements by urine in man. In: "Trace element metabolism
    in Animals-2", Hoekstra, W.G., Suttie, J.W., Ganther, H.E.,
    and Mertz, W. (eds.), p. 473-475, University Park Press,
    Baltimore.
Dunn, M.J., and Walser, M., 1966, Magnesium depletion in normal man.
    Metab. Clin. Exp. 15: 884-895.
Eichner, E.R., and Hillman, R.S., 1971, The evolution of anemia in
    alcoholic patients. Am. J. Med. 50: 218-232.
Evans, G.W., and Johnson, P.E., 1978, Copper- and zinc-binding
    ligands in the intestinal mucosa. In: "Trace Element Metab-
    olism in Man and Animals-3", Kirchgessner, M. (ed.), p. 98-
    105, ATW, Weihenstephan.
Evans, G.W., Myron, D.R., and Wiederanders, R.E., 1969: Effect of
    protein synthesis inhibitors on plasma ceruloplasmin in the
    rat. Am. J. Physiol. 216: 340-342.
Flink, E.B., 1976, Magnesium deficiency and magnesium toxicity in
    man. In: "Trace Elements in Human Health and Disease",
    Prasad, A.S. (ed.), p. 1-21, Academic Press, New York, San
    Francisco, London.
Flink, E.B., Stutzman, F.L., Anderson, A.R., Konig, T., and Fraser,
    R., 1954, Magnesium deficiency after prolonged parenteral
    fluid administration and after chronic alcoholism, complicated
    by delirium tremens. J.Lab. Clin. Med. 43: 169-183.
Forth, W., and Rummel, W., 1973: Iron absorption. Physiol. Rev.

53: 724-792.

Ganther, H.E., Hafeman, D.G., Lawrence, R.A., Serfass, R.E., and
    Hoekstra, W.G., 1976: Selenium and glutathione peroxidase in
    health and disease - a review. In: "Trace elements in Human
    Health and Disease", Prasad, A.S. (ed.), p. 165-234, Academic
    Press, New York, San Francisco, London.

Grassmann, E., 1976, Zur Verwertung verschiedener Eisenverbindungen
    bei der Ratte. Zbl. Vet. Med. A 23: 292-306.

Grassmann, E., 1977, Zur Verwertung suboptimaler Zulagen verschiedener
    Eisenverbindungen durch die Ratte. Zbl. Vet. Med. A 24: 817-
    826.

Grassmann, E., and Kirchgessner, M., 1973, Zur Eisenverwertung bei
    unterschiedlicher Kupferversorgung. Arch. Tierernährg.
    23: 261-271.

Grassmann, E., and Kirchgessner, M., 1974, On the metabolic avail-
    ability of absorbed copper and iron, in: "Trace Element Metab-
    olism in Animals-2", Hoekstra, W.G., Suttie, J.W., Ganther,
    R.E., and Mertz, W. (eds.), p. 523-526, University Park Press,
    Baltimore.

Grassmann, E., von Krziwanek, S. and Kirchgessner, M., 1978, Zum
    Einfluß verschiedener Proteingehalte in der Diät auf die
    Aktivität einiger Metalloenzyme. Arch. Tierernährg. 28: 451-
    458.

Hartoma, T.R., Sotaniemi, E.A., Pelkonen, O., and Ahlqvist, J., 1977,
    Serum zinc and serum copper and indices of drug metabolism in
    alcoholics. Europ. J. Clin. Pharmacol. 12: 147-151.

Heaton, F.W., Pyrah, L.N., Beresford, C.C., Bryson, R.W., and Martin,
    D.F., 1962, Hypomagnesemia in chronic alcoholism. Lancet 2:
    802-805.

Held, H., 1977, Effect of alcohol on the heme and porphyrin synthesis
    interaction with phenobarbital and pyrazole. Digestion 15:
    136-146.

Hellman, E.S., Tschudy, D.T., and Bartter, F.C., 1962, Abnormal elec-
    trolyte and water metabolism in acute intermittent porphyria;
    transient inappropriate secretion of antidiuretic hormone.
    Am. J. Med. 32: 734-746.

Kalbfleisch, J.M., Lindeman, R.D., Ginn, H.E., and Smith, W.O.,
    1963, Effects of ethanol administration on urinary excretion
    of magnesium and other electrolytes in alcoholic and normal
    subjects. J. Clin. Invest. 42: 1471-1475.

Kiessling, K.-H., Pilström, L., Bylund, A.-C., Piehl, K., and Saltin,
    B., 1975, Effects of chronic ethanol abuse on structure and
    enzyme activities of skeletal muscle in man. Scand. J. Clin.
    Lab. Invest., 35: 601-607.

Kirchgessner, M. and Grassmann, E., 1970, Absorption von Kupfer aus
    den Cu(II)-L-Aminosäure-Komplexen. Z. Tierphysiol., Tierer-
    nährg. u. futtermittelkde. 26: 3-7.

Kirchgessner, M., Roth, H.-P., and Weigand, E., 1976a., Biochemical
    changes in zinc deficiency, in: "Trace elements in Human
    Health and Disease", Prasad, A.S. (ed.), p. 189-225, Academic

Press, New york, San Francisco, London.

Kirchgessner, M., Grassmann, E., Roth, H.-P., Spoerl, R., and Schnegg, A., 1976b, Trace element deficiency and its diagnosis by biochemical criteria. In: "International Atomic Energy Agency: Nuclear techniques in animal production and health", p. 61-79, Wien.

Kirchgessner, M., Schwarz, F.J., Grassmann, E., and Steinhart, H., 1978, Interactions of copper with other trace elements. In: "Copper in the Environment", Nriagu, I.O., (ed.), Wiley Interscience, New York (in press).

Lim, P., and Jacob, E., 1972, Magnesium status of alcoholic patients. Metabolism 21: 1045-1051.

Lin, S.C., 1978: Ca^{2+} ATPase and its interaction with morphine. Fed. Proc. 37: 771.

Loh, T.T., and Juggi, J.S., 1975, Changes in the blood and iron metabolism in rats with alcoholic fatty livers. Proc. Soc. Exp. Biol. Med. 150: 537-540.

Luisier, M., Vodoz, J.F., Donath, A., Courvoisier, B., and Garcia, B., 1977, Carence en 25-hydroxyvitamine D avec diminution de l'absorption intestinale de calcium et de la densité osseuse dans l'alcoolisme chronique. Schweizerische Medizinische Wochenschrift 107: 1 29-1533.

Lund, B., Sorensen, O.H., Hilden, M., and Lund, B., 1977, The hepatic conversion of vitamin D in alcoholics with varying degrees of liver affection. Acta Med. Scand. 202: 221-224.

Lundin, P., Lundvall, O., and Weinfeld, A., 1971: Iron storage in alcoholic fatty liver. Acta Med. Scand. 189: 541-546.

Lundvall, O., Weinfeld, A., and Lundin, P., 1969, Iron stores in alcohol abusers. Acta Med. Scand. 185: 259-269.

Martin, J.B., Craig, J.W., Eckel, R.E., and Munger, J., 1971, Hypokalemic myopathy in chronic alcoholism. Neurology 21: 1160-1168.

McCollister, R.J., Flink, E.B., and Lewis, M., 1963, Urinary excretion of magnesium in man following ingestion of ethanol. Amer. J. Clin. Nutr. 12: 415-420.

Mekhjian, H.S., Sury, T., and Baba, N., 1975, Chronic ethanol intake blocks the intestinal transport of sodium in man. Gastroenterology 68: A 96/953.

Morrison, S.A., Russell, R.M., Carney, E.A., and Oaks, E.V., 1978, Zinc deficiency: a cause of abnormal dark adaptation in cirrhotics. Amer. J. Clin. Nutr. 31: 276-281.

Orkin, L.R., and Chen, C.-H., 1977, Addiction, alcoholism, and anesthesia. South. Med. J. 70: 1172-1174.

Reinhold, J.G., 1975, Trace elements - a selective survey. Clin. Chem. 21: 476-500.

Ross, D.H., Lynn, S.C., and Lee Gardenas, H., 1976: Cellular adaptation to opiates. Fed. Proc. 35: 385.

Roth, H.-P., and Kirchgessner, M., 1974, Zum Aktivitätsverlauf verschiedener Dehydrogenasen in der Rattenleber be unterschiedlicher Zinkversorgung. Z. Tierphysiol., Tierernährg.

u. Futtermittelkde. 33: 1-9.
Schechter, P.J., Giroux, E., and Sjoerdsma, A., 1978: Zinc distribu-
    tion in human serum, with data on alterations in pregnancy and
    hepatic cirrhosis. In: "Trace Element Metabolism in Man and
    Animals-3", Kirchgessner, M. (ed.), p. 343-345, ATW, Weihen-
    stephan.
Simon, M., Bourel, M., Genetet, B., Fauchet, R., Edan, G., and
    Brissot, P., 1977, Idiopathic hemochromatosis and iron over-
    load in alcoholic liver disease: differentiation by HLA
    phenotype. Gastroenterology 73: 655-658.
Smith, W.O., and Hammarsten, J.F., 1959, Intracellular magnesium in
    delirium tremens and uremia. Amer. J. Med. Sci. 237: 413-417.
Somogyi, J.C., and Kopp, P.M., 1976, Alkoholismus und Ernahrungstatus.
    Bibliotheca Nutritio et Dieta 24:17-31.
Spencer, H., Osis, D., Kramer, L., and Norris, C., 1976, Intake, ex-
    cretion, and retention of zinc in man. In: "Trace Elements
    in Human Health and Disease", Prasad, A. (ed.), p. 345-361,
    Academic Press, New York, San Francisco, London.
Sullivan, J.F., and Burch, R.E., 1976, Potential role of zinc in
    liver disease. In: "Trace Elements in Human Health and Di-
    sease", Prasad, A.S. (ed.), p. 67-85, Academic Press, New
    York, San Francisco, London.
Sullivan, J.F., Lankford, H.G., and Robertson, P., 1966, Renal ex-
    cretion of lactate and magnesium in alcoholism. Amer. J.
    Clin. Nutr. 18: 231-236.
Tephly, T.R., Wagner, G., Sedman, R., and Piper, W., 1978: Effects
    of metals on heme biosynthesis and metabolism. Fed. Proc.
    37: 35-39.
Thurman, R.G., 1977, Hepatic alcohol oxidation and its metabolic
    liability. Fed. Proc. 36: 1640-1646.
Vallee, B.L., Wacker, W.E.C., Bartholomay, A.F., and Robin, E.D.,
    1956, Zinc metabolism in hepatic dysfunction. I. Serum zinc
    concentrations in Laennec's cirrhosis and their validation
    by sequential analysis. N. Engl. J. Med. 255: 403-408.
Vallee, B.L., Wacker, W.E.C., Bartholomay, A.F., and Hoch, F.L.,
    1957, Zinc metabolism in hepatic dysfunction. II. Correla-
    tion of metabolic patterns with biochemical findings. N.
    Engl. J. Med. 257: 1055-1065.
Van Thiel, D.H., Gavaler, J.S., Little, J.M., and Lester, R., 1977,
    Alcohol: Its effect on the kidney. Metabolism 26: 857-866.
Vodoz, J.F., Luisier, M., Donath, A., Courvoisier, B., and Garcia,
    B., 1977, Diminution de l'absorption intestinale de 47-calcium
    dans l'alcoolisme chronique. Schweizerische Medizinische
    Wochenschrift 107: 1525-1529.
Walser, M., 1969, Renal excretion of alkaline earths. In: "Mineral
    Metabolism", Comar, C.L., and Bronner, F. (eds.), p. 235-320,
    Academic Press, London, New York.

III. Nutrition, Growth and Human Development

MALNUTRITION DURING PREGNANCY AND LACTATION,

PERINATAL NUTRITION PROBLEMS

Mamdouh Gabr

Prof. of Pediatrics, Fac. of Med. Cairo University

Cairo - EGYPT

Nutritional privileges for pregnant and lactanting mothers have been recognized by most nations for hundreds of years and were verbally transmitted from generation to generation[1]. In Egypt, pregnant mothers are advised to eat chicken daily during the first forty days after delivery. In North China, she must eat 100 eggs during the first month of lactation. Fenugreck, which is rich in protein, is a popular food for pregnant mothers in Egypt, Abyssenia and Sudan.

A voluminous amount of literature is at present accumulating regarding the biological, metabolic and socieconomic aspects of perinatal nutrition. As more information is gained, gaps in our knowledge appear to require further research for better clarification.

MATERNAL NUTRITION AND INFANT DEVELOPMENT

Certain aspects of the deleterious effects of maternal malnutrition during pregnancy on fetal and infant development are now well documented. Field studies in several countries have demonstrated significant differences in anthropometric measurements of the newborn as related to maternal nutritional status. The interaction between nutrition and other environmental and biological factors frequently interfered with proper interpretation of the results. These interacting factors include racial and genetic background, age, health, education. nutritional habits and past nutritional status of the mother; parity, multiple pregnancies and fertility pattern; socioeconomic conditions especially as realted to sanitation, infection, availability of health services; climate and possible health and nutrition of the father.

Although many recent publications have accounted for most of
these variables, further studies on larger and better matched populations using up to date statistical techniques are still warrented
in many areas of the world.

Such studies should particularly consider the possible effects
of marginal cases of maternal malnutrition as compared to well nourished groups. In a study which was carried out in Cairo, 15 years
ago, a significant but slight difference was observed in birth
weight, height and skull circumference between infants of well nourished and moderately malnourished mothers[2]. Since the differences
were slight, it was concluded that moderate malnutrition did not
seriously affect the health of the infant. This should be re-evaluated on the light of present knowledge suggesting that differences
as little as 200-300 gms in birth weight might affect the well being
of the newborn[3].

The postulation that maternal malnutrition might affect fetal
development through various generations would justify longitudinal
studies in this respect. Some animal experiments have shown that
correction of malnutrition in one generation of mothers resulted in
normal offspring of the second generation, to be followed by growth
retardation of the third generation.

More studies on placental transfer of nutrients in the malnourished mother are needed.

MATERNAL MALNUTRITION AND
ORGAN DEVELOPMENT

The effects of maternal malnutrition and organ develpment has
been the subject of extensive studies. Most of these studies by
necessity were carried out on experimental animals with the well
recognized limitations in their applications to human. Recent knowledge on intrauterine development of the human brain viz early neuronal multiplication, to be followed later by dentritic growth and development of synaptic network; the special vulnerability of specific
areas of the brain; the various biochemical changes involving neuro-
-transmitters and neuro-endocrine system as well as the development
of newer tests and techniques to detect changes in mental performance, psychomotor behaviour and other congnitive functions, are
stimulating further research in this most important field. More
knowledge is needed on the effect of the type, degree, onset and
duration of nutritional insult to the fetus on future neurological
development.

Studies on the effect of maternal malnutrition on the development of other organs such as the liver, kidney, lungs and endocrine

glands as well as the functional capabilities of the newborn are
needed.

Although the incidence of congenital anomalies are reported to
be much higher with maternal malnutrition, more studies are required
to exclude other non-nutritional factors operating specially intrauterine infections.

MATERNAL MALNUTRITION AND
IMMUNO-COMPETENCE OF THE INFANT

It has been reported that small-for-date infants of malnourished
mothers show defects in the defense mechanisms similar to those of
the older child with P.E.M.[4]. On the other hand, it has been reported that milk of lactanting mothers suffering from malnutrition is not
defective in certain immunological factors which are of prime importance in the defense mechanism of the gastro-intestinal tract during
the neonatal period[5]. The rapidly growing field of immunology neces-
sitates more specific studies on the effect of maternal malnutrition
on immunocompetence of the infant and of the fetus.

MATERNAL MALNUTRITION AND
MATERNAL HEALTH

The relationship between maternal malnutrition and toxemias of
pregnancy is still controversial. Although better maternal nutrition
is associated with marked decrease in maternal morbidity and mortality, several investigators still advocate that low energy protein
diet lowers the incidence of eclampsia[6].

MATERNAL MALNUTRITION AND LACTATION

There is ample evidence that malnutrition of the lactating
mother influences the quantity of breast milk[7]. Certain constituents, such as lipids, retinal, ascorbic acid, thiamin and riboflavin
are also affected[8]. Protein content of breast milk is reported to
be satisfactory even under conditions of severe nutritional deprivation[9,10]. Present knowledge on the effect of maternal malnutrition
on the other protein constituents of milk apart from lactalbumin and
caseinogen is scanty. The possible interference with immunological
factors, mostly proteins, in human milk of poorly nourished lactating mothers has been referred to previously.

Marginal degrees of maternal malnutrition are more liable to be
reflected on the breastfed infant than on the fetus. In such societies where moderate degrees of malnutrition prevail, birth weight is
only slightly affected while deviation in the weight of the breastfed
infant from Western Standards appear as early as the 3rd month. A

study recently carried out on lactating mothers in a poor urban Cairo
district revealed that figures for underweight infants compared to
Western standards were 5% at birth and 24% at 3 months. More studies
on lactation mothers with various degree of malnutrition are to be
encouraged.

MATERNAL MALNUTRITION AND
POPULATION DYNAMICS

Reflecting of maternal malnutrition on population dynamics are
great. Small for date babies of malnourished mothers have a higher
perinatal and infant mortality rate. It is commom belief by socio-
logists that acceptance of family planning will not occur until a
significant reduction in infant mortality has been achieved[11]. Whether this child survival hypothesis is correct or not for certain cultures and certain environmental conditions is difficult to say. A
recent survey by the high council of Census and Statistics in Egypt
revealed that 25% of mothers with six surviving children wanted more
children while on the other hand, 60% of mothers with four living
children did not want more. This percentage was higher in urban than
rural areas. At any rate even if the child survival hypothesis is
correct, a time lag would exist between the initial effect of increased population following better nutritional status of the mothers
and the later decrease in population growth following voluntary pur-
poseful fertility control by the parents. During this period, efforts to improve maternal nutrition should be integrated with family
planning measures.

Other aspects of the influence of maternal nutrition on population dynamics include effect of malnutrition on fertility and vice
versa, the effect of estrogen containg pills on breast milk, the loss
of blood associated with intrauterine device in anaemic mothers etc.
Further research is certainly required to solve many such problems.

REMEDIAL PROGRAMS FOR PREVENTION
OF MATERNAL MALNUTRITION

Knowledge of several aspects necessary for the success of effective remedial programs to combat maternal malnutrition is still
lacking and contraversial.

Protein and Energy Requirements

Energy requirement has been established at 80.000 Kcal (335 M.
j.) for pregnancy and 100.000 Kcal (415 M.j.) during the first six
months of lactation. Daily requirements of good quality protein have
been estimated to be 6 grms and 17 grms for pregnant and lactating

mothers, respectively[12]. Reports from various developing countries
indicate that these values might be too liberal[13]. It is argued that
adptation mechanisms and restriction of physical exercise during pregnancy cut down these requirements[14]. On the other hand, evidence has
been presented from developed countries that higher requirements for
protein are necessary for optimal intrauterine growth[15]. The whole
issue of protein and energy requirement during pregnancy and lactation should be re-evaluated taking into consideration several factors: climate, adaptation genetic selection, amino-acid requirements,
more refined methods of measuring dietary intakes and energy expenditure and development of better methods to assess protein status
including tests of function. Research on optimal nutritional requirements during pregnancy and lactation is needed especially in developing countries, where the cost of intervention programs represent an
economic burden.

At-risk concept and
Maternal Nutrition

In countries where malnutrition is prevalent, the limitations of
available resources necessitate the adoption of a certain priority
approach. Local studies in each community should be carried out to
identify the most appropriate at-risk factors related to maternal nutrition. Criteria for selection of at-risk factors include: meaning
fulness i.e. occurrence with sufficient frequency and strongly associated with risk, simplicity, reliability, feasibility, sensitivity to
change and cost[16]. The at-risk factor might be biological, socioeconomic, environmental and might operate on the community level, family
level or the individual level. Under very socioeconomic conditions
all mothers in a community might be at risk. Examples of at risk
indicators of maternal malnutrition in developing countries include
family income, family size, sanitary conditions, availability of
health services, maternal age, education, parity, history of previous
pregnancy, weight and height. More elaborate indicators such as
weight gain during pregnancy are rarely applicable in these communities where health services are scarce. In more privileged communities,
more soffisticated techniques such as ultrasonic scanning, biochemical
changes in serum, tissue enxymes, hair root examination etc.[17], might
be resorted to. Refinement, simplification and rendering such techniques less costly is a challenge to research workers.

Integrated approach:

An integrated approach is essential for the development of an
effective practical comabatting program. The relationship between
planning and maternal nutrition has been previously referred to. Up-
grading of health services in general and MCH services in particular
has been found to be more effective than dietary supplementation in

several studies. Occupational and field reserach of these and other
aspects are essential before rationalization of such programs on
nationwide or community basis.

Finally it must be stressed that ethical requirements must be
particularly strict in all research studies on maternal fetal malnutrition to safe guard the growth of the fetus and child.

REFERENCES

[1]Platt, B.S., Proc. Nutr. Soc. 13, 94, 1954.
[2]Kamal. I, Gaz, Egypt Ped. Ass., 10, 1, 1962.
[3]Rush, D., Sterin Z., Christakes G. and Susser M., "The prenatal
    project in Nutrition and fetal development", M. Winnick, ed.,
    Wiley-Interscience Publishers, New York, 1974, pp. 95.
[4]Chandra, quoted by Hallman, N. in proceeding of IPA Seminar of
    feeding the preschoold child, Montreaux August, 1975.
[5]Reddy V., Antimicrobial factors in human milk, Current topics in
    Pediatrics, 15th International Congress of Pediatrics, New Delhi,
    India, 1977, pp. 169.
[6]Kido K., Yamada N., Hayshi S., Taii S., Noda Y., Mukiano, S. And
    Nishimura T., Abstracts, 10th International Congress of Nutri-
    tion, Kyoto, Japan, 1975.
[7]Harfouche, J.K., J. Trop. Ped. 16, 3, Monograph, 10, 1970.
[8]Belvady B. and Gopalan L., Indian J. of Med. Res. 45, 518, 1960.
[9]Jelliffe, D.B., Brit. Med. J. 2, 1131, 1952.
[10]Walker, A.R.D., Arvidson U.P. and Draper W.L., Trans. Roy. Soc.
    Trop. Med. Hyg., 48, 395, 1954.
[11]Wishik, S.M., Vender Vynckt, S., P.A. G. Bulletin, 5, 11, 1975.
[12]FAO/WHO, Energy and Protein requirement, Joint FAO/WHO Committee,
    WHO Technical report Series nº. 522, 1975.
[13]Greenfield, N., Abstrcts, 10th International Congress of Nutrition,
    Kyoto, 1975, PP. 179.
[14]Abdou, I, Amer, A.K., Bulletin of the Nutrition Institute of the
    United Arab Republic, 1, 21, 1965.
[15]Calloway, D.H., "Nitrogen balance during pregnancy in Nutrition and
    fetal development", M. Winnicked, Wiley, Interscience publishers,
    New York, 1974.
[16]Guidelines in the at risk concept on the Nutrition and health of
    young children, Am. J. Clin. Nutr., 30:242, 1977.
[17]Metcoff, J., Costileo J.P., Crosby W.M. Sandstood, H., Leucocytes
    as predictors of human fetal growth, Joint Speciality Session,
    ASCN/AFCR/ASCI/AAP, San Francisco, April, 1978.

MATERNAL FERTILITY AND NUTRITION

IN RELATION TO EARLY CHIDHOOD SURVIVAL

F.T. Sai

University of Ghana

Accra - GHANA

INTRODUCTION

Maternal nutrition has been accepted for a long time as in some way having an influence on the foetus. The importance of maternal nutrition also for the immediate postnatal life of a child has, to a certain extent, been recognized. In both instances, however, the mechanisms through which this maternal nutrition or lack of good nutrition during pregnancy and the immediate post-natal period act have not been completely clarafied. For a time, the fetus was considered a very good parasite. Opinion no longer holds that such is the case. Increasingly, the evidence is accumulating that, although the fetus is to a great extent a parasite, it is not a very complete parasite, and therefore is unable, given shortages of nutrients in the mother, to arrive with its full complement of nutrients or even to arrive with a full weight. Low birth weight has itself now been associated to a great extent with maternal nutritional influences. Thus, what happens during pregnancy, especially during the latter part of pregnancy, is of considerable importance for the early beginnings of a child. These early beginnings which were not considered to be of such great consequence in later life, are beginning also to appear as having a very important relationship to what happens to the child during the rest of its life, as measured by most development parameters.

Prenatal influences to some extent continue during the postnatal period. Lactation, for example, has its origins in the prenatal period, and breastfeeding which follows on lactation is of great importance during the first half-year or so of the child's life — longer in most developing countries. In the neonatal period it is of crucial importance since in most LDCs, there is practically no other way of feeding a very young infant than by the breast.

557

Fertility of the mother is another important variable in the early start of the child. Attention has recently been focused on the age at which childbearing starts and ends, intervals between births, and total fertility, and their relationship to pregnancy outcome and early childhood survival. The possible interactions between these fertility variables and maternal nutrition with pregnancy outcome still need further study. An attempt is made in this paper to assemble some of the evidence on maternal factors influencing pregnancy and its outcome and early childhood survival, and to relate these to some nutritional and other variables that affect the early development of a child.

MATERNAL AGE AND PARITY

The age at menarche has been decreasing with improvement in nutritional and general health status while with the same improvement menopause is delayed. This therefore, increases the total fertile period of a woman. In parts of Africa the women in the 40 to 50 year age group who have some education and have married into the upper classes of society, reached their own menarche at between 15 and 17 years of age. Their children have been reaching their menarche at 11 to 13 years, similar to the situation in Europe and the United States of America now. Such early onset of puberty makes teenage pregnancies much more likely, yet the evidence suggests that irrespective of social class and the amount of care available, the risks to life and health of the mother and child are higher for women under the age of 18, and particularly high for those under the age of 16 even in the more advanced countries. In the less developed countries, the evidence is pretty strong that those under the age of 18 and particularly those under the age of 16 are unlikely to have finished their own growth, so if they get pregnant the chances are that their own growth is stunted or the influences that this incomplete growth plays on the child will be much more magnified. Those having their first child are more likely to have pregnancy complications if they are under the age of 18 than if they were in the age group above 18. Such complications rise very sharply when the women are aged 35 and over too, as demonstrated by Nortman (1974). In the over-35 age group, the likelihood of various nutritional complications is also very much greater.

In addition to maternal age, a factor of some considerable importance is the frequency with which a mother has her children. In developing countries, women who have many children, whose births are badly spaced are unlikely to have their blood nutrient levels and general physical resources restored adequately before they have the next child. They therefore themselves are a very serious risk group and in various studies it has been shown that children who arrive at intervals closer than two years apart are likely to have more early childhood complications and difficulties than children who are spaced

at intervals beyond two years. When comparisons have been made by
maternal age and parity, it has been found that for any given age,
particularly young mothers, parity over four or five is more likely
to lead to serious trouble for both mother and child than parity under
four. In may developing countries, the death rates of women during
the fertile period is higher than the rates for men in the same age
group, unlike the situation in the developed countries. These death
rates by themselves pose very severe and serious consequences to the
early start of the child. In many parts of Africa, if the mother dies
in childbirth, the chances of the child surviving its first month are
very seriously jeopardised, and this is simply due to an inability to
feed rhe child properly. Infant and childhood mortality and morbidity
are themselves very closely related to birth order and spacing. Omran
(1974), among others, has reviewed the literature and has come to the
following conclusions:

Mothers who are very young have more of their children dying in
both the neonatal and post-neonatal periods than mothers aged between
20 and 26. There is also a greater neonatal and infant wastage in
those having their first child and in those having more than four or
five children than those having two or four children. What influences
actually come to play must be considered as multiple. First of all,
there is the general economic situation in which the mothers find
themselves. Generally very high fertility rates are associated with
poorer socio-economic and nutritional status. However, there is a
fertility influence on its own quite apart from socio-economic status.
How this is mediated is not too well understood.

MATERNAL HEALTH AND NUTRITION

Pregnancy constitutes a great physiological stress for the
mother. Ths child is, to a certain extent, a parasite which has to
receive all its nutrients from the mother. In those countries where
the nutritional status is marginal, a pregnancy may therefore tilt
the balance and precipitate the woman into an overt state of malnutrition. There are communities in Latin America, Africa and Asia
where women may carry a baby to term with little or no gain in
weight, and where the women actually lose weight while lactating.
Such women are literally milking the flesh off their bodies for their
children. Signs of vitamin deficiency, such as cheilosis, bleeding
gums, macrocytic anaemia as well as iron deficiency anaemia and
goitre, are all much commoner among the pregnant than ordinary women
in many developing countries. When pregnancies occur at frequent
intervals, the mothers are unable to make up for the deficiency, and
many of them are nutritionally depleted by the time they have their
fifth and subsequent children, a situation which has been described
as the "maternal depletion syndrome". The grand multipara of tropical countries already looking very old at 30 is no fiction.

There is ample evidence, also, that pregnancy reduces the resistance of women to infections and infestations of various kinds.
In malarious areas, for example, this may pose a serious threat to
health and pregnancy outcome. Gilles (1969), working in Nigeria,
showed that malaria immunity, which is built up during the first
nine years of life and normally retained in adult life, may be so
depressed during pregnancy as to cause the pregnant woman to have
clinical attacks of malaria with its attendant complications such as
abortions and premature delivery or low for age babies. Parasitis-
ation of the placenta may be partly to blame. But malaria also increases the need for folates and the role that this may play in
marginally nourished communities needs further study. Iron deficiency
anaemia has been shown recently to pose problem of inability to re-
sist infection. Iron deficiency is highly prevalent in pregnancy.
To what extent does this threaten the fetus faced with infections?
How much of a risk is this for the neonate?

Early maternal nutrition is also very important, since many
studies have shown a positive correlation between low maternal stature
and low birth weight. Poor stature is generally associated with
chronic malnutrition or malnutrition at some stage during earlier
development. This is also linked with poverty and adverse socio-
-economic circumstances. There are studies which indicate that
among less well-off communities, birth weight of children showed a
correlation with social class. In a study comparing birth weights of
children in two hospital in Accra, Hollingsworth (1960), showed a
statistical difference between the birth weights of children born in
the Ridge Hospital (at that time catering for senior public servants
and the elite of the society), as compared to children born in the
Kirle Bu Hospital, which was catering for the general public. Many
studies in other parts of the world, particularly in India, have also
shown this general trend that women from the poorer social classes
have children whose birth weights are on the whole, lower than those
from the higher social classes. This difference is only partly accounted for by the greater incidence of prematurity among chidren
from the lower-socio-economic classes. Some of it is also due to
low for age babies — a situation partly correctable by supplementary
feeding in pregnancy.

SOME SPECIFIC NUTRIENT DEFICIENCIES
IN PREGNANCY AND LACTATION

1) Calorie deficiency in pregnancy

Calorie deficiency, when very severe, as exists in famines, is
not very consistent with a satisfactory pregnancy and, in fact, there
is a depression of fertility during famines. However, in most communities, that severity of calorie deficiency is an occasional hap-

pening only; but there is in many communities calorie decifiency of
sufficient level to influence the in-utero development of the child.
Even though human beings do adapt to calorie deficiency to an extent,
yet in very poor communities, pregnant women may find it absolutely
impossible to conserve energy to the extent required to maintain their
own weight as well as accommodate an increase of the order of 5-10
kilos, including the laying down of fat against lactation. In such
situations, therefore, the children arrive very much smaller than
children of well fed mothers. Studies from INCAP seem to confirm
that the final birth weight of infants in marginal communities can
be improved by feeding extra calories in the last trimester. The
actual limits of malnutrition within which the positive mechanisms
will work are unknown and need study.

2) Iodine deficiency

There is no direct evidence of what happens to the child whose
mother has iodine deficiency, but indirect evidence on epidemiological
grounds would appear to suggest that such children also arrives with
a certain amount of iodine deficiency or thyroid decifiency. There is
an increased amount of cretinism among children in goitrous areas
compared to children in non-goitrous regions. In such regions both
the prevalence and severity of goitres correlate closely with parity
and particularly with closely spaced births.

3) Iron deficiency

Recent evidence would appear to show that iron is a very important element in resistance to infection and, if that is the case, then
iron deficiency which is of such common occurrence in tropical countries, must he considered as a very hazardous situations for the neonate. The evidence is still very small but it should be further
investigated.

4) General malnutrition in relation to brain development and mental performance

This has increasingly occupied researchers in the last decade.
INCAP studies have concluded that maternal malnutrition during pregnancy and before has a very serious consequence to brain development
and the further mental developemnt of the child after birth. This is
still subject to a good deal od debate but the evidence is strong.
Therefore, the question of feeding women adequately during pregnancy
and even before becomes very important indeed.

5) Malnutrition and Lactation

The evidence is rather conflicting about the importance of maternal malnutrition to the quantity and quality of milk produced. For a long time some people have been stating that women in poorer communities are able to lactate and breastfeed satisfactorily while women in better socio-economic circumstances are in general not able to breastfeed satisfactorily. Unfortunately, both lactation and breastfeeding are subjecte to influences which are not merely nutritional but psychological, social and cultural, and it has not been very easy to isolate the separate influences and be able to determine the extent to which malnutrition per se affects breast milk and the output of the breast. What is more, the milk contents synthesised by the breast behave differently from those filtered. Wray (1977) among others, has attempted an analysis of the finding of the latest literature. He has quoted work by Gopalan, by Chavez and by Edozien. Edozien found that the volume of milk produced in the reasonably nourished African woman is of the order of seven hundred to eight hundred cc's a day, but when the calorie intake is less than 2000 calories per day, the volume tends to range from 400 to 600 cc's a day. Wray also quotes studies by Gopalan in which he found among hospitalizes women that when they were provided with a hospital diet of 2900 calories, milk production increased by about 100 cc's per day when protein intake increased from 61 grams to 99 grams a day. Edozien found that where mothers were receiving 3000 calories per day, and producing 700 or 800 cc's of milk, he has able to provoke increases of the order of 200 cc's per day, by increasing the daily protein intake from 25 grams to 100 grams with diets that were carefully isocaloric. These findings, however, have not been generally accepted. There is a major debate going on right now as to whether calories by themselves would not increase the flow of milk. This needs to be carefully studied.

The quality of milk to a certain extent also reflects the quality of the material diet, although in general grosso constituent terms the quality of milk of a malnourished woman, judged by protein fats and sugars, is not very different from that of a well nourished one. Recent evidence would appear to indicate that where the mothers are relatively short in vitamin C, the milk that comes out is also dificient in vitamin C. Arroyare has shown this to be the case with vitamin A. He has also shown an improvement with consumption of vitamin A fortified sugar as reported at this Congress. The question of deficiency of other vitamins requires study. In developing countries the qualitative and quantitative adequacy of breast milk means life or death to the very young child. Infantile beriberi, which has been reported in some parts of the world, is directly due to a deficiency of vitamin B in the mother.

6) Maternal Nutrition, Lactation and Fertility

Although malnutrition itself has no great influence on the ability to conceive, it may have an influence on the ability to carry a
child to term. Lactation and breast feeding, on the other hand, have
been shown to have a very major influence on fertility. Lactation
through the production of prolactin is able to inhibit ovulation for
a period of time. This has been shown to be one of the major means
of fertility regulation in poorer countries. Breast feeding itself,
especially breast feeding on demand, through constant nipple stimil-
ation, has a feedback mechanism which leads to prolonged prolactin
production and consequently, depression of the ovary, and through
that, continuation of amenorroea or, at least, an-ovulatory cycles.
In many poor communities, this has been found to be one of the most
important means of prolonging the interval between births. Since
the shorter the interval the higher the intra-uterine or neonatal
risks to the child; this in itself must be considered as a major
nutritional influence on early childhood survival, as well as on
fertility generally.

SOME PREVENTIVE MEASURES

As the title of the symposium has altered slightly and the Chair-
man has set the tone it is necessary to draw attention to preventive
measures to alleviate maternal malnutrition and its consequences to
perinatal morbidity and mortality:

a) General and long term approaches: General socio-economic up-
   lift with special programmes to improve the status of women
   including relevant and general education as well as training
   and employment for out of school youth; improvement in general health care.

b) Specific Measures:

   i) Nutrition intervention programmes for needy women.

   ii) Identification of special risk groups and their surveillance.

   iii) Comprehensive antenatal care.

   iv) Realistic nutrition education and training for lactation
   and breast feeding.

   v) Education and services for family planning.

These specific measures are related to different periods of the
mother's reproductive activity but they need to be considered as part
of a package. The importance of family planning to both the nutritional health of mother and fetus and to pregnancy outcome and early

child survival deserve stressing. A nutrition intervention programme
for women which does not carry a component of family planning is an
incomplete programme.

RESEARCH

Moghissi and Evans (1977) have done this field a useful service
by bringing togehter into one compact volume much of what is known
today about nutritional impacts on women in relation to reproduction.
Naturally most of the information comes from developed countries and
the studies relate to reasonably nourished communities. Work like
the 7 year longitudinal study of INCAP is needed in many more centres
of developing countries.

Such work should study the influence of malnutrition of moderate
severity continued over varying periods of time on the ability to conceive and carry a conception to term, the nutritional status of the
child at term and its ability to survive its early period; what are
the prospects for long term normal development of such children?

Another study is to relate the above to the birth order of children, the interval between births and maternal age. It is also necessary to find out the extent to which various infections afflicting a
woman during pregnancy interact with the other factors mentioned above
and how they influence the fetus and the young child.

There is much more needed on the effects of specific nutrient
deficiencies of varying degrees of severity on pregnancy outcome and
early childhood survival in many developing countries.

There is little doubt that studies of such nature will advance
our understanding of the complex mechanisms involved in the mother
child with respect to nutrition. They will also help to sharpen the
focus of public health programmes seeking to protect and promote the
health of mothers and infants.

REFERENCES

Gilles HM, Lawson JB, Sibelas M and Vollera AN (1969). Malaria, Ana-
     emia and Pregnancy, Ann Trop. Med. Parasitol. 63,2
Hollingsworth MJ (1960). West Afr. Med. J. 9, 256.
Nortman D (1974). Parental Age as a Factor in Pregnancy Outcome and
     Child Development. Reports in Population and Family Planning
     No. 16. Pop. Council, Aug. 1974.
Omran AR (1974). "Health Rationale for Family Planning in the Physi-
     cian and Population Change", Lucille Bloch, ed. World Federation
     of Medical Education.
Wray JD (1977). Maternal Nutrition, Breastfeeding and Infant Surviv-

val.   Paper prepared for Conference on Nutrition and Reproduction, Bethesda USA, Feb. 13-16 1977.

RELATIONSHIP BETWEEN MATERNAL NUTRITION

AND INFANT MORTALITY

Aaron Lechtig

Div. of Human Development, (INCAP)

P.O. Box 1188, Guatemala, GUATEMALA, C.A.

INTRODUCTION

High infant mortality rates (IMR) are a major health problem in many countries of the world. Maternal malnutrition has been implicated as an important cause of the high IMR reported in many developing nations. The objetive of this paper is to explore the relationship between maternal nutrition and infant mortality.

For this purpose we will first explore the historical trends in the infant mortality rates of the developed countries. We shall then undertake cross-sectional analysis relating nutrient availability to IMR. Differences in infant mortality associated with social class within countries will then be studied and finally we shall review our field results in order to document to what extent maternal nutrition is a determinant of infant mortality.

RELATIONSHIP BETWEEN SOCIOECONOMIC DEVELOPMENT
AND INFANT MORTALITY RATES

During the last four to five centuries, infant mortality rates in Europe fluctuated between 150-250/1,000 live births[1], reflecting the same situation seen in developing countries today.

However, infant mortality rates began to fall dramatically in the late 19th century, first in Sweden and then in England, France, Italy and the U.S.A. This appears to have been the result of both improved sanitary practices and increased standards of living, and it was relatively independent of improvements in medical care.

the same industrialized countries where infant mortality rates decreased in the twentieth century had a simultaneous improvement in gross national product per capita.

Figure 1, based on cross-sectional data, shows that dietary energy per capita per day — a measure of food availability — and IMR, are associated. Similar results have been shown previously with the proportion of low birth weight (LBW) babies[2]. The threshold appears to be around 2,800 calories per capita, above which IMR and the closely related LBW rates cease to decline significantly. For the same level of economic developement there is a wide range in IMR. This variability may be partially due to variations in the efficiency of translating economic growth into improved nutrition and health. For example, in Sweden, the drop in mortality occurred at a lower dietary energy per capita than in the U.S.A.

In Figure 2, 123 countries have been divided into three groups according to dietary energy per capita per day. It is clear that no country with less than 2,400 calories per day has "low" IMR while most of the countries with more than 2,800 calories presented "low" IMR.

In sumary, from these independent analysis we can deduce a clear association between socioeconomic characteristics and infant mortality rate.

It should be pointed out that poor sociaconomic conditions entail economic, cultural and biological deprivation. Lower class women are shorter, work more during pregnancy and have generally poorer health. They are also more likely to have smaller pelves, and poorer diets during pregnancy. Low SES mothers are also more likely to marry young, to be multiparous and to have illegitimate births. In addition, impoverished women are likely to show less than optimal care both for themselves during pregnancy and for their children. Each of the above-named factors has been shown to be associated with a high risk of infant loss. Our next task is to explore to what extent specific improvement in maternal nutrition may produce a decrease in IMR.

EFFECT OF MATERNAL NUTRITION ON INFANT SURVIVAL

Although there is no published data on investigations to assess specifically the relationship between maternal nutrition and infant mortality, the hypothesis of an effect of maternal nutrition on infant mortality rate seems reasonable and is supported by several studies. For instance, birth weight is consistently associated with infant mortality[4,5]. The majority of the racial mortality differential in the United States can be attributed to the higher proportion of low weight at birth of the black neonates[6,7], a difference that falls

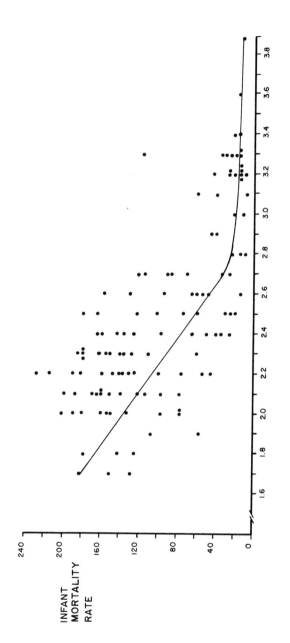

DIETARY ENERGY PER CAPITA PER DAY (THOUSAND CALORIES)

1 IMR. = infant deaths/1000 live births. Computed from the world population data sheet (Pop. Ref. Bureau Inc., 1975)

Fig. 1 – Relationship between dietary energy per capita and infant mortality rate[1].

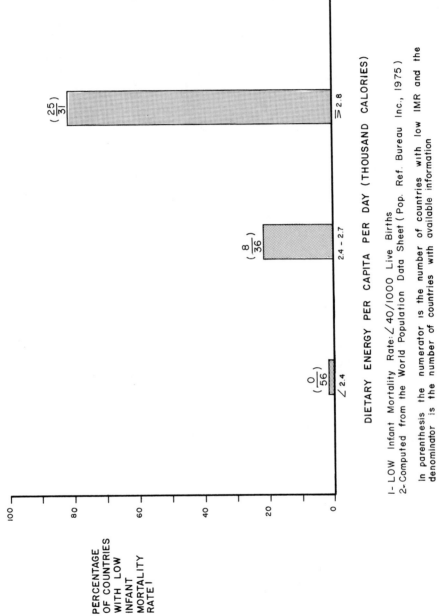

DIETARY ENERGY PER CAPITA PER DAY (THOUSAND CALORIES)

1- LOW Infant Mortality Rate: ∠40/1000 Live Births
2- Computed from the World Population Data Sheet (Pop. Ref. Bureau Inc, 1975)

In parenthesis the numerator is the number of countries with low IMR and the
denominator is the number of countries with available information

Fig. 2 – Relationship between dietary energy *per capita* and percentage
of countries with low infant mortality rate[1,2] ($p < 0.001$).

within the range of the effect of maternal nutrition on birthweight
demonstrated through dietary surveys[8] or nutritional supplementation[9]. Thus, maternal nutrition seems to be related to infant mortality through low birth weight (LBW≤ 2.5 kg). The next part of the
hypothesis, that lower SES women have poorer diets during pregnancy
has been demonstrated in developing nations where the social class
gap is wide[10]. In addition, placental size is smaller in low SES
women, a factor that may contribute to fetal malnutrition[11]. In
other words, low socioeconomic status may lead to poor maternal nutrition, high prevalence of LBW babies and consequently high IMR.
It has been argued that the accelerated drop in infant mortality
rates which occured in England during World War II could only be attributed to the wartime food distribution program which favored pregnant mothers[12]. Further, maternal malnutrition during the mother's
infancy and childhood could also produce an effect on infant mortality perpetuated through generations.

    We have explored the interrelationship between maternal nutrition
and infant mortality in two different studies: the first one in an
urban population of low social class from Guatemala City and the second one in four rural villages of eastern Guatemala.

Urban Study

    In this study the design corresponded to a case-control retros-
pective study with the main purpose of identifying simple risk indicators of infant death. For this purpose we studied the records of
101 consecutive infant deaths from low social class during 1975 in a
hospital of Guatemala City. These were compared with 199 children
(control group) who survived the first year of life, were also of low
social class and were being followed in the same hospital. Of 42
variables examined, 14 showed significant differences between the
study and the control groups.

    Table 1 presents the relative risk associated with these vari-
ables as well as their sensitivity and specificity as indicators of
risk of infant death. Of the 14 risk indicators, 6 concern the nutritional status of either the infant or the mother. These are:
weight for age (≤ 80%), breast feeding (≤ 6 months), weight for
height (≤ 90%), height for age (≤ 87%), low birth weight (≤ 2.5kg),
and maternal arm perimeter (≤ 24cm). Two additional indicators probably affect infant survival by means of impaired maternal nutritional
status. These are birth interval (≤ 30 months) and maternal age
(≤ 19 years).

    A risk scale was built on the basis of 7 of the indicators pre-
sented in Table 1, the possible score ranging between 0 and 7. The
high risk population group, composed of those children with high
score (from 5 to 7) in this scale, comprised 86 per cent of the

TABLE 1

URBAN STUDY — RISK INDICATORS OF INFANT MORTALITY

INDICATOR	NUMBER OF CASES		Relative[1] Risk
	Deaths	Control	
1. Hemorrhage during pregnancy[2]	101	199	23.2**
2. Weight for age $\leqslant$ 80%[2]	101	197	21.1**
3. Breast feeding $\leqslant$ 6 months[2]	101	199	19.6**
4. Weight for height $\leqslant$ 90%	31	191	13.9**
5. Umbilical cord rolled in neck	101	199	12.6**
6. Height for age $\leqslant$ 87%	70	191	7.6**
7. Birth weight $\leqslant$ 2.5 kg[2]	101	199	6.9**
8. Gestational age $\leqslant$ 37 weeks[2]	101	199	6.0**
9. Preceding child dead	64	147	3.7*
10. Birth interval $\leqslant$ months[2]	64	147	3.0*
11. Arm perimeter $\leqslant$ 24 cm	101	199	3.0*
12. Absence of perinatal medical care	101	199	2.9*
13. Maternal age $\leqslant$ 19 years[2]	101	199	2.6*
14. Age of menarche $\leqslant$ 13 years	101	199	1.9*
Score of 5-7 in risk scale[3]	64	147	25.2**

*   p < 0.05; ** p < 0.01.

1   Computed increment of probability of death in high risk group.

2   Components of risk scale

3   Score range: 0-7; high risk score: 5-7.

infant deaths and had a relative risk of dying during the first year
of life 85 times higher than children with low score (from 0 to 4).
Of the 7 components of this risk scale, 5 (weight for age, breast
feeding, birth weight, birth interval and maternal age) concern the
nutritional status of the baby, the mother or both. In conclusion,
the results of the urban study bring support to the hypothesis that
maternal nutrition is related to infant survival.

Rural Study

     We have explored the interrelationship among these variables as
part of a study in four villages of Guatemala[9]. These are communities where chronic malnutrition and infectious diseases are highly
prevalent, a situation unfortunately common to most of the rural populations of the Third World.

     In 1969 INCAP provided a system of curative medical care in
these comunities. The infant mortality rates have been reduced from
160 per 1,000 in 1968 to about 50 per 1,000 in 1975.

     The study design and the principal examinations made in mothers
and preschool children are presented in Table 2. Two types of food
supplements are provided: atole* and fresco**. Two villages receive
atole while the other two receive fresco. Atendance at the supplementation center is voluntary and, consequently, a wide range of supplement intake is observed. Table 3 presents the nutrient content
for both atole and fresco.

     As the home diet is more limiting in energy than in proteins[9],
ingestion of supplmented calories was selected as the criteria to assess supplement intake. We stress that while energy is the main limiting nutrient in this population, other populations may present
very different nutritional situations. Three additional independent
variables were also included in the present analysis: maternal height
(an indicator of the nutritional history of the mother during the age
of growth); socioeconomic score of the family; and birth weight (an
indicator of fetal growth). The socioeconomic score is a composite
indicator reflecting the physical conditions of the family house, the
mother's clothing and the reported extent of teaching various skills
and tasks to preschool children by family members. The principal
outcome variable was IMR in the cohort of children born from January
1, 1969 to February 28, 1975. Table 4 presents the limits were defined on the basis of reported literature (i.e. low birth < 2.5 kf)
or were based on results of prior analyses predicting birth weight.

---

 * The name of a gruel, commonly made with corn
** Spanish for a refreshing, cool drink

TABLE 2

STUDY DESIGN FOR FOUR VILLAGES[*]

INFORMATION[**]	WHEN COLLECTED
Obstetrical history	Once
Clinical examination[1]	Quarterly
Anthropometry[1]	Quarterly
Surveys[1]	
Diet	Quarterly
Morbidity	Fortnightly
Attendance at feeding center[1]	Daily
Amount of supplement ingested[1]	Daily
Socieconomic status[1]	Annually
Birth weight	At delivery
Infant death	First year age

[**] Two villages received atole, a protein-calorie supplement, and two fresco, a calorie supplement.

[*] Pregnancy was diagnosed by absence of menstruation; these surveys were made fortnightly.

[1] In mothers and preschool children

Figure 3 explores the relationship between socieconomic score (SES), maternal height, caloric supplementation during pregnancy and birth weight with the proportion of infant deaths in the four villages combined. In all four groups there is a lower proportion of infant deaths in the "high" category of each variable. However, the difference between the proportion of infant deaths in teh "high" and "low" categories is statistically significant only with maternal height and with birth weight.

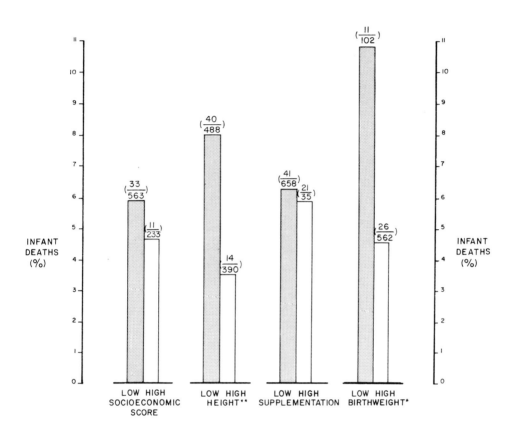

x^2 test: *p $\angle$ .05
**p $\angle$ .01

In parenthesis the numerator is the number of infant deaths and the denominator is the total number of live births

Fig. 3 – Percentage of infant deaths by levels of socioeconomic score, maternal height and food supplementation during pregnancy.

TABLE 3

NUTRIENT CONTENT PER CUP

(180 ml)

	ATOLE	FRESCO
Total calories (Kcal)	163	59
Protein (g)	11	--
Fats (g)	0.7	--
Carbohydrates (g)	27	15.3
Ascorbic acid (mg)	4.0	4.0
Calcium (g)	0.4	--
Phosphorus (g)	0.3	--
Thiamine (mg)	1.1	1.1
Riboflavin (mg)	1.5	1.5
Niacin (mg)	18.5	18.5
Vitamin A (mg)	1.2	1.2
Iron (mg)	5.4	5.0
Fluor (mg)	0.2	0.2

Next, we studied to what extent each of these apparent associations with infant mortality held after controlling for the remaining three independent variables. Given the small sample size, we
might not be able to measure the magnitude of the relationship between each of the variables presented in Figure 4 and proportion of
infant deaths. In cosequence, we explored mainly the consistency
or replicability of the direction of these relationships across eight
mutually independent comparisons.

Figure 4 shows a comparison between two categories of caloric
supplementation during pregnancy within each category of socioeconomic score, maternal height and birth weight. It can be seen in
this Figure that in six of the eight independent comparisons infant
mortality was lower in the high supplemented group than in the low
supplemented group.

In order to control for constant maternal factors, either measured or not measured, we explored the relationship between caloric
supplementation during pregnancy and infant death within pairs of

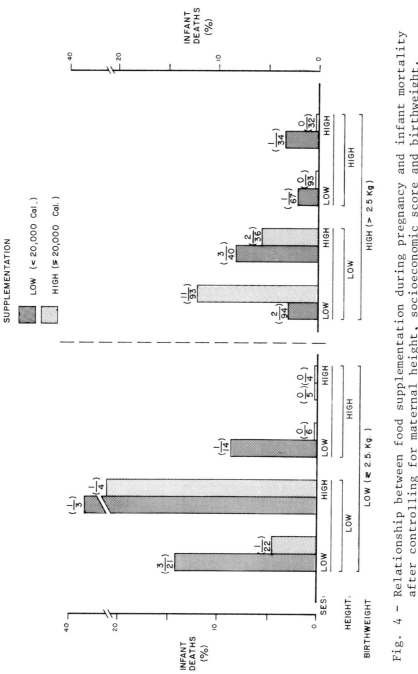

Fig. 4 – Relationship between food supplementation during pregnancy and infant mortality after controlling for maternal height, socioeconomic score and birthweight.

TABLE 4

LIMITS USED TO FORM DICTOMOUS VARIABLES

| | CATEGORY | |
VARIABLE	Low	High
1. Supplemented calories during pregnancy	< 20,000	≥ 20,000
2. Maternal height (cm)	≤ 149	> 149
3. Socioeconomic score	≤ mean + 1SD of four villages	> mean + 1SD of four villages
4. Birth weight	≤ 2.5 kg	> 2.5 kg

siblings of the same mother. In this analysis, the proportion of mothers with decreased caloric supplementation during the latter pregnancy was higher in the groups in which the latter child died than in those in which the latter child survived.

In consequence, in spite of the small sample size of the study groups, at present we believe that these results are compatible with the hypothesis that maternal nutrition is causally related to infant mortality. This conclusion arises from the following facts: 1) maternal height, an indicator of nutritional history of the mother during ages of growth, is consistently associated with infant mortality, and 2) caloric supplementation during pregnancy, an indicator of maternal nutritional status during intrauterine life of the baby, is also associated with infant mortality in these populations.

The causal may be composed by the following steps: 1) maternal malnutrition may lead to smaller placental size and decreased nutrient supply from the mother to the fetus. This would result in developmental retardation during intrauterine life and therefore in decreased ability to survive during postnatal life. 2) maternal malnutrition may also produce suboptimal lactation performance which will contribute to the infant malnutrition, growth retardation and, in consequence, may limit the infant's potential to survive in its environment. Usually, this gradual deterioration of the child's development may increase susceptibility to infections of the gastrointestinal and respiratory tracts which in turn would worsen the health status of the baby and end up as the "final" cause of death. There is available evidence supporting the plausibility of several parts of this hypothesis.

Our conclusion from these findings is that both short-and long--term maternal nutrition status may be causally related to infant mortality, This does not mean that other factors, namely medical care and environmental sanitation are not important determinants or infant mortality in these populations. Actually, it is probable that in these rural populations in which medical care is available, infant mortality is the end result of a complex interaction among several factors including maternal nutrition. For instance, we have estimated that in the four study villages the medical care system using paramedical personnel and strict quality control systems is mainly responsible for a decrement in IMR from 160/1000 in 1968 to about 85/1000 in 1969. The rest, from 85/1000 to 47/1000 may be ascribed to the program of food supplementation or, in other words, to the improvement of maternal and infant nutrition and to improved medicl care. Some interactions are also probably occuring between both programs, food supplementation and medical care. Several studies have shown higher infant survival from mothers who have had more antenatal visits when compared to those who have had less, or none (see Table 1). If the mother's use of medical care facilities is associated with maternal height or supplement intake, this could explain differences in survival associated with maternal height or caloric supplementation. However, the present sample size does not permit us to solve this problem and there is no study available that provides an answer to these questions.

In conclusion, mechanisms for translating SES into variations in infant mortality exist at several levels. The main maternal factors are malnutrition and illness which lead to delivery of poorly viable infants. These effects, aggravated by poor medical care, lead to infant death. Based on both the literature reports and our own findings we believe it reasonable to assume that both and short and long-term maternal nutritional status are important determinants of infant survival, growth and development.

The actions to decrease infant mortality rates should be specifically adapted to the needs of each population group, and planners should take account of the socio-political constraints that influence maternal nutritional status. As immediate actions simplified health care porgrams with a strong nutritional component and using paremedical personnel may be very effective. However, in the long-term, the most effective actions in developing countries will be those oriented to a comprehensive attack of the causes of underdevelopment-To be successful, this approcah will require positive socio-political changes focused on social objectives and rational economic methods. The social changes mentioned above are justified not only because the ultimate goal of development is to improve the quality of human existence but also because the quality of human life is the key to development.

REFERENCES

[1] Lechtig, A., H. Delgado, R. Martorell, D. Richardson, C. Yarbrough and R.E. Klein. Effect of maternal nutrition on infant mortality. in: "Nutrition and Human Reproduction" (Conference on Nutrition and Human Reproduction held at the National Institute of Health, Bethesda, Md., February, 1977). W. Henry Mosley, ed. Plenum Press, New York, 1978, p. 147-174

[2] Lechtig, A., S. Margen, T. Farrell, H. Delgado, C. Yarbrough, R. Martorell and R.E. Klein. Low birth weight babies: world wide incidence, economic cost and program needs. (Chapter II). in: "Perinatal Care in Developing Countries" (Based on a workshop held at Gimo, Sweden, jointly sponsored by the World Health Organization and the 5th European Congress of Perinatal Medicine). G. Rooth and L. Engström, lds. Uppsala, Sweden, 1977, p. 17-30.

[3] Morris, J.N. and J.A. Heady. Social and biological factors in infant mortality. V. Mortality in relation to the father's occupation, 1911-1950. Lancet 1: 554-560, 1955.

[4] Chase, H.C. Infant mortality and weight at birth: 1960 United States birth cohort. Am. J. Pub. Hlth, 59: 1618-1628, 1969.

[5] Lechtig, A., H. Delgado, R. Martorell, D. Richardson, C. Yarbrough and R. E. Klein. Socioeconomic factors, maternal nutrition and infant mortality in developing countries. Paper presented at the VII World Congress of Gynecology and Obstretics, Mexico, D.F., October 17-24, 1976.

[6] Bergner, L. and M.W. Susser. Low birth weight and prenatal nutrition: an interpretative review. Pediatrics 46:946-966, 1970.

[7] Habicht, J-P., A. Lechtig, C. Yarbrough and R.E. Klein. Maternal nutrition, birth weight, and infant mortality. "Size at Birth (Ciba Foundation Symposium 27", in: K. Elliot and J. Knight (lds.) Associated Scientific Publishers, Amsterdam, 1974 p. 353-377.

[8] Lechtig, A., J-P. Habicht, E. De León, G. Guzmán and M. Flores. Influencia de la nutrición sobre el crecimiento fetal en poblaciones rurales de Guatemala: aspectos dietéticos. Arch. Latinoamer. Nutr., 22: 101-115, 1972.

[9] Lectig, A., J-P. Habicht, H. Delgado, R.E. Klein, C. Yarbrough and R. Martorell. Effect of food supplementation during pregnancy on birth weight. Pediatrics 56: 508-520, 1975.

[10] Lechtig, A., H. Delgado, R.E. Lasky, R.E. Klein, P. Engle, C. Yarbrough and J-P. Habicht. Maternal nutrition and fetal growth in developing societies: socioeconomic factors. Am. J. Dis. Child. 129: 434-437, 1975.

[11] Lechtig, A., C. Yarbrough, H. Delgado, R. Martorell, R.E. Klein and M. Béhar. Effect of moderate maternal malnutrition on the placenta. Am J. Obstet. Gynecol., 123: 191-201, 1975.

[12] Baird, D. Environmental and obstetrical factors in prematurity with special reference to experience in Aberdeen. Bull. WHO 26: 291-295, 1962.

PREDICTING FETAL MALNUTRITION

Jack Metcoff

Univ. of Oklahoma Health Sciences Center

Oklahoma City, Oklahoma USA

INTRODUCTION

Human fetal malnutrition is associated with reduced neuroblast formation, cell numbers, myelin formation and dendritic arborization in the developing brain. Postnatally, neurologic defects and impaired mental development and school performance are common. Congenital anomalies are eight times more frequent and there is little or no catch-up growth of brain or stature. These all lead to net reduction of competitive competence in childhood and adult life. The cost of a society in terms of lost potential may be enoumous. Fetal malnutrition may compromise 5 - 8% of all pregnancies in developed countries like the U.S.A., and 9 - 28% of all live births with gestational period more than 37 weeks in developing countries. World wide, fetal malnutrition of the growing fetus and prevent the consequences of fetal malnutrition, prenatal diagnosis should identify those mothers carrying undernourished babies before the rapid phase of brain growth and cell hyperplasia is completed. If some constellation of nutrients and other remediable factors adversely affecting fetal growth can be identified at the same time, appropriate interventions could be initiated to improve intrauterine nutrition and growth of the fetus.

HYPOTHESIS

We hypothesize that maternal nutrition regulates fetal growth. Non-nutritional factors also conditions fetal growth. These include: prepregnancy factors such as maternal age, height, weight, race, number of pregnancies, socioeconomic status, education; and events associated with pregnancy, such as weight gain, cigarette smoking,

581

activity, complications, (including toxemia, infections, hypertension,
diabetes, etc.), and, of course, the length of gestation and the sex
of the baby. If the role of maternal nutrition is to be assessed,
these non-nutritional conditioning factors must be taken into account.

The process by which the organism utilizes food determines the
nutritional status at any moment. "Nutritional status" is an operational term which implies evalutation of the response to nutrients or
lack of nutrients. The precise definition of nutritional status
changes with the purpose of the evaluation and the modalities and responses being measured. In the present study we use the term nutritional status to define some relation between nutrients in the body
fluids which influence cell functions and modify fetal growth. An
appropriate balance of nutrients will support effective cell functions, including differentation, replication, growth and metabolism
of maternal and fetal cells. Inappropriate levels or imbalances of
nutrients in the maternal microenvironment, would limit growth and
development of fetal cells. The result would be fetal malnutrition
and its consequences.

To test this hypothesis, we use maternal polymorphonuclear
leukocytes as an indicator for nutrient regulation of maternal and
fetal cells. Leukocytes are rapidly replicating cells which are
constantly maturing in the bone marrow and are released into the
maternal circulation, where they have a short half-life before being
replaced by new generations of cells. The circulating leukocyte has
most of the metabolic characteristics of other nucleated cells. Like
fetal and placental cells they derive their energy primarily from metabolism of glucose. The metabolism of the circulating neutrophil is
altered by pregnancy and impaired by malnutrition.

OBJETIVES

Thus the prime objectives of our project are:

(1) To predict fetal growth from maternal variables measured once
     at midpregnancy.

(2) To identify mothers likely to have malnourished babies, early
     enough in pregnancy to modulate fetal growth.

(3) Operationally, we propose to test the hypothesis that the nutri-
     ent microenvironment regulates the growth and metabolism of all
     rapidly replicating cells in the mother, using leukocytes as a
     cell model, and maternal diet, hair roots, plasma nutrient and
     amino acid levels as determinants of the nutrient microenviron-
     ment.

DESIGN

The overall design of the data collection procedure is the following: patients are entered into the study between 16 and 24 weeks
of pregnancy after being oriented to the project and having agreed
to participate by signing the informed consent form. Complete demographic information, physical examination, anthropometry, and obstetrical information are recorded. Twenty-four hour diet recall and
a 3-day food record are obtained by the project nutritionist; blood
is drawn for leukocyte, nutrient and amino acid analyses; hair is
pulled for assay of protein, DNA and RNA in the hair root and zinc
in the hair shaft. About 20 weeks later, when the mother is admitted to the hospital for delivery of her baby, the project is notifi-
ed. At birth, the baby is examined by a pediatrician of the project.
Anthropometric measures with Dubowits assessment of gestational age
are obtained twice, first at about 24 hours and secondly, at about
48 hours of age.

PREDICTION OF ADJUSTED BIRTH MEASURES

As the first step, we defined relationships between maternal
nutrition and fetal growth in our population sample and used those
relationships to develop prediction equations. Later, the midpreg-
nancy data of individual mothers are entered into the equation which
predicts the various measurements her baby should have when it is
born. For example, we attempt to predict birth measurements such as
weight, lenght, head circumference, arm muscle area, etc. The prediction of the birth weight her baby will have, is made at midpreg-
nancy from maternal nutrient levels, leukocyte metabolism and other
characteristics measured at that time. The birth measurments are
adjusted by multiple regression analysis to account for individual
differences in gestational age, maternal size, etc.

ADJUSTING FOR NON-NUTRITIONAL FACTORS
INFLUENCING BABY SIZE

Each mother and baby differ in some way from the mean of the
entire sample. Partial regression coefficients were developed from
mother/baby pairs having uncomplicated pregnancies. These coefficients were used to adjust the raw for all birth size measurements
of all babies.

MATERNAL VARIABLES MEASURED
AT MID-PREGNANCY

The data on the independent maternal variables used to predict
baby size ar birth are obtained at a single time around mid-preg-

nancy (on the average, at 22 weeks), and include four groups of variables: 12 maternal characteristics, 16 nutrient measures, 18 plasma free amino acids and 10 measures of leukocyte metabolism. Baby size is predicted as soon as the measurements are completed.

CHARACTERISTICS OF THE SAMPLE

Maternal

     At the time of this data analysis, 756 mothers have been entered prospectively in our study; 584 of these mothers have delivered babies. Data are available for 576 of these mother/baby pairs.

     Mean raw data values are available for age, race, height and non-pregnant weight for 292 uncomplicated pregnancies and 284 pregnancies having some complications. The raw data for both groups of mothers are similar.

     Week of gestation when samples were obtained, weight gain to that interval and fundal heights also were similar. Total weight gain by women with uncomplicated pregnancies was slightly better.

Baby

     Raw data for duration of gestation, sex, birth weight, length and head circumference for the two groups of babies also were comparable.

     Other anthropometric measures on the 2 groups of babies also were similar, except for thigh muscle area, which was larger in the babies from uncomplicated pregnancies.

EFFECT OF PREGNANCY COMPLICATIONS
ON BIRTH SIZE

     Although the raw data are similar it is possible that pregnancy complications affect fetal growth.

     To determine the effect of complications on baby size standard scores (T scores) were computed for each adjusted measure in 573 babies (out of 576) having complete data. Two hundred eighty-nine (289) babies with complete data had uncomplicated pregnancies or deliveries. Two hundred eighty-four (284) pregnancies had one or more complications, such as infections, fetal distress, toxemia, abnormal glucose tolerance, etc. For example, there were 143 babies whose mothers had some sort of infection during pregnancy. Infections were associeated with significant reductions of birth weight, head circumference, chest circumference, arm and thigh circumference and muscle area with a reduced estimate of gestational maturity by Dubowitz assessment. These data emphasize the need to consider the influence of pregnancy complications on baby size at birth if the role of maternal nutrition per se on fetal development is to be evaluated.

PRODUCT: MOMENT CORRELATIONS
294 UNCOMPLICATED PREGNANCIES

A Product: Moment Correlation Matrix was developed between the 57 maternal variables measured at mid-pregnancy and various measures of baby size at birth. Numerous statiscally significant correlations were found. Some statistically significant (P<.05) correlations for uncomplicated pregnancies were found between the adjusted birth weights and arm-muscle area of the baby with some maternal variables, including smoking, leukocyte and nutritional variables. The correlation coefficients have been multiplied by 100 to remove decimal points. While statistically significant, the correlations are not very strong. Some of the variables are intercorrelated, for example, hair-root protein and DNA and several of the amino acids. Simple correlations such as these are not likely to have predictive value, since more of them account for more than 15% of the variance in the birth measurements.

RELATION BETWEEN MATERNAL LEUKOCYTE METABOLISM
AND NUTRIENT STATUS AT MID-PREGNANCY

Multiple regression analysis was used to determine if the metabolism of the maternal leukocyte is regulated by maternal nutrition and cell metabolism. In 145 mothers who have delivered babies and had complete data for all variables, a set of 33 independet nutrient variables, including 13 plasma nutrients, 18 plasma free amino acids, and one day diet protein and calorie intakes, were significantly correlated with 8 of 10 measures of leukocyte metabolism (R = 0.55 - - 0.65). These multiple correlation coefficents, derived from a principal component analysis, confirm a relationship between the metabolism of the mother's circulating leukocytes and her nutritional status.

RELATION BETWEEN MEASURES OF FETAL GROWTH
AND MATERNAL CHARACTERISTICS, NUTRITIONAL STATUS
AND LEUKOCYTE METABOLISM AT MID-PREGNANCY

Finally, prediction of baby size at birth was derived from all the mid-pregnancy measures.

Our hypothesis that fetal growth is related to maternal nutrition and leukocyte metabolism at mid-pregnancy is supported by the significant multiple correlation coefficeints encountered.

Principal component scores (PCS) are calculated for each of the 4 sets of variables independently. All subjects used in the analysis must have complete data for each set of variables. Principal component scores are a linear combination of all variables within a data set, such as the nutrients, are orthogonal and thus reduce the problem of multicolinearity. Linear multiple regression analysis selects one of two highly correlated variables and rejects the other as a redundant variable. The rejected variable may not be less important biologically than that selected by the computation process. The PCS retains all variables, assigning a relative weight to each.

With the leukocyte measures, plus hair root DNA and protein, or with the nutrient and amino acid values as independent variables, R values of 0.42 and 0.49 respectively, were found ($p<0.05$). When leukocytes, nutrients and amino acids, together with other maternal characteristics comprising a set of 57 independent variables were used, R was 0.75 and accounted for more than 50% of the variance in adjusted birthweight, head circumference, thigh muscle area; and 45 and 48% of the variance in lenght and arm muscle area, respectively.

These prospective observations give strong support to our hypotheses: first, that the nutritional microenvironment of the mother regulates the metabolism of her rapidly dividing and growing cells; and second, can account for a biologically significant proportion of the variability in fetal growth. The significant multiple correlation coefficients indicate that a combination of some maternal characteristics, nutrient levels and leukocyte metabolism measured at mid-pregnancy can predict the birth size of the baby.

This research was supported by Contract AID-TA-C/1224 from the Agency for International Development, Washington, D.C.

BREASTFEEDING PATTERNS IN RURAL BANGLADESH

Sandra L. Huffman

Dept. of Pop. Dynamics, School of Hygiene and Public
Health, The Johns Hopkins Univ., 615 North Wolfe Street
Baltimore, Maryland - USA

BACKGROUND

The often cited advantages of breastfeeding for women in the
developing world include breastmilk's anti-infective properties,
nutritional superiority to cow's milk, low cost, freedom from contamination, contraceptive effects, and its phychological benefits
(Jelliffe and Jelliffe, 1971). Although these advantages are considerable, the incidence and duration of breastfeeding vary widely,
depending on the country and location within it. Generally, in more
developed areas, breastfeeding is less common; rural women traditionally breastfeed more extensively than urban women.

The proportion of women breastfeeding their infants has been
observed to range from less than 10% in urban areas of the United
States to up to 100% in some rural areas in countries of Asia and
Africa. Average durations of breastfeeding have been observed to be
3.7 months in urban Chile (Perez et al., 1971), 5.5 months in Colombia (Oberndorfer and Mejia, 1868), 10.2 months in the Philippines
(Osteria, 1973), 14.0 months in Taiwan (Jain et al., 1970), 15.1
months in Egypt (Kamal et al., 1969), 18.0 months in Rwanda (Bonte
and Balen, 1969), 24 months in Senegal (Cantrelle and Leridon, 1971),
and 26 months in Indonesia (Singaringbum and Manning, 1976).

Along with incidence and duration, patterns of breastfeeding
also differ with regard to the frequency and length of suckling.
In a Mexican study, the degree of suckling was estimated to represent 15% of the day, or a total of three and one-half hours (Chavez,
1975). The number of breastfeedings per day varied between 12 and
13 during the first eight months, decreasing to 8-10 at eighteen
months. As might be expected, as the number of feeds per day in-

creased, the total number of minutes involved in suckling also increased. Rao (1959) observed that the minimum frequency of breastfeeding in India was six times per day, with the frequency decreasing
as the baby's age increased. Berman et al. (1972), reported that
among Eskimos, infants were nursed whenever they acted hungry or
restless. Nursing customarily continued for about five minutes at
each breast and occurred every 2-3 hours during both day and night.
In two Filipino studies, it was found that infants were fed 7-8 times
per day with each feeding period lasting 15-30 minutes (Popkin, 1975).

     The practice of breastfeeding in developing countries has often
been termed prolonged or extended, implying that it is greater than
what should occur. However, this type of breasfeeding is essential
for the majority of infants involved. Breastmilk constitutes a major
share of the overall protein intake of children even up to the age
of 2-3 years, and thus contributes significantly to the nutritional
well-being of the child (Rao and Gopalan, 1969). Blankhart (1962)
observed that, among New Guinean populations, breast-milk represented
a large proportion of the energy that the infant consumed: at 2-5
months 80%-100% of the calories came from breastmilk, at 8-11 months
28%-90%, and at 13-23 months 17%-54%.

     What factors are involved in the initiation and termination of
breast-feeding? In countries of the Indian subcontinent, breastfeeding is usually started 2-3 days after delivery. During the
interim, babies in South India are usually given water, dilute with
cow's milk, honey, or dates (Rao, 1959). In a field study of several
villages in Matlab, Bangladesh, Lindenbaus (1966) reported that, for
the first 3-4 days postpartum until milk production began, babies
were given honey, mustard oil, or cow's milk diluted with water —
all fed to the child on the tip of the finger.

     Infants in developing countries are usually fed on demand.
Since infants sleep with their mothers, night feedings often occur.
The practice of breastfeeding is so extensive that non-puerperal
women are at times also enlisted to breastfeed. Wieschoff (1940)
related reports from physicians in several areas of Africa, India,
Indonesia, Polynesia, North and South America, stating that grandmothers and neighbors suckled infants of women who had died.

     The durantion of breastfeeding is determined by several factors.
Pregnancy is reported to be one of the greatest determinants of
weaning (Rao, 1976; Swaminathan et al., 1973; Kamal et al., 1969;
Oberndorfer and Mejia, 1968; Gopalan and Belavady, 1961). In some
countries, breastmilk is considered inferior when the woman is pregnant. Often during the later months of pregnancy the amount of
milk produced is greatly reduced because of the high quantities of
circulating endogenous steroids evident during pregnancy (Vorherr,
1974). Other reasons given for weaning are the infant's age, insufficiency of milk, disease or death of mother or child. The

mother's working status may also affect the timing of weaning. Women in urban areas often have shorter durations of breastfeeding than those of rural areas, influenced by differences in cultural practices and occupational patterns. Popkin (1975) noted a significant decline in breastfeeding when Filipino mothers left their own barrio to work in another. He reported that for most this meant an increase in supplementation and decrease in the frequency of breastfeeding rather than a complete cessation of breastfeeding.

Among the group of women who do breastfeed, what characteristics are likely to be associated with variations in suckling patterns? As suggested above, changes in occupational activities may lead to variations in the frequency of breastfeeding. Other factors which may affect breastfeeding behavior are education and socioeconomic status. Increased socioeconomic status has been observed to be associated with decreased durations of lactation among Taiwanese women (Jain et al., 1970).

Maternal health status may affect breastfeeding practices through behavioral and physiological factors. Mothers who are ill may decrease their frequency of breastfeeding because of fear of passing the illness to the infant via the breastmilk. Sick mothers may also reduce their tendency to breastfeed because of weakness and discomfort and subsequent inability to breastfeed as freely as when they are well.

The infant's health status may also be an important factor in suckling patterns. A weak but significant correlation between changes in child health and variability in milk consumption was observed in Mexican studies, with sick infants suckling less frequently (Martinez and Chavez, 1971). Antonov (1947) observed that children of low vitality exhibited lower intensities of suckling than healthy infants. In many countries, cultural prohibitions to breastfeeding while the child is ill are common.

Infant age and nutritional status are likely to affect suckling behavior as reported by Thomson and Black (1965) and Gopalan (1962). The infant's nutritional status may affect suckling behavior indirectly through an effect on maternal behavior. Graves (1975) studied behavior of both well and malnourished infants in West Bengal, India. She found that mothers of malnourished children were less attentive to them and less likely to respond to the child's crying. Crying can indicate hunger or some other form of distress. Breastfeeding is often used as a method of comforting the infant as well as for feeding purposes. Thus, it may be that mothers of malnourished infants, by responding to crying less often than other mothers, are also likely to exhibit lower frequencies of breastfeeding.

Sex of the infant may also be a determining factor in suckling

patterns. Fomon et al.(1971) noted the quantities of formula consumed by bottlefed infants and observed that food consumption by
males was greater than by females. Martinez and Chavez (1971) noted
no differences in total volume of breastmilk consumption by sex, but
did observe that males gained more weight per unit of food consumed.

The present study describes the patterns of breastfeeding among
women in a rural area of Bangladesh where more than 90% of population
resides in rural areas.

METHODOLOGY

The data examined in this study are derived from a larger scale
investigation of the association between maternal nutritional status,
breastfeeding patterns, and postpartum amenorrhea in rural Bangladesh.
The methodology of data collection has been described in greater
detail previously (Huffman et al., 1978a).

This study employed both cross-sectional, semi-prospective, and
longitudinal methods of data collection. The population chosen for
study included residents of Matlab Thana, a rural area forty miles
south of Dacca, Bangladesh. The Cholera Research Laboratory (CRL)
has been maintaining a field research station and hospital in Matlab
since 1963. An ongoing vital registration system for the 265,000
residents (88% Muslim, 12% Hindu) has been in operation since 1966.
This system includes a registration of births allowing for exact
information on children's ages.

The original study, which was designed to examine the determinants of postpartum amenorrhea, was conducted in several phases.
Since the purpose was to focus on the return of menses postpartum,
only specified women were reinterviewed in subsequent phases. Phase
I collected information on all women in 86 villages of Matlab with
live births from February to September 1974. There were approximately 2500 women interviewed in Phase I. Information collected in this
phase included breastfeeding behavior, menstrual and pregnancy
status, and contraceptive usage. Socioeconomic information, including mother's education and a measure of wealth, was obtained from a
1974 census of the area.

In Phase II women who were still breastfeeding but who were not
pregnant were reinterviewed. The interviews, which were conducted
2-3 months after Phase I, obtained further information on menstruation, breastfeeding, morbidity, and diet. Anthropometric measurements of the mothers and infants were also taken.

A sample of 216 women found to be amenorrheic in Phase II were
selected for a longitudinal study of breastfeeding patterns which
was initiated in March 1976. These women were parity 2-4, aged

20-34. Only Muslim women were chosen for this phase because of methodological problems involved, since the field workers employed for
the study were primarily Muslim who for cultural reasons were unable
to enter Hindu homes. In this phase, women were visited at monthly
intervals by female field assistants who collected information on
morbidity, diet, and activities of the mothers. Anthropometric measurements were also obtained. Suckling time during the eight-hour
time period was measured by timing the number and frequency of
breastfeedings. This study continued for a period of one and one-
half years or until the respondent became pregnant or stopped breastfeeding. Because of incompleteness of some data after twelve months
of interview, some results only include twelve months of observation.

In Phase IV of the study, women who had not resumed menses as
of Phase II were reinterviewed. These interviews, which were conducted in August-September 1976, collected the same information as
in Phase I. Table 1 provides the ages of the index children at each
phase, as well as the information collected at each interview. Out
of the initial 2500 women, we have completed information on 1419
respondents. This excludes women who were lost to follow-up and
those who were eliminated from reinterview because of having resumed
menses as of Phase II. This presents biases which will be discussed
later.

RESULTS

For women in this study with births in February through September 1974, the median duration of breastfeeding was thirty months.
This includes women whose infants had died. Table 2 illustrates
that the probability of stopping breastfeeding because of infant
death is highest during the first year. At one year postpartum, 22%
of the women had stopped breastfeeding due to the death of their
infant. If infant death is excluded as a reason for stopping breastfeeding, 98% of all women with live infants were still breastfeeding
an infant aged one year, and 91% were still breastfeeding children
of two years of age. Less than 30% of the women with surviving infants had terminated breastfeeding by two and one-half years postpartum.

Although infant death is the primary reason for discontinuing
breastfeeding during the first year after birth, subsequent to that
time other reasons, including pregnancy, insufficient milk, and
infant's age, become prevailing factors influencing the likelihood
of continuing breastfeeding. Table 3 illustrates this point. Pregnancy is the primary reason for stopping breastfeeding. In the later
ages, insufficient milk also is an often reported reason for discontinuation of lactation. Among other reasons reported for stopping
breastfeeding are the child's becoming too old to breastfeed or
stopping willingly, maternal or infant illness, and the mother separated from the child through divorce or adoption.

TABLE 1

AGES OF CHILDREN AND INFORMATION COLLECTED AT STUDY PHASES

PHASE	I	II	III	IV
Child's Age at date of Interview	10–18 months	13–21 months	17–25 months	22–30 months
Information Collected	Breastfeeding Practice, Menstrual Status, Pregnancy Status, Contraceptive Status	Same as I Morbidity Diet Anthropometrics	Same as II Timing of Suckling Activities	Same as I
Date of Interview	Aug.–Oct. 1975	Nov./75–Jan./76	Aug.–Sept. 1976	March/76–Sept./77

TABLE 2

PROBABILITY OF BREASTFEEDING BY INFANT's AGE
AND REASON FOR DISCONTINUATION

INFANT's AGE IN MONTHS	n_x	CUMULATIVE PROBABILITY OF STOPPING BREASTFEEDING BECAUSE OF INFANT DEATH Q_{xd}	CUMULATIVE PROBABILITY OF STOPPING BREASTFEEDING FOR OTHER REASON Q_{xo}	CUMULATIVE PROBABILITY OF CONTINUING BREASTFEEDING FOR ALL REASONS P_x *
0	1414**	.00	.00	1.00
3	1216	.14	.00	.86
6	1171	.17	.01	.83
9	1130	.19	.01	.80
12	1082	.22	.02	.77
15	1026	.24	.02	.74
18	966	.25	.03	.72
21	904	.26	.05	.69
24	689	.26	.09	.65
27	348	.27	.13	.60
30	60	.27	.23	.50

* $Q_{xd} + Q_{xo} = 1 - P_x$

** Excludes five cases whose timing of cessation of breastfeeding was unknown

TABLE 3

PROPORTION OF WOMEN STOPPING BREASTFEEDING
BY INFANT'S AGE AND REASONS FOR STOPPING

CHILD'S AGE (completed months)	REASONS FOR STOPPING BREASTFEEDING				
	Child Death	Pregnancy	Insufficient Milk	Other*	ALL n
0– 6	.97	–	.01	.02	256 (1.00)
7–12	.80	.09	.02	.10	93 (1.00)
13–18	.57	.27	.02	.14	56 (1.00)
19–24	.15	.53	.14	.18	88 (1.00)
25–30	.09	.42	.25	.23	64 (1.00)
n	373	97	33	54	557**
	(.67)	(.18)	(.06)	(.10)	

* Includes infants becoming too old or stopped willingly, infant sick, mother sick, mother divorced or separated from child.

** Excludes five cases where timing of cessation of breastfeeding was unknown

Of the women who stopped breastfeeding for reasons other than infant death, 53% stopped due to pregnancy. Although a high proportion of women stop breastfeeding because of pregnancy, Table 4 illustrates that the majority of women continue to breastfeed through the sixth month of pregnancy. Even during the ninth month of pregnancy, nearly 20% were still breastfeeding. It should be remembered that most of these women were breastfeeding children aged over two years.

The bias in this study of not including women found to be menstruating as of Phase II may cause our estimates of the median duration of breastfeeding to be slightly high. These women are likely to become pregnant sooner than those who were still amenorrheic at Phase II. Thus, a higher proportion of women may have stopped breastfeeding because of an earlier onset of pregnancy.

Another reason commonly given for discontinuation of breastfeeding is an insufficiency of milk. Eighteen percent (18%) of the women with living children stopped breastfeeding because of insufficient milk. However, this finding needs to be examined closely, since 59% of these women were pregnant when they stopped breastfeeding. During pregnancy, milk production is reduced because of the inhibitory effect of high amounts of circulatory estrogen on the release of prolactin (Vorherr, 1974). Therefore, it is not surprising that pregnant women would report insufficient milk as a reason for stopping breastfeeding. Although other researchers have often reported insufficient milk as a reason for stopping breastfeeding, whether the association has been due to the onset of pregnancy has not been previously examined.

Table 5 illustrates a breakdown of breastfeeding and non-breastfeeding women according to education and familial wealth. Women with some education were more likely to have stopped breastfeeding for reasons other than infant death than those with no education. Wealth, as measured by familial ownership of certain household goods was also related to continuation of breastfeeding with women in the lowest one-third in ownership of goods exhibiting a greater likelihood of continuing breastfeeding than those in the upper one-third. An interesting finding in this table is that the women whose infants had died were more likely to be poorer and less well educated, suggesting that socioeconomic status probably affects the infant's survival probability. Breastfeeding duration may, therefore, be indirectly affected by socioeconomic status in the opposite direction to that described above, since death of the infant is a principal reason for stopping breastfeeding.

Maternal nutritional status has been reported to be associated with lactation performance. Some have suggested that malnourished

TABLE 4

PROBABILITY OF BREASTFEEDING DURING PREGNANCY
BY BY NUMBER OF MONTHS PREGNANT

NUMBER OF MONTHS PREGNANT	NUMBER OF WOMEM n_x	CUMULATIVE PROBABILITY OF BREASTFEEDING
1	295	1.00
2	266	.98
3	228	.96
4	170	.88
5	110	.78
6	60	.64
7	23	.45
8	8	.31
9	0	.19

mothers are more likely to terminate breastfeeding than their well
nourished counterparts (Wray, 1977). We do not have weights on women
whose infants died in the early postpartum period. However, of those
356 women, only four reported to have stopped breastfeeding for
reasons other than infant death. It is possible that they were more
malnourished and may have been poor milk producers, but from the
other data in the study, the theory that chronic malnutrition as observed among women in Bangladesh causes the cessation of milk production is doubtful. Breastfeeding is such a pervasive characteristic of the society that even among women with infant deaths, 5% continued breastfeeding another child, generally the next older sibling
in the family.

SUCKLING PATTERNS

Another manner of examining breastfeeding patterns is to compare
the amount of time women spend suckling their infants. For the approximately 200 women studied in the final phase of this research,
suckling frequency was observed for a period of eight hours at monthly intervals. At the initiation of this phase, the child's age
ranged from 17-25 months. After eighteen months of observation, ages
ranged from 35-43 months.

TABLE 5

PROPORTION OF WOMEN ACCORDING TO BREASTFEEDING STATUS,
EDUCATION, AND FAMILIAL WEALTH

| | BREASTFEEDING | NOT BREASTFEEDING DUE TO | | All |
	Proportion	INFANT DEATH Proportion	OTHER REASONS Proportion	n
Education*				
No education	.61	.27	.12	1053 (1.00)
Some education	.63	.21	.16	338 (1.00)
				1391+
Familial Wealth**				
Low	.60	.29	.11	547 (1.00)
Medium	.62	.26	.12	461 (1.00)
High	.63	.20	.17	383 (1.00)
				1391+

*X^2 = 6.0   d.f. = 2   p < .05

**X^2 = 11.9   d.f. = 4   p < .02

+excludes 28 cases of missing data

Retrospective data on feeding patterns in the early postpartum period were collected for these women. Forty-one percent of the mothers had begun breastfeeding during the first 24 hours after birth. All but 3% had initiated breastfeeding by the third postpartum day, when colostrum is replaced by the production of milk. Women reported giving their infants mustard oil or honey in the early postpartum period until they had initiated breastfeeding.

On average, women in the study breastfed their children 5-6 times in the observed eight-hour period. Because of the often intermittent pattern of breastfeeding, a feeding period was defined as one which occurred within an interval of thirty minutes. In 62% of the feedings, women used both breasts when suckling their children. Suckling was initiated by the child in 96% of the cases, either by his/her coming to the mother to suckle, or reaching for a breast while being held by the mother, or crying until the mother offered her child a breast.

A seasonal trend in suckling time was evident, with women reducing their suckling time during the months of November-February, and increasing again in the following months. This trend was independent of the infant's age. The months of November-February represent the post-harvest season when demands of women's time increase due to activities related to the processing of the rice crop. When we examined the pattern of crop-related activities women reported to have been involved in during the proceding month, we found that a peak month for crop-related activities was November.

Maternal nutritional status was found to be unrelated to the frequency of breastfeeding. Women in the lowest quartile of weight were similar in the pattern of breastfeeding to those in the highest weight quartile. Although there was a slight seasonal variation in maternal weight, with women losing weight during the months of July to September, corresponding to the preharvest season when food availability is limited, there was not an observable independent decline in weight as the duration postpartum increased. It must be noted, however, that our data do not include information on women's weight change patterns prior to one and one-half years postpartum. It is possible that women lose weight during the earlier period of lactation because of increased metabolic demands. However, these findings suggest that after this time, weight changes are affected by seasonal patterns rather than by duration effects.

Although maternal nutritional status appears unrelated to breastfeeding frequency, infant nutritional status seems to affect the frequency of suckling. Mothers with malnourished infants (defined as weight for height less than 80% of the Harvard Standard) exhibited a consistently higher mean suckling time than mothers of more well-nourished children. This finding which was contrary to that expected may be due to the consumption of poorer diets among

malnourished children. Their heavier reliance on breastmilk for
nourishment may cause them to suckle more frequently. Examinations
of mean suckling time by maternal morbidity, infant morbidity, and
child's sex did not reveal significant differences.

CONCLUSIONS

Matlab Thana, a rural area in Bangladesh, was the site of this
multiple phase study involving 1419 women who gave birth to live infants from February through September 1974. In this study virtually
all of the women breastfed their infants. For all women the median
duration of breastfeeding was observed to be thirty months, with
infant death the principal reason for terminating breastfeeding.

Among women with living infants, over 75% were breastfeeding at
thirty months postpartum. Major reasons stated for discontinuing
breastfeeding included pregnancy, insufficient milk, and infant's
age. Although pregnancy was the principal reason for discontinuing
breastfeeding among women with live infants, over 50% of the women
who became pregnant continued to breastfeed through the sicth month
of pregnancy. Almost 60% of the women who cited insufficient milk
as the cause of discontinuing breastfeeding were pregnant when they
stopped breastfeeding. Although poorer women were more likely to
have stopped breastfeeding because of infant death, among women with
living infants, richer women were more likely to have stopped breastfeeding than poorer women.

A group of 200 mothers, a subsample of the women described above,
were followed longitudinally for up to eighteen months. For these
women, whose index children were initially aged one and one-half
years or older, a seasonal trend in total suckling time was observed.
Women exhibited decreased suckling during the time of the major
harvest season. During this season, an increased activity pattern
was observed which may partially explain the reduced time spent suckling. Total suckling time was inversely associated with mother's
education, familial wealth, and infant nutritional status. No association was found between suckling time and maternal nutritional
status, infant illness, maternal illness, or sex of child.

RECOMMENDATIONS

Our data illustrate that women breastfeeding their children from
one and one-half to three years postpartum appear to maintain body
weight. This may be due to metabolic adaptation caused by the lactation process, coupled with the problable low quantities of milk
produced. Although weight is maintained, supplementation of these
women sould lead to increased output of milk or to replenishing of

maternal stores lost through the demands of pregnancy and early
lactation.

Of particular nutritional concern are the women who breastfeed
while pregnant. We found that more than half of the women who became
pregnant continued to breastfeed into the third trimester of pregnancy. These women are especially vulnerable to the consequences of
malnutrition. In addition to their own health risk, the health of
their suckling child and unborn infant is jeopardized. Supplementation of lactating women could benefit the child by increasing its
consumption of a valuable protein source. The fetus would be aided
through the direct transfer of maternal nutrients necessary for fetal
growth.

Nutritional programs attempting to supplement breastfeeding women
and their infants should be carefully designed to support the current
patterns of breastfeeding in Bangladesh. Recent research has shown
that maternal nutritional status does not significantly affect the
duration of postpartum infertility (Huffman, et al., 1978). Feeding
programs for lactating women are therefore unlikely to result in increased fertility, as previously expected, as long as the mothers
rather than the infants are consuming the supplements.

Supplementation programs aimed ar infants and children, however,
need to be concerned with adverse effects on breastfeeding — both in
terms of causing the eventual cessation of milk production as well as
on perhaps influencing an earlier resumption of postpartum menses
because of the reduced suckling stimulus. Such programs should be
specially designed to supplement breastfeeding rather than replace
it. These projects need to be aware of the adverse effects of possible increases in fertility among women whose infants receive supplementation. This does not imply that infant feeding programs
should not be initiated but rather that contraceptive services should
be made available simultaneously.

In Bangladesh where the annual rate of population growth approximates 3%, the general need for a strong family planning program is
widely recognized. Because of the lengthy pattern of breastfeeding,
care must be taken in selecting which contraceptives will be offered
to breastfeeding women. There are several studies indicating that
estrogen-progesterone combination oral contraceptives may decrease
the quantity of breastmilk (Gupta, et al., 1977; Buchanan, 1974).
Oral contraceptives may detrimentally also affect the composition of
breastmilk (Jelliffe, 1978) and may result in the excretion of exogenous steroids in breastmilk. The effects of injectable contraceptives upon lactation are still unclear with some investigators reporting some diminution of breastmilk (Parveen, et al., 1977) and others
reporting no deleterious effects (Badraoui, et al., 1978).

In many low income countries where some degree of economic development has occurred, the practice of breastfeeding has diminished

(Carballo, 1977). In these countries health planners are faced with the difficult task of reeducating women to the value of breastfeeding. Fortunately, in Bangladesh, breastfeeding is still an integral part of the culture. However, as we have observed in other countries, the traditional practice of breastfeeding can quickly change with modernization influences. With adequate foresight, development experts can help Bangladesh to retain the advantages of prolonged breastfeeding as the process of modernization continues.

BIBLIOGRAPHY

Antonov, A.N. 1974. "Children born during the siege of Leningrad in
    1942." J. Ped. 30: 250-259.
Badraoui, M.H.H. et al. 1977. Effects of some progestational ste-
    roids on lactation. In: "Fertility Regulation During Human
    Lactation", Parkes, Alan S., et al. (eds.). Galton Foundation,
    Cambridge, England.
Berman, M.L., et al. 1972. Effect of breastfeeing on postpartum mens-
    truation, ovulation and pregnancy in Alskan Eskimos. Amer. J.
    Obst. Gyn. (111 (1): 524-534.
Blankhart, D.M. 1962. Measured food intakes of young Indonesian
    children. J. Trop. Ped. (June): 18-21.
Bonte, M. and Balen, H. van. 1969. "Prolonged lactation and family
    planning in Rwand." J. Biosoc. Sci. 1:97-100.
Buchanan, R. 1975. Breastfeeding - Aid to infant health and fertility
    control. Pop. Reports 14: 149-167.
Cantrelle, P., and H. Leridon. 1971. Breastfeeding, mortality in
    childhood and fertility in a rural zone of Senegal. Pop.
    Studies 25 (3): 505-533.
Carballo, M. 1977. Social and behavioral aspects of breast-feeding.
    In: Fertility Regulation During Human Lactation", Parkes, Alan
    S., et al. (eds.). Galton Foundation, Cambridge, England.
Chavez, A., C. Martinez, and H. Bourges. 1975. Role of lactation
    in the nutrition of low socioeconomic groups. Ecol. Food
    Nutr. (May).
Fomon, S.J., et al. 1971. Food consumption and growth of normal
    infants fed milk based formulas. ACTA Paed. Scand. Suppl.
    223.
Gopalan, C. 1962. Effect of nutrition on pregnancy and lactation.
    Bull. Wld. Hlth. Org. 36: 203-211.
Gopalan, C., and B. Belavady. 1961. Nutrition and lactation. Fed.
    Proc. 20 (111) Suppl. 177-184.
Graves, P.M. 1975. Nutrition, infant behaviour and maternal char-
    acteristics: A pilot study in West Bengal, India. Amer. J.
    Clin. Nutr. 29 (3): 305-319.
Gupta, A.N., V.S. Mathur and S.K. Garg. 1977. Effect of oral con-
    traceptives on the production and composition of human milk.
    In: "Fertility Regulation during Human Lactation", Parkes,

Alan S., et al. (eds.). Galton Foundation, Cambridge, England.

Huffman, Sandra L., A.K.M. Alauddin Chowdhury and Zenas M. Sykes. 1978. Lactation and Fertility in Rural Bangladesh, submitted to Demography.

Huffman, Sandra L., A.K.M. Alauddin Chowdhury and W. Henry Mosley. 1978. Postpartum amenorrhea: How is it affected by maternal nutritional status? Science 200: 1155-1157.

Jain, A.K., T. C. Hsu, R. Freedman, and W.C. Chang. Demographic aspects of lactation and postpartum amenorrhea. Demography 7(2): 255-271.

Jelliffe, D.B., and E.F.P. Jelliffe. 1971. Human milk, nutrition, and the world resource crises. Science 188: 557-561.

Jelliffe, D.B. and E.F.P. Jelliffe. 1978. The volume and composition of human milk in poorly nourished communities: A review. Amer. J. Clin. Nutr. 31(3): 492-515.

Kamal, I., F. Hefnawi, M. Ghoneim, M. Tallat, N. Younis, A. Tagui, and M. Abdalla. 1969. Clinical, biochemical, and experimental studies on lactation. Amer. J. Obst. Gyn. 105(3): 315-323.

Lindebaum, S. 1967. Infant care in rural East Pakistan. Cholera Research Laboratory, Dacca, mimeo.

Martinez, C. and A. Chavez. 1971. Nutrition and development in infants of poor rural areas. I. Comsumption of mother's milk by infants. Nutr. Rep. Intern. 4(3): 139-149.

Oberndorfer, L., and W. Mejia. 1968. Statistical analysis of the duration of breastfeeding (a study of 200 mothers of Antioquia Province, Colombia). J. Trop. Ped. 4(1): 27-42.

Osteria, T. Lactation and postpartum amenorrhea in a rural community. 1973. Acta Med. Phil. 9 (Series 2, 4): 144-151.

Parveen, L., Chowdhury, A.Q. and Chowdhury, Z. 1972. Injectable Contraception (Medroxyprogesterone acetate) in Rural Bangladesh. Lancet (ii) 946-948.

Perez, A., R. Potter, and G.S. Masnick. 1971. Timing and sequence of resuming ovulation and menstruation after childbirth. Pop. Studies 25(3): 491-503.

Popkin, B. 1975. Income, time, the working mother, and child nutriture. Discussion Paper No. 75-9. Institute of Economic Development and Research, School of Economics, University of Philippines.

Rao, K.S., M.C. Swaminathan, P.S. Swaru, and V.N. Patwardhan. 1959. Protein malnutrition in South India. Bull. Wld. Hlth. Org. 20: 603-639.

Rao, V.K. and Gopalan. 1969. Nutrition and family size. J. Nutr. Diet. 6: 258-265.

Rao, N.P. 1976. Determinants of nutritional status. Presented at International Development Council Workshop on Household Studies in Singapore, August 3-7.

Singarimbun, M. and C. Manning. 1976. Breastfeeding, amenorrhea, and abstinence in a Javanese village: A case study of Mojolama. Studies in Fam. Plan. 7(6): 175-179.

Swaminathan, M.C., K. Vijayaraghavan, and D.H. Rao. 1973. Nutritional status of refugees from Bangladesh. Ind. J. Med. Res. 61(2): 278-284.

Thomson, A.M. and A.E. Black. 1965. Nutritional aspects of human lactation. Bull. Wld. Hlth. Org. 52: 163-177.

Vorherr, H. 1974. "The Breast: Morphology, Physiology and Lactation". Academic Press, New York.

Wieschoff, H.A. 1940. Artificial stimulation of lactation in primitive cultures. Bull. Hist. Med. 8(10): 1403-1415.

Wray, Joe D. 1977. Maternal nutrition, breast-feeding and infant survival. In: "Nutrition and Human Reproduction", W.H. Mosley, ed., Plenum, New York.

Supported in part by an interim grant by the Ford Foundation, a research grant (HD-11118) from the National Institute of Child Health and Human Development, and research funds of the Cholera Research Laboratory.

BREASTFEEDING IN HUNGARY

Mária Barna

Central Municipal Hospital for Infectious Diseases

Budapest, HUNGARY

Mother's milk is the best suitable food for infants because of its ideal composition. However, its function is more than feeding: it protects against the gastrointestinal and respiratory infections and allergic reactions.

Some papers have stated that all the immunologic protection of the fetus originated transplacentally and though there were some immunoglobulins in mother's milk, first of all in colostrum they could not absorb and had no clinical consequence. At this time, on this basis, the defensive role of the mother's milk was discredited. Recent investigations have demonstrated that the mother's milk contains secretoric IgA in a big quantity, which gives a local protection to the muco-cutaneous membrane of the gastrointestinal tract. The newest investigations performed in Hungary too, have drawn attention to importance of the cellular elements of the mother's milk. The newborn gets the most part of its immunologic protection from the mother's milk by way of cellular transmission of β-lymphocites. Presumably the protection of β-lymphocites has a more important role than that of the immunglobulins of the mother's milk.

Considering the last 10 years, in Hungary the number of births increased 14%; the infant mortality decreased almost 2/3. However, the number of breastfed infants gradually decreased during this period. Whereas in 1968, 49.2% of infants under 3 months consumed only mother's milk and the percentage of artificially-fed infants was 6.8%, in 1977 only 31.5% was breast fed alone and the number of artificially-fed infants increased to 16%.

In László Municipal Hospital for Infectious Diseases, infants under 3 months suffering of Coli enteritis or respiratory infections

were artificially fed at admissions during the period of 1975-1977,
which has drawn attention to the fact that, first of all, the artificially-fed babies got ill or they were admitted to the hospital,
respectively.

In 1975 we examined the results of a representative survey on
infant feeding of 8019 children in order to analyze the relationship
between economic activity, age and educational level of the mother,
family backgrouns, type of residence, birth-weight of the child,
etc., and breastfeeding.

In our sample, 88.5% of mothers were economically active and
11.5% were dependents.

In Hungary, maternity leave lasts 20 weeks. After the termination of maternity leave the mother is entitled to remain at home
until her child reaches the age of 3, and by this her working relation does not cease, nor is it shortened. During this pediod she
has the right to a child care allowance which is equal to 800 Ft per
month for the first child, 900 Ft for the second child and 1000 Ft
for the third child. If the mother does not make use of the child
care allowance and after the termination of maternity leave she goes
back to work, in the case of her child falling ill the mother may
stay at home and get a sick-pay for this period without any limitation of time, up to the age of 1 year. The breastfeeding woman has
the right to interrupt the work every day for one hour and a half
until her child reaches the age of 6 months and for three quarters
of an hour up to the age of 10 months, to feed him or her. She may
use this time during her work, before the working time or at the end
of it. These allowances make it possible for economically-active
mothers to breast-feed.

But according to the results of the survey, the economic activity and the occupation of the mother still has a negative effect
on breastfeeding. Among the economically active mothers the number
of whose who suckled her 1, 3, and 7 month old infants was 8-15 %
less than among the dependent mothers.

According to our investigations, breastfeeding seems to be independent of the educational level of the mother and father.

The relationship between mother's age and breastfeeding gave an
interesting result. On the one hand, among the mothers above the
age of 40, the number of those who never suckled her child was 3
times greater than among the younger ones. On the other hand, there
were 13-20% more 3 month old infants of mothers above the age of 40
who were breastfed, compared to the infants of the young mothers.
And this rate was about 17-21% among the 7 month old babies.

We examined the reason why babies have never sucked: 42.8% of the mothers had no milk, and 13.2% were ill.

We controlled the relationship between family background and
breastfeeding. If the husband lived together with the family, the
percentage of children who was never breastfed was 3.4%; if he did
not live together, it was 19.2%, and the number of children under
1 month who were not breastfed was 12% more than among a good family
backgroung, which emphasizes the role of the environment.

The type of residence seems not to have a very important effect
on breastfeeding, although most of the children who were never
breastfed and those who were not breastfed at the age of 1 and 3
months were living in the capital, in Budapest, at the time of the
survey. This suggests that the rate of the breastfed babies is lower
in the great cities than in the countryside.

No wonder that there is a relationship between the birth weight
of child and breastfeeding. The rate of infants born with less than
2500 g who were never breastfed was 18.4%, while this rate was only
1.9% for those born with more than 2500g.

In Hungary, 170-240 000 liters (340-480 000 pints) of mother's
milk were collected every year. Half of the female milk collected
is given to the infants' departments of hospitals, and the other half
is given to infants with low weight or to sick babies whose mothers
are not able to breastfeed them, also free of charge.

According to the data of our survey the low birth-weight was
more frequent among the mothers above 30 years old, but there was no
relationship between the economic activity and the lower weight.
There were less low birth-weight or premature babies among the children of highly educated mothers. The mothers of high educational
level went to the medical specialist 9-10 times during pregnancy,
which is the ideal frequency. Only 29 (0.4%/ of the 8019 mothers
did not go to the advisory dispensary.

During pregnancy, as well as after delivery, all women are under
the care of the Public Health Service. It is desirable that the
pregnant woman should go to medical examination, therefore, the total
amount of maternity allowance is paid only to women who were at least
three times at a control during pregnancy, and the first time was in
the period of the first trimester of pregnancy. Medical control and
provision are carried out within the framework of the Public Health
Service and the health visitor of this Service goes to see the pregnant woamn in her flat or working place. In case of bad family conditions, if the mother has no place to live during her pregnancy, she
might be placed in a home for mothers. After delivery, while she is
breastfeeding her child, she might be taken with the child to a home
for infants.

To the question: what was the reason for giving complementary food, the answer was the following:

1. mothers had no milk in 13.7% of the cases;
2. mothers had few milk in 63.7% of the cases;
3. mothers had no time in 0.3% of the cases;
4. mothers wanted to give not only mother's milk in 11.4% of the cases.

In 85.5% of the cases the pediatrician, the health visitor or other health personnel advised the mother to begin to give supplement, and only in 14.5% of the cases it was the mother's decision.

But giving the complementary food preparation was proceeded by test sucking only in 28.8% cases.

According to the opinion of the majority of mothers, the necessity of breastfeeding lasts to the age of 8-9 months; only 6-13% of them did not know how long the period of breastfeeding should be.

We asked the breastfeeding women what advice they would give their daughters concerning the feeding of her child. The answers were corresponding to the general view in 88.9% of the mothers; 5.9% of the mothers would advice a longer period and 1.6% a shorter breastfeeding period.

In order to get information about onset, circumstances and practice of breastfeeding I examined 5 new-born sections of hospitals in Budapest. My experiences were the following: in Hungary there are practically no deliveries at home; 99.2% of deliveries occurred in obstetrical departments and 0.8% of child-birth took place out of institutions. In the latter case deliveries occured at home or during transportation to the hospital, because of the speed of the process. In such cases the mother and her child are carried to the nearest obstetric institute. Naturally, hospitalization is free of charge for everybody.

Our new-born sections are crowded. The bed utilization rate was 100% in all of them, with the exception of one department. The new-borns are cared for by pediatric sisters. During their studies, the pediatric sisters learn a subject called the "healthy infant", for 1 year. His is taught in 56 hours, 22 lessons being spent for teaching of nutrition including 8 lessons on breastfeeding. The pediatric sisters' qualification is appropriate. Their task is to persuade the mother to breastfeed, to prepare the mother and the infant for breastfeeding, to help the mother by their advice and to give them practical assistance during the breastfeeding period. They can perform their task only partly because of the crowddedness. But I met a very undesirable view among the doctors, that breastfeeding causes an aesthetically unfavourable effect on the breast.

In one of the 5 new-born sections every infant got milk preparation from the second day until the beginning of the excretion of the
mother's milk production, and later too; the quantity of the sucked
milk was estimated by the way of 2 measurements.

In another new-born section, on the basis of four measurements
it was stated that the mother's milk was not enough for 50% of infants. Here the supplementary food was pumped, boiled mother's milk.
This is an incorrect practice because if the babies get pumped
mother's milk too early they will not be hungry, will not suck, the
milk of the mother will stagnate and will cease flowing. In the
other new-born sections the supplement was rare.

It is a further question, that because of the crowdedness, the
new-borns are carried home on the 4-5th day after birth, and the
mothers did not learn to suckle or to pump their milk.

In this way, it is the responsibility of pediatricians and
health visitors for the breastfeeding increases. The obstetrical
institute sends information on childbirth immediately to the dispensary, by residence, and so within 48 hours after the mother's
return home the health visitor goes to see the mother and her infant
in their flat and within one week the pediatrician does so too.

In 1975 infants up to the age of one year were presented 11.4
times at the advisory dispensary and the health visitor went to see
the infants in their home 15.2 times, averages for the country.

At the same time a survey in one district of Budapest demonstrated that the frequency of the medical examinations increased by
18% in a year but the task of 33 health visitors was provided by only
22 nurses because 11 of them remained at home to use the child care
allowance. The overload of pediatricians and health visitors may be
one of the reasons that the mothers resort to the baby-food preparations when they first believe that the sucking was unsuccessful.

In several textbooks it has been uniformly emphasized that
breastfeeding is highly recommended at least up to three months and
is useful up to 6-8 months.

In Hungary the exaggerated propaganda of the drug factories has
no important role in the declining trend of breastfeeding. On our
milk products it is indicated that one should use it only in case
of the lack of mother's milk and for substitution. The infant-feeding encyclopedia, developed by a Hungarian drug factory, has 43 pages
which deal with nutrition, 14 pages with breast feeding and only 2
pages with milk products. I confess that I was the author of the
infant feeding part of this little book and it is given to all the
mothers after deliveries. I am sure, that it is a propaganda for
breastfeeding and it was financed by producers of artificial food.

The time-table of infant-feeding constructed also by the drug factory emphasizes the breast feeding.

The health workers are responsible in a high degree for the worse rate of breastfeeding among the young infants. It would be necessary to draw attention emphatically on the importance of breast-feeding in fundamental and postgraduate training too.

In addition I should like to make the following proposals:

1. The economically active mothers should get all the possibil-ities of getting the necessary free time to breastfeed their infants.

2. During pregnancy, at the advisory dispensary, the mothers should be prepared for breastfeeding. It would be advisable if the Red Cross organized the mother's school for breast-feeding.

3. It would be desirable if, in the new-born sections, the nurses helped the mothers more intensively by practical advice, assisting the process of breastfeeding, particularly in the case of the first child.

4. It would be suitable propaganda material for this aim, to draw attention that the mothers should not make use of baby-food preparations without medical advice. It would awake mothers consciousness of the fact that the milk-products are only substitutes.

5. The decreasing number of premature or low birth weight babies may increase the rate of the breastfed babies.

6. It would be desirable to improve the work of the pediatri-cians and the health visitors by fulfilling the vacancies.

MAIN NUTRITION PROBLEMS DURING

THE WEANING PERIOD AND THEIR SOLUTION

P.M. Shah

Inst. of Child Health, J.J. Group of Gov. Hospitals

Bombay, INDIA

In a vast majority of situations nutrition problems start from
the latter half of infancy to the second year of life. The pattern
of malnutrition differs to some extent in urban and rural children.
The main nutritional problems frequently encountered during the weaning period in rural and urban communities of India are:

1. Energy-protein malnutrition;
2. Iron deficiency anaemia;
3. Vitamin A deficiency, and
4. Rickets.

The magnitude, distribution and genesis of these problems in a
developing country, like India are described briefly below.

1. ENERGY-PROTEIN MALNUTRITION

The prevalence of energy-protein malnutrition in rural and urban
communities in and around Bombay is presented in Table I.

More than half the number of infants in urban and non-tribal
rural areas and only 37.0% of the infants in tribal villages were
weighing above 80% of the reference weight. However, prevalence of
severe malnutrition, weight of 60% and below, was not much different
in these areas. A large number of children in tribal villages were
mildly or moderately malnourished. Malnutrition is more frequently
seen during the second year of life. The prevalence of severe malnutrition was higher in urban and tribal rural communities. One
fourth of the children in their second year were weighing above 80%
of the reference weight in non-tribal villages and urban areas, and

TABLE 1

NUTRITIONAL STATUS OF CHILDREN DURING INFANCY
AND SECOND YEAR OF LIFE

AGE	n	NUTRITIONAL STATUS PERCENTAGE OF HARVARD REFERENCE WEIGHT				
		Above 80	71-80	61-70	51-60	Below 50
Rural						
Non-Tribal						
Infants	166	57.8%	21.1%	13.3%	5.4%	2.4%
1-2 years	190	23.7%	40.0%	25.2%	8.9%	2.2%
Tribal						
Infants	135	37.0%	30.3%	23.7%	4.5%	4.5%
1-2 years	232	12.3%	34.1%	33.7%	14.6%	5.3%
Urban						
Infants	258	56.2%	24.4%	12.4%	5.8%	1.2%
1-2 years	272	25.8%	25.4%	31.6%	12.1%	5.1%
Urban Slums						
Infants	1226	65.88%	23.47%	6.72%	3.29%	0.74%
1-2 years	2491	35.01%	28.66%	22.80%	10.20%	3.33%

only 12.3% had that nutritional status in tribal villages. Malnutrition increases during the second year and prevalence is higher in tribal villages.

Breastfeeding and delayed introduction of solids

In a longitudinal study at the WHO-aided Project in the 23 villages around Palghar, interesting and useful information was gathered on the changes in the nutritional status of the under-fives[1]. All the children in the villages were covered and followed up. It was observed that 50.6% of the children who had weight above 80% of the reference[2] in June, 1973, deteriorated in their nutritional status

TABLE 2

CHANGES IN NUTRITIONAL STATUS OF UNDER-FIVES
FROM JUNE 1973 TO MAY 1974 AT THE WHO-AIDED
PROJECT, PALGHAR[1]

WEIGHT GROUPS IN JUNE, 1973	n	IMPROVE- MENT	NO CHANGE	DETERIO- RATION
Above 80%	399	–	197	202
(%)	(100.0)	–	(49.4)	(50.6)
71 - 80%	651	100	412	139
(%)	(100.0)	(15.4)	(63.2)	(21.4)
66 - 70%	414	152	182	80
(%)	(100.0)	(36.7)	(44.0)	(19.3)
61 - 65%	176	99	48	29
(%)	(100.0)	(56.4)	(27.4)	(16.2)
51 - 60%	177	107	62	8
(%)	(100.0)	(60.4)	(35.0)	(3.6)
Below 50%	36	22	14	–
(%)	(100.0)	(62.1)	(37.9)	–

Note: The children in weight grades above dotted line did not re-
      ceive nutrition supplements except seven.

in May, 1974 (Table 2). On further analysis, it was found that a
majority of those who lost weight were children in the age group of
1-2 years. During the previous year, they had been breast fed and
had normal nutrition. Delayed introduction of semi-solids and solids
was the factor causing malnutrition. Moreover, their mothers started
going to work 5 to 6 months after delivery and the infants were
looded after by young mother-substitutes who did not feed them well
enough and which lead to deterioration in the weight of 6 to 24 months
old children. In a cohort study it was observed that only one third
of the ruban and 1.8% of rural children were given semi-solids during
6 to 8 months of age while 10% of urban and 50.9% of rural nine
months-old were not given any semi-solids. Introduction of semi-
solids is governed by beliefs and customs. 80% of the urban and 96%
or rural mothers believed in 'hot' and 'cold' foods[3].

TABLE 3

GROWTH OF THE INFANTS DURING THEIR FIRST YEAR OF LIFE[5]

INITIAL NUTRITIONAL STATUS/BIRTH WEIGHT	No. OF INFANTS 1973/75	NUTRITIONAL STATUS							
		Normal by				Severe malnutrition by			
		3rd mth.	6th mth.	9th mth.	12th mth.	3rd mth.	6th mth.	9th mth.	12th mth.
2.5 kg and below	289	34	105	91	44	13	4	5	14
(%)		(11.8)	(37.7)	(31.5)	(15.6)	(4.5)	(1.4%)	(1.7%)	(4.8%)
3.1 kg and above	191	118	126	105	82	–	–	–	1
(%)		(61.7)	(65.9)	(52.3)	(42.8)	–	–	–	(0.5%)

Mother-substitutes and young child nurture

At the Kasa Project in tribal villages about 80% of the mothers were working and 57.4% of their children were looked after by child mother-substitutes who themselves were under 6 years of age. 48.4% of children who were cared by mother substitutes were below 2 years of age. It was observed that 28.6% of the children below 2 years were fed thrice and 54.5% four times a day. The incidence of severe malnutrition, weight 60% and below of reference weight, was 55.5% when the age of the mother-substitutes was between 6 to 8 years of age. When the children were looked after by the mothers or elderly mother-substitutes, the prevalence of severe malnutrition was 8.5% and 21.3%, respectively[4].

Foetal malnutrition and subsequent nutrition during infancy

Malnutrition during intrauterine life continues during infancy and second year of life. At the WHO-aided Project, Palghar out of 289, newborns weighing 2.5 kg and below, only 37.0% and 15.6% were having normal nutrition by sixth and twelfth month, respectively. 4.8% of the previous group were weighing 60% and below as compared to 0.5% of the latter group (Table 3)[5].

Inter-relationship between maternal fetal and infant nutrition

It was observed[6] that the newborns of the women who had pre-pregnancy weight of 48 kg or above and had high antenatal weight gain of more than 6 kg, weighed 2.9 kg as compared to those of the women whose pre-pregnancy weight was 38 kg or below, weighed 2.6 kg at the time of birth irrespective of low or high antenatal weight gain. The newborns of the women whose pre-pregnancy weight was 48 kg and had low weight gain during pregnancy, weighed 2.8 kg. The interesting results were the weights of the same infants at 3 and 6 months of age. Only infants of those mothers whose pregnancy weight was high and gained high weight during pregnancy were weighing 5.8 kg and 7.2 kg, as compared to other three groups of infants who weighed 4.5 kg and 5.6/5.7 kg, respectively (Table IV). All these infants were breastfed.

These observations suggest that maternal nutrition particularly chronic malnutrition in women contributes not only to low birth-weight (fetal malnutrition) but also malnutrition during early infancy[6]

TABLE 4

RELATIONSHIP OF MATERNAL NUTRITION,
ANTENATAL WEIGHT GAIN AND WEIGHT OF
INFANT DURING FIRST SIX MONTHS[6]

ANTENATAL WEIGHT AT 20th WEEK	ANTENATAL WEIGHT GAIN	INFANT'S MEAN WEIGHT IN kg AT		
		Birth	3 months	6 months
Low weight group (40 kg or below at 20th week, pre-pregnancy weight 38kg or less)	Low wt. gain of 3 kg or less	2.6	4.5	5.6
	High wt. gain or above 6 kg	2.6	4.5	–
High weight group (50 kg or above at 20th week pre-pregnancy weight 48 kg or above)	Low wt. gain of 3 kg or less	2.8	4.5	5.7
	High wt. gain above 6 kg	2.9	5.8	7.2

B R E A S T F E D

Faltering of growth in a community

277 under-fives in 20 villages who were either not gaining weight for three months or loosing weight during two months prior to the study (45.5 percent) had been investigated. About one fourth of these children were weighing 65% of less of the reference weight and 51.2% were in 60% and less weight group. 28.9% of these under-fives with faltering of growth were in their second year of life and 28.8% were in their third year. 54.9% were girls. 64.0% of the mothers of 277 children were economically active. 44.9% had history of upper respiratory tract infection, 36.8% recurrent diarrhea and 25.2% had round worm infestation. Urine was normal in all and two had tuberculosis proved on tuberculin testing and MMR[7].

Family size

There was a significant relationship between size of family and

nutritional status of young children. The percentage of children
weighing 70% or less of the reference weight was 46.2% in small and
62.5% in large families in rural community and 12.5% and 20.0%, respectively in urban families[8].

Child Spacing

An inverse relationship was observed between the period of spacing and prevalence of moderate or severe malnutrition. When spacing
of children was 1 year or less, there were 1.9 times as many moderate
and severely malnourished children as when the period between births
was 3-to-4 years in the urban as well as rural groups[9].

Diluted Supplementary Feeds and Bottle Feeding

Feeds when prepared out of milk purchased from a vendor or buffalo's and cow's milk are excessively diluted. Many a times excessive dilution continues for a prolonged period. In a cohort study on
feeding practices in nine month old infants in Bombay city and villages it was observed that 46.8% of the urban and 13.0% of rural
infants received diluted milk. In 42.5% of the urban situations
dilution was advised by the doctors. The grandmothers influenced
dilution in 17% of urban and 21.7% of the rural infants. In 40.5%
of the urban and 60.5% of the rural mothers thought of diluting the
milk on their own[3]. Parents are often ignorant about how to recon-
stitute the milk from milk powder for infant feeding. They prepare
a very diluted formula. All the milk powder tins mention that one
measureful of powder should be added to one ounce of water. In practice, the parents do not have an ounce measure at home and above 80%
of the feeding bottles used in the community were "medicine" or
"tonic" bottles, as those empty bottles were available at home.
They take one cupful or at times glassful of water to which they add
one measure of milk powder, making a four to eight times diluted
formula. The graduate and post-graduate mothers and at times even
the doctors have committed similar mistakes[10]. Parents from poor or
middle economic group get misguided by the example of an infant from
an economically well-placed family who is put on supplementary milk
powder. Milk powder is considered a prestige food. Moreover, commercial milk powders were not within the reach of rural parents.
Only 28.0% of urban mothers knew proper technique of reconstitution
of the milk. Above 4 out of 5 urban infants and 1 out of 8 rural
infants were fed on the bottle when supplementary feeds were given[3].

Severe malnutrition sets in when the babies are fed with the
bottle and teat. Unhygienic conditions and improper sterilization
in bottle feeding play havoc. These factors contribute to gastroenteritis. The mother and grand-mother attributes it due to concentrated milk formula which small infant is not "tolerating". They

further dilute the milk which results in marasmus. Hence, a vicious
cycle of bottle feeding, diarrhea further dilution of the formula and
marasmus continues.

Sale Promotion and Advertisements

The advertisements do more harm than good in promoting nutritional status of the infants. The parents give importance to the
protein contents of a particular preparation and administer few tea-
spoons per day to their children hoping to get magic returns.

Lactation Failure

The incidence of complete lactation failure in the mothers of
three months old cohorts infants was 9.0% in Bombay city and 3.9% in
the rural community. Breast conditions, emotional problems, severe
malnutrition and associated chronic diseases like tuberculosis were
the leading factors contributing to the lactation failure. 87.0% of
the infants of these mothers have malnutrition[11].

Another study in the city had revealed that 9.7% of the mothers
never breastfed their infants and 56.5% of them introduced supplementary feeds during the third month and 79.1% by tenth month. 82.7%
of the mothers had insufficient milk and 8.9% had to be away during
the day[12]. When the 9 months old cohorts were followed, nearly third
(32.8%) of the urban and about half of the rural infants did not
receive any supplementary feeds[3].

2. IRON DEFICIENCY ANEMIA

During weaning period, the prevalence of nutritional anemia runs
parallel with that of energy-protein malnutrition and in realities
all the factors which cause EPM lead to nutritional anemia.

3. VITAMIN A DEFICIENCY

1.6% of the infants and 4.2% of 1-2 years olds had signs of
Vitamin A deficiency in tribal villages of the Kasa Primary Health
Centre[13].

4. RICKETS

Survey carried out in urban and rural communities in and around
Bombay revealed that 17.1% of the urban and 25.5% of the rural children between 12-18 months had clinical rickets. However, only 1/6th

of these had biochemical or radiological evidences of rickets[14].

9.4% and 24.4% of the nine months old cohort infants in Bombay and villages around Palghar had clinical rickets[3]

SOLUTION OF THE NUTRITIONAL PROBLEMS

Even though poverty remains very much in the background in caus-ing malnutrition during weaning period, it is not the limiting fac--or. It is observed that in the same ethnic socioeconomic groups, there are young children who have normal nutrition. This observation opens a door for optimism and indicates a role of the medical pro-fession. It is surprising that nobody has ever studied why those children are normal. All studies have aimed to find out the cause of abnormal conditions.

The studies mentioned above in rural as well as urban communi-ties have revealed that the following factors, in sequence of their importance, lead to malnutrition during weaning period:

1. Delayed introduction of solids;
2. lessened frequency of feedings;
3. low birth weight;
4. chronic maternal malnutrition;
5. recurrent infections like diarrhea and upper respiratory infection;
6. excessively diluted supplementary feeds;
7. spacing between children, and
8. lactation failure.

Majority of these causes mentioned here could be tackled through proper education of the mothers or mother-substitutes on nutritional care of the malnourished pregnant women and timely treatment of in-fections. However, it is imperative that all the children and women in the reproductive age group in a community are covered. This is a Herculian task as the problem of looking after the most vulnerable and needy in a developing country is that of reaching the unreached and unreachables. The clinic approach will not solve the problem in countries like India where half the number of women for the number of men are economically active[15]. In some countries where the women are not working, the under-five clinic may be instrumental in approaching all the children.

The solution will be to reach all in the vulnerable group through someone from the local community preferably those who have apptitude, interest, physical capacity and few years of schooling background. Alternative integrated approaches to combat malnutri-tion were tried out in India at number of projects even before the outside world knew anything about the China's barefoot doctors[16].

The accomplishments are very promising.

Various nutrition intervention programmes have frequently not yielded rewarding results due to the lack of a system of identifying the needy through medical criteria, a poor appreciation of the local etioecological factors causing malnutrition, a difficulty in approaching the inaccessible needy and an absence of involvement of a local community. Equally important factor is a lack of collaborative efforts on the part of the various departments and agencies to integrate the health services with nutrition programmes. These are the germs of the disease and the cure will be possible only if the system of operation is altered drastically, if not, revolutionized.

To arrest malnutrition, all children in the weaning period and preferably up to five years and the pregnant women should be put on surveillance. The Community Health Workers (Part-time Social Workers) from the local villages should be involved in community diagnosis[17] of malnutrition in children and pregnant women through innovative and appropriate technology[18]. Monthly weighing of children and of the pregnant women is very important. Through a simplified technology, they can identify the grade of malnutrition and select severely malnourished for therapeutic nutrition supplementation or suggest to the community for sponsorship of supplementary feeding[19]. The surveillance for at-risk children and pregnant women provides the list of those who need special care[20,21]. At the WHO-aided project the following five indicators have been worked out for at-risk system:

1. Weight of 65% of the reference or less;
2. fails to gain weight for three successive months;
3. losses weight during two successive months;
4. weighs less than 1.5 kg at birth, and
5. had an illness such as measles, whooping cough or acute gastro-enterites.

The local community should be involved in managing such cases. The main thrust of the programmes should be on nutrition and health education emphasizing an introduction of semi-solids and solids by 4th month of age, increasing frequency of feeds during weaning period, encouraging breast feeding up to two years and if supplementary feeds are given then correct dilution of the formulae should be taught to the mothers. The mother-substitutes should be given equal importance for nutrition and health education.

The following at-risk pregnant woman who is going to deliver low birth weight baby should be kept on nutrition surveillance:

1. Weighs 38 kg or less before pregnancy or weighs 42 kg or less at 34th week of pregnancy;
2. is less than 145 cm in height;

3. is severely pale;
4. has a child from a previous delivery who weighed less than 2 kg at birth;
5. had swollen legs while pregnant or having swollen legs at time of examination;
6. has high blood pressure;
7. is below 18 or above 30 years old and primpara, and
8. there is doubt about her having twins.

The Community Health Workers should manage common illness and carry out timely health care of children and women[22]. They should enlist and collect children for immunization.

Under financial constraints, the intervention programmes with the specific nutrition supplementation should be organized for the selective population. Preference should be given to at-risk regions like tribal or chronically draught affected areas and urban slums. Importance may be given to the jobless-period of the parents and intervention programmes with supplements be organized during the at-risk season. Benefits must go to those who are medically selected as at-risk rather than all in the community. The nutrition supplements should be of local foods which while preparing are neither time nor fuel consuming inexpensive and educative to parents and mother-substitutes. These nutrition supplements should not substitute for a regular meal. The local community must be involved in feeding the beneficiaries in a group which will help in controlling the sharing of supplements by others.

The severely malnourished children and severely anaemic pregnant women should be given ferrous sulphate tablets orally to reduce the incidence of anaemia. Those who have clinical anaemia should be treated by the Community Health Workers[22].

The Community Health Workers should administer oral massive vitamin A six monthly to severely malnourished children in one-to-five year age group and should treat the children who have signs of vitamin A deficiency. Pregnant women should be given oral doses of 600,000 units of vitamin A during 9th month in areas where prevalence of the deficiency is high.

Rickets is not a public health problem. However, these workers should be trained in identifying clinical rickets and refer the cases to multipurpose health workers and doctors.

All children should be protected against infectious diseases. Timety health care including oral rehydration with fluids and electrolytes should be given to the children who have diarrhea, and heavily infested cases should be dewormed[22]. The spacing of the children and size of the family should be the focal points of education for planning of the families by various health workers. Pro-

grammes on enviromental sanitations, actively involving community are
of extreme importance in preserving the satisfactory nutritional
status of children.

There is no doubt that overall development of the community will
result in better nutritional status of vulnerable groups of the population. However, that could be achieved by the determination of
planners and administrators to coordinate and implement various development programmes. That task should not be left for a medical
officer of the health centre. The health staff should have a wider
view of the development and they should encourage and advise collaborative efforts for overall community involvement. There are some
isolated examples of doctors who have accomplished a lot in overall
development in a small region, but all these projects could not be
implemented in a typical governmental set-up of management with many
constraints. Hence, such experiments are less likely to be replicat-
ed on a large scale elsewhere in the country, eventhough they are
outstanding when viewed in isolation. Until overall development in
agriculture, animal husbandry, education, communication, cottage
industries and other departments is achieved, malnutrition in children and in women is to be corrected by the health manpower through
some alternative approach which can fit into the management and
economic structure of the country[17].

REFERENCES

[1] Shah, P.M., and A.R. Junnarkar. Domiciliary treatment of protein-
        calorie malnutrition. Second Year Progress Report, WHO-aided
        Project, Institute of Child Health, Bombay, (1974).
[2] Stewart, H.C. and S.S. Stevenson. Physical growth and development,
        in: "Text Book of Pediatrics". W.E. Nelson, Ed., 8th ed. B.
        Saunders, Philadelphia, (1964).
[3] Oberai, S. Feeding practices in nine months old infants: A Cohort
        study. Thesis submitted to Bombay University, 1978.
[4] Shah, P.M., Walimbe, S.R. and Dhole, V.S. Wage-earning mothers,
        Mother-substitutes and Care of the young children in rural
        Maharashtra. The paper presented at the International Congress
        of Pediatrics, New Delhi, Oct. 1977. (Under publication).
[5] Shah, P.M., Shah Kusum, P. and Junnarkar, A.R. Domiciliary treat-
        ment of Protein-Calorie Malnutrition. Third and Fourth Year
        Progress Report. WHO-aided project, Institute of Child Health,
        Bombay, 1978.
[6] Shah, Kusum P. and Shah, P.M. Factors leading to severe malnutri-
        tion. A Relationship of Maternal Nutrition and Marasmus in the
        Infants. Indian Pediat. 12, 64-67, 1975.
[7] Shah, P.M., and Bhalerao, V.Y. Faltering of growth in rural under-
        fives. (Under publication).
[8] Mudkhedkar, S.N. and Shah, P.M. The impact of Family size on
        Child Nutrition and Health. Indian Pediat. 13, 1073-1079, 1975.

[9] Mudkhedkar, S.N. and Shah, P.M. The effect of spacing of children on the Nutrition and mortality of under-fives. Indian Jour. Med. Res. 64, 453-458, 1976.

[10] Shah, P.M. Factors leading to severe malnutrition. B. Iatrogenic and commerciogenic factors leading to marasmus, Indian Pediat. 12, 67-68, 1975.

[11] Shah, P.M., D.J. Karkik, Shah, Kusum P. and Shah, B.P. Lactation failure in the urban and rural communities: A cohort study. (Under publication).

[12] Karkal, M. Socio-cultural and Economic aspects of Infant feeding. Indian Pediat. 12, 13-19, 1975.

[13] Shah, P.M., and Rane, A.V. Selective approach for prevention of Malnutrition for blindness with oral massive vitamin A (Under publication).

[14] Mankodi, N.A., Monkykar, A., Siddhaye, S., and Shah, P.M. Rickets in preschool children in and around Bombay. Trop. Gr. Med. 26, 375, 378, 1974.

[15] United Nations Publications. Demographic Data of 1970, in:"Demographic Year Book", New York, 1971.

[16] Indian Council of Medical Research. Alternative approaches to health care, ICMR, New Delhi, 1977.

[17] Shah, P.M. Community diagnosis and community management of malnutrition. Food and Nutrition, 1978.

[18] Shah, P.M. Community participation and Nutrition. The Kasa Project in India, UNICEF's Assignments Children, 35, 53-71, UNICEF, New York.

[19] Shah, P.M. Early detection and prevention of protein-calorie malnutrition. 2. ed. Popular Prakashan, Bombay, 1978.

[20] Shah, P.M., Junnarkar, A.R. and Khare, R.D. Communitywide surveillance of "at-risk" under-fives in need of special care. Jour. Trop. Ped. & Env. Child Health, 22. 103-107, 1976.

[21] Shah, P.M. "Weightage and practicability of management of at-risk factors affecting the health of young children in 'at-risk' factors and health of young children", Jelliffe, D.B. ed., Cairo, 1978.

[22] Shah, P.M. and Shah, Kusum P. "Timely Health Care of children and mothers". 1. ed. Popular Prakashan, Bombay, 1978.

NUTRITIONAL PROBLEMS DURING THE WEANING PERIOD

S.G. Srikantia, M.D.

National Institute of Nutrition

Hyderabad - 500007, INDIA

The period in a child's life when he shifts progressively from exclusive breastfeeding to a mixed type of food, ultimately to the adult type, is a crucial one. This transitional period — the weaning period, has been recognized as being potentially dangerous in several cultures. The process of weaning is perhaps of minor importance in populations and communities with good nutrition and good standards of infant feeding, but is of major concern in communities where incomes are low, knowledge about weaning practices is not always correct and where environmental sanitation and standards of personal hygiene are far from satisfactory. Depending upon the culture, this crucial period lies between the 4th and 24th months of life.

Breastfeeding practices show marked variations from one country to another and in the same country from one socio-economic group to another. The time at which weaning is started, the nature of the weaning food that is given and the frequency of feeding also show considerable variations. It is these variables that bring in the important dimension of nutrition into the practice of weaning.

Although on a world-wide basis, a majority of infants are still being satisfactorily breastfed, there appears to be a declining trend in the extent and duration of breastfeeding, with a consequent increase in the numbers of bottle-fed infants. This has been found to be so in both the developed and the developing countries, and has become a cause for major concern from the public health point of view. Among important reasons which have contributed to this changing trend are the increasing level of maternal education, altered pattern of family life with more and more nuclear, smaller, mobile and isolated families, the changing role of women as earners to supplement family income, the increased work potential due to in-

dustrialization and the easy availability of alternatives to breast-
mil, coupled with high pressure advertising techniques. Among the
developing countries, the situation with respect to breastfeeding
does not appears to be uniform. While in several of them, even among
the rural communities, early weaning is being practices, in several
others, including India, women belonging to the poor income groups
— both rural and urban — successfully breastfeed their infants for
prolonged periods of time (Table 1). Lactation failure is virtually
unknown among this group[1, 2] — an observation similar to that made
over two decades ago[3].

The time of weaning is important from the nutritional angle.
Both very early and delayed weaning are associated with adverse nutritional consequences. Very early weaning is often associated with
the introduction of either too little or too much of the weaning
food. Early weaning, with inadequate amounts of milk substitutes,
can lead to various grades of growth retardation including marasmus,
due not only to the high cost of the weaning food, but also the lack
of knowledge as to how much has to be given. In addition to this
direct effect of inadequate intake of calories and protein on nutritional status, another important factor that influences nutritional
status is the diarrhea which accompanies weaning, particularly in
rural communities, for a variety of reason.

Early weaning with semi-solid foods which have a high calorie
density can lead to overfeeding, with consequent risk of childhood
obesity, persisting even into late obesity.

Delayed weaning has its nutritional repercussions too. It is
not often recognized in many rural communities that breast milk
alone is insufficient to meet the nutritional needs of an infant
beyond the age of 5 to 6 months, and that additional foods are necessary aroung this age. Many women in rural India, for example,
believe that as long as there is some breast milk, the child does
not need additional food, with the result that almost one fourth of
all infants are exclusively breastfed, even at the end of 12 months.
Such a situation leads to growth retardation, and very late weaning
can lead to markedly stunted growth and development of kwashiorkor.

Weanling diarrhea is one of the most important and most frequently encountered problems during infancy with widespread nutritional repercussions. In various parts of the world, the peak incidence of weanling diarrhea varies considerably and is related to
the time of weaning. When breastfeeding is prolonged, weanling
diarrhea comes late, and with early weaning, it comes early. Weanling diarrhea is an important cause of not only infant mortality but
also of growth retardation. This is not simply due to the infection
per se, which is known to bring about metabolic alterations, but also
to feeding practices which are followed when an infant or young
child has diarrhea. Food is often withheld in such situations and

TABLE 1

BREAST FEEDING PRACTICE AMONG RURAL AND URBAN POOR
IN PARTS OF SOUTH INDIA

AGE (months)	URBAN POOR BREAST FED			RURAL POOR BREAST FED		
	Number	Solely %	Partially %	Number	Solely %	Partially %
< 1	29	93.1	0	34	100	0
1	38	89.5	7.9	50	98	2.0
2	36	94.4	2.8	60	98.3	1.7
3	43	93.0	4.7	56	98.2	1.8
8	46	76.1	21.7	52	80.8	19.2
11	46	63.0	32.6	59	42.4	57.6
18	41	24.4	56.1	59	8.1	86.4
23	34	8.8	64.7	58	1.7	89.7

Source: Bhavani Belavady – Personal Communication 1977.

this enforced semi-starvation aggravates the adverse effects of in-
infection on growth. This is particularly true when there are recurrent attacks of diarrhea.

     The relationship between the nutritional status of infants and
young children on the one hand and the incidence of diarrheal disease
on the other has been a matter of some controversy. Frequency of
diarrheal disease and duration of an attack have been reported to
tun parallel to the degree of malnutrition as judged by nutritional
anthropometry[4, 5]; other workers have not found this relationship.
However, before one concludes that they are, in fact, cause and effect, it is necessary to consider the possibility that both the extent of growth retardation and increased frequency of diarrheal
attacks may be due to a common factor, which independently influences
both.

     It was earlier believed that the differences in the incidence
of diarrhea seen throughout the world between breastfed and bottle-
fed infants was due to the bacterial contamination introduced from
outside. It is now recognized that human milk contains a number of
antibacterial factors and that these may have a positive beneficial
role. Among recognized antibacterial factors are the immunoglobulins, particularly secretory IgA, lysozyme, lactoferrin, complement
and specific antibodies against a number of microorganisms including
strains of E. coli, salmonella, polio and coxsackie virus. Colostrum
is particularly rich in secretory immunoglobulins. Breast milk also
contains competent macrophages and lymphocytes. There is enough
evidence that these factors have an active role. For example, fresh
human colostrum can control diarrhea due to E. coli infection[6], the
concentration of E. coli in stools parallels that of IgA[7], and live
polio vaccine is often less effective when given close to a breast
milk meal[8]. The nutritional status of the mother is known to modify
the concentrations of some of the nutrients in breast milk, but it
is gratifying to note that the concentrations of all these antibacterial factors in milk samples obtained from women belonging to the
poorest sections of the population, who were undernourished and had
deficiency signs, were essentially similar to those seen in samples
obtained from well nourished women belonging to the well-to-do
groups[9, 10] (Table 2). This has considerable public health significance, and in countries where the risk of infection is high, but for
the prolonged breast-feeding, the infection rate would perhaps have
been even higher.

     It has long been held that contaminated water supplies are
sources of water borne diarrheal diseases in childhood and on this
basis it has been argued that the provision of protected water supplies will help to combat these diseases. Results of studies carried
out almost two decades ago, as also some recent studies, have suggested that the availability of water is perhaps as important, if
not more so, in influencing the quantum of diarrheal disease since

TABLE 2

ANTIBACTERIAL FACTORS IN BREAST MILK

GROUP	IMMUNOGLOBULINS: mg/100 ml			LYSOZYME mg/100 ml	LACTO-FERRIN mg/100 ml	COMPLEMENT mg/100 ml
	IgA	IgG	IgM			
Colostrum						
Well-nourished	336	5.9	17.1	14.2	420	33
Under-nourished	374	5.3	15.3	16.4	520	25
Mature Milk						
Well-nourished	120	2.9	2.8	24.8	250	16
Under-nourished	118	5.8	4.7	23.3	270	17

it can be an important determinant of personal hygiene. Rates of
shigella infection have been found to fall when water availability
close to home increased[11]. These findings do not minimize the importance of providing safe drinking water, but do point to the importance of water availability.

Among problems associated with weanling is the nature of supplements that can be introduced as weaning foods. It is clear that
predominantly milk-based products cannot be used in developing countries both because of cost and because of low milk production. They
may, however, have a supplementary role. Recent evidence also sug-
fest that weaning foods do not need to contain very high concentra-
trions of protein. What is needed appears to be the development and
popularization of weaning foods — multimixes, based upon locally
available, culturally acceptable, traditional, inexpensive vegetable
foods, for use even in rural areas if an impact has to be made on
nutritional problems during the weaning period. Because of changes
in socio-cultural practices, which leave little time for the mother
to prepare weaning foods freshly everyday, there is a need for time
saving, almost ready-to-serve weaning foods. In fact, the results
of a recent study carried out in some rural communities in India
showed that following nutrition education, which included weaning
as an important component, it was possible to transfer the knowledge
to a great majority of mothers, but that only 20% or less put this
knowledge to use. Of those who did not, half the number gave economic constraints as the reason and the other half, lack of time to
prepare special foods as the reason[12]. This, coupled with the infrequent feeding of the adult type of diet, which has a low calorie
density, was an important reason for inadequate growth during the
weaning period. These constraints need to be removed by the development of suitable weaning foods.

REFERENCES

[1] Bhavani Belavady (1977). Personal communication.
[2] Prema, K., Nadamuni Naidu, A. and Neelakumari, S. Lactation and
       fertility (To be published).
[3] Gopalan, C. and Bhavani Belavady. Nutrition and lactation. Fed.
       Proc. 20: 177, 1961.
[4] Mata, Leonardo J., Kronmal, R.A., Garcia, B., Butler, W., Urrutia,
       J.J. and Murillo, S. Breast feeding, weaning and the diar-
       rhoeal syndrome in a Guatemalan Indian village. In: "Acute
       Diarrhoea in Childhood" Ciba Foundation Symposium, K. Elliott
       and J. Knight (eds.), 42: 311, 1976.
[5] Gordon, J.E. Synergism of malnutrition and infectious disease.
       In: "Nutrition and Preventive Medicine" WHO Monograph Series,
       No 62, p. 193, G.H. Beaton and J. M. Bengoa (eds.), Bengoa,
       1976.
[6] Larguia, A.M., Urman, J., Stoliar, O.A., Ceriani, J.M., O'Donnell,

A., Buscaglia, J.C. and Martinez, J.C. Fresh human colostrum
for the prevention of E. coli diarrhoea - A clinical experience.
Env. Child Health 23: 289, 1977.

Michael, J.G., Ringenback, R. and Hottenstein, S. The Antimicrobial activity of human colostrum antibody in the newborn. J.
Inf. Dis. 124: 445, 1971

Warren, R.J., Lepow, M.L., Bartsch, G.E. and Robbens, F.C. The
relationship of maternal antibody, breast feeding and age to
the susceptibility of newborn infants to infection with at-
tenuated polio virus. Pediatrics, 34: 4, 1964.

Vinodini Reddy, Bhaskaram, C., Raghuramulu, N. and Jagadeesan, V.
Antimicrobial factors in human milk. Acta Pediat. Scand. 66:
229, 1977.

Jagadeesan, V. and Vinodini Reddy. C_3 in human milk. Acta Pediat.
Scand. 67: 237, 1978.

Hollister, A.C., Dorothy Beck, M., Gittelsohn, A.M. and Hemphill,
E.C. Influence of water availability on shigella prevalence in
children of farm labour families. Am. J. Pub. Hlth. 45: 354,
1955.

Indian Council of Medical Research, Annual Report, National Institute of Nutrition, p. 184, 1977.

NUTRITIONAL PROBLEMS IN THE WEANING PERIOD

WITH SPECIAL REFERENCE TO CHILE

Enrique Yáñez

Inst. of Nutrition and Food Technology

Univ. of Chile, Santiago, CHILE

There is wide agreement among pediatricians and nutritionists that human breast milk represents the optimal food for the human infant, but the substitution of breast feeding continues, often with dire consequences. The American Academy of Pediatrics has stated that "breast feeding is recommended for all full-term and vigorous preterm infants because human milk is nutritionally sound and because breast feeding tends to facilitate a close mother/child relationship". Despite this and many other official statements there is a strong trend towards a short breast feeding period. At best, only 30 percent of American children are breast-fed for any length of time and women around the world seem to be nursing for less and less time. In the United States only 25 percent of infants are breast-fed at age one week and this figure declines to 5 percent at age 6 months. The decline in breast feeding has been marked and progressive during the last decades in industrialized countries. This fact is clearly illustrated by data from Sweden collected from 1944 to 1970.

Unfortunately the same phenomenon is being observed in developing countries where the practice of bottle-feeding has increased with economic development. This problem has been observed in Chile for the last 3 or 4 decades and has acquired dramatic proportions. In 1940 about 85 percent of our children reached the age of 6 months with exclusive breast feeding, while in 1974 only 11 percent of the infants in urban areas and 19 percent in rural areas were being breast-fed at the same age.

This trend has been well documented by several studies conducted in Chile. Plank & Milanesi found in women under age 25 that of 53.1 percent of all infants were weaned before age 6 months. Jones et al.

found that between 0.1 month of age 56.4 percent of all mothers were breast-feeding their babies. This percentage decreased gradually to 9.9 percent for age 5-6 months. Jelliffe has pointed out that women in developing countries have followed the example of their counter-parts in industrialized countries thus giving rise to serious nutri-tional problems in young infants. Studies from Chile have shown that women from high income groups can bottle-feed their babies without increased infant morbidity and mortality, while those from low income groups pay the price of increased morbidity and mortality due to gastrointertinal infections, malnutrition and other conditions.

In the case of Chile there has been a displacement of protein-calorie malnutrition to younger age groups. This is a well-known fact for Chilean pediatricians and nutritionists. In 1940, twenty five percent of the children hospitalized with severe malnutrition were less than 1 year of age. This figure increased dramatically to 72 percent in 1967 and to 88 percent in 1974. This has happened in spite of other advances achieved by the country during the same period. Therefore, a clear association can be established between early weaning and increase in malnutrition. Moreover, the analysis of the 1977 figures shows that 43 percent of all protein-malnutrition cases corresponded to children 0 to 6 months of age.

Plank & Milanesi stated that postneonatal death rates were sig-nificantly higher among infants who started bottle-feeding in the first 3 months than among those exclusively breast-fed during that time. The early onset of malnutrition in Chile causes a high infant mortality rate. This was 120.3 per thousand live births in 1960 and has decreased slowly to 79.3 in 1970 and 54.7 in 1976. This is a disappointing result for a country that has had a nation-wide program of free distribution of milk in operation for several decades. Cur-rently, every child under 6 months receives 3 kg of full-fat powdered milk. From 6 months to 24 months the child receives 2 kg of milk monthly. From 2 years to 6 years of age the child is given bimonthly 3 kg of a protein-rich mix (FORTE-SAN or similar) containing about 20 percent protein.

It has been pointed out that in areas where standards of hygiene are poor and hospital facilities inadequate, the protective value of breast-feeding is beyond dispute. This has been confirmed in Chile by the experience of Plank & Milanesi. At 4 weeks and 3 months mor-tality rates in bottle-fed babies were twice those in babies re-ceiving breast milk alone. Breast-fed babies who also received cow's milk fared no better than those given cow's milk alone, indicating that if a baby is to benefit from breast-feeding he must be given breast milk exclusively for at least 3 months. In other words, in that study there was no evidence that the continuation of breastfeed-ind afforded any protection once bottle-feeding began. The Chilean study also showed that as income rose, mothers tended to switch from breast-feeding to bottle-feeding. This explained why mortality rates rose in families with higher incomes within the low socio-economic stratum. An important fact closely related to morbidity and mortality is the high frequency of diarrhea and gastrointestinal infections in bottle-fed infants due to contamination of the formula or bottle in poor hygiene households. In a study conducted in a slum area of Santiago a high percentage of the milk bottles were contaminated with pathogenic bacteria.

The economic consequences of declining breast-feeding are rarely taken into consideration. Poor families of developing countries cannot afford buying expensive foods developed by commercial companies. In those countries like Chile where the Government has taken the initiative to provide free milk to infants the cost of the Program is extraordinarily high and weighs heavily in the national budget.

The food used as a substitute for the mother's milk must provide nutrients qualitatively and quantitatively comparable to breast-milk. Special consideration must be given to meeting protein and energy requirements because among the nutritional deficiency of this period these factors rank first and they may result in the severe types of protein-calorie malnutrition — marasmus and kwashiorkor. Of common occurrence are also iron deficiency anaemia, vitamin A deficiency, rickets and less commonly vitamin C and thiamin deficiencies.

The type of weaning food used will depend on the economic capability of the family. In Chile mothers from high income groups feed their babies preferentially commercial formulas available in the market. While those of middle and low income strata use the full-fat powdered milk distributed freely by the National Health Service to all children up to 2 years of age. Special care must be taken in order to make that mothers prepare the bottles properly, as regards to hygiene and concentration. This must provide the adequate caloric density and nutrient levels needed to meet the requirements.

The notable advance in the industrial production of nutrients allow control of xerophthalmia and blindness by addition of vitamin A to skimmed milk powder in infant feeding; the prevention of rickets through fortification of milk with vitamin D; reduction in infant morbidity and mortality and assurance of good growth of infants by increasing utilization of industrially processed vegetable products as human food; and the successful development and utilization of complete infant feeding formulas. The addition of iron to powdered milk to prevent iron deficiency anaemia has been investigated extensively in my Institute and the results will be communicated in another Symposium of the Congress.

NUTRITION PROBLEMS DURING THE WEANING PERIOD -

THE UTILIZATION OF LOCAL FOODS

J.B. Fashakin, M.D.

Dept. of Food Science & Technology

Univ. of Ife, Ile-Ife, NIGERIA

As we all know, the problems of nutrition during the weaning period are complex and sometimes misunderstood. Moreover there seems to be no real agreement over the true definition of weaning. However, in this paper, weaning period is presumed to mean the transitional period following lactation and the full dependence of the child on non-milk products. During this period the child is exposed to various food commodities which depend on the resources within the locality under consideration.

It is therefore important that in the consideration of a new weaning food, much attention must be paid to the availability of raw materials, the purchasing power of the population and finally the quality of the end product in meeting with the daily requirements of the consumers that it is intended for.

In most developing countries, particularly in the villages and rural areas, children are weaned on cheap maize gruels or other starchy preparations from roots and tubers. The primary aim of such feedings is not to meet with the daily requirements of the infants and children but merely to make use of the available resources. Investigations into the nutrient composition of these traditional weaning foods showed that many of them are grossly inadequate in most nutrients except the provisions of calories. This has led to widespread reports of protein-calorie malnutrition within such age groups. In improving the qualities of these traditional weaning foods therefore, it is important to supply more available proteins.

In Nigeria for example, food crops vary from one geographical location to another. Moreover, because of the monsoon rainfall in some parts of the year, there are periods of plenty and periods of

scarcity. It seems to me to be appropriate to bear these kinds of
facts in mind when formulating weaning foods from locally produced
raw materials. Similarly in Nigeria, the protein rich legumes such
as sorghum, millet, cowpeas and groundnuts are predominantly grown in
the Savannah regions of the Northern States while corn, roots and
other tubers thrive in the Southern thick mangrove forests.

Traditionally therefore, in the Southern parts of Nigeria, children are weaned on a food preparation from fermented corn known as
"OGI". This preparation has been reported[1, 2] to be deficient in protein and thus being the etiological factor in the causation of protein-calorie malnutrition among the children in these regions. In
proposing an adequate weaning formula the Federal Institute of Industrial Research developed a process for incorporating a full fat
soya flour into Ogi and gave it the name Soy-Ogi which contained 30%
soya flour with about 70% fermented corn. This product was shown[3, 4]
to be nutritionally adequate both in nutrient composition and amino-
acid pattern when compared with other commercially imported weaning
foods. The authors further showed that the formula was effective in
ameliorating the common symptoms associated with the incidence of
protein-calorie malnutrition.

Unfortunately soya beans are not widely grown in Nigeria. There
are, however, other sources of protein rich pulses that are popularly grown which have not hitherto been fully utilized in the formula-
tion of weaning foods. Among these pulses, groundnuts are probably
the most viable. Although groundnuts are grown on a large scale in
Nigeria, over 90% of the production is exported as groundnut-cake to
the developed countries where they are used as animal feeds. The
inherent problem associated with the application of groundnuts is
the presence of mycotoxins especially Aflatoxin. Several authors[5, 6, 7] have shown that this mycotoxin is a carcinogen with high
degree of potency even at small doses.

However, the growth of moulds on groundnuts, which is responsible for the presence of the mycotoxins can be avoided under low
moisture content and proper storage of groundnuts. Moreover there
is no direct evidence to link the consumption of groundnuts in human
with the incidence of liver cancer. In fact a product containing
42% groundnut protein known as 'Arlac' was found to be highly acceptable in many African countries including Senegal, Uganda, Kenya
and Nigeria.

EXPERIMENTAL PROCEDURE

In our experiments, corn was steeped for a period of 72 hours
to encourage fermentation and the growth of facultative microorganisms. After soaking, the material was milled. Then the starch was
separated from the husk using a sieve of about 3 mm aperture. The

product was then centrifuged to obtain a partially dry flour.

Similarly, groundnut seeds were dry-blanched with sand to remove the outer coat which would have contributed a bitter taste to the end product. The resultant kernels were minced and extracted with n-hexane in order to concentrate the protein. The product was dried, milled and sieved to achieve a fine particle size not greater than 3 mm in diameter. The groundnut flour thus prepared was mixed with the corn flour, as prepared above, at a ratio of 3:1, before the end product was spray-dried. The resultant product was named 'Nut-Ogi'. This product was mixed with the basal* diet to constitute one of the experimental diets.

The other experimental diets included:

(i)   Soy-Ogi as generously donated by the Federal Institute of Industrial Research, Oshodi, Nigeria;
(ii) Cerelac, a commercially imported weaning food;
(iii) Sample A, which was a refined Soy-Ogi, as prepared by Nestle Food Company intended for marketing in Nigeria.

All the diets were diluted to an isocaloric level (Table 1).

Twenty four weaning albino rats were placed in separate metabolic cages and fed on the basal diet for a period of 7 days. The animals were then weighed and separated into six equal groups when the experimental diets were introduced. The animals were maintained on these diets for a period of 21 days.

RESULTS AND DISCUSSIONS

The nutrient composition of the various diets are shown in Table 2. Similarly the weight changes of the animals during the experiment are shown in Table 3. This is further illustrated in Fig. 1

According to these results the commercial product known as cerelac gave the highest weight gain by the animals. This was closely followed by the product labelled sample 'A', which is also a potential commercial product by Nestle Food Company.

The performance of 'Nut-Ogi' a combination of groundnut protein and corn flour was not quite as good as that of the commercial products. However, it may be observed that the 'Nut-Ogi' and the 'Soy-

---

* The basal diet contained in g/kg: corn flour 800; sugar 60; vegetable oil 100; salt mix (8) 30; vitamin mixture 10; and cod liver oil 5.

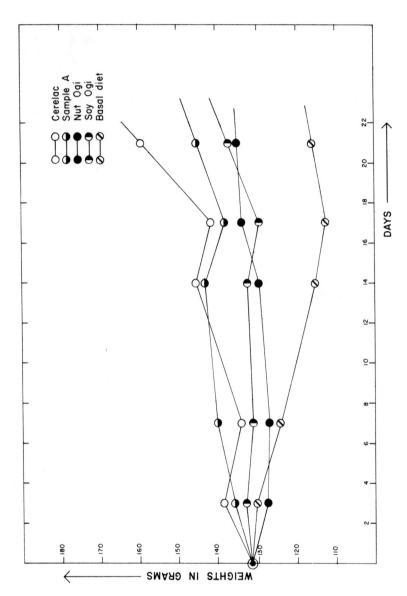

Fig. 1 – Increase in growth rates over the 21 days experiment.

TABLE 1

THE ISOCALORIC MIXTURES OF BASAL DIETS
AND EXPERIMENTAL DIETS

DIETARY SAMPLES*	INITIAL PROTEIN CONTENT %	WEIGHT OF FOOD (g)	WEIGHT BASAL DIET (g)	FINAL PROTEIN CONTENT %
I Nut Ogi	13.0	355.0	145.0	10.0
II Soy Ogi	20.0	250.0	250.0	10.0
III Cerelac	10.9	255.0	45.0	10.0
IV Sample A	9.5	500.0	-	10.0

* Diatery Sample I: mixture of groundnut protein and corn flour.
  Dietary Sample II: mixture of soy protein and corn flour.
  Dietary Sample III: a commercial baby food as described in the
                      text
  Dietary Sample IV: refined mixture of soy protein and corn
                      flour.

TABLE 2

NUTRIENT COMPOSITION OF DIETARY FOODS EXAMINED

DIETARY SAMPLES	PROTEIN %	FAT %	MOISTURE %	ASH %	THIAMINE (mg %)	ASCORBIC ACID (mg %)
Nut Ogi	13.70	9.95	8.90	1.60	0.16	17.00
Soy Ogi*	20.00	6.3	4.7	3.0	-	-
Cerelac	10.90	7.80	2.10	1.86	0.19	17.00
Sample A	9.50	8.50	4.10	2.60	0.28	16.00

* Taken from Akinrele et al[4].

TABLE 3

THE AVERAGE WEIGHT CHANGES IN GRAMS
DURING THE   EXPERIMENTAL PERIOD

DIETARY GROUPS OF ANIMALS	T I M E		I N		D A Y S	
	1	3	7	14	17	21
I Nut Ogi	132.0	129.0	127.9	131.0	135.8	137.7
II Soy Ogi	132.0	134.8	133.4	134.3	131.5	138.6
III Cerelac	132.0	139.5	136.6	146.0	144.5	164.2
IV Sample A	132.0	137.8	141.7	145.3	141.0	148.0
V Basal Diet	132.0	131.8	127.1	116.3	116.0	119.9

Ogi' ranked very closely.

It is presumed that the relative performance of these products
reflect on the amino-acid patterns and the different ratios of the
essential to the non-essential amino-acids. While soya bean lacks
methionine, it is rich in lysine. Groundnut protein on the other
hand lacks methionine and lysine (see Table 4).

It has also been reported[9] that 89% of the total sulphur amino-
acids requirements by animals could be met by cystine. Furthermore
other reports[10] also showed that in diets with a poor amino acid
balance, the amount of methionine suggested by the FAO reference
protein is too high for infants and may have undesirable effects.
This would seem to indicate here that the deficiency of lysine in
the groundnut protein is more crucial than that of methionine in the
soya protein. The situation may be further aggravated by the low
level of lysine and thyptophan in the corn flour.

In Table 5 the amount of food consumed by the animals under
these experimental diets can be compared. The results are consistent
with other similar results as we have observed in the weight changes
and also in Table 6 the net protein ratios. Since the NPR measures
the total nitrogen retention in these weaning rats it is reflected
on the utilization of the dietary proteins. The results therefore
indicate that unless the limiting amino-acid in this groundnut is
supplied the utilization of the protein is seriously inhibited.

It is true that the food industries have a role to play in for-

TABLE 4

THE ESSENTIAL AMINO-ACIDS IN THE PROTEIN CONCENTRATES
(in mg/g)

AMINO-ACIDS	FAO PATTERN (1965)	GROUNDNUT PROTEIN	SOY PROTEIN	MAIZE
Isoleucine	129.0	136.0	121.0	105.0
Leucine	172.0	186.0	192.0	295.0
Lysine	125.0	106.0	160.0	65.0
Phenilalanine	114.0	138.0	121.0	102.0
Tyrosine	81.0	94.0	79.0	138.0
Methionine	61.0	37.0	37.0	44.0
Threonine	99.0	104.0	96.0	91.0
Tryptophan	31.0	30.0	32.0	15.0
Valine	141.0	133.0	126.0	116.0

TABLE 5

THE AVERAGE FOOD CONSUMPTION
OVER THE EXPERIMENTAL PERIOD
(in   gm)

DIETARY GROUPS ANIMALS	TIME IN DAYS		
	7	14	21
I Nut-Ogi	31.8	68.2	97.7
II Soy-Ogi	32.4	65.0	94.4
III Cerelac	32.9	104.4	182.0
IV Sample A	31.0	93.5	116.8
V Basal Diet	33.8	89.9	124.2

TABLE 6

THE PROTEIN EFFICIENCY RATIOS (PER)
AND NET PROTEIN RATIOS (NPR)
OF THE DIFFERENT DIETARY SAMPLES

DIETARY SAMPLES	PER	NPR
I Nut-Ogi	0.60	2.27
II Soy-Ogi	0.70	2.85
III Cerelac	1.70	3.12
IV Sample A	1.40	2.84

mulating new baby weaning foods. It is considered easy to supplement
groundnut protein with lysine in order to increase its efficacy as a
potential protein source in weaning foods. Such a product still be
cheaper than most available commercial products. Provided that the
product is processed in a manner that conforms with the Laws and Food
Legislation of the country, groundnut protein is a viable protein
source in weaning foods.

REFERENCES

[1] Akinrele, I.A. A Biochemical Study ot the Traditional Method of
        Preparation of Ogi and its effects on the nutritive value of
        corn. Ph. D. Thesis, Univ. of Ibadan, 1966.
[2] Collis, W.R.F., I.S. Dema and A. Omololu. The Ecology of Child
        Nutrition in Nigeria villages. Trop. Geogr. Med. 14: 140,
        1962.
[3] Akinrele, I.A., O. Adeyinka, C.C.A. Edwards, F.O. Olatunji,
        J.A. Dina and O.A. Koleoso. The Development and Production of
        Soy-Ogi. F.I.I.R. Research Report no. 42. Lagos, Nigeria,
        Federal Ministry of Industries, 1970.
[4] Akinrele, I.A. and C.C.A. Edwards. An Assessment of the Nutritive
        Value of a Maize-Soya Mixture, 'Soy-Ogi', as a weaning food in
        Nigeria. Br. J. Nutr. 26: 177, 1971.
[5] Butler, W.H. and J.M. Barnes. Toxic Effects on Groundnut Meal
        containing Aflatoxin to rats and guinea pigs. Br. J. Cancer
        17: 699, 1964.
[6] Canaghan, R.B.A. and K. Sargent. The toxicity of certain Ground-
        nut Meals to Poultry. Vet. Record 73: 726, 1961.
[7] Newberne, P.M. Carcinogenicity of Aflatoxin - contaminated peanut

meals. in: "Mycotoxins in Foodstuffs", G.N. Wogan (ed.), M.I.T. Press, Cambridge, Mass., 1965.

[8] Bernhart, F.W. and R.M. Tomarelli. A Salt Mixture Supplying National Research Council Estimates of the Mineral Requirements of the Rat. J. Nutr. (U.S.), 89: 495, 1966.

[9] Rose, W.C. and R.L. The Amino-acid requirement of man. XIII. The sparing effect of cystine on the methionine requirement. J. Biol. Chem. 216: 763, 1955.

[10] Scrimshaw, N.S., R. Bressani, M. Behar and F. Viteri. Effect of amino acid supplementation of corn-masa at high levels of protein intake on the nitrogen retention of young children. J. Nutr. (U.S.) 66: 485, 1958.

[11] Hegsted, D.M. and B.O. Juliano. Difficulties in assessing the nutritional quality of rice protein. J. Nutr. (U.S.), 104: 772, 1974.

Supported by grant-in-aid and Funds for Research and Teaching, Department of Food Science & Technology, University of Ife, Ile-Ife, Nigeria.

NUTRITION PROBLEMS IN THE WEANING PERIOD -

THE CONTRIBUTION OF THE FOOD INDUSTRY

George A. Purvis, Ph.D.

Gerber Products Co.

Fremont, MI, USA

The infant food industry has a responsibility to provide the most favorable nutrition for infants compatible with the need of each individual infant regardless of the location or situation of that infant. Advantages of existing feeding circumstances should be utilized, and every effort made to compliment and not compete with those beneficial foods. Breast-feeding provides the best and most contemporary example. Breast-feeding is recognized as the preferred feeding for infants. After the infant reaches an age and size that nutrients from breast milk are no longer adequate, there is a need for supplementation with other suitable foods. This is the place for the contribution of the food industry. Supplementary foods, judiciously selected in reasonable quantity, can provide support for breast feeding for an extended period. Normal infants, under normal feeding circumstances, provide the basis for this statement, and special needs, both the infant's and mother's, require separate consideration.

The challenge of the food industry for supplementary foods which can become weaning foods can be detailed and discussed separately. First a nutritional need must be defined for a large enough segment of the population to make large scale, efficient manufacture feasible. When the need has been defined, foods must be developed or modified to fulfill that need. These must be compatible with local custom, ingredients and practice if they are to be successful.

The local requirements for foods must be recognized — not only nutritional but cultural, regulatory and food availability. The foods provided for infants, to be successful, have to fit the eating habits of the family. The parents select food for the infant, therefore the foods must be compatible with familiar foods. The transi-

tion fo family foods also demands that introductory foods can be used
to teach the infant favorable eating habits including a variety of
available foods.

Local regulations and standards for foods must be recognized
and inspections, reports and conversation arranged to assure compliance. Individual nations, and in many cases local regulatory
agencies, have regulations which require separate interpretation and
compliance. In most cases the regulations have good reason for existance, however lack of uniformity frequently causes a barrier to
distribution between jurisdictions. Codex Alimentarius has made
substantial advances in reducing these barriers, as will be noted in
a later section.

Availability of foods and ingredients must be ascertained to
meet the rigid specifications that are expected and demanded for
infants and children. The microbiological, microphysical and chemical standards are, as they should be, very stringent for infant
foods. Sources must be located or developed for high quality ingredients.

Safety of foods developed must be evidenced or established
either by preparation procedure, freedom from contaminants, and practical usage in the home. Many areas do not have access to a safe
and clean water supply, and in these cases the form of food must
compensate either by manufacturing process to preclude the addition
of water or by treatment of the food by cooking to make the water
safe. Growing and storage conditions are in some cases not sufficiently advanced to yield satisfactory ingredients, and steps must
be taken to improve conditions or locate alternate sources of supply.

Economic feasibility must be present to make manufacture of
foods a self-sufficient operation. Efficiency in manufacturing requires a sufficient volume to permit mass production, efficient
operation and ultimately a price that can be afforded by consumers.
The ultimate success of an operation demands that industry can have
a sufficient return to make a venture profitable.

Packaging is a critical consideration that must be adjusted to
individual conditions and foods. The basic purpose of packaging is
to protect the food. Other functions include establishment of
serving size, provision of information concerning use, and an attractive presentation appropriate to the food. The protection requirements are often demanding due to conditions of heat, humidity,
insect infestation, exposure to rodents and extensive transportation
and storage. It is a challenge to provide an appropriate package
at a cost that is not excessive.

The final challenge is to, with the preceding challenges·, provide a food for infants that is nutritionally sound. The foods must

not only be versatile enough to fulfill the nutritional needs for a
wide variety of individual infants, but also must be appropriate for
use by infants at the stage of development when they will be used.
The individual likes, sensitivities and needs of infants must be
fulfilled.

There are a number of existing product modes that need to be
discussed, with the extent of applications for each.

The most extensive form utilized in developed nations are canned
baby foods, either in metal cans or glass jars. An extensive variety
of hermetically sealed sterile products is possible through this
form. A disadvantage is cost and the security for extensive manufacturing, transportation and storage facilities.

Foods based on cereal represent another, more universally used
food form. Infant foods based on cereal are predominately precook-
ed, dry, enzyme-treated foods that are well tolerated by infants.
Cost of preparation is advantageous, and usage offers excellent
potential. Foods based on cereal offer an excellent media for fortification. A nutritional disadvantage is that variety of food usage
is not possible — cereal is in most products the only ingredient.

Infant formula is an extensively used manufactured food, most
of which are designed to simulate human milk. The advantageous use
of infant formulas is for mothers who cannot breastfeed or who do not
have sufficient milk to fulfill the nutritional needs of infants.
Vitamins and iron are added to most infant formulas to make them
complete nutritionally. The technology is available to manufacture
infant formulas that are ready-to-feed; in liquid form to be diluted
with water and in dry form to be prepared with water.

Fortified complete foods have been developed in many areas that
use local sources of food, and are in the form generally used locally. Several examples of low cost manufactured foods have been manufactured; Incaparina is a stapel cereal with vegetable protein concentrate (INCAP), Superamine is predominantly peas and lentils with
added NFDMS (Algeria), Falfa is a mixture of a cereal, a legume and
a vegetable protein concentrate (Ethiopia). These combination foods
with the application of advanced manufacturing technology, provide
an important basis for extended manufacturing development.

Formulated local foods represent an excellent source of development for the food industry. The use of available tropical fruits has
proven compatible with technology of manufacture for canned baby
foods, and provides an excellent example of extended technology.
More extensive evaluation of locally available foods is worthy of
investigation from a standpoint of nutritional quality in addition
to the compatibility with local preference — ex.: Guava is an excellent source of vitamin C.

A brief discussion of Codex Alimentarius is pertinent to a discussion of processed foods for infants. Codex is sponsored by FAO/WHO and interested governments participate in deliberations to propose regulations for foods, ingredients, processes and practices. Proposed regulations have been completed for canned baby foods, foods for infants based on cereal and infant formula. The proposed regulations provide an excellent basis for the food industry since they lend uniformity to regulation, provide guidelines for preparation of foods and assure a uniform quality food for all infants and children. Codex also reduces barriers to trade since the definition of products, ingredients and practices is the same regardless of location.

The projection of products, production and needs rest only with the imagination of scientists and technologists and the needs of infants and children. Innovative processes to increase safety of food, improve further nutritional quality and utilize untapped food resources are continually being investigated. The control of processes and techniques is constantly being improved to minimize the extent of treatment and maximize safety (for example extrusion heating). Dry food technology is evolving which will make many foods available at lower cost than they have been in the past. Packaging to protect food quality with less weight and a lower expenditure of energy.

The ultimate objective of the infant food industry is to provide the most nutritious foods for support of other appropriate methods of infant feeding. The challenge is to provide foods compatible with breast feeding to make the individual child's transition to family foods favorable and nutritionally advantageous.

THE EFFECT OF NUTRITION ON GROWTH

OF CHILDREN IN BAGHDAD

Adnan Shakir, M.D.

Baghdad University

Baghdad, IRAQ

INTRODUCTION

In this paper the dietary patterns, weights, heights, skeletal maturity, and the age at menarche of Baghdad children, belonging to two different socio-economic groups is presented. The results were compared between the different groups and those of British children. The objective is to show the effect of nutrition on the genetical potential of growth in its various aspects on the different groups.

The children studied were mainly of Arab ancestry, with variable Kurdish, Turkish and Persian contribution. All were born in Baghdad and their ages were determined from their birth certificates. They lived in different parts of the city and belonged to two different socio-economic classes. Their separation into the different social classes was defined according to their father's occupation, and the district and type of house they lived in. The Class A children represent the privileged section of the community, the Class B children represent the under-privileged.

The Dietary Patterns

The 24 hour recall method was used and 400 girls of each class were interviewed and the results compared to the dietary pattern of London children (Wadsworth and Cameron, 1968). The frequency of consumption of meat, milk and its products, and starchy roots (including potatoes) was higher in the London group than in Baghdad Class A children. Fresh fruits, eggs and legumes were consumed about equally in both groups. The frequency of consumption of meat, milk and its products, starchy roots, fruits and eggs was significantly higher in

Class A than Class B Baghdad children, but the consumption frequency
of bread, sugar and legumes was higher in Class B than Class A children.

Skeletal Maturity

The study for skeletal maturity of the hand and wrist included
711 children, aged from birth to the age of 6 years, of both sexes.
There were 240 boys and 179 girls belonging to Class A, and 156 boys
and 136 girls of Class B.

The skeletal maturity of each child was estimated by the method
of Tanner, Whitehouse and Healy, (1962). Each of the Baghdad children's skeletal maturity score was converted into a "bone age", this
being the chronological age of an average British child of the same
sex, having the same total bone score. In unbiased sample of British
children the mean value of bone age minus chronological age should
not be significantly different from zero. The difference between
bone age and chronological age was calculated for each one of the
Baghad children, and the mean difference was calculated for each
yearly age interval for each of the sexes and each of the different
social class. The Student's test was used for estimating the statistical significance of the difference between these means and zero,
which indicated the significant advancement or retardation of the
Baghdad children's skeletal maturity from that of the British standards.

It was clear that up to the age of about 1.5 years the skeletal
development of both Class A and B Baghdad Children was more advanced
than that of their British counterparts. The boys in Class A continue to be more advanced than the British boys up to the age of 3 years
(significantly so at the 1-2 years age interval). At the age interval of 3-4 years they become more or less equal to the British standard, although this retardation is not statistically significant.

The Class B boys, although more advanced than their British
counterparts in the first year of life, slow down, approach and equal
the British boys at 1-2 year age interval. After the age of 2 years
they fall below the British standards, and this retardation becomes
statistically significant from the age of 4 years onward.

The Class A girls, like the boys of the same class, are more advanced than their British counterparts up to the age of 3 years (being significantly advanced at the 1-2 year age interval), but after
the age of 3 years fall to below the British standard (but not to a
statistically significant level) and remain so till about the age of
4-5 years. After this age their advancement increases.

The Class B girls are above the British standard in the first

year of life, then fall to the level of this standard at the age of
2 years, after which they fall still further below their British
counterparts, and this becomes statistically significant at the age
of 5 years.

The differences in the skeletal maturity of Class A and Class B
boys and girls at yearly age intervals showed that both boys and
girls of Class A are more advanced in their skeletal maturity than
their counterparts in Class B at all age intervals. This advancement
is statistically significant in the first 3 years of life and in the
5-6 year age interval, Class A girls being more significantly advanced over Class B girls than Class A over Class B boys in the 1-3
year age interval.

Weights and Heights

The same sample used for the skeletal maturity study were measured at the same time of the hand and wrist x-ray was taken.

The weights were measured using either the "Supreme" infant beam
balance (Weylux), or the UNICEF beam balance. The supine length was
taken for children under the age of 3 years and the height was measured in older children. For these measurements the Harpenden range
of stadiometers were employed.

The mean heights and weights of the children studied were compared with those of the British standards (Tanner, Whitehouse and
Takaishi, 1966). The height of both Class A boys and girls is almost
equal to the British standards up to the age of 4 years, but thereafter they start to fall a little below the standard in the first
6 months of life, but start to fall to below their British counterparts from then onwards, but always remain above the 3rd centile of
the British standards.

The weight of Class A boys and girls start a little above the
British standard and remains almost at the same level of the standard
till about the age of 4-5 years, when boys and girls start to differ.
The girl's weight remains almost the same as the standard up to the
age of 6 years, whereas the boys mean weight starts falling behind,
and is below the standard at the age interval of 5-6 years, so at
that age both Class A and B boys will have nearly the same mean
weight for age.

The weight of Class B boys and girls is above the British standard for about the first 3 months of life, but starts to fall to below
the standard from then onwards, though always remaining above the 3rd
centile of the British standard. The girls show more fluctuation in
their weight curve than the boys.

The Age at Menarche

The study included 1725 girls in Class A and 2122 girls in Class B. With the aid of the statistical technique of logit analysis, the mean age at attainment of menarche and the standard error of the two groups was determined; (Table 1).

The girls in Class A menstruate 0.37 years earlier than the girls in Class B, and this difference (over four months) is statistically significant at the 1% level.

The mean age at menarche of the well-to-do Baghdad girls (Class A), is somewhat later than in the highest classes of West European girls, which is around 13.0 years. They compare well with the well--to-do girls in Kampala, East Africa, and urban Kerala, India (Tanner, 1969).

Conclusion

The various differences in the growth indicators mentioned between Class A and Class B children reflects the effect of the difference between their dietary intake and the effect of the various socio-economic factors of growth. The trend in height growth and skeletal maturity is generally the same in both classes, being advanced in early life and later becoming retarded, (similar to African children), but the extent to which this advancement is different and is due to the unfacourable environmental influence in the under--privileged Class B. Weight difficiency comes first, and weight gain is affected early by the adverse environmental factor, namely under--nutrition. The weight of Class A boys and girls is almost the same as the British standards, indicating that the nutrition of Class A children is favourable, and the behavior of height and skeletal maturity is therefore influenced purely by the genetic characteristics. Thus the boys in Class A show an advanced skeletal maturation till the age of 4-5 years, after which both their skeletal maturation and height growth fall below the British standards. The Class A girls show the same trend, but their skeletal maturation and height growth, both fall below the British standards about 6 months earlier than that of boys.

The difference in skeletal maturation between Class A and Class B was greater in girls than in boys. Class A girls being more significantly advanced than Class B girls. This is contrary to what is always said, that the effect of class difference is less on girls than on boys. This explained on the Basis that due to local cultural habits of the poor classes, boys are better cared for than girls. This is well shown in the weight and height charts. The weight, and to a lesser extent the height, of Class B girls are much below those of Class A girls, while the weight and height (especially the weight)

TABLE 1

AGE AT MENARCHE IN BAGHDAD

SOCIO-ECONOMIC CLASS	NUMBER	MEAN AGE AT MENARCHE AND STANDRAD ERROR (years)	CHI SQUARE FOR ACCURACY OF FIT
Class A	1725	13.59 + 0.062	17.3(*)
Class B	2122	13.96 + 0.049	7.4
Total	3847		

(*) The fit for the logit is good, except for three very early
     menstruating girls in Class A

of Class B boys approaches the level of Class A boys after age of 5
years.

    The difference in the rate of maturation between Class A and Class
B girls continues throughout childhood, and is reflected by the difference in the age of menarche between the two classes.

BIBLIOGRAPHY

Shakir, A. (1971): The Age at menarche in girls attending schools
    in Baghdad. Human Biology, 43, 265-270.
Shakir, A. (1973): The dietary pattern of children attending schools
    in the city of Baghdad. Nutrition (London), XXVII, 5, 330-343.
Shakir, A. (1974): Skeletal maturation of the hand and wrist of
    young children in Bahdad. Annals of Human Biology, Vol. 1,
    No. 2, 189-199.
Tanner, J.M. (1969): Growth and Endocrinology of the Adolescent,
    in: "Endocrine and Genetic Diseases of Childhood". L. Gardner,
    ed. Saunders, London.
Tanner, J.M., Whitehouse, R.H., and Healy, M.J.R. (1962): Part I
    & II. Paris: Centere International de l'Enfance.
Tanner, J.M., Whitehouse, R.H, and Takaishi, M. (1966): Archives of
    Disease of Childhood, 41, 454-471; 613-635.
Wadsworth, G.R. and Cameron, M.E. (1968): Rapid determination of
    dietary patterns. Nutrition (London) 22, 188-191.

NUTRITIONAL STATUS, PHYSICAL ACTIVITY AND

EXERCISE PERFORMANCE IN PRESCHOOL CHILDREN

Anna Ferro-Luzzi

National Institute of Nutrition

Rome, ITALY

In spite of the interest that the dynamic and reciprocal interelations between nutrition, physical activity, body composition and physical fitness have always evoked, their recognized public health relevance and hypothetical socio-economic importance, it is surprising how incomplete and often conflicting is still at present our knowledge of the mechanisms involved in their regulation.

There is general consensus that in man, as in the laboratory animal, overt serious malnutrition is not compatible with optimal physical performance and is usually associated, among others, with decrease of spontaneous physical activity and structural or biochemical alterations in the systems involved with the uptake and delivery of oxygen. It is well known, for example, that one of the earliest consequences of experimental Protein Energy Malnutrition (PEM) is represented by the decrease in diameter of the muscle fibre, reflecting the rapid mobilization of muscle protein for the benefit of other prioritarian tissues[1,2]. Glycogen, as well as creatinphosphate and ATP are also lost, and the mitochondrial density of the muscle cell is decreased, leading to deterioration of the muscle functions and a decline of oxygen consumption capacity of the organism[3-5].

Our knowledge about the influence of chronic exposure to marginally deficient nutrition on spontaneous physical activity, body composition, aerobic power, endurance, etc., is much more controversial and restricted. This paucity is very striking, given the formidable prominence of the phenomenon of mild-moderate malnutrition. Over 200 million children below 5 years of age are reportedly affected by it, today in the world[6], and it would be obviously of top importance to know to what extent is the motor and functional deve-

lopment of these children affected, and whether the deterioration
would totally revert upon rehabilitation, or would permanently affect
the adult's work capacity, thus setting into action the spiral of
malnutrition, low work capacity, low productivity, poverty, malnutrition, etc.

Unfortunately, but not unexpectedly, the number of existing
studies that combine dietary intake with energy expenditure and exercise response measurements in children is exceedingly small, mostly
as a consequence of the difficulties encountered in collecting valid
informations on the diet and spontaneous physical activity of unres-
tricted young children.

As physical activity in childhood and adolecesce may be a very
important factor in determining the potential for exercise in adult
life, an acceptable indirect approach for the estimation of mean
daily net energy available for physical activity consists in deriv-
ing it from dietary informations. The presence and nature of environmental stimulation participates in determining to what extent is this
available energy indeed spent in motor activities, thus maintaining
or developing the present physical fitness of the child and his future
work capacity, or is it channelled towards the energy stores of the
body, leading to hypokinesia and obesity.

This paper attempts to relate different levels of nutritional
status and dietary intakes to the level of spontaneous habitual
physical activity as well as to the capacity for exercise at earlier
ages. The material for this analyses derives from data collected in
various studies on dietary intakes and growth performances of children of preschool age exposed to widely different styles[7-9]. For two
of the examined groups some information on exercise performance had
also been collected[9], while for the other two, a detailed study of
the aerobic capacity of the adults belonging to the same community
had been performed by other authors[10].

The location of the investigation sites is indicated in Fig.1.
The way of life and general environment has been described in detail[9-11].

The group of <u>Rome</u> children includes a representative sample of
3-6 year old children attending the nursery schools of the County of
Rome. These children represent the reference group, as they belong
to an area of the Country with high standard of life, that previous
investigations have shown to be free from nutritional problems[12-14].

The three other study groups present various degrees of dietary
deficits.

The groups from <u>Cosenza</u> includes children whose ethnic background is comparable to that of the Roman children, but who are living

Fig. 1 - Geographical distribution of the study groups.

in one of the most underdeveloped areas of Italy. The average yearly
income pro capite in the Cosenza Province arranged on the basis of
their incomes[15]. The area is characterized by low standard of life,
socio-economic deprivation, high levels of un-or under-employment,
high rates of migration, cultural emargination, poor hygienic conditions. The school age child of this area had been found, in previous
investigations, to grow at a significantly slower rate than the Roman
child and it had been assumed that nutrition had a prominent role in
the causation of the observed growth retardation although no direct
dietary evidence was available at that time[16].

The other two groups of children are New Guineans who have been
studied as part of an I.B.P. multidisciplinary investigation on Human
Adaptability in a Tropical Ecosystem[17]. They are 3-6 year olds and
represent 80% of the population of that age of two separate villages.

Kaul[11] is a village on a small volcanic island off the North
coast. The environment is typical tropical, with heavy rain fall,
high temperature and humidity all the year around. Malaria is endemic, but the general health is moderately good by European standards.
The staple food is the taro in its many varieties.

The other group, Lufa[11], is formed by the people from a highland
cluster of villages, situated at an average altitude of 1800 meters
on sea level. The climate is temperate, with cool nights and large
temperature excursions during the day. Sweet potato is the main crop
and the staple food. Malaria is absent and the general physical
appearance is rather good. Cases of clinical PEM may be occasionally
observed in the village, but were all traced back to disrupted family
conditions.

Dietary intakes and anthropometric measures were obtained using
techniques reported elsewhere[7-9].

Tests of physical fitness were carried out to investigate whether shorter, lighter, nutritionally at risk and possibly hypokinetic
children of the under-priviliged areas were functionally, as well as
somatically, handicapped as compared to the faster growing, well
nourished children of privileged sections of the population.

Exercise performance has been tested on a subsample of 362 three
to six year old Italian children, 245 in Rome and 117 in Cosenza.
The modified step test as suggested by Cermak[18, 19] was adopted as
suitable for measuring exercise performance in so young subjects
under field conditions. This test measures the response of the
circulatory system to a constant medium intensity work load. The
load consists of 5 minutes stepping at a constant rate of 30 steps/
/min on a bench 25 cm high. Heart rate is continuously monitored
during the whole duration of the work load and during the following
5 minutes of recovery. Heart beats are counted by direct auscultation

with a modified phonendoscope strapped to the chest of the child
(Fig. 2). Heart response is taken as an indicator of work performance. The index CE, Cardiac Efficiency, is obtained by dividing the
amount of work performed (expressed in kpm) by the total pulse count
during the stepping and the recovery phase. Cardiac Efficiency is
thus described by the amount of mechanical work performed per one
beat of the heart. High levels of this quotient are indicative of
an efficient circulation.

Tests of physical performance were not carried put on the New
Guinea children.

Table 1 summarizes the number of subjects studied in each location and the type of survey they underwent.

RESULTS

The anthropometric data, expressed as absolute differences from
the values recorded for the Roman child, thus emphasizing the gradient in developmental performance of the four groups, are presented
in Fig. 3. The retardation in growth of the New Guinean children as
compared to the Roman children is quite marked and is maintained
with advancing age. Cosenza children, although performing better
than the New Guineans, are nevertheless shorter and lighter than the
Romans, the difference being just over 300 g and 0.5 cm at the age
of 3, and increasing with age to reach about 1.2 kg and 2.6 cm at
6 years of age.

In Table 2, where weight for height for age is presented as
percent deviation from standards of reference[20], [21], it may be seen
that the Roman children are heavier for each cm of their height,
although the difference becomes significant only when the two New
Guinean groups are compared to the Roman children ($P<0.01$).

Fat content and arm muscle area values[22], shown in Fig. 4, are
consistent with the height and weight data, the New Guinean children
having smaller skinfolds and muscle areas.

The dietary intakes of subgroups of the study children are
compared and shown in Table 3. The mean daily energy intakes range
from 5.52 MJ/day in Kaul to 6.77 MJ/day in Rome. Protein intakes
show an even wider range of distribution, as the Roman child receives a diet containing more than twice (58 g/day) the amount of protein consumed by the Kaul child (23 g/day). Lufa has a slightly
higher protein intake, and also Cosenza, with 48 g/day, ranges considerably lower than the Roman child. Carbohydrates are present in
the highest amounts in the New Guinea diets, as could perhaps have
been anticipated, while fats are lowest in Lufa (16 g fat/day).

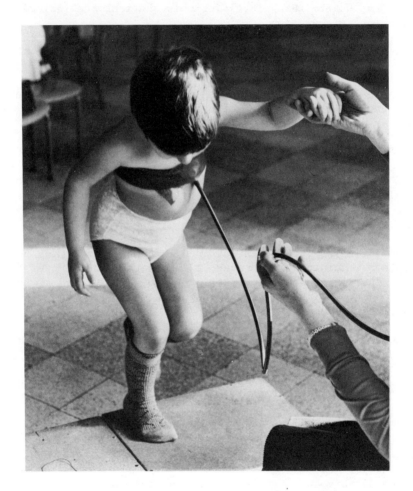

Fig. 2 - Execution under field conditions of the modified
         step-test (18, 19) by a 4-year-old child.

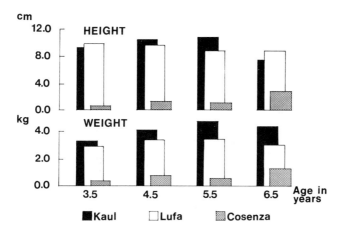

Fig. 3 – Absolute differences of weight (kg) and height (cm) of Kaul (black columns), Lufa (white columns) and Cosenza (shaded columns) children as compared to Rome children.

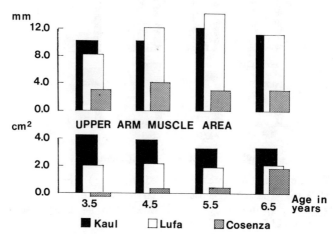

Fig. 4 - Absolute differences of skinfold thicknesses (sum
         of triceps, biceps, subscapular and suprailiac, mm)
         and muscle area (cm²) of Kaul (black columns), Lufa
         (white columns) and Cosenza (shaded columns) child-
         ren as compared to Rome children.

TABLE 1

NUMBER OF SUBJECTS STUDIED IN EACH AREA

LOCATION	NUTRITIONAL STATUS SURVEY	DIETARY	EXERCISE PERFORMANCE SURVEY
	no.	no.	no.
KAUL	208	74	--
LUFA	147	89	--
COSENZA	725	225	117
ROME	2684	437	245

On the average, protein contributes 7% of the total energy intake in Kaul and 8% in Lufa, while this value raises to 14% in both Italian groups.

Expressing the results in terms of body weight, to account for the recorded differences in body size and to standardize the results, provides an accurate dietary profile of the four groups. As may be seen in Table 4, protein intakes show a wide range and a well difined regular trend with a step wise increase from the lowest values recorded in Kaul, 1. g/kg, to the highest in Rome, 3.3 g/kg. Energy intakes expressed in terms of body weight show a rather surprising picture, as both New Guinean groups are seen to subsist on energy levels (389 and 380 kJ/kg) comparable to the values recorded for the Roman children (364 kJ/kg) while in Cosenza children receive only 317 kJ/kg.

Heart responses to exercise of the two groups of Italian children are shown in Fig. 5. heart rates at rest do not differ in either group or sex, while 5 minutes stepping induces a significantly smaller response in the Conseza children of both sexes as compared to the Roman children. A sex difference is also present, the girls having a faster heart rate than boys. No difference is present during the recovery phase between the Rome and the Cosenza child, while sex differences persist with higher rates for the girls. In either cases, the normalization of the pulse rate after the cessation of the exercise is very rapid, and the values recorded during the 2nd minute of recovery are already near basal with a slow almost imperceptible decline thereafter.

The CE of the two groups pooled together, is shown in Table 5. The index is significantly higher in males and, for both sexes, shows

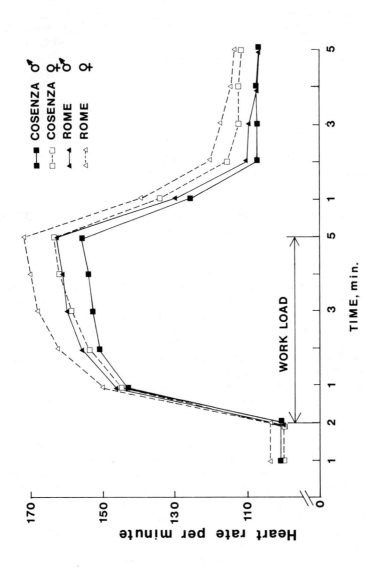

Fig. 5 - Heart rates of Rome boys ▲ ── ▲ and girls △ -- △ , and Cosenza
boys ■ ── ■ and girls □ -- □ , at rest, during 5 minutes of
medium intensity work load, and 5 minutes recovery.

TABLE 2

SOMATIC DEVELOPMENT (AS MEAN PERCENT OF REFERENCE*)
OF 3-6 YEAR-OLD CHILDREN FROM FOUR CONTRASTING
ENVIRONMENTS (SEXES ARE COMBINED)

AGE yrs.	GROUP	no.	% WEIGHT FOR HEIGHT FOR AGE		% TRICEPS FOR AGE	
			Mean	SD	Mean	SD
3.5	KAUL	58	88	8	70	15
	LUFA	32	94	9	74	15
	COSENZA	140	100	12	93	25
	ROMA	494	103	12	101	26
4.5	KAUL	66	87	7	74	11
	LUFA	36	90	9	68	13
	COSENZA	257	100	11	95	27
	ROMA	913	102	12	106	30
5.5	KAUL	37	85	8	70	15
	LUFA	33	91	7	66	14
	COSENZA	291	101	13	99	34
	ROMA	1142	103	14	110	36
6.5	KAUL	47	85	5	66	10
	LUFA	46	91	8	69	13
	COSENZA	37	98	12	95	34
	ROMA	135	103	13	110	36

*) Tanner, Whitehouse and Takaishi (1966);
   Tanner and Whitehouse (1975).

a regular and significant increase with age, perhaps in relation with
the shift occuring with growth to an increasing participation of
changes in stroke volume in response to the work load. Grouping the
children on the basis of their geographical location, Rome versus
Cosenza (Table 6) shows that the cardiac performance is slightly,
even if not significantly, better in the growth retarded, nutritionally at risk children from Cosenza. When the children are grouped
on the basis of their family's income, or in rural versus urban, or
on the basis of their level of spontaneous activity, as described

TABLE 3

MEAN DAILY ENERGY, PROTEIN, FAT AND CARBOHYDRATE INTAKES OF 3-6 YEAR-OLD CHILDREN
FROM FOUR CONTRASTING ENVIRONMENTS (SEXES ARE COMBINED)

| | no. | ENERGY | | | | PROTEIN (g) | | CARBOHYDRATE (g) | | FAT (g) | |
| | | (MJ) | | (kcal) | | | | | | | |
		Mean	SD	Mean	Mean	SD	Mean	SD	Mean	SD
KAUL	74	5.52	1.62	1318	23.3	9.9	247	60	30	13
LUFA	89	6.01	1.41	1435	28.6	10.3	302	65	16	14
COSENZA	225	5.68	18.5	1358	48.2	17.5	211	73	42	21
ROMA	437	6.77	2.03	1618	58.4	18.2	227	70	59	21

TABLE 4

MEAN DAILY PROTEIN AND ENERGY INTAKES PER kg BODY WEIGHT
OF FOUR GROUPS OF 3-6 YEAR-OLD CHILDREN

| | PROTEIN | | ENERGY | | |
| | (g / kg) | | (kJ / kg) | | (kcal / kg) |
	Mean	SD	Mean	SD	Mean
KAUL	1.64	0.71	389	131	93
LUFA	1.81	0.64	380	84	91
COSENZA	2.68	0.95	317	104	76
ROMA	3.26	124	364	117	87

by parents and teachers in a questionnaire specially developed for
the purpose, we fail to discriminate between either their energy
intakes or their heart responses to exercise.

DISCUSSION

     Growth performance is regarded as a reasonably sensitive indica-
tor of nutritional status, and the smaller child is currently assumed
to be suboptimally nourished. On the basic of this assumption, both
New Guinean groups may be considered as malnourished as compared to
the Roman child; and also the growth performance of the Cosenza
children does not match that recorded for our reference group, Roman
children. The smaller absolute value for arm muscle area of the New
Guinean groups as compared with the Roman ones, could be expected on
the basis of their smaller body size. It should be noted however
that the relationship between arm muscle area and body weight indicates that if weight is held constant, the New Guinean children have
an amount of muscle which is practically similar to that of either
Italian children (Fig. 6); and the same applies to the relationship
between arm muscle area and height (Fig. 7).

     If skinfolds are taken as proxie of body fat content, these
findings suggest that the smaller, lighter children from New Guinea,
subsisting on a significantly poorer diet, are in fact also the more
'dense', having less adipose tissue and more lean tissue per unit
body mass than the well nourished, privileged Italian peers from
Rome. It also means that part of the extra weight recorded for the
Roman child may be attributed to the presence of greater fat deposits. The same observation holds true, albeit with smaller differences, when the two Italian groups are compared. In other words,

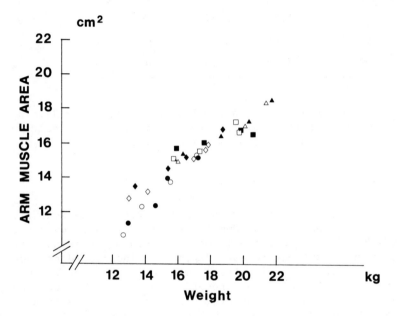

Fig. 6 - Relationship between weight and arm muscle area (age specific average values) of <u>Kaul</u> boys ●, <u>Kaul</u> girls ○, <u>Lufa</u>
boys ◆, <u>Lufa</u> girls ◇, <u>Cosenza</u> boys ■, <u>Cosenza</u> girls □,
<u>Rome</u> boys ▲ and <u>Rome</u> girls △ .

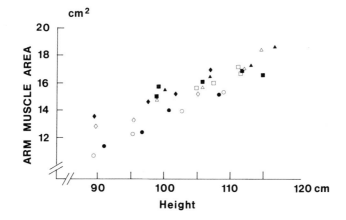

Fig. 7 – Relationship between height and arm muscle area (age specific average values) of <u>Kaul</u> boys ●, <u>Kaul</u> girls ○, <u>Lufa</u> boys ◆, <u>Lufa</u> girls ◇, <u>Cosenza</u> boys ■, <u>Cosenza</u> girls □, <u>Rome</u> boys ▲ and <u>Rome</u> girls △.

TABLE 5

CARDIAC EFFICIENCY INDEX OF PRESCHOOL ITALIAN CHILDREN

| AGE yrs. | CARDIAC EFFICIENCY INDEX | | | | | |
| | BOYS | | | GIRLS | | |
	no.	Mean	SD	no.	Mean	SD
3.5	15	0.451	0.040	17	0.426	0.034
4.5	57	0.519	0.080	61	0.478	0.053
5.5	94	0.579	0.093	93	0.541	0.102
6.5	15	0.601	0.091	10	0.568	0.099

TABLE 6

ENERGY INTAKE, BODY FAT AND CARDIAC EFFICIENCY INDEX
OF 3-6 YEAR-OLD CHILDREN

| GROUP | no. | ENERGY INTAKE (kj / kg) | SUM OF FOUR SKINFOLDS (mm) | CARDIAC EFFICIENCY INDEX | |
				Mean	SD
Boys					
COSENZA	61	310	26	0.567	0.09
ROMA	120	377	29	0.543	0.09
Girls					
COSENZA	56	293	30	0.519	0.10
ROMA	125	356	35	0.505	0.08

these children would be, for the same height and weight, relatively more muscular, or less fat, than Roman children. Muscle mass is very important, as experimental evidence has largely demonstrated, representing the major bodily reserves of protein that may be promptly mobilized in conditions of dietary protein shortage to provide amino acids to the liver[1,2]. A large muscle mass thus, beside representing a good store for emergency dietary situations, is also an indicator that the protein requirement of the body has been satisfactorily met by the dietary intake and that its function has been preserved.

The net amount of energy available to the study children for
physical activity was calculated from the values of their energy intakes after allowing for the cost of growth[23] and of maintanance
(B.M.R. x 1.5)[24] and assuming no relevant changes in their body energy
stores. As may be seen in Fig. 8, the net energy available to the
two New Guinean groups equals 1.15 and 1.23 MJ/day for Kaul and Lufa,
which compares favourably with the value of 1.26 MJ/day suggested and
advocated by FAO/WHO for this age group[23]. The net energy available
to the Roman child, 1.09 MJ/day, is not very dissimilar, while it
comes as a surprise that the children in Cosenza have to face their
every day life with little more than 0.21 MJ/day. At the light of
the information on the amount of energy available to the children
for physical activity, and on their body composition, the greater
body fat content of the Roman child may be reasonably interpreted as
being generated by the deposition as fat of at least a part of the
net 1.09 MJ available to him daily, instead of its utilization for
higher output of physical activity. On the other hand, most of the
energy available to the New Guinean children appears to have been
channelled into higher levels of physical activity, possibly stimulated and facilitated by the more demanding habitat and different
cultural attitudes, with very little diversion towards its conservation in the body stores, as shown by the smaller values of their
physical fitness is available. However a detailed study of the response to submaximal exercise on bycicle ergometer of the young adults
of the same villages, most of them parents or relatives of the study
children, was carried out by Cotes and coworkers[10]. The results of
that study indicated that both groups, but especially the Lufans, had
excellent exercises performances, superior to that of most western
populations with similar levels of energy expenditure and consuming
far better diets. The energy expenditures of these New Guinean
adults were in fact indicative of moderate physical activity, and
the time allocation study confirmed that their way of life was un-
exacting[11]. One could speculate whether their excellent exercise
performance could be due to special genetic endowment or be the
result of past, rather than present, high levels of physical activity
ad significantly and positively affected the adult's response to
exercise. These conclusions evidently lack of the necessary supportive evidence, and more observations are warranted before a cor-
recte interpretation may be reached. The results of the fitness
tests on the Italian children indicate that the children with poorer
diets and with smaller body size have a response to exercise not
dissimilar to that of the better fed faster growing children of the
more privileged areas of the Country, suggesting that the degree of
under-nutrition and risk of malnutrition they were exposed to was
compatible with normal and satisfactory levels of physical fitness.
These findings are not unique, as come evidence has accumulated in
the recent literature suggesting that low levels of energy or food
intake do not necessarily always results in a deterioration of physical performance. Some of the studies which failed to show a de-

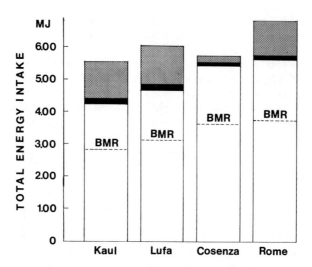

Fig. 8 - Comparison between the amount of energy available for phy-
        sical activity (shaded area) to four groups of 3-6 year old
        children, after deduction of the energy cost of growth
        (black area) and of maintenance (B.M.R. x 1.5) (white area)
        from the measured value of total energy intake.

crease in aerobic power or exercise response in individual with direct
or indirect evidence or marginal energy intakes are those on Ethiopian
children[25], on Tunisian shcool children[26], on Terai Nepalese peas-
sants[27]. This epidemiological evidence finds support in experimental
results obtained on growing rats, showing that if the nutritional deprivation is of a moderate degree, the growing organism is able to
deploy advantageous mechanisms enabling it to reach desirable forms
of adaptation compatible with good physical performance, health and
longevity[28-30].

Great caution should however be exercised in drawing conclusions
from findings of unmodified or improved physical performance in
growth retarded, undernourished subjects. Genetic or environmental
factors other than diet may affect the quality of the response to
exercise. The test itself provides information only on one aspect
of physical performance; there are no direct data available on the
fitness of the New Guinean children and their physical activity has
been assessed only by indirect means; the impact of early malnutrition on physical performance may become evident only much later in
life. On the other hand, a small body size may be the expression
of different genetic potential or may derive from the impact of
environmental factors other than nutrition, or finally, may indicate
a successful form of adaptation to different ways of life and nutrition, compatible with normal and satisfactory physical performance.
The results of the present study prompt us to question our present
ability to identify, diagnose and fully understand the mild-moderate
form of malnutrition and its consequence on physical activity and
fitness, and it is hoped that they may represent a plea and a challenge for more research on the subject.

REFERENCES

[1]Waterlow, J.C. amd Stephen, J.M.L. The effect of low protein diets
    on the turnover rates of serum, liver and muscle proteins in the
    rat, measured by continuous infusion of L-(^{14}) lysine. Clin.
    Sci. 35:287-292 (1968).
[2]Waterlow, J.C. and Alleyne, G.A.O. Protein malnutrition in child-
    ren. Advances in knowledge in the last ten years. Adv. Protein
    Chem. 25:117-227 (1971).
[3]Barac-Nieto, M., Spurr, G.B., Maksud, M.G. and Lotero, H. Aerobic
    work capacity in chronically undernourished adult males. J. Apl.
    Physiol. 44:209-215 (1978).
[4]Gold, A. and Costello, L.C. Effects of semi-starvation on rat
    liver, kidney and heart mitochondrial function. J. Nutr. 105:
    208-214 (1975).
[5]Waterlow, J.C. Adaptation to low protein intakes. in: "protein-
    -calorie malnutrition". R.E. Olson (ed.), Academic Press, New
    York, 1975, pp. 23-25.
[6]United Nations World Food Conference. Fd. Nutr. 1:1-17 (1975).

[7]Ferro-Luzzi, A., Norgan, N.G. and Durnin, J.V.G.A. Food intake, its relationship to body weight and age, and its apparent nutritional adequacy in New Guinean children. Am. J. Clin. Nutr. 28: 1443-1453 (1975).

[8]Ferro-Luzzi, A., Norgan, N.G. and Durnin, J.V.G.A. The nutritional status of some New Guinean children as assessed by anthropometric, biochemical and other indices. Ecology Food Nutr. 7: 115--128 (1978).

[9]Ferro-Luzzi, A., D'Amicis, A., Ferrini, A.M. and Maiale, G. Nutrition, environment and physical performance of preschool children in Italy. Biblthca Nutr. Dieta, 27: 85-106 (1979).

[10]Cotes, J.E., Anderson, H.R. and Patrick, J.M. Lung function and the response to exercise in New Guineans, role of genetic and environmental factors. Phil. Trans. R. Soc. Lond. B. 268:349-361 (1974).

[11]Norgan, N.G., Ferro-Luzzi, A. and Durnin, J.V.G.A. The energy and nutrient intake and the energy expenditure of 204 New Guinean adults. Phil. Trans. R. Soc. Lond. B. 268:309-348 (1974).

[12]Ferro-Luzzi, G. amd Proja, M. Lo stato di nutrizione delle comunita' scolari Italiane. 9: Lo stato di nutrizione dei bambini delle elementari romane. Quad. Nutr. 22: 269-292 (1962).

[13]Mariani, A., Lancia, B., Migliaccio, P.A. and Sorrentino, D. Growth, nutritional status and food consumption of Roman school children. in: "Auxology: human growth in health and disorder". A. Gedda and P. Parisi (eds.) Academic Press, London, 1978, pp. 49-61.

[14]Ferro-Luzzi, A., D'Amicis, A., Ferrini, A.M. and Maiale, G. Nutrition and growth performance in an Italian rural environment. in: "Auxology: human growth in health and disorder". A. Gedda and P. Parisi (ed.), Academic Press, London 1978, pp. 199-205.

[15]Tagliacarne, G. "Il reddito prodotto nelle Provincie Italiene nel 1974 (e confronto col 1971/1972/1973)". Quad. Sint. Econ., Angeli, Roma, 1975.

[16]Ferro-Luzzi, G. and Termine, L. Lo stato di nutrizione delle comunita' scolari Italiene. 9: Lo stato di nutrizione degli scolari nella Provincia di Cosenza. Quad. Nutr. 23: 1-16 (1963).

[17]Harrison, G.A. and Walsh, R.J. A discussion on human adaptability in a tropical ecosystem. An I.B.P. human biological investigation of two New Guinea Communities. Phil. Trans. Soc. Lond. B. 268: 221-400 (1974).

[18]Cermak, J. Möglichkeit der Wertung der Funktionstüchtigkeit des Kreislaufsystems der Schuljugend mil Hilfe der eigenen Modifikation der Step-Testes mit durchlaufender Messung der Pulsfrequenz. Schweiz. Z. Sportmed. 1: 1-8 (1969).

[19]Cermak, J. Cermakova, J., Zudova, Z., Cerna, M. and Tuma, S.T. Die Unterschiede in der Funktionstüchtigkeit des Kreislaufsystems von Schülern der Experimental-Schwimmklassen in Vergleich mit gleichalten keinen Sportreibenden Schülern. Schweiz. Z. Sportmed. 1: 9-19 (1969).

[20]Tanner, J.M., Whitehouse, R.M. and Takaishi, M.T. Standards for

height, weight, height velocity and weight velocity in British children. Arch. Dis. Childh. 31: 454-471 and 613-635 (1966).

[21] Tanner, J.M. and Whitehouse, R.M. Revised standards for triceps and subscapular skinfold in british childre. Arch. Dis. Childh. 50: 142-145 (1975).

[22] Gurney, J.M. and Jelliffe, D.J. Arm anthropometry in nutritional assessment. Nomogram for rapid calculation os muscle circumference and cross-sectional muscle and fat areas. Am. J. clin. Nutr. 26: 912-915 (1973).

[23] FAO/WHO Energy and protein requirements. FAO Nutr. Meeting Rep. Series 52, Rome, 1973.

[24] FAO/WHO Energy and protein requirements. Recommendations by a joint FAO/WHO informal gathering of experts. Fd. Nutr. 2:11-19 (1975).

[25] Areskog, N.H. Short-time exercise and nutritional status in Ethiopian boys and young males. Acta Paediat. Scand. Suppl. 21 (1971).

[26] Parizkova, J. and Merhautova, J. The comparison of somatic development, body composition and functional characteristics in Tunisian and Czech boys of 11 and 12 years. Hum. Biol. 42: 391-400 (1970).

[27] Weitz, C.A. and Lahiri, S. Factors affecting the work capacity of native and migrant groups living in a jungle area of Nepal. Hum. Biol. 49: 91-108 (1974).

[28] Borer, K.T. Weight regulation in hamsters and rats: Differences in the effect of exercise and neonatal nutrition. in: "Huger: Basic mechanisms and clinical implications." D. Novin, W. Wyrwicka & G. Bray (eds.) Raven Press, New York, 1976.

[29] Parizkova, J. and Poledne, R. Consequences of long-term hypokinesia as compared to mild exercise in lipid metabolism of the heart, skeletal muscle and adipose tissue. Europ. J. Appl. Physiol. 33: 331-338 (1975).

[30] Parizkova, J. "Body fat and physical fitness." M. Nijhoff B.V. Publ. The Hague, 1977.

MALNUTRITION AND PHYSICAL

AND MENTAL DEVELOPMENT

Fabio Ancona Lopes

Pediatric Dept., Botucatu Medical School

São Paulo, BRAZIL

This work is concerned with clinical and experimental investigations about malnutrition, physical and mental growth.

In the experimental aspects, malnutrition has been studied during the fetal and post-natal phase, and the influence of nutritional recovery on biochemical determinants that follow intrauterine and post-natal malnutrition have been determined.

In the clinical aspects, we studied the relation between social and economic status and intelectual deficiency in school children.

During the last decade it became well established that malnutrition affects people's growth. Presently, the studies in the field have been intensified with respect to cerebral development.

Some models of experimental malnutrition have been established either modifying the quality and/or quantity of the animal's food, or the time during which the restriction is imposed. The problem was more precisely defined and got more serious, when it became clear that malnutrition imposed during the proliferative phase of growth slowed the rate of cell division, resulting in reduction of the ultimate number of cells (Winick and Noble, 1966). These changes become irreversible after the normal time for cell proliferation was surpassed. In contrast, malnutrition imposed during the period of cellular enlargement causes reversible changes, the cells returning to their normal size on subsequent rehabilitation.

Concomitant with the effects of malnutrition on cell1 growth, there are also very important alterations in lipids and proteins incorporation. Consequently, areas where myelinization is more rapid, are more vulnerable to the effects of early malnutrition.

It is already known that the intrauterine phase and the first period of post-natal life are very important times for an adequate utilization of all growth potential.

We studied firstly[1], the results of proteic-caloric restriction, in rats, during the entire gestational period. We observed mean values of body weight significantly lower in all malnourished off-springs. In spite of the pre-natal period not being the one of greater cerebral vulnerability, in rats, we also observed reduction of total lipid, cholesterol and protein contents (mg/hemisphere).

Following the same variables, we studied[2] the effect of nutritional recuperation starting at birth, to know whether the damages observed on the newborns would remain or not during post-natal life. The malnourished offsprings were fed by a normal foster mother. For this experiment we chose two periods of post-natal growth: the 28th and 90th days of life.

The statystical analysis revealed:

- Mean values of body, brain and hemisphere weights significantly lower in the malnourished group (mg/hemisphere).

- Mean values of water content not significantly different in all studied groups.

We conclude that the animals could not recover from the nutritional injury imposed in the intra-uterine phase despite of all the time of recupartion in post-natal life.

In another experiment[3] we studied the effects of protein-caloric restrition, in rats, during the breast feeding period. For this purpose, the offsprings number was elevated to twelve and the animals were impeded from feeding for five hours per day. At the end of 28th day, one group was sacrified. The other group was subsequently recovered with "ad libitum" food until two months of life.

Another group[4] was submitted to proteic-caloric restriction (4g diet/day) for thirty-three days after the normal breast feeding period. At this time one group was sacrificed and the other received "ad libitum" diet until the third month of life.

The group of adult[5] animals (three months of life) received proteic-caloric restriction (8g diet/day) until five months of life. At this time one group was sacrificed and the other received "ad libitum" diet until the sixth month of life.

The control groups of all experiments were sacrificed with the same age as the malnourished groups.

The results obtained were:

- No alteration was observed in qualitative histological aspects
  in all groups[3-5].

- Brain weight was significantly decreased in all groups[3-5].

- Lipid concentration (mg/g) was significantly decreased in
  younger groups[3,4]. In adult animals the damage was also
  significant but of smaller intensity.

- Protein concentration (mg/g) was significantly decreased in
  Group[3] and group[4].

The nutritional recovery experiments showed that protein contents (mg/g) reached normal levels in all groups.

However, lipid recovery in adult animals was more evident than
in younger animals.

Therefore, the effects of malnutrition were in all situations
more severely in younger than in adult groups and the biochemical
modifications observed were more persistent for lipid contents than
for protein content.

Many hypothesis could be discussed for the still scarce results
on lipids[3-5] and cholesterol contents[2]:

- Slowed enzymic maturation with consequent reduction of their
  activity (Davison and Dobbing, 1966; Chase, 1967).

- Slowed cellular division rate (Winick and Noble, 1966; Fishman
  et al., 1971).

- Lipid utilization for energetic metabolism (Trindade et al.,
  1978; Kfouri et al., 1977).

- Decreased pool of cerebral lipids consequent to smaller local
  intake and/or greater use.

About the proteins, we can examine the following hypothesis:

- Decreased cellular number in the cerebrum (Winick and Noble,
  1966).

- Decreased concentration of the many cerebral enzymes (Rajalak-
  hsmi and col., 1965; Swaisman, 1972) that would maintain the
  cerebral amino acid pool (Reddy et al., 1971). These processes

could lead to alterations in protein synthesis.

In the clinical aspects, we studied the relation between social and economic status and intellectual deficienty in school children. Many studies about the prevalence of mental retardation in advanced countries show that 80% of affected children have slight alterations in the intellectual standards. These children belong mostly to the economically less privileged sectors of community.

There are not many studies of this kind in underdeveloped countries. It has been discussed by W.H.O. that the prevalence of mental retardation in these countries could have a prepoderant component of cases with slight changes, since severe mental retardation is usually concomitant with important clinical complications, it would lead to great mortality in these populations without medical care.

In the present work we studied 536 children from 6 to 12 years of age of two schools of the Botucatu (São Paulo) region. The centra city in the region has 50.000 inhabitants. One of the schools is situated in the city and the other in the peripheral region.

The school children were submitted to collective psychometric tests with the "Raven progressive Matrixes".

The children that obtained equal or lower than the 25th percentile were submitted to individual tests, and those sellected as intellectualy deficient (percetile < 5) were submitted to "Wichsler's Tests for Children" (W.I.S.C.). The social economic status of the children's family was stablished by the utilization of the following indications: family head occupation (load 4), total family income (load 3), habitation and utensils (load 3) and head family instruction (load 3).

The results showed: association between precarious social-economic status and peripheral localization of the school and high prevalence of intellectual deficiency in this group (29% in the peripheral school and 8,7% in the city school).

The genral prevalence of intellectual deficiency was higher than that related for advanced countries (17,5%).

There was also a predominance of I.Q.'s in the borderline and lower-normal levels in the peripheral school. An adverse family environment could be detected in most cases of intellectual deficiency.

REFERENCES

[1] Campana, A.P.; Nobrega, F.J. Moraes, S.; Yida, M. Status sócio-
    -econômico e deficiência intelectual em escolares. J. Ped.
    43 (1): 17-25, 1977.
[2] Kfouri, J.R.N.; Trindade, C.E.P.; Nóbrega, F.J.; Tonete, S.S.Q.;
    Montenegro, M.R.G. Metabolismo (Lipídeos e Proteínas) e histo-
    patologia do cérebro de ratos recém-nascidos, jovens e adultos
    submetidos à desnutrição proteico-calórica, tipo marasmo. J.
    Ped. 42 (1): 28-39, 1977.
[3] Kfouri, J.R.N.; Trindade, C.E.P.; Nóbrega, F.J.; Tonete, S.S.Q.;
    Montenegro, M.R.G. Metabolismo (lipídeos e proteínas) e histo-
    patologia do cérebro de ratos recém-nascidos, jovens e adultos,
    submetidos à desnutrição proteico-calórica (tipo marasmo) e
    recuperação nutricional. J. Ped. 42 (1): 19-27, 1977.
[4] Tonete, S.S.Q.; Coelho, C.A.R.; Nóbrega, F.J. Desnutrição fetal
    experimental em ratos: Efeitos sobre o peso corporal, peso ce-
    rebral, teor de lipídeos totais proteinas e colesterol no cére-
    bro. J. Ped. 44 (4): 213-221, 1978.
[5] Tonete, S.S.Q.; Nóbrega, F.J. Metabolismo do cérebro de ratos jo-
    vens e adultos submetidos à desnutrição fetal. J. Ped. 45 (1):
    18-30, 1978.

NUTRITION, PHYSICAL ACTIVITY, AND FITNESS.

Per-Olof Astrand, M.D.

Dept. Physiology III, Karolinska Institute

Lidingövägen 1, S-114 33, Stockholm, SWEDEN

In my comments I would like to make some sort of a cocktail of
physiology (energy output; the substrate for working skeletal muscles), nutrition (intake of energy and nutrients), psychology (motivation for exercise), education (what should we teach).

ENERGY OUTPUT

In a sense the function of the individual cell is very much dependent on the function of the whole body and vice versa. A relatively constant environment is a vital condition for the undisturbed
operation of individual cells. Thus, the composition of the cell-
-bathing fluid must be kept extremely constant. For this purpose,
special service systems were developed, and I will particularly
mention the oxygen transporting system including the heart and the
circulation.

No cells are able to vary their metabolism to the same extent
as skeletal muscles. During maximal exercise, muscles metabolize up
to 100 times as much energy as at rest. Metabolism, measured per
gram of tissue, is then not greater than for nerve cells. However,
nerve cells are always on a high metabolic level, whether a person
is asleep or engaged in intensive thought. The service systems
developed to keep inner environment more or less constant must therefore be dimensioned to accommodate the demands arising in conjunction
with muscular effort. This applies especially to respiration and
circulation. Otherwise, the human body could only operate at rest.
But the "machinery" has to be turned on every now and again in the
muscles and the important service organs are to be kept in proper
working order. The only way to develop and maintain an optimal

function of the locomotor organs and the circulation is to perform
muscular work, and that activity must not be too rare!

It should be pointed out that the resting metabolic rate is
relatively high. For a person with a body weight of 75 kg the 24-hr
value may be about 7 MJ (1700 kcal). As a comparison the extra energy
demand for walking 35 to 40 km would be on the same level. It is not
relevant always to express the energy content of Danish pastry, beer,
chocolate etc. in terms of how many kilometers the consumer must walk
in order to spend the energy!

Unfortunately, our methods for measurements of the energy output
over long periods of time (e.g. for 24 hours) are not accurate enough
for an exact estimation of the body's energy turnover. When measuring
the energy intake by food consumption variations in the body water
content over the period of observations can simulate a relatively
large energy surplus or deficiency (a 1-kg increase in the body
weight could be 1 kg of water = no change in energy content; if the
1 kg was fatty tissue the gain in energy would be about 28 MJ or
7000 kcal). Tegelman (1974) applied this method for a estimation
of energy intake during two one-week periods in four subjects. The
energy output was calculated from (i) records of major physical activi
ties and then applying standard values of energy costs from the liter
ature; (ii) determination of the heart rate - energy output relation
ship during standardized activities, then continuous recording of the
heart rate and interpolation of the actual energy output from the
heart rate - energy output curves; (iii) a combination of methods
(i) and (ii). His conclusion was that the heart rate - energy output
relationship, method (ii), might best quantitate the total energy
metabolism when the subjects varied their levels of habitual physical
activity. When the rate of physical activity was reduced the energy
intake did not fall as much (see Table 1). (The noted maximal day-
-to-day variation in body weight was in each case 0.7 to 1.0 during
the inactive week and 1.0 to 3.3 kg during the active week. These
fluctuations complicated the calculation of the energy intake. There
fore, it was still an open question in this particular study: which
was the actual level of energy turnover?)

It should be emphasized that a measurement of the aerobic power
can be very accurate by recording the oxygen uptake (1 liter of oxygen
corresponds to an energy yield of 19.7 to 21.1 kJ or 4.7 to 5.05 kcal
(the exact figure depending on the proportion of utilization of fat
and carbohydrate). However, the apparatus for the measurements will
inevitably disturb the subject and that factor will limit its application. A realistic figure for maximal aerobic power for young
healthy persons is about 55 kJ (13 kcal) $\times$min^{-1} in women and 75 kJ
(18 kcal) $\times$min^{-1} in men. However, the individual variations are
large, and in top athletes up to 155 kJ (37 kcal) $\times$min^{-1} has been
noticed. Normally, one cannot work at 100% of the maximal aerobic
power for more than about 10 min (see Astrand and Rodahl, 1977,

TABLE 1

Mean values for daily energy intake estimated from food consumption (A) and energy output estimated by various methods (B; see text) in four subjects during one physically active week and another week dominated by inactivity. (From Tegelman, 1974).

ACTIVITY	METHOD							
	A		B: (i)		B: (ii)		B: (iii)	
	MJ	kcal	MJ	kcal	MJ	kcal	MJ	kcal
Active week	11.9	2,850	15.9	3,800	14.0	3,350	15.0	3,600
Inactive week	11.4	2,700	13.5	3,200	11.0	2,600	11.8	2,800

p. 316). As examples of 24-hour energy expenditures a 300 km (about 190 miles) brought the metabolism to about 59 MJ (12,000 kcal). The energy cost of a day including a 85 km (53 miles) ski-race was calculated to 25 to 30 MJ (6,000 - 7,000 kcal); i.e. the energy content of 1 kg of fatty tissue.

It is rather popular to present tables on figures giving the energy cost of various activities. They can give some guidance, but they are far from accurate. As a rule of thumb the energy cost of walking at moderate speed is 2 kJ (0.5 kcal) x kg^{-1} x km^{-1} and for jogging or running (at various speeds) 4 kJ (1 kcal) x kg^{-1} x km^{-1} (kg refers to the body weight). However, there is a tendency that we ignore the effect on energy cost by type of surface, gradient, wind, etc. (see Fig. 1). In swimming different strokes vary in efficiency, and in the same stroke a swimmer with poor technique may need three times as much energy for a given distance than the elite swimmer.

NUTRITION

The "fuel" for exercising skeletal muscles is carbohydrate and fat (free fatty acids, FFA). Considering the shortage of protein in part of the world it is an advantage that protein is normally not yielding energy for contracting muscles. During mild exercise the muscles oxidize free fatty acid and glycogen in a ratio of about 1:1. The heavier the exercise, the more dominating becomes the energy yield from glycogen. With emptied glycogen stores in the muscles it is nor possible to work hard. By a diet rich in carbohydrates for a few days, the glycogen stores can reach supernormal values. The procedure is particularly effective if the glycogen stores are first exhausted by prolonged, heavy exercise prior to the diet. Such a scheme can help anyone interested in competitive sport in events

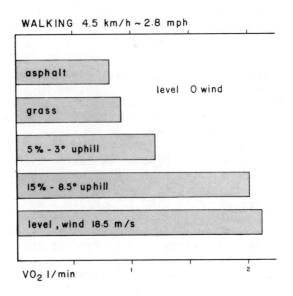

Fig. 1 – The energy expenditure of walking under different
        conditions. (Data from Pugh, see Astrand and
        Rodahl, 1977, p. 582.)

calling for endurance, in hiking etc., to improve his performance.
By various mechanisms the blood sugar is normally not available as
a substrate for the muscles, it is saved for the nerve cells.   Otherwise hypoglycemia would be a threat during any exercise!

A glucose intake during streneous exercise can improve endurance
for the glucose can be utilized by the working muscles and the emptying of the muscle glycogen is postponed. (However, for each glucosyl
unit glycogen will yield 3 ATP per mol but glucose only 2 ATP net.).
For a more detailed discussion, see Astrand and Rodahl, 1977, chapter
14.

A very important effect of training is certainly the induced
increase in maximal oxygen uptake. Another positive effect is the
increase in oxidation of FFA and the reduced energy yield from glycogen. This change in fuel utilization is not only evident at a
given metabolic rate but also when working at a given percentage of
the maximal aerobic power. This adaptation is certainly a glycogen-
-saving mechanism. A reduced lactate production in the trained
individual may be one factor behind this difference in metabolic
pattern. The modification in the mitochondrial enzyme profile induced
in trained muscles may also be of importance. A reduction in the
diffusion distance between capillaries and the interior of the muscle
cells will facilitate the FFA uptake. There is, namely, an increased
capillary density in the training muscle. The cross section area of
the Type I muscle fibers (slow twitch fibers) does not increase in
endurance training; the opposite may occur. An increased oxidation
of FFA will elevate the concentration of citrate which in turn will
inhibit the activity of the enzyme phosphofructokinase retarding
the glycogenolysis and lactate formation. Another effect of endurance
training is an enhanced potential of the Type II muscle fibers (fast
:witch fibers) to work aerobically. (For details, see Holloszy and
Booth, 1976; Saltin et al., 1977).

Henriksson (1977) noted after 2 months of training a 27 percent
higher SDH activity in the trained leg than in the untrained one
(quadriceps femoris muscle). The maximal oxigen uptake wall 11 and
4 percent respectively higher in the posttraining test than before
training. At a submaximal two-leg exercise at an average of about
70 percent of the maximal oxygen uptake (for one hour) the subjects
apparently worked harder with the trained leg. However, the degree
of utilization of free fatty acids was higher in the trained than
in the untrained leg indicating a difference in the oxidative capacity.

Studies indicate that irrespective of what factors trigger the
adaptations on the cellular levels causing effects mainly in those
muscles which are being trained, there are no, or only modest, effects in nontrained muscles.

Another example of effects of physical training on the aerobic metabolism is illustrated by Fig. 2 from studies by Henriksson and Reitman (1977). They noticed a 19 percent increase in maximal aerobic power in their 13 subjects who had trained for endurance for 8 to 10 weeks. The activities of SDH and cytochrome oxidase has increased 32 and 35 percent resepctively (muscle samples from vastus lateralis). Within two weeks posttraining, the cytochome oxidase activity had returned to the pretraining level and after six weeks the SDH activity has back to the control level. However, the maximal oxygen uptake was still high or 16 percent above the pretraining level. The authors conclude that an enhancement of the oxidative potential in skeletal muscle is not a necessity for a high maximal oxygen uptake. As pointed out, changes in the enzyme pattern may be of great importance in the utilization of various substrates in the muscles which may have particular consequences in prolonged exercise.

OBESITY, DIET AND EXERCISE

Within a wide range of energy expenditures through various degrees of physical activity, there is an accurate balance between energy output and intake so that the body weight is maintained constant. In the daily activity is very intensive and prolonged, the spontaneous energy intake is often less than the output, with a reduction in body weight as a result. A daily energy expenditure below a threshold level often leeds to an energy surplus and consequently obesity. In this case, satiety is not reached until more energy has been taken in than has been expended. It appears conclusively that obesity is to a large extent the result of reduced activity with a maintenance of an "old-fashioned" appetite center set for an energy expenditure well above the one typical for a sedentary individual. This is true for children as well as for adults. The reason for their different attitudes toward physical exertion is not clear. When studying very obese individulas on the bicycle ergometer, it has been noted that obese subjects complained of fatigue and felt exhausted at low work loads, when the blood lactic acid levels were not significantly elevated and would be easily tolerated by normal individuals. Obese individuals are characterized by their response to external rather than to internal cues in their eating. (References see Garrow, 1974; Mayer, 1968).

It is particularly important to stimulate young individuals to regular physical activity, for such activity in the long run effectively counteract obesity by keeping the individual within the range where spontaneous energy intake is properly regulated by the energy output. Obesity in infancy may increase the number of fat cells, causing predisposition for subsequent overweight. The treatment of obesity is particularly difficult with an increased number of fat cells.

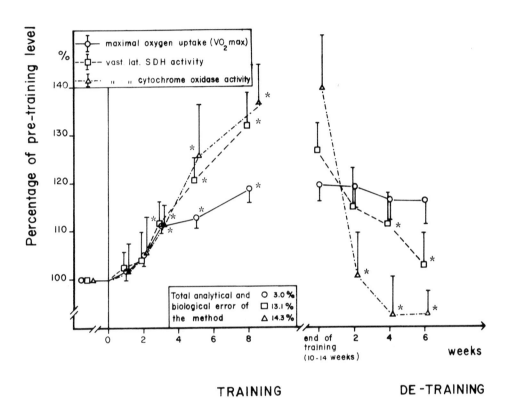

Fig. 2 - Changes in maximal oxygen uptake and oxidativve enzyme activities (vastus lateralis) during training and de-training
periods respectively. Asterisks denote significant difference (p >0.05) from pre-training and after training period,
or end of training and de-training period. (From Henriksson and Reitman, 1976).

It is pointed out by Garrow (1974) that the rat has an astonish-
ing ability to regulate its energy intake over long periods of time,
but it applies only if the food available is monotonous, if the rat
has been fed ad libitum with this diet, and if flavouring agents,
especially sweeteners, are excluded. These conditions are not relevant for the human species. For primitive man there was available
a relatively small selection of naturally occuring animal and plant
foods. Appetite, which may at one time have been a reliable guide
to a correctly balanced diet, is now merely a sensation which can
be manipulated in many ways by food manufacturers. In experiments,
subjects have been over or underfed by various methods (e.g. by
supplementing food by an intragastric tube, or by meals with differing energy density but indistinghishable in taste and volume). The
correlation between the state of energy balance and the voluntary
energy intake is very low, and the individual's ability to regulate
the food intake, at least over a short term, is very poor. Man of
today is facing two aspects of modern society: (1) a food industry,
social and cultural impacts favouring a relatively large energy intake, and (2) a reduced energy demand in jobs and during leisure
time.

DIET AND EXERCISE

With the exception of children during growth, of women during
pregnancy, and sometimes of convalescents, the energy intake should
not exceed the energy expenditure. The energy requirements vary
naturally with the individual's physical activity. On the other
hand, the need for most of the nutrients is comparatively independent
of the individual's activity level; therefore the less active the
individual the higher is the content of the essential nutrients
required per energy unit in order to obtain the desired optimal
nutritional level (Wretlind, 1967). In a homogenous population, a
dietary tradition is usually quite similar. Blix (1965) found that
a linear relation exists between daily energy supply and supply of
many nutrientes (protein, calcium, vitamin A, thiamin, iron). Through
the centuries people obtained their choice of food to an energy output
of 12.5 MJ (3,000 kcal) or more, which also gave all the nutrients
needed. For many of the nutrients mentioned, the energy intake should
actually exceed about 10.5 MJ (2,500 kcal) to ensure an adequate
supply. In other words, the diet in Sweden, and no doubt in many
other countries, seems to be adjusted to persons with an energy
requirement of at least 10.5 MJ (2,500 to 3,000 kcal) (Wretlind, 1967).

This diet is, however, not suitable for the large number of "low
energy consumers" who actually do exist. This unsuitability may
explain the rather common disturbances in the state of health and
wellbeing associated with malnutrition even in countries with plenty
of food available. it has been noted that in Sweden, as well as in
many other countries, the percentage of energy from protein has

remained remarkably constant from the end of the nineteenth century
to the present day, or between 11 and 12 percent; the proportion of
fat in the energy supply, however, has greatly increased from about
20 to 40 percent. Per capita, the amount of energy of protein and
carbohydrates has decreased, but the amount of fat has increased
during this period of time (Wretlind, 1967).

There are two general ways of improving the nutritional conditions of the low energy consumers: (1) Change their food habits so
that their diet consists of a higher content of essential nutrients
per energy unit than it does now. A dietary habit should be developed
so that the requirements of those who consume only 6.3 to 8.4 MJ
1,500 to 2,000 kcal) per day can be satisfied. (2) Low energy consumers should be stimulated to become high energy consumers by taking
part in regular physical activity in one form or another. By an increase in energy output, they can, without the risk of obesity, eat
more and automatically get a greater supply of essential nutrients.

Now for the philosophy and the teleology: Some years ago I had
a chance to visit Botswana and go out into the Kalahari desert to see
little Bushmen, the last remaining stone-age people. They still
follow the life of true hunters and food-gatherers. Gathering sufficient food means that the men trudge along great distances in their
hunting efforts, and the women and children also walk to collect
berries, melons, and various plants. This process of walking, stopping, and squatting to dig, and walking again, is exhausting and gets
worse as the day goes by and the load becomes heavier. When the women
have gathered sufficient and turn for home, they still have to collect and carry firewood for the night. During most of the year, game
and edible plants are not found in great profusion and, in order tc
get enough to eat, the Bushmen have to work for hours every day. The
driving factor behind this habitual physical activity is actually
the feeding center, plain hunger. I did not see adult Bushmen running, but their walking was fast. They were well trained!

Our bodies are, of course, in their "construction" exactly like
those of our stone age ancestors. However, most of us do not have
to work physically to get enough food do satisfy our satiety centers.
Technological and agricultural development was not physiologically
anticipated. In the hypothalamus or other parts of the brain there
isn't, unfortunatelly, a built-in center forcing us, driving us, to
regular exercise. We do look forward to a meal, to eat is one of
the primitive and basic enjoyment of life; as we go through the
dishes, our appetite disappears. With exercise, the situation is
just the opposite - very few people look forward with pleasure to
a bout of training. Some pleasure comes during the exercise, but
most of it comes afterwards - one perhaps feels pleased to have once
again conquered physical laziness that is typical for homo sapiens
from puberty onwards. Getting more exercise is like a New Year's
resolution. Giving up smoking, losing weight, changing one's eating

habits, etc., belong to the same category. But finding some excuse
to postpone putting a resolution into effect is never any problem!

I don't think that the Bushman particularly liked and enjoyed
his many hours of physical effort. But he had to put his muscles
to work in order to survive. We are not different in attitude, but
we can get enough to eat without much muscular effort. We must
intellectually learn and understand how the human body functions and
how to treat it well to attain its optimal function and optimal
health!

Regular physical activity is essential for the optimal function
of many vital organs and tissues and for general function. I am less
sure about its direct and significant effects in the area of preventive medicine, but I can mention much indirect evidence to justify
a wholehearted promotion of physical exercise. Information about
diet and exercise should be tied together. My present prescription
for exercise is as follows:

Daily – at least 60 minutes of physical activity, not necessarily
        vigorous, nor all at the same time. During your daily rout-
        tine, moving, walking, climbing stairs, etc., whether one
        minute 60 times per day, 12 minutes 5 times a day, or any
        combination totalling 60 minutes, will burn up approximately
        1.2 MJ (300 kcal).

Weekly – at least two or three periods of 30 minutes of intermittent
        or sustained activity at submaximal rate of work (brisk walk-
        ing, jogging, cycling, swimming, cross-country skiing, gym-
        nastics etc.) are necessary for maintaining good cardiovas-
        cular fitness, and will consume an additional 3 MJ (750 kcal)
        per week. Note that the recommendations are "measures" in
        tie, not in distance.

IN SUMMARY

Some daily physical activity is good for the function of the
body and also for the health! The individual who is habitually active
may eat more without the risk of becoming obese and, if on a traditional diet, this increased food intake ensures an adequate supply
of essential nutrients. The habitually inactive one runs the risk
of malnutrition. Living in highly developed countries with plenty
of food available is no guarantee for good health.

(Part of this article was published in "Why Obesity", Näringsforskning
22, No. 15, 1978).

BIBLIOGRAPHY

Astrand, P.-O. and Rodahl, K.: "Textbook of Work Physiology";
McGraw-Hill, New York, 1977 (2nd ed.)

Blix, G.: A study on the relation between total calories and single
nutrients in Swedish food; Acta Soc. Med. Upsala. 70: 117, 1965

Garrow, J.S.: "Energy Balance and Obesity in Man"; North-Holland
Publ. Co., Amsterdam/American Elsevier Publ. Inc., New York,
1974.

Henriksson, J.: Training Induced Adaptation of Skeletal Muscle and
Metabolism during Submaximal Exercise; J. Physiol., 270: 661,
1977.

Henriksson, J. and J.S. Reitman: Time Course of Changes in Human
Skeletal Muscle Succinate Dehydrogenase and Cytochrome Oxidase
Activities and Maximal Oxygen Uptake with Physical Activity and
Inactivity; Acta Physiol. Scand., 99: 91, 1977.

Holloszy, J.O., and F.W. Booth: Biochemical Adaptations to Endurance
Exercise in Muscle; Ann. Rev. Physiol., 18: 273, 1976.

Mayer, J.: "Overweight Causes, Cost and Control"; Prentice-Hall,
Inc., Englewood Cliffs, N.Y., 1968.

Saltin, B., J. Henriksson, E. Nygaard, P. Andersen, and E. Jansson:
Fiber Types and Metabolic Potentials of Skeletal Muscles in
Sedentary Man and Endurance Runners; Annals N.Y. Acad. Sci.:
301: 3, 1977.

Tegelman, R.: A study of different methods of analysing total energy
metabolis; Näringsforskning, 18: 117, 1974 (in Swedish).

NUTRITION, PHYSICAL ACTIVITY

AND PHYSICAL FITNESS

J. Paŕízková

Charles University

11807 Prague I, CZECHOSLOVAQUIA

Physical fitness and performance have been for some years of interest, also from the nutritional aspect, to sports medical officers and occupational and sports physiologists. In recent years, however, the interaction between physical activity, muscular work, resulting physical fitness and nutrition is the focus of nutritionsts -physiologists as well a clinicians.

Food intake is a decisive factor in the energy supply of the organism and its activity; muscular work ensuing from physical activity determines most of the above-basal energy output. The balance and optimum relationship between these factors play an important role also in the development of general physical fitness and performance, health status and its prognosis.

Physical fitness is nowadays one of the criteria of adequacy of somatic development in industrialized countries and in developing countries, during the period of growth and development, which is associated also with an adequate nutritional status of children and adolescence. There is no need to stress that good nutritional status is indispensable for a desirable work capacity from the occupational aspect and thus also for economic productivity.

A desirable health status which depends closely on physical fitness, however, does not display an unequivocal relationship with nutrition; assessment of the nutritional optimum in individual instances is still very difficult. Inadequate as well as excessive nutrition interferes with physical fitness and performance, health status and longevity; malnutrition in developing countries leads to a high mortality, reduces resistance to infections in childhood and later life. In industrialized countries conversely overnutrition and in

697

unbalanced diet are generally accepted to be one of the most important risk factors of so-called diseases of civilization which account
in these countries for a high proportion of the mortality.

    Relations between diet and muscular work have various aspects.
At present one of the aspects elaborated in greatest detail is e.g.
the problem of utilization of different substrates during muscular
work of different grades and types (e.g. a short dynamic anaerobic
load, or a prolonged aerobic load). Many investigations were made
in experimental animals as well as men pertaining to different organs
and tissues (in particular skeletal and heart muscle and liver resp.)
Thus evidence was provided that during an anaerobic load combustion
of carbohydrate predominates and that during prolonged aerobic combustion of fatty substance is accentuated due to the increased aerobic capacity of the organism (increased capacity to supply different
tissues with oxygen is an essential prerequisite of increased lipid
utilization during work). These processes were investigated in
blood, directly in muscle tissue and other tissues of the organism;
the consequence for body composition were also investigated, i.e. of
the ratio of lean body mass and fat. Detailed analyses were also
made of the relations between physical fitness and the ability to
perform a certain type of work after the administration of a certain
substrate (e.g. glucose), as shown already in the classical work of
Christensen, (Astrand and Rodahl, 1977) and others. Great attention
was devoted to the ability to accumulate under certain circumstances
increased amounts of glycogen and the ensuing increased ability to
perform for some time aerobic work under a steady state of oxygen
supply. Detailed investigations were made of enzymatic processes in
muscle and heart tissue, etc. Findings pertaining to these processes
are summarized best in biochemical studies assembled at the occasion
of symposia on the biochemistry of exercise. So far we do not possess, however, much information what happens in other organs and
tissues than those which are actually engaged during muscular work
although this activity is reflected in the whole organism.

    Water and electrolyte economy and its optimum adaptation during
a load is the basis prerequisite of satisfactory performance; these
phenomena were investigated not only from the aspect of work performance in certain occupations but in particular in top sportmen where
concrete recommendations were made e.g. preparation of beverages with
a certain ionic composition which are to compensate the sequelae of
theincreased muscle load and to maintain the milieu interieur in a
state redering optimum performance possible. All these aspects can
be covered under the heading of an adequate supply of all substances
needed for the long-term establishment of the desirable nutritional
status of the organism before, during and after physical performance.

    The required actual nutritional status depends, however, also
on other factors of the external environment. The load acting on
the organism and its response under optimal conditions in different

environments takes a different course when the organism is exposed to increased or reduced temperature and humidity, various pressures or the gravitationless state etc. Action of a muscular load and hypokinesia resp. were studied in detail under conditions of flight in a gravitationless state where the right type of diet in relation to physical fitness influences substantially the activity in the cosmos. From this aspect an important part is played also by the degree of adaptation to various grades of muscular activity and different diets. It is beyond doubt that in this perspective sphere of human activity a very important part is played by a high standard of fitness attained by prolonged training and adaptation to different types of loads, similarly as a diet suitable as regards quantity, composition, preparation and mode of ingestion, meal frequency, etc. This problem concerns so far only industrial countries, similarly as the problem of increased risk of cardiovascular diseases as a result of overeating and concurrent hypokinesia.

The opposite pole in the sphere of relations between motor acti- vity, fitness and nutrition is the problem of malnutrition in deve- loping countries which among others substantially restricts the deve- lopment of physical fitness, work performance and thus also economic productivity of the country. As regards health and its relationship with nutrition this is at present the most urgent world problem; its solution, however, by far does not depend on research but mainly on the social and economic position of the country. Results of research can hasten the solution of selected partial problems in developing countries but hardly the overall food shortage and unequal distri- bution.

The relations netween nutrition, adaptation to a load and phy- sical fitness play a very important role also in the development of top performance in Olympic athletes in different sports contents. Optimal diets for sportsmen were of interest already in ancient Greece and since then various dietary regimes were elaborated for the basic types of sports activities. So far it did not prove pos- sible to resolve this problem in a satisfactory and final way. Se- veral laboratories in different parts of the world are still concern- ed with intensive research of these problems but there is so far no definite recipe as regards quantity, quality, meal frequency, etc. for sportsmen of different disciplines and age, although many ins- tructions are available. This is due in particular to the great variability of the human organism where optimal values as regards food intake are not unequivocal neither in physically fit subjects nor in subjects of the normal population under varying environmental conditions.

The nutritional individuality of man conditioned among others also by the motor stereotype which in individual cases differs mark- edly and is one fo the unresolved key problems of physiology and clinical nutrition. As the importance of early stages of ontogenesis

is stressed from the aspect of later development of the organism in
other disciplines (medicine, pedagogics, etc.) it is important to
mention also selected problems in conjunction with relations between
nutrition and fitness and performance. There is evidence that e.g.
constitutional factors of the organism play in this respect a very
important part. The nutritional individuality is manifested at the
very onset of life, as apparent during the neonatal period and early
childhood. Here we have, however, to consider the question to what
extent only genetic constitutional factors play a part and to what
extent environmental factors incl. diet and motor regime are involved, e.g. during the foetal period, i.e. during pregnancy. It seems
that it is possible already during the foetal period to alter by
certain interferences the future programming of the organism as
regards its future spontaneous food intake on the one hand and his
motor activity on the other. Both determine along with other factors
also the development of physical fitness and performance during subsequent ontogenesis. It was revealed for instance that a certain
locomotor regime of the mother during pregnancy in investigations
of experimental laboratory animal models changes some parameters
(such as the size and structure of the heart muscle), which may have
an impact on the future development of physical fitness and performance of the cardiorespiratory apparatus. At the same time also
changes are manifested in selected parameters pertaining e.g. to
the lipid metabolism in different tissue (liver, small intestine,
etc.) which may have certain consequences in later life. For instance the lipid concentration was higher in female offspring of mothers subjected during pregnancy to an increased work load on a treadmill. Also lipid synthesis in the above organs differed in offspring
of mothers with a different locomotor regime during pregnancy.
Although the interpretation of mechanisms similary as the importance
of these changes are not clear so far, from these facts it is obvious
that it is possible to influence during the foetal period the subsequent metabolic development of the organism and this may be manifested also in the physical fitness and development of health status.

A graded food intake (investigated on the model of laboratory
animals kept during the suckling period in nests with a different
number of animals) influences significantly the grade of spontaneous
motor activity and this in turn was manifested also in the growth
curve, spontaneous food intake, body composition and other selected
parameters of the lipid metabolism in different organs. (Pařízková
and Petrásek, 1978).

A part is played not only the total food intake of the young
but also by the composition of the diet of the mother during early
life of the young: the offspring of mothers on a low-calorie diet
which contains also little protein (5 caloric %) and who up to the
age of 42 days have a similar low intake are characterized not only
by restricted growth (smaller body weight associated with lower
weight of skeletal muscles), a relatively elevated caloric intake

(g of food per 100 g body weight) but also by an elevated spontaneous
motor activity during the realimentation period when a normal diet is
available ad lib. Other metabolic parameters in these animals also
have a different character (Table 1 - Pařízková and Petrásek, in
press).

Consistent with previous results of investigations of children
from Tunisia where we found a relatively and absolutely higher level
of functional and motor development in children who has a reduced
food intake during childhood (who however did not suffer from extreme
malnutrition), leading to more delicate body build, it seems that
food restriction does not lead necessarily to physical and functional
deterioration but may be useful for the development of physical fitness from certain aspects, as well as health (Pařžková, 1977). Work
has been published on the longer survival of laboratory animals fed
only a certain ratio of the spontaneous food intake (cca 60 %), as
well as work on humans who had a very plain and modest diet in particular during early childhood and adolescence (Hejda, 1966). These
conclusions seem to be important in particular from the aspect of
developing countries where from the economic aspect it is impossible
to contemplate on a diet which in the near future could lead to the
attainment of standards of physical development as encountered in
industrialized countries. However, this goal is not entirely desirable among other reasons because such somatic development is associated with obesity of civilization affecting in particular the cardiovascular apparatus. Apparently it is essential to ensure an optimum
energy balance i.e. an adequate food intake as regards quantity and
composition on the one hand and an adequate motor regime on the other
hand from the very beginning of life incl. the foetal period. The
definition of the optimum should arise from an analysis of experimental data from developing as well as industrial countries. Obviously it is essential to ensure nutrition throughout ontogenesis,
in particular the maternal diet during pregnancy and in children
during early childhood. On the other hand, in industrial countries
the idea is abondoned, supported by some experimental findings that
a large volume of a high energy diet is the best road to physical
functional development and satisfactory development of the health
status. To ensure the required energy balance we need also a satisfactory energy expenditure which is difficult to ensure in particular
in large urban agglomerations of industrial countries.

In our country physical training is organized from an early age
i.e. from the first months of life and preschool age to advanced age.
In this ways an effort is made to ensure despite an ample and often
unbanlanced food intake an optimal energy balance and thus a high
standard of fitness and performance, a satisfactory health status
and long active life span in all shperes of human activities.

TABLE 1

CHANGES IN LIPID METABOLISM IN MALE RATS (125 days old)
WITH DIFFERENT NUTRITION AT THE BEGINNING OF LIFE
(Parízkova and Petrasek, 1978)

	CONTROLS		TEMPORARY PROTEIN-ENERGY DEFICIENCY	
	$\bar{x}$	SD	$\bar{x}$	SD
Body weight (g)	401.1	45.1	316.5	32.5
Concentration of total lipids – liver (g%)	5.69	0.16	6.31	0.26
Liposynthesis in the liver (%)	15.47	3.16	6.48	0.81
Concentration of fatty acids – liver (g%)	3.63	0.21	4.22	0.43
Synthesis of fatty acids – liver (%)	7.80	3.38	4.47	0.61
Concentration of cholesterol (liver) – mg/g	2.26	0.12	2.51	0.19
Cholesterogenesis liver (%)	0.85	0.16	0.72	0.09
Cholesterolaemia (mg%)	67.4	5.2	68.8	6.1

BIBLIOGRAPHY

Astrand P.O., Rodahl K.: "Textbook of work physiology". McGraw-Hill
    Book Company, New York etc., 1977. (2nd ed.)
Hejda S.: Z. Altersforsch. 21: 159, 1968.
Pařízková J:: "Body fat and physical fitness". Martinus Nijhoff B.V.
    Medical Division, The Hague 1977
Pařízková J., Petrásek R.: "Nutrition and Metabolism", in press 1978
Pařízková J., Petrásk R.: prepared for publication.

EFFECT OF MALNUTRITION ON BRAIN GROWTH-

HISTOLOGICAL AND AUTORADIOGRAPHICAL STUDIES

Tomoichi Kusunoki, M.D.

Kyoto Prefectural Univ. of Medicine

Kyoto, JAPAN

There is increasing evidence that infants who are born with in-trauterine growth retardation have a higher incidence of central nervous system problems than infants born with appropriate growth for gestational age. This evidence has also been confirmed by our retrospective study on the causes of microcephaly, in which as many as 36 cases of 140 microcephalies, I mean, 26% were the products of intrauterine growth retardation. Severe maternal toxemia, metabolic and endocrinoligic disorders, and chronic renal disease of the mother, placental anomalies, and fetal infections are the main etiologic factors of intrauterine growth retardation. Most of these factors also often give rise to nutritional deprivation for the infant either directly or indirectly.

In 1977, we reported that most neurons in the mouse cerebrum are produced at the matrix layer, where active cellular proliferation takes place from the 10th to 16th days of gestation. And maternal malnutrition during this period caused about 14% prolongation of the generation time of the matrix cell, the precursor of neurons, thus causing a decrease in neuron production of fetal brain. From these results, it is suggested that malnutrition during the first and second trimester may cause similar results in the human embryo.

Recently many reports revealed that early malnutrition results in the distortions of brain chemical composition and function of growing child. However, most results have been obtained biochemically by determining the DNA content in the cerebrum and cerebellum as a whole, after homogenization.

Our experimental studies were undertaken to clarify what kind of cells and which part of the cell in child brain were most susceptible to early malnutrition.

As you know, histogenesis of central nervous system can be
divided into three stages. The first of them is the stage of orga-
nogenesis, when neural groove closes and becomes neural tube, subsequently brain vesicles are formed. The 2nd is the stage of neuron
production. Howard et al and Dobbing et al demonstrated biochemically that all neurons in the neocortex were produced during 10 weeks
from the 10th to the 20th week of gestation. We also confirmed their
data histologically. The 3rd is the stage of differentiation of neuron, that is, dendritic and anoxic arbolization, synaptogenesis and
myelination. A major part of differentiation of the cortical neurons
may be said to be postnatal events; at this stage, a specific neuron,
i.e. granule cell in the cerebellum are produced.

Thus, it may be apparent that severe malnutrition during early
postnatal life affects the cellular proliferation, especially production of cerebellar microneurons, dendrite growth, synaptogenesis and
myelination.

Postnatal Neurocytogenesis
and Undernutrition

In the cerebrum of full term infant, we can not find any specific
regions where active cellular proliferation is taking place. However
in the cerebrum of pre-term infant, two specific regions, i.e. vesti-
gial subependymal layer and ganglionic eminence, are found at the
surface of the lateral ventricles. An autoradiograph of subependymal
layer of 7-day-old mouse sacrificed 2 hours after tritiated thymidine
(^{3}H-thymidine) injection shows many labeled cells are found in this
layer. This phenomenon indicates the very active cellular proliferation is taking place in the subependymal layer and ganglionic eminence of preterm infant. The cells formed in this layer of mouse are
microneurons and glias. As for the human cerebrum, Rackic and Sidman
reported that the neuron production at the ganglionic eminence lasted
until 34 weeks of gestational age.

Contrary to the cerebrum, neuron production is still very active
in the cerebellum of full-term infant.

In the full-term newborn infant the cerebellar cortex has a thick
layer, that is the external granular layer at the surface of cortex.
This layer disappears by 12 months of age. Cerebellar microneurons
such as granule cell, basket cell and satellite cell are produced at
this layer.

The autoradiograph of cerebellar cortex of a 10-day-old mouse
sacrificed 2 hours after injection of ^{3}H-thymidine, a radioactive
precursor of DNA shows that a number of labeled cells are found

exclusively at the external granular layer (EGL). This indicates
that the cell in the EGL synthesize DNA very actively for the cell
division. By means of autoradiography, we demonstrated that more
than 50% of granule cell in the mouse cerebellum were produced after
10-days of age. There is a close similarity between the morphological maturation of cerebellar cortex of a full-term newborn and that
of a 10-day-old mouse. Thus it may be conjectured that more than
50% of granule cell are formed after birth even in human.

So, we examined the effect of early undernutrition on cell proliferation kinetics at the EGL. We found that brain weights of
undernourished animals were lower even after 60 and 100 days after
birth, and the cerebellar size of the undernourished mouse was apparently smaller than that of control. The external granular layer
had already disappeared from the surface of cerebellar cortex of the
control mouse, whereas one or three cell rows of this layer was still
found at these regions of the cerebellum of undernourished animal.
This indicates dismaturity rather than compensatory cell production.

Now, the effect of undernutrition on the kinetics of cell proliferation in the external granular layer was examined in this manner.

Twenty five mice in the undernourished control groups were injected with ^{3}H-thymidine on day 10 of age, and were sacrificed hourly
or at two hours interval. Sagital sections of cerebellums were then
processed for autoradiography. The percentage of labeled mitosis
increased abruptly to reach 100% by 4 hours after ^{3}H-thymidine injection. This high value lasted for several hours and then decreased
rapidly. 14 or 16 hours after ^{3}H-thymidine injection, the percentage
of labeled mitosis increased again. Generation time of the prolifer-
ating cell at the EGL of undernourished mouse was about 18.0 hours,
whereas that of the control was 15.5 hours. This result indicates
that early undernutrition decreases cell proliferation of cerebellum
and results in fewer production of microneurons.

Dendrite Growth and Undernutrition

Neurons in the neocortex show the rapid growth of dendrite during
the first 12 months of age; dendrite arbolization of the neurons of
full-term neonate is very poor comparing with that of 3 years old
child. In the mouse and rat, dendrite of neuron in the cerebral
cortex becomes apparent after 7 to 10 days after birth.

To study the effects of early undernutrition on dendrite growth,
we compared the number and length of dendrite of cortical neuron in
the control and undernourished mouse, by calculating the number of
branching and lengh of dendrite using special ocular glass in which
concentric circle at the radius of 30μ, 60μ and 80μ was drown. The
number of dendrite which crossed each circles were counted.

15 days after birth, we found no difference in dendrite growth between the control and undernourished group. However, differences became apparent when examined on day 20. When examined on days 30 and 60 of life, dendrite growth of undernourished mouse were retarded significantly at the distance of 60μ and 80μ from cell body. When examined on 10 day of age, decreased dendrite growth in undernourished animals was still significant at 60μ point. However, considerable catch-up growth was found at 30μ and 80μ points. Recently Salas in Mexico, and Araya in Chile also reported same results on suckling rat. The decreased dendrite growth reflects decreased number and function of synapses.

Thus I believe that you can understand the very very important role of early nutrition from the viewpoint of our precise histological studies of child brain growth.

NUTRTION, ENVIRONMENT AND MENTAL DEVELOPMENT - SPECIAL REFERENCE TO

INFANT FEEDING, GROWTH AND EVELOPMENT IN NORTHERN JAPAN

Hiroshi Wako, M.D., and Tomiji Hatakeyama, M.D.

Dept. of Peditc., Iwate Medical Univ, School of Med.

Morioka, JAPAN

The relationship between severe protein energy malnutrition (PEM) and mental development has been extensively studied during critical periods of development. This information is especially important in urbanized areas where mild to moderate PEM is not uncommon. Therefore we have decided to do our own field research in Northern Japan, where mild to moderate PEM is still common in remote rural areas. Considerably less is known about the impact of lesser levels of malnutrition on intellectual and behavioral development. This type of malnutrition occurs in the background of low socioeconomic status, reduced education, poor sanitary conditions, and is further complicated by recurrent infections.

Children in Iwate Prefecture of Northern Japan continue to have the higher infant mortality which, on an average, however, has tremendously decreased in Japan, as a whole in recent years. (Fig. 1 and 2).

Effect of Lysine and Threonine Supplementation
on Brain Function in Monozygotic Twins

Under these circumstances, we have attempted an epidemiologic study in humans, using a comparative longitudinal study of the physical and intellectual development of monozygotic twins, 2 to 3 years of age. This was carried out over a period of 1,300 days, from 1967 to 1971. In cases of twins younger than 3 years; one set of twins, 3 years and 6 months old were selected from the undernourished population. The twins from these areas had a low intake of protein and fat. Rice comprised about 85% of all dietary intake. A small amount of fish and eggs was eaten only twice a week. The survey of the

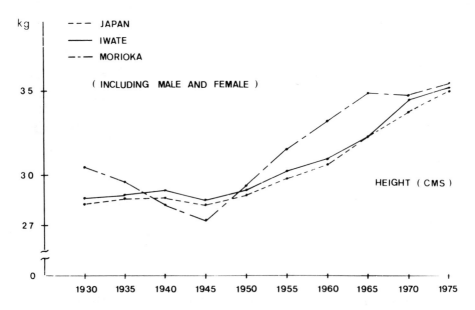

Fig. 1 - Distance growth curves in 12 year of age
from 1930 to 1975 (height (cms).

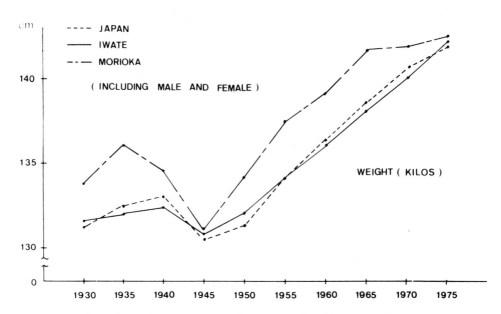

Fig. 2 - Distance growth curves in 12 year of age
from 1930 to 1975. (weight (kilos).

actual food intake in the 6 pairs of twins selected was very important
in evaluating their nutritional status, but we did not conduct a
detailed dietary survey because of psychological influence of the
survey on the twins and their parents. However, a dietary survey was
performed on children whose living conditions were similar to those
of the subjects.

These groups served as a control. An outline of the results is
summarized in Table 1. Each pair of the monozygotic subjects lived
at home and kept to their normal routine, except that one child of each
each pair received 2 tablets daily, one containing 0.1g of lysine
and other 0.05g of threonine, at morning and evening mealtimes respectively; the other child receiving placebo tablets at the same
times. The children were weighed and measured regularly. Tests for
perception, kinethesis, sociability, learning, manipulatory capacity
and psychic function, were performed at the beginning of the experiment. D.Q. was estimated by the methods Ushijima et al. currently
used in Japan. The same tests were conducted after 300 days, 540
days, 780 days, 920 days and 1,300 days. Suzuki-Binet's intelligence
test for intelligence quotient (I.Q.) was conducted after 980 and
1,300 days. The test showed a significant rise in D.Q. in the
lysine - thereonine administered twins. Statistical examination of
the difference between the rise in D.Q. of the controlled twins and
that of the group supplemented with the amino acids, revealed that
D.Q. of the supplemented twins rose significantly higher than that
of the controlled twins ($p > 0.01$). They have been kept the parallel
I.Q. till recent examination.

Recent Trends of Infant Feeding in Japan

During the development of pediatrics as a speciality, no facet
was more controversial, more variable or more in need of accurate
scientific evaluation than that of infant feeding.

It has been recognized again that breast feeding is essential
for the infant up to 6 months. However, we have had no Japanese
growth standards by breast and formula feeding in a strict sense.
We attempted to investigate growth standards according to feeding
methods in Northern Japan.

The superiority of breast milk has been established, such as
in regard to infections, or due to feeding problems and allergy to
cow's milk protein. Moreover, the immunological importance of
colostrum for newborns has become clear.

Formula feeding is defined as not being reared by breast milk,
until 6 months old. Infants who did not receive formula milk, from
3 days after birth, were assumed to be in the breast feeding category.

TABLE 1

D.Q. and I.Q. SCORES OF TWINS IN EACH TEST

MONOZYGOTIC TWINS		BEFORE Exp. (I) D.Q.	(II) 300 Days D.Q.	(III) 540 Days D.Q.	(IV) 780 Days D.Q.	(V) 920 Days D.Q. (I.Q.)	(VI) 1300 Days (I.Q.)
Case 1	Exp.	125	139	149	125	135 (140)	(140)
(1y, 1m)	Contr.	131	137	135	117	119 (119)	(123)
Case 2	Exp.	79	106	112	106	101 (96)	(96)
(1y, 1m)	Contr.	81	87	100	100	92 (92)	(92)
Case 3	Exp.	93	105	120	103	106 (105)	(106)
(2y, 1m)	Contr.	96	94	115	100	100 (99)	(103)
Case 4	Exp.	77	94	–	101	123 (124)	(124)
(2y, 3m)	Contr.	84	86	–	91	115 (101)	(118)
Case 5	Exp.	67	85	101	101	105 (105)	(103)
(2y, 6m)	Contr.	69	81	87	87	99 (89)	(82)
Case 6	Exp.	81	82	81	97	108 (108)	(107)
(3y, 6m)	Contr.	86	88	87	87	108 (101)	(101)

Nutritional education for mothers of selected areas has been given by our pediatric staff and regional "health nurses".

According to Japanese statistics in infant feeding in 1970, by Ministry of Welfare and Health, the percentages of receiving breast milk, formula milk, and mixed feeding were shown to be 31%, 28.1% and 40.9% in 3 months, respectively.

Breast feeding has been decreasing remarkably in urbanization and industrialization.

Not only in developed countries, but also in developing countries, breast feeding has become very important for human infants.

This is important for physical growth and also for psychological or behavioral development.

Though growth curves showed an almost similar curve, the body weight seemed to show a slight increase in formula feeding, and these were statistically not significant.

The incidence of rickets are increasing in industrialized areas and towns, and are decreasing in farm villages where the sunbath is adequately taken.

The changing face of rickets is mainly due to the changing pattern of infant feeding and their environmental situation.

Breast Feeding for Human Infants

From the view point of a nutritional environment in infant feeding, we investigated behavioral development in children who had been carefully brought up for six years, in our Well Baby Clinic, by different feeding methods; breast, mixed and formula feeding and it was noted that breast milk feeding was superior to other feedings, as in regard to "behavioral personality", as tested by a combination of psychological tests currently used in Japan.

In children brought up by breast feeding, there were less problems in nervousness and emotional stability. In addition to these findings, breast feeding is economically superior to other feeding methods, regarding the cost and days for therapy in common infant diseases, that were observed in rural areas. In breast feeding, the cost and duration of therapy are one third as compared with those in formula feeding. Thus breast feeding for infants should be recommended again from the nutritional and behavioral standpoints. The biochemical uniqueness and environmental influence for infants of human milk should be reemphasized.

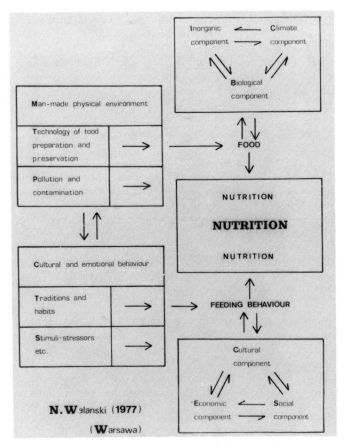

Fig. 3 - Nutrition, environment and behavioral development.

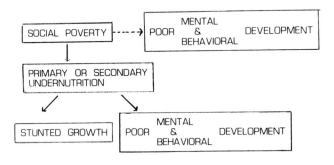

(MODIFICATION OF CRAVIOTO'S SCHEME)

Fig. 4 - Nutrition, environment and behavioral development.

Nutritional Deficient State and
Endocrinological Function -
- In case of Vitamin E Deficient State

The relation between malnutrition and endocrinological function
has been sometimes controversial and complicated.

By our ultramicroscopic observation of anterior pituitary and
hormonal secretion in experimental vitamin E deficient rats, we could
confirm that there were definite relationships between them.

In this respect, vitamin E seems to have an important function,
especially in protecting the highly unsaturated fatty acids, derived
in the body from linoleic and linolenic acid. We, therefeore, observed the pituitary changes in vitamin E deficient rats by ultramicros-
cope and ascertained the remarkable enlargement of GT cells, which
are similar to the findings, due to gonadectomy in orchiectomized or
ovariectomized rats, although there were a little differences, as in
regard to the decreased gonadtropic secretion granules in castrated
rats.

The secretions of FSH and LH by radioimmoassay were changed
significantly by the vitamin E deficient state.

It is conceivable that the actual requirement of vitamin E is
higher than usually found in the brain, especially for a developing
brain. A requirement of vitamin E in the brain should be further
studied.

These kinds of phenomena of endocrinological function and malnutrition should be further investigated.

These human and animal experiments strongly suggests that environmental and nutrition status may act synergistically on the development of the central nervous system and behavior, especially in
early infancy. (Fig. 3 and 4).

BIBLIOGRAPHY

Cravioto, J., Delicardie and Birch, H.G., Pediatrics., 38, 319(1966)
Oiso, T. et al., J. Nutri. Sci. Vitaminol., 19, 157-163, 1973.
Prescott, T.W., Read, M.S. and Coursin, D.B., eds. Brain Function
    and Malnutrition; Neurophysiological Method of Assessment.
    John Wiley & Sons, New York-London-Sydney-Toraito(1975).
Falkner, F. and Kretchmer, N., ed. Nutrition, Growth and Development
    S. Kargar, Basel-Munchen- Paris- London- New York-Sydney(1975).
Winick, M.; Nutrition and Mental Development Medical Clinics of
    North America 54, 1413(1970).

PSYCHO-SOCIAL ASPECTS OF

NUTRITION PROBLEMS

Jacques le Magnen

Lab. de Neurophysiologie Sensorielle et Comportementale
Collège de France
11 Place Marcelin Berthelot, 75231 Paris Codex 05, FRANCE

Ingestion of foods is an evident prerequisite of metabolic processes of nutrition. For this simple reason, generally ignored, the
study of feeding behavior belongs to sciences of nutrition. Feeding
behavior is not determined by cognitive and voluntary factors. In
human beings as in animal species, hunger and satiety, food preferences and aversions are physiological processes. For some time,
the observational and experimental study of feeding behavior has demonstrated the regulatory nature of this process. Neuroendocrine
mechanisms operate so that the control of feeding is integrated in
the overall regulation of body energy and body composition. Most of
nutrition problems and diseases result in man from defects of this
physiological and regulatory control of eating behavior rather than
metabolic disorders. Regardless of the availability of food sources,
under-, over- or malnutritions result in fact from under-, over- or
imbalanced feeding. Obesity or extreme leanness, protein deficiencies, alcoholis, and so on... are unknown in animals under the same
condition of food disposal. They are consequences of specific diseases of human feeding behavior.

The question therefore is to know the exact nature of the physiological and regulatory control of food intake, and to understand how
it is upset in man by psycho-social or other factors.

I - HUNGER AND SATIETY

Though some uncertainties still do persist, general mechanisms
of hunger and satiety or no hunger are to-day elucidated. The hunger
arousal of eating is elicited in conditions of depleted glycogen and
fat stores, and by a neuroendocrine pattern involved in their resto-

ration. Satiety and therefore feeding frequency are dependent on
repleted energy stores and maintained, by the neuroendocrine pattern
involved in their utilization in the current energy metabolism. In
rats, and also in man, different combinations of stimulation to eat
are provided by the depletion of carbohydrate and fat stores along
the 24 hr dark-light cycle[1, 2, 3]. In man, during light time, the
re-storage of both glycogen and fats and the current utilization only
of carbohydrates provide a maximal stimulation to eat and a minimal
satiety. During the dark and sleeping time, the mobilization and
current utilization of the two energy metabolites proce a minimal
stimulation to eat and a maximal persitency of satiety. In other
words, lipogenesis and glycogenesis add their effects to realise the
metabolic conditions of hunger. Lipolysis and oxidation of fats
counteract the stimulation to eat resulting from the depletion of
carbohydrate stores.

     That, ultimately, hunger is stimulated by glucopenia due to the
shortage of the glucose hepatic supply into the blood is highly probable but is not as yet definitevely proved. In man, as in animals,
hunger is experimentally elicited, by insulin of 2-deoxy-D-Glucose
(2-DG) -induced glucopenia, respectively associated with hypo- and
hyperglycemia. In acute natural hunger after food deprivation or
following a high level of energy expenses such as muscular exercise,
the role of a glucoprivic condition is clear. However, in chronic
and steady state condition of feeding, the same role of glucoprivic
state in the onset of food acceptance has not yet been fully demonstrated. The location either in the liver or in the brain of targets
or receptors of this hunger signal is also still debated.

     Regardless these still uncertain detailed mechanisms, it is now
clear that the regulation of body energy balance and of body weight
is merely insured because... and in as much as metabolic expenses
govern food intake. It will be seen below how such a feedback mechanism may be disturbed by human food habits.

II - PALATABILITY OF FOODS

     The effect of the hunger arousal is to give rise to the oral
food acceptance. Oro-pharyngeal sensory activities of foods trigger
the hunger-dependent eating response. This response to be oro-sensory activity of the food, named "palatability" of that food, is mainly
involved in determining the size of a meal, that is to say the amount
eaten from the onset of eating until its end by the onset of satiety,
or satiation. The palatability of various foods is learned[4, 5, 6].
It has been demonstrated that, from a genetically determined background of preferences and aversions (for sweet and bitter substances,
for example), a conditioning of palatability occurs. In this conditioning, the post-ingestive nutritive, non-nutritive or toxic effects
of foods act as positive or negative reinforcers or unconditioned

stimuli. Through this physiological learning (extensively studied in the particular case of conditioned aversions), the oro-sensory activity of each foods conditioned stimulus to eat allows and explains eating responses, anticipating their further metabolic consequences.

This homeostatic mechanism by which body energy expenses and specific needs of nutrients directly govern feeding is still functioning in human beings; but, it is obscured and biased by the psycho-social and cognitive human nature.

These biases may be overviewed as follows:

1) In man, the free alternation of hunger and satiety driven by the metabolic pattern is basically impaired by the social habit of a meal feeding. In such a pattern of three or four meals a day, the duration of satiety is precluded to be an efficient modulator of the cumulative intake. Whatever the size of a meal, man takes his next meal at the socially imposed time. Man is not free to eat at all times.

   In such a feeding schedule and in a steady state condition regarding energy expenses, the metabolic driving of hunger is altered. In rats placed in a human-like feeding regimen, it was shown that neuroendocrine basis for hunger becomes conditioned to the time and contingent stimuli of the habitual meal[7, 8, 9]. It seems that in man, through this conditioned hunger, the intake of food is no longer dependent on the current metabolic depletion, but anticipates later expenses. Thus, the balance between such an anticipated intake and the actual metabolic expenditures is apparently delayed and only operated by long term corrections. Regulation of body energy balance and of body weight, operative within the 24 hrs in ad lib fed rats is achieved in humans (when it is) from day to day, and rather from week to week, or more.

2) Food prohibitions and allowances, rejections and acceptances are transmitted by education like a language from the socio-cultural background. Most of them, by counteracting the spontaneous learning of food palatability are non-regulatory in nature. It is well known that they lead to tragic conditions of under and malnutrition.

3) Irrational believes added to... or substituted for the traditional sources recently appeared. They are new taboos. It is the case with the reluctance for industrially prepared foods.

4) The limit of science of nutrition is also a cause of uncertain allowances. It is, for example, presently impossible to give a scientifically based recommendation about the delay protein and even caloric need. More generally, it is vain and pehaps dangerous of attempting to substitute a cognitive and rational direction of

feeding behavior for its physicological ones.

5) The management of an over-palatability of foods is an art but it leads to eat to be fed but for a chemo-sensory self-stimulation. It may cause obesity; rats, fed overpalatable human foods, become obese[10, 11, 12].

6) The eating of some caloric foods or non-caloric materials is no longer stimulated by nutritional needs. New pharmacological and psychological needs of CNS stimulants associated with stresses of industrial civilization supports excessive or abusive consumptions. Rats stressed bu tail pinching are hyperphagic and are rendered obese[13, 14].

Remedies of these various sources of a non-regulatory feeding behavior in humans are difficult. The strategy of their study and application should be inspired by the principle that man, as animal, cannot eat against his natural hunger and satiety and should not be pushed to eat against his physiologically occuring preferences and aversions. Man eats what he likes and as he likes. The aim of a new and delicate feeding education should be to protect man from conditions leading him to artificially like and dislike contraty to the natural determinants of his food intake.

REFERENCES

[1] Le Magnen, J. & Devons, M. Metabolic correlates of the meal onset in the free food intake or rats. Physiol. & Behav., 5, 805-814, 1970.

[2] Le Magnen, J., Devos, M., Gaudilliere, J.P., Louis-Sylvestre, J., & Tallon, S. Role of a lipostatic mechanism in regulation by feeding of energy balance in rats. J. Comp. Physiol. Psychol. 84: 1-23, 1973.

[3] Le Magnen, J. Interactions of glucostatic and lipostatic mechanisms in the regulatory control of feeding. in: "Hunger: Wyrwicka, E. & Bray, G. ed., Raven Press (NY) 1976.

[4] Le Magnen, J. Advances in studies of the physiological control and regulation of food intake. in: "Progress in Physiological Psychology", E. Stellar & J.M. Spragues ed., NY Academic Press, 4, 204-261, 1971.

[5] Le Magnen, J. Hunger and Food palatability in the control of feeding behaviour. in: "Food Intake & Chemical Senses", Y. Katsuki, S. Takagi, Y. OOmura & M. Sato, ed., Univ. Tokyo Press, 1977.

[6] Booth, D.A. Satiety and appetite are conditioned reactions. Psychosom. Med., 39: 76-81, 1977.

[7] Woods, S.C., Vasselli, J.R., Kaestner, E., Szakmary, G.A., Milburn, P. & Vitiello, M.V. Conditioned insulin secretion and meal feeding in rats. J. Comp. Physiol. Psychol., 91: 128-133, 1977

[8] Balagura, S., Harrell, L.E. & Roy, E. Effect of the light-dark

cycle on neuroendocrine and behavioral response to scheduled feeding. Physiol. & Behav., 15: 245-248, 1975.

[9]Balaguara, S. & Harrell, L.E. Neuroendocrine conditioning: conditioned feeding after alterations in glucose utilization. Am. J. Physiol., 228: 329-396, 1975.

[10]Sclafani, A. & Springer, D. Dietary obesity in adult rats: similarities to hypothalamic and human obesity syndromes. Physiol. & Behav. 17: 461-471, 1976.

[11]Sclafani, A. & Gorman, A.N. Effects of age, sex and prior body weight on the development of dietary obesity in adult rats. Physiol. & Behav., 18: 1021-1026, 1977

[12]Rolls, B.J. & Rowe, E.A. Dietary obesity: permanent changes in body weight. J. Physiol. (London), 272 (1) 2 P, 1977

[13]Antelman, S.M., Rowland, N.E. & Fisher, A.E. Stimulation bound ingestive behaviour: a view from the tail. Physiol. & Behav. 17: 743-748, 1976.

[14]Rowland, N.E. & Antelman, S.M. Stress-induced hyperphagia and obesity in rats: a possible model for understanding human obesity. Science, 191: 310-312, 1976.

PSYCHO-SOCIAL ASPECTS OF

NUTRITION PROBLEMS

Mario Campagnoli, M.D., M.P.H.

National Office of Mental Health, Sec. of Pub. Health

Buenos Aires, ARGENTINA

Usually, the interrelationships nutrtion-psychology are considered from the viewpoint of what is commonly called "Nutrition, learning
and behavior", that is: the consequences of poor nutrition on intelectual and cultural development.

Whereas this approach has revealed to be of utmost importance,
its counterpart, namely the influence of psycho-social factors upon
nutrition — both hyper and hyponutrition — seem to have been neglected in recent times and considered unimportant as compared with
hunger in a global scale. Notwithstanding, as it was aptly emphasized by Mme. Simone Veil, Ministry of Health of France[1] "...illnesses
of deficiency have their counterparts in illnesses of excess; the
diseases of poverty are matched by diseases of opulence". Man "...
must revolutionize his food habits, overcome his anxieties...".

We have coined the expression "Psycho-social Nutrology" as a
name for this "new" scientific discipline which should concern itself
with the study and consideration of psycho-social factors that act,
on a positive or negative manner, on human nutrition.

The XXIX World Health Assembly reccomended to WHO "The application of psycho-social knowledge to the betterment of health care,
particularly in cases afflicted by greater needs"[2].

Psycho-social aspects of Nutrition include:

1) The influence of nutritional problems and nutritional pathology
   upon society.

2) The influence of psycho-social factors upon normal and pathological

nutrition of individuals.

3) Nutrition and nutritional pathology considered in its whole, sanitary-epidemiological scope.

FIELDS OF ACTIVITY

The psycho-social approach to Nutrology finds its main areas of application in the following fields:

a) Normal Nutrition

1) Nutrition Education at all levels.

2) Group Nutrition.

3) Dietary habits, patterns, folklore, taboos.

4) Diet and Nutrition Surveys.

5) Sanitary planning in Nutrition.

6) Integration and interrelations of experts in these fields.

7) Implementation of research in the matter. "Change of approach".

8) Nutritional Surveillance.

9) Food assistance and distribution.

b) Nutritional Pathology

1) Pathological Nutrition at the endemo-epidemic level: diabets; obesity; cardiovascular diseases; consequences of "Hyper" nutrition in general.

2) Undernutrition; vitamin-mineral deficiencies; calorie-protein malnutrition and other "hypo" nutropathies.

3) Psycho-social behavior among patients of nutritional diseases.

4) Lay societies: diabetics, obese and the like. "Endogroups".

5) Special cases: pregnancy; children development and growth.

The psycho-social medical approach should not be confused with the so-called "socialization of medicine". The main goal of the former is the consideration of the patient as an individual person although taking primarily into account the socio-psycho-epidemiologic factors influencing upon his clinical condition and vice-versa. As it was aptly recently said[3] a new "bio-psycho-social model" should provide "a basis for investigation, a framework for teaching and a design for action in the real world of health care".

REFERENCES

[1]Nutrition Notes, 12 (4), December, 1976.
[2]WHO Chronicle, 30: 365-367, 1976.
[3]Engel, George L., Science 196: 129-135, 1977.

NUTRITION AND THE BRAIN:

PSYCHOSOCIAL CORRELATIONS

Miguel R. Covian

Dept. of Physiology, Fac. of Medicine

14100 Ribeirão Preto, São Paulo, BRAZIL

MALNUTRITION IN ANIMALS

Many investigators have demonstrated[1] what malnutrition during the first years of life causes decresed intellectual ability with consequent decreased learning and adaptation to the environment. To what degree are these deficiencies caused by malnutrition and/or other psychological and social factors associated with poverty? According to studies carried out on people in underdeveloped countries and on animals, malnutrition is the most important factor. Nutritional restrictions retard brain development during its earliest stages. In rats and guinea pigs, the brain reaches 80% of its adult size at the time of weaning, when body weight is only 20% of its adult size. An 8% decrease in total brain weight and 53% in body weight has been observed in rabbits that suffered from malnutrition during the first 21 days of life. The animals exhibited irritable behavior and became easy victims to respiratory ane eye infections[10]. Another observation is that total brain weight, plus DNA and protein content of the brain in rats that were malnourished during the earliest stages of life do not return to normal values when the animals are fed ad libitum after the initial period of deprivation[12]. In general, experiments carried out on animals which suffered malnutrition during infancy show behavioral deficiencies in adult life in terms of emotional stability, motor ability and learning. It is well known, however, that maternal and social deprivation during infancy also elicits these symptoms in animals. Female rats suffering from malnutrition but allowed to enjoy maternal care and social interaction did not develop motor or learning deficiencies, nor greater emotional instability during adulhood[6].

Undernourished mice exhibited greater motor activity, made fewer space discrimination mistakes, and a greater number of mistakes in solving maze problems[9].

MALNUTRITION IN MAN

    The research carried out on animals points to the role that
chronic, although not severe, undernourishement may be playing in
human behavior and mental ability. Malnutrition has more severe
effects when it reaches the organism during the period of maximum
growth ("vulnerable period"), which, for the brain, takes place during
the first 2-3 years of life. Good nutrition is ideal for all ages,
but the first three years are crucial. The human brain reaches 70%
of its final weight during the first year, and its growth is complete
by the end of the second year, while body growth continues until the
18th year of age. A comparison was made of children who were undernourished at the time of the experiment and a group of children in
similar socioeconomic circumstances, but in good nutritional condition.
Ages varied between 10 months and three years. Brain growth (measured by head circumference) and IQ were significantly lower in the
undernourished group than in the controls. No improvement was observed up to 7 years of age[11]. Children who survived a severe malnutrition episode in infancy showed lower learning ability in school[2].
However, these results are not considered to be conclusive by other
investigators[4]. Brain damage is irreversible, i.e. normalization of
nutrition does not improve the condition of the brain. The number of
children subsisting on a protein-deficient diet today the world over
is estimated at 350 million, or 7 out of 10 children under 6 in the
world population[8]. In general, protein deficiency is the major malnutrition factor, but vitamins also play an important role.

    Two hypothesis on the vulnerability of the developing brain have
already been testes: that of Winick and Noble[13] which maintains that
malnutrition during the period of growth of any tissue, when cell
division takes place, will result in a reduction of the number of
cells size, will produce only a reduction in size, which is reversible with good nutrition. The other hypothesis, by Dobbing[3], affirms
that the brain is more vulnerable to restricted nutrition during the
period of brain growth-spurt, i.e. the period of rapid growth of the
sigmoid growth curve. The period of brain cell division forms the
first part of brain growth-spurt, while the second part concerns
growth in terms of number and lenght of prolongations, establishment
of synapses and myelinization. which are not directly related to cell
division but are vulnerable during the rapid growth period. The
number of neurons an adult will have is already reached before the
growth-spurt. In the guinea pig, this process occurs by the 42nd
day of pregnancy; in the rat, 2 days after birth; in man, at 25 days
of gestation. However, the greatest number of cells is represented
by the glia, which starts its spurt after cessation of neuronal
division and at the time when the dendrites grow and synaptic contacts
are established. The two hypothesis have been tested when restricted

nutrition occured after the period of neuronal division. Thus, we may conclude that malnutrition affects almost exclusively the glia, the growth of neuronal processes and the synapses.

PSYCHOSOCIAL PROBLEMS

The solution of the worldwide nutritional problem does not depend only on the disciplines directly related to nutrition, but rather on the integration of several scientific disciplines and political and social sciences. Two thirds of the world population are estimated to suffer from absolute physical hunger, and three quarters are possibly suffering from some form of hunger. Two opposing courses of action exist in the face of world hunger[7]: 1) Eliminate the new poor population; 2) Eliminate poverty. The first solution is Malthusian in character and rests on the policies of birth control and family planning, directed and financed by the centers of power. The second is a revolutionary and human one, requiring modifications of the social structure which will extinguish the blatant differences between power and subjection, wealth and poverty, domination and submission. World population is currently estimated at 3500 million (3 bilion and a half), and is projected at 6000 million (6 billion) by the year 2000. The population explosion is not only due to a higher birth rate, but also to a decreased mortality rate. How will all those people be fed, when difficulties already exist with today's population? According to those who favor limited world population, nutritional insufficiency is due to the scarcity of arable land. The "Bariloche Plan", however, declares that: "According to FAO estimates, current food production is sufficient in practice to satisfy the basic needs of every inhabitant of the earth. We also know that physical limits do not restrict, so far, food production, since only 43% of potentially arable land is utilized, with yields which are much lower than theoritically possible, even by conservative calculations"[5]. Thus the possibility of an agricultural revolution in the future exists, especially when we consider that man only utilizes 23,000 classes of plants (six hundred in agriculture) out of the half million available to him[7].

Obviously, if the sociopolitical factors hindering food production and distribution were to be modified, the worldwide problems of hunger and malnutrition could be solved during our time.

BIBLIOGRAPHY

[1]Abelson, P.H. Malnutrition, learning and behavior. Science, 164: 17, 1969.
[2]Cravioto, J. Mental performances in school age children. Findings after recovery from early severe malnutrition. Amer. J. Dis. Chil., 120: 404-410, 1970.

[3] Dobbing, J. Vulnerable period in developing brain. in: "Applied Neurochemistry". Davison, A.N., Dobbin, J. eds., England Blackwell Scientific Publications, Oxford, 1968.

[4] Frisch, R.E. Does malnutrition cause permanent mental retardation in human beings? Psychiat. Neurol. Neurochir., 74: 463-479, 1971.

[5] Herrera, A.O. Un monde pour tous-Le modèle mondial Latino-Américaine. Presses Unviersitaires de France, Paris, France, 1977.

[6] Koos Slob, A., Snow, C.E. and Natris-Mathot, E. Absence of behavioral deficits following neonatal undernutrition in the rat. Develop. Psychobiol., 6(2): 177-186, 1978.

[7] Mara Zavala, D.F. Explosion demográfica y crecimiento. Imprenta Universitaria, Caracas, Venezuela, 1970.

[8] Praag, Van, H.M. Emotional and nutritional deprivation. Psychiat. Neurol. Neurochir., 74: 419-420, 1971.

[9] Randt, C.T., and Derby, B.M. Behavioral and brain correlations in early life nutritional depriviation. Arch. Neurol. 28: 167-172, 1973.

[10] Schain, R.J. and Watanaba, K.S. Effects of undernutrition on early brain growth in the rabbit. Exo. Neurol. 41: 366-370, 1973.

[11] Stoch, M.B. and Smythe, P.M. Does undernutrition during infancy inhibit brain growth and subsequent intellectual development? Arch. Dis. Chil. 38: 546-552, 1963.

[12] Swaiman, K.F. The effect of food-deprivation and re-feeding on brain activity in inmature rat. Brain Res., 43: 296-298, 1972.

[13] Winick, M. and Noble, A. Cellular response in rats during malnutrition at various ages. J. Nutr. 89: 300-306, 1966.

ECOLOGY AND NUTRITION:

AN OBSERVATION IN THE HUMID TROPICS

Nelson Chaves

Depto. Nutrição CCSUFPe - Cidade Universitária

50.000 - Recife, Pernambuco - BRAZIL

For the purposes of this paper, the definition of "Tropics" is the following: the rigorously tropical climate is the rainy one of the jungles, with an annual average temperature of 21° C ($\pm 70^{\circ}$ F) and an elevated pluviometric index[1]. Grographically, the tropics are in the band situated between the Tropics of Cancer and Capricorn, which are parallel to the Equator, at a distance of 26° on either side. The tropical climates are divided into: 1) humid, that of the jungle; 2) semi-arid, of the fertile lands; 3) arid, that of the deserts; and 4) semi-arid, of the savannahs[2].

This study was carried out in an area of the humid tropics, the Zona da Mata-Sul of Pernambuco, with a population which was estimated, in 1975, to total 565,284 inhabitants[12], and where, over 4 centuries ago, the sugar-cane agro-industry was installed, initially under the slave system, then, after the freedom of slaves, under the patriarchal system of the sugar mill, and, lastly, with the coming of the industrial plants, under the capitalistic regime. This micro-region is situated in the Brazilian Northeast, which covers an area of approximately 1,548,672 km^2, 18.20% of the national territory[10], with an estimated population (in 1977) of 33,642,000 inhabitants[11]. The Northeast region, with thermal averages between 20° and 28°C, is divided into the humid tropics (Zona do Litoral - Mata) and semi-arid tropics (Agreste and Sertão).

Seing as how in this region the sugar-cane agro-industry was installed and developed, in a well-defined economic cycle, especially in the States of Pernambuco and Alagoas, whose sugar production represents 88% of the Northeast's total production[15], we studied the ecological conditions and the socio-economic aspects of said agro-industry.

    In the slavery and patriarchal period, subsistence agriculture and cattle and goat raising were relatively developed. There were the dwellers, who worked in the sugar plantations and practiced subsistence agriculture and small animal husbandry; they planted manioc, corn, beans and fruits, which assured them of a reasonable diet and a modest source of economic revenue.

    With the arrival of the industrial plants, however, the situation changed: the large property owners, in order to rid themselves of the social obligations imposed by measures taken by the government in the sense of providing greater assistance to the rural worker, substituted their old dwellers by paid workers, who started to live in the cities and villages next to the mills where they worked, in unsatisfactory living conditions. The subsistence agriculture and animal husbandry practically disappeared: dietary monotony became established and nutritional deficiencies began to appear.

    The installation of the sugar-cane agro-industry in São Paulo and Parana, where the ecological conditions are much more favorable and the consumer market is greater; and the utilization of primitive agricultural technology and the gradual impoverishment of the soil caused by the massive and uninterrupted devastation of the forests, led to the appearance of a severe crisis in the sugar agro-industry in Pernambuco, indicated by the closing down of several plants, by unemployment and under-employment, by hunger and by the migration of the field population to the large urban centers of the Northeast, especially Recife. This disordered urbanization caused the appearance of severe problems in the cities receiving these migratory fluxes.

    On the other side, the precarious hygienic conditions due to the deficiency of basic sanitation and water supply have contributed towards the aggravation of the already poor health status of the populations of this area, where the incidence of intestinal parasitosis is very high, including among them, schistosomosis mansonica, a real social scourge, brought by the African slave at the time of the colonization. Associated with other parasitosis, the schistosomosis was disseminated and is responsible for the high incidence of anemia and the health deterioration of the populations studied.

    With the installation of the industrial plants there also took place the establishment of the routine practice of dumping the residues from sugar industrialization — molasses — in the river, which caused massive killing of the fish and crustaceans, with repercussions in the economic and nutritional status of the populations. It should be remembered that the molasses from the plants can be industrialized, as was demonstrated by Prof. Oswaldo Lima, preparing from it the torula, rich in nucleic acids, proteins and

vitamins of the B-complex. In spite of this, the continuous destruction of the existing river foods (fish and crustaceans) proceeds, along with another rich source of food: the industrial residues of the sugar cane.

One should also consider the low productivity of the rural
worker in the Northeast, caused by his precarious health status
and also by the restrictions of ecological nature imposed by a hot
and humid climate. We know that perspiration is of extreme importance in the mechanism of thermal regulation. An individual doing
intense work under the sun perspires profusely, being capable of
losing from 1 to 3 liters of perspiration/hour; in the desert, this
loss can reach 10 to 12 liters. With intense exercise, under a hot
and humid climate, an individual can lose 4 liters of perspiration/
hour. For the evaporation of 1 liter of perspiration the organism
loses about 580 kcal, which causes a large aqueous and energetic
loss and can lead to real dehydration[13]. Seeing as how the perspiration contains several organic and inorganic substances, the abundant perspiration, together with the aqueous loss, leads to the
loss of chemical substances. Among the inorganic substances, the
most important is sodium, in the chloride form. After an excessive
loss of salt there occurs an alarm reaction from the adrenal cortex;
if the saline loss continues, there follows an adaptation and exhaustion phases. The adrenal responds to the stress by increasing
the aldosterone secretion in order to avoid a greater sodium loss.
With the continuation of work and perspiration the adrenals become
exhausted, causing fatigue, commonly known as "tropical fatigue".
Among the organic substances there are the lactic acid, pyruvic
acid and amino-acids. Nitrogen is also lost, due to its high elimination through perspiration (about 150 mg/day, being possible
to reach 500 mg/day)[19].

For the determination of the nitrogen balance one usually considered the elimination of urinary and fecal nitrogen. Lately, the
importance of the nitrogen eliminated through perspiration has been
emphasized. In intense muscular work, like the one carried out by
the rural worker cutting sugar cane, perspiration becomes abundant,
and the elimination of nitrogen through perspiration increases, as
a result of the protein catabolism for energetic purposes. If protein ingestion is low, as happens in the sugar cane area in Pernambuco, the organism becomes more deficient in proteins. A 70kg man,
under normal conditions, loses 3.7 gm of nitrogen per day through
urine, feces and perspiration, which is equivalent to a protein
loss of 23 gm per day; if he perspires intensely, this proteic
loss is accentuated, draining his organic reserves[19]. Scrimshaw[17]
makes references to the observations by Huang and collaborators,
who studied the obligatory nitrogen losses in eight individuals in
hot climates. The authors observed that when these individuals
were submitted to elevated temperatures there was a fall in urinary nitrogen, but increased nitrogen loss through perspiration.

He also mentions the studies by Way and Morgen about nitrogen balance in individuals with proteic depletion, stressing the importance of the nitrogen loss through perspiration.

These facts serve too well as an argument against the recommendations proposed by the FAO/WHO Groups in 1973 for proteic and
caloric allowances, when applied to poor populations who live and
work in tropical regions with hot and humid climate.

This complex set of factors, in which the ecological conditions are associated with modifications of a social order, has
deep repercussions in the living conditions of the rural worker
and in his nutritional and health status, having contributed towards the devaluation of man in the area of Mata-Sul of Pernambuco, where malnutrition is endemic and progressive.

A study carried out by the Nutrition Institute, presently the
Nutrition Department of the Center for Health Sciences of the Federal University of Pernambuco (20), revealed a low caloric ingestion (much lower than the amount necessary for the basal metabolism of an active individual), reduced protein ingestion (especially of animal origin), and reduced ingestion of several other
nutrients in the city of Ribeirão(PE).

Other studies[3, 4, 7, 9, 14, 16] revealed the same deficiencies,
intestinal multi-parasitosis, and a high index of schistosomosis
mansonica (always superior to 50%). Evident signs of endemic malnutrition were found in children and adults, by means of anthropometric measures (reduced height and weight), anemia, precarious
dental state, weight loss, precocious aging and a certain level of
adanimia. About 2/3 of the infant population presented malnutrition of the I, II and III degrees, with a high incidence of marasmus and kwashiorkor.

A nutritional survey carried out by the Interdepartmental Committee on Nutrition for National Development (ICNND)[18] showed weight
and height deficiencies in children and adults of both sexes when
compared to the North American norms and to those of São Paulo.
In Palmares (PE), male individuals with an average of 35.9 years of
age presented an average height of 157 cm and an average weight of
52.5 kg; in female individuals with an average of 33 years of age
the average height was 149.3 cm and the average weight 49.2 kg.
In Catende, in male individuals with an average of 26.9 years of
age, the average height found was of 157.5cm and the average weight
52.7 kg; in female individuals with an average of 35.4 years of age,
the averages were 150.9 cm for height and 50.4 kg for weight.

A later study, carried out in Catende and Agua Preta revealed
the average height for the adult man to be 1.61 cm and for the woman, 1.51 cm (non published data). The simple visual exam of the

population shows a high frequency of women, between 20 and 30 years
of age, presenting endocrine deficiency due to malnutrition, as
shown by low stature, reduced pelvic waist, mammary hypoplasia (even
in women with previous pregnancies).

A study was carried out to verify the influence of nutrition
on hormonal secretion in the cities of Agua Preta and Ribeirão,
situated in the full sugar area of the State of Pernambuco[5, 6].
During the first stage of the work, carried out in Agua Preta, 9
children of both sexes were selected, between the ages of 1 and 6
years of age, with 2nd degree malnutrition. The signs of severe
malnutrition were evident, such as hepatomegalia, conjuntival xe-
rosis, discolored hair, etc. The serum level of the growth hormone was high, as well as that of insulin; there were reduced urinary levels of 17-cetosteroids and 17-hydroxycorcosteroids. Proteins, especially serum albumin, were reduced. With the dietary
treatment, the hormone levels became normal and the children presented improvements in the nutritional status, as shown by biochemical, clinical and anthropometric data. Figures 1-7 show
a synthesis of the study[5].

In the Ribeirão study 10 children from 1-6 years of age, with
II and III degree malnutrition were studied. Significant increases
in the levels of growth hormone and reduced insulin were also found.
After the diet-therapy treatment there was nutritional rehabilitation and the hormone levels returned to normal. Twelve days after
the initiation of the treatment the levels of serum albumin increased, but this increase was more evident in the children with
III degree malnutrition. Gamma-globulin levels were reduced.
Tables 1 and 2 reveal the alterations encountered[6].

In the Zona da Mata-Norte of Pernambuco, the nutritional and
health status is also quite precarious. Costa[8], based on anthropometric measures done in children from 6-12 months of age residing
in the municipality of Ferreiros (area of the sugar cane agro-industry) showed the occurrence of endemic malnutrition.

The growing state of endemic malnutrition of the populations
studied, as revealed by the studies carried out in different time
periods, is the result mainly of the socio-economic conditions imposed by the sugar cycle, when the rural worker was not adequately cared for. If we consider that the total population of the
Zona da Mata de Pernambuco (Mata Umida and Mata Seca), in 1975,
was estimated at 1,054,661 people[12], with 65-70% malnutrition, we
shall reach the conclusion that the situation is severe, with
their socio-economic recuperation being extremely difficult. The
low wages, the growing unemployement and under-employment, the decreasing immunity, disease and the high mortality coefficients,
reduce the work capacity and decrease the life average. In addition, there is a predomination of a very young dependent popula-

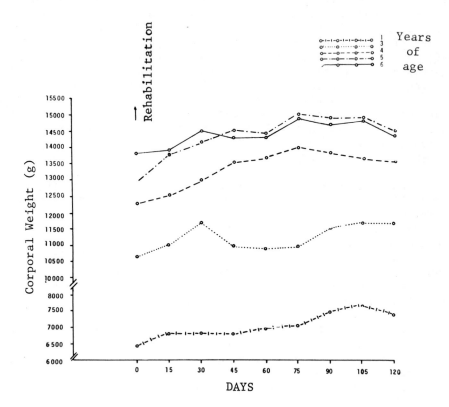

Fig. 1 - Corporal weight (gms) of children with malnutrition - Água Preta

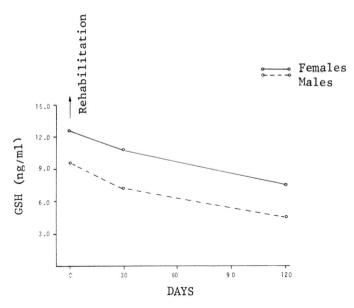

Fig. 2 - Serum levels of the growth hormone (GSH) in children
         with malnutrition: before, during and after nutritional
         rehabilitation - Água Preta.

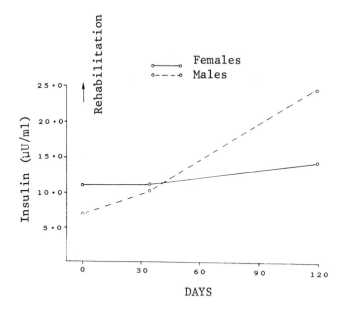

Fig. 3 - Serum levels of insulin in children with malnutrition:
         before, during and after nutritional rehabilitation -
         - Água Preta.

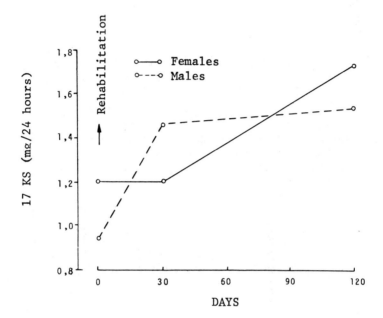

Fig. 4 – Urinary levels of 17 cetosteroids (17 KS) in children
         with malnutrition; before, during and after the nutrition-
         al rehabilitation - Água Preta

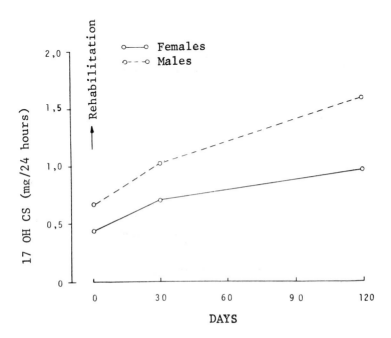

Fig. 5 – Urinary levels of 17 hidroxycorticosteroids (17-OH CS)
in children with malnutrition: before, during and
after the nutritional rehabilitation – Água Preta.

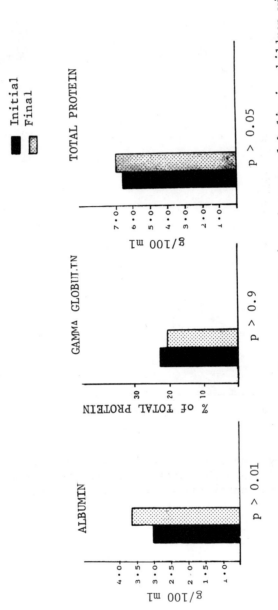

Fig. 6 – Average plasma values of total protein, albumin and gamma-globulin in children with 2nd degree malnutrition: before and after nutritional rehabilitation – Ribeirão.

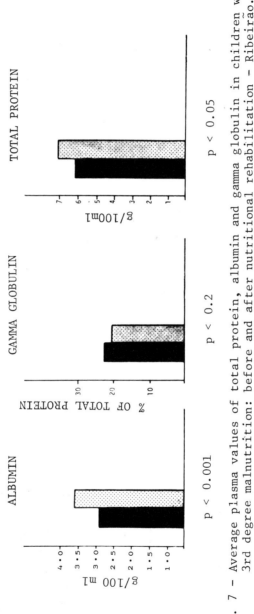

Fig. 7 – Average plasma values of total protein, albumin and gamma globulin in children with 3rd degree malnutrition: before and after nutritional rehabilitation – Ribeirão.

TABLE 1

CLINICAL–NUTRITIONAL SIGNS FOUND IN MALNOURISHED CHILDREN (II and III degree), BEFORE AND AFTER NUTRITIONAL REHABILITATION – RIBEIRÃO

	III DEGREE (nº of cases)		II DEGREE (nº of cases)	
	0 days	120 days	0 days	120 days
Hepatomelagia	3	3	3	–
Papillary atrophy	5	5	5	1
Irritability	1	–	–	–
Paleness	5	5	5	1
Hair easily plucked	5	5	5	4
Discolored hair	5	5	5	1
Angular lesion	3	–	4	–
Peripheral redness	4	1	2	1
Interdentary papillary hypertrophy	5	5	3	1
Apathy	3	3	5	1
Ankle edema	1	–	1	–

TABLE 2

HEIGHT AND AVERAGE SKINFOLD OF MALNOURISHED CHILDREN:
BEFORE AND AFTER NUTRITIONAL REHABILITATION – RIBEIRÃO

III DEGREE nº	HEIGHT (cm)		SKINFOLD	
	0 Days	120 Days	0 Days	120 Days
10	70.00	74.00	7.20	12.00
11	68.00	71.00	7.40	12.00
12	80.00	85.00	6.40	12.00
13	90.00	94.00	4.00	10.00
14	73.00	74.00	7.00	14.50
x̄	76.20	79.60	6.40	12.10
S.D. ±	8.95	9.65	1.39	1.58

II DEGREE nº	HEIGHT (cm)		SKINFOLD	
	0 Days	120 Days	0 Days	120 Days
15	80.00	82.00	6.40	11.00
16	64.00	69.00	4.20	8.00
17	75.00	79.00	7.00	11.50
18	71.00	73.00	5.00	9.00
19	84.00	88.00	9.20	15.00
x̄	74.80	78.20	6.36	10.90
S.D. ±	7.79	7.46	1.93	2.70

x̄ – arithmetic average

S.D. ± – standard deviation

tion, which aggravates the social problem and requires great amounts of investments. In Ribeirão, for example, 45% of the population is under 15 years of age, while 8.5% is over 50 years of age[20]. It is a population with practically no old people and with an infant population of close to 50%.

REFERENCES

[1] Bates, M. "Les tropiques; l'Homme et la Nature entre le Cancer et le Capricorne". Payot, Paris, 1953. p. 84 (Bibliothèque Géographique).

[2] Ibidem, p. 91-92.

[3] Batista Filho, M. et al. Inquérito nutricional em área urbana da Zona da Mata do Nordeste brasileiro, Água Preta, Pernambuco. O Hospital, Rio de Janeiro, 79 (5): 707-723, maio 1971.

[4] Chaves, N. Environmental factors and malnutrition, in "Newer Horizons in Tropical Pediatrics", S. Gupte, Jaipee Brothers, Delhi, pp. 152-175, (1977).

[5] ———, et al. Influência da Nutrição sobre a secreção hormonal, Estudo in Água Preta, Revista Bras. de Pesquisas Med. e Biol., São Paulo, 8(5-6): 353-362 (1975).

[6] Ibidem, Estudo em Ribeirão, Pernambuco, Brasil. 8(5-6): 353-362 (1975).

[7] Chaves, N. A nutrição, o cérebro e a mente, O Cruzeiro, 110 pp. (1971).

[8] Costa, E. Consumo protéico-calórico e desenvolvimento somatométrico em crianças de 6-72 meses de idade, Tese Univ. Fed. Pernambuco, Inst. Nutrição, 74 pp. Recife (1975).

[9] Freitas, L.P. da C.G. de, Inquérito da morbilidade e mortalidade por sarampo em Água Preta, Pernambuco, Tese Univ. Fed. Pernambuco, Inst. Nutrição, 75 pp., Recife (1975).

[10] Fundação Instituto Brasileiro de Geografia e Estatística, "Anuário Estatístico do Brasil, 1976, Rio de Janeiro, vol. 37, p. 20 (1977).

[11] Ibidem, p. 83.

[12] Ibidem, p. 98.

[13] Houssay, B.A. et al. "Fisiologia Humana", 4th. ed., El Ateneo, Buenos Aires, p. 651-656 (1969).

[14] Lago, E.S. Sarampo e desnutrição; estudo bioquímico transversal, Tese Univ. Fed. Pernambuco, Inst. Nutrição, 66 pp. Recife (1974).

[15] Melo, M. Lacerda de, O açucar e o homem; problemas sociais e econômicos do Nordeste canavieiro, Série Estudos e Pesquisas 4, Instituto Joaquim Nabuco de Pesquisas Sociais, Recife, p. 19 (1975).

[16] Puffer, R.R. and Serrano, C.V., Patterns of mortality in childhood, Report of the Inter-American Investigation of Mortality in Childhood, WHO Scientific Publication 262, Washington, D.C., 470 pp. (1973).

[17] Scrimshaw, N.S.   Through a glass darkly: discerning the practical implications of human dietary protein-energy inter-relationships, Nutrition Reviews, New York, 35(12): 321-337, (Dec. 1977).

[18] U.S.   Interdepartmental Committee on Nutrition for National Development, Northeast Brasil, nutrition survey, March-May 1963. Washington, D.C., p. 210. 1965.

[19] U.S.   National Academy of Sciences, National Research Council, Food and Nutrition Board, "Recommended dietary allowances", 8th ed.   Washington. D.C. p. 39.   (1974)

[20] Universidade Federal de Pernambuco.   Instituto de Nutrição. Pesquisa Nutricional da Zona da Mata.   Recife, Imprensa Universitária, 1968. 133p.

NUTRITION AND PRIMARY HEALTH CARE SERVICES

Derrick B. Jelliffe, M.D.

Prof. of Pub. Health & of Pediat., Univ. of California,

Los Angeles, California, 90024, U.S.A.

Nutrition is at the heart of effective primary health care, especially for the physiologically vulnerable (pregnant and lactating
mothers and their young children), and for the culturally or economically disadvantaged, as with groups regarded as socially inferior and
for confused new slumdwellers. The level of nutritional status is
any community depends on many factors which can be summarized in the
"community nutrition level" (CNL) equation (Fig. 1).

From this, it can be seen that many aspects of primary health
care — or more exactly, community improvement — affect nutrition.
For example, minor increases in income may make nutritious foods more
available. Simpleseeming agricultural extension guidance — for example, with self-seeding tropical dark green leafy vegetables — can be
a significant help, as these are often villagers' vitamin and mineral
supplements. Improved waste disposal and a less contaminated water
supply decrease the risks of infections, which in themselves are
nutritional problems as they raise nutrient needs and increase losses.
The same, of course, applies to the provisions of basic health services to prevent infections (such as by immunization) and to make economical basic treatment easily and speedily available (such as by oral
rehydration). Likewise, the provision of whatever child-spacing advice or services as are felt to be needed by, and acceptable to, the
community decrease the likelihood of "maternal nutritional depletion"
and of babies denied an opportunity for biological infant feeding,
particularly breast feeding.

IMPROVING COMMUNITY NUTRITION

The International Union of Nutritional Sciences (IUNS), in

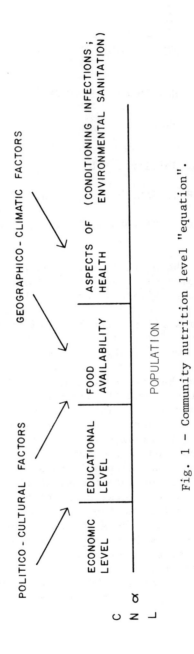

Fig. 1 – Community nutrition level "equation".

parallel with WHO, FAO, UNICEF, and other bodies, has come to realize
that the improvement of nutritional status depends very largely on the
activities by the community itself. Both through general measures
which can include "community self-assessment", increasing food availability, improving family finances, protecting the environment, etc.
All have been undertaken by village communities in various parts of
the world[1]. In a more classical nutritional sense, they need to be
concerned with improving the feeding of mothers and young children
using locally available resources[1,2,3,4]. Some of the "principles"
in such activities have been outlined in the publications mentioned,
among which the following may be reiterated: focus on priority
problems (usually protein-energy malnutrition and deficiencies of
iron or vitamin A); local causation; identification of main "risk
factors" responsible in the particular community; methods based on
local resources, foods, kitchen facilities, meal patterns and cultural
attitudes; joint preventive and curative activities; integration with
other village level improvement activities (such as through agricultural extension); making available nutritional information and blocking nutritional misinformation.

IMPROVED FEEDING OF MOTHERS
AND YOUNG CHILDREN

     Recent scientific studies clearly emphasize the fact that the
nutrition of pregnant and lactating mothers cannot be considered apart
from their young offspring. In other words, the nutrition of the
fetus, the young baby ("extero-gestate fetus") and the infant during
weaning ("transitional") cannot be considered separately. This is
so much so that the expression "maternal-channelled infant feeding"[5]
has been suggested to ensure that this inter-relationship is appreciated. In other words, modern knowledge further endorses the wisdom
of feeding the mother to nourish the young child.

(i) Fetus - During pregnancy, the fetus is entirely dependent on the
            mother for nutrition, mainly via the placental circula-
tion, but also to some extent from amniotic fluid swallowed during
the last months in the womb. Inadequate maternal nutrition during
pregnancy can lead to low birth weight, with small, feeble newborns
with inadequate stores of nutrients (including body fat). In developing countries, this is a huge public health problem, probably involving many millions of babies each year[6,7]. In addition, an inadequate
maternal diet in pregnancy leads to deficient stores for use in subsequent lactation (particularly the subcutaneous far normally laid
down in healthy, well-fed women), and in poorly nourished parts of
the world this can lead to "maternal nutritional depletion" becoming
worse each reproductive cycle (e.g. goiter, anemia or general malnutrition).

(ii) Extero-gestate Fetus - The human neonate is born relatively
immature and for the first 6-9 months
is dependent on the mother in a similar way to the fetus in the
uterus. Nutritionally, the breast takes the place of the placenta.
All over the world, but certainly in developing countries, the scientific significance of human milk and breast feeding has become
continually endorsed nutritionally, as a form of active protection
against infection, as a natural form of child spacing and as the
best use of local resources and of cutting down on the drain of
foreign currency for the purchase of higly expensive and nutritionally
totally inferior substitutes in the form of commercially processed
infant formulas[8]. In fact, the inroads of the promotion of such
products is a major public health hazard and needs to be borne in
mind in national policy decisions. Are such products necessary? Or
should their importation be controlled, banned or circumscribed by
laying down locally appropriate criteria?[9] It is sometimes assumed
that this is exclusively an urban problem. This is not so. Although
obviously more central in urban slums, the outreach of the transistor
radio and other forms of mass media introduce the persuasion of profit
dominated industries into areas where their products are both inappropriate and unaffordable.

(iii) Transitional - The so-called weaning period is inherent to all
mammals, including humankind. As is well-known,
the danger is increased by the fact that this is the period when
resistance to infections acquired from the mother in utero is dimi-
nishing and the psychological problems of separation from the mother
may also be contributing. This is obviously a higly hazardous period
on which primary care services should be focussed. At the same time,
the background for prevention is earlier — in a sense, the severe
malnutrition in the transitional (particularly in the second year
of life) is due to causes operating then, but with seeds sown in
fetal life and in the extero-gestate fetal stage.

     In countries anywhere, especially in developing areas of the
world, the biological method of feeding young children during the
transitional period is on a firm basis of sound nutrition earlier on.
During the transitional (or weaning) period itself, the aim should
be to devise weaning mixtures based on local cultural practices,
foods and cooking procedures. Whether these be prepared specially
for children or mixed together from foods already cooked for the
family, the concept of "multimixes" has been found to be very
useful[10, 11]. This is based on the recognition that most communities
have envolved food mixtures which are the excellent ways to combine
traditional foods long used and prized in the particular culture
(e.g. various legumes or pulses and cereal grains). The ideal is to
try to ensure the preparation of such home-made weaning foods. However, in some circumstances it may be practical and preferable to try

to prepare such mixtures at a central point in a community or region, using appropriate technology[9].

From the opposite pointe of view, the usefulness of processed weaning foods for young children imported from abroad needs to be scrutinized carefully — often it will be found that these are highly expensive nutritionally inappropriate.

CONCLUSION

Modern scientific research clearly shows that the optimal method of infant feeding for developing countries is exactly the same as industrialized countries. Universally the three main planks of scientifically guided, biological infant feeding need to be "dyadic"[12] — that is as an interaction between the mother and her young offspring:

(a)   to feed the pregnant and lactating mother with a mixed diet of locally available foods,

(b)   to breast feed alone for four to six months,

(c)   to introduce the least costly weaning foods based on the concept of "multimixes" from four to six months onwards, preferably prepared from locally available foods, but with continuing lactation into the second year of life.

As can be seen, all of these are often achievable with resources available to communities themselves and, indeed, need to be based on a full and clear understanding of the methods and practices that have evolved over millenia in different cultures.

Conversely, communities need to be protected from the disruptive ill-effects of uncontrolled, harmful advertising of infant foods which have neither relevance nor possibility of being used appropriately because of poverty or poor home hygiene. In other words, community education with regard to infant feeding needs to be based on the improvement (where necessary) of traditional methods and practices, and on shielding from the harmful incursion of processed infant foods.

Centrally, a policy needs recognizing which assists in improving infant (and maternal) feeding based on local resources, but also is designed to select processed infant foods, both manufactured nationally and imported from industrialized countries. In these choices, the following guidelines may be helpful:

1.   Processed infant foods need to be geared to the real nutritional needs of priority ages of young children and mothers in the area.

2. Processed infant foods need to be promoted (or not) with
    regard to their economic feasibility, within the constraints
    of the family budgets and the national income of the communi-
    ty concerned.

3. Processed infant foods must not be advertised and promoted
    if and when they are culturally, socially, and economically
    inappropriate for the particular community.

4. Processed infant foods must not be promoted if this will lead
    to the destruction of an existing satisfactory pattern of
    infant feeding, particularly breast feeding, which cannot be
    replaced by the proposed substitute.

Much of the malnutrition seen in mothers and young children in
developing countries is the result of inequitable distribution of
income and other resources, so that adjustment in this regard is a
primary goal. At the same time, tremendous improvements in nutrition
could be achieved through wide-spectrum primary care services, designed to improve the quality of life as a whole and to ensure biological
infant feeding, by trying to improve the diets of mothers during pregnancy and lactation and by realizing the absolutely key and vital role
of breast feeding in achieving optimal maternal and child health[8].

REFERENCES

[1] IUNS-UNICEF (1977) Community Action Family Nutrition Programs.
        Indian Council of Medical Research, New Delhi, India.
[2] IUNS Nutrition Abstracts (1971) Nutrition Programs for Pre-school
        Children: Guilines. Report of an IUNS Working Conference,
        Zagreb, Yugoslavia.
[3] Guidelines on the At-Risk Concept and the Health and Nutrition of
        Young Children. Amer. J. Clin. Nutr., 30: 242, 1977.
[4] IUNS (1978) Practical Approaches to Combat Malnutrition. Cairo,
        May, 1977.
[5] Jellifee, D.B. and Jellife, E.F.P. (1976) Maternal-channelled
        infant Feeding. Food and Nutrition, in press.
[6] Petros-Barvazian, A. and Behar, M. (1978) Low birth weight — what
        should be done to deal with this global problem. WHO Chronicle
        32: 231.
[7] Sterky, G. (editor) (1978) Birthweight Distribution — an indicator
        of Development. SAREC/WHO Worshop, Sigatuna, Sweden.
[8] Jelliffe, D.B. and Jelliffe, E.F.P. (1978) Human Milk in the Modern
        World. Oxford University Press: Oxford and New York.
[9] IUNS Position Paper (1978) Weaning Foods. Rio de Janeiro, Brazil.
[10] Jelliffe, E.F.P. (1974) A New Look at Weaning Multimixes in the
        Commonwealth Caribbean. J. Trop. Pediat. Env. Chld. Hlth. 20:14.
[11] Cameron, M. and Hofvander, Y. (1976) Manual of Feeding Infants and
        Young Children. (second edition) WHO: Geneva.

[12]IUNS Paper for the International Year (1979) of the Child (IYC) (1978) Dyadic Infant Feeding: Improving the Nutrition of Mothers and Young Children.

NUTRITION AND AGEING

Itsiro Nakagawa

Dept. of Nutr. & Biochemistry, Inst. of Pub. Health

Tokyo - JAPAN

Decrepitude is not a phenomenon to be seen only in advanced age, but begins in infancy. It is generally considered that growth is a theme for pediatricians and decrepitude a theme for geriatricians. However, growth and decrepitude cannot be differentiate from each other, and should be studied simultaneously. Therefore, ageing should be studied from the viewpoint on two fields.

From the anthropometric and biochemical viewpoint, pubertal growth is spurted earlier in urban children when compared with rural children, and the difference between urban and rural children became remarkable at 12-13 years of age. However, the difference of these parameters disappeared by the end of growth period.

The difference in the amount of protein intake is one of causes which affect pubertal growth. However, it is questionable that such acceleration is desirable for good health and long life span. We conducted ten similar experiments with a total of 385 female rats. We examined the effect of protein nutrition on growth and life span. The effect of the amount and composition of protein (10, 18,27 and 36% casein and pure amino-acid mixture diet natural L-amino acid mixture corresponding to 18% casein of which amino acids were estimated by an automatic amino acid analyzer) on growth and life span of rats was examined by using littermate females. Littermates were evenly distributed in the different groups. The amount of diet intake was the same in all the groups. The increases in both tail length and body weight of rats fed a 10% casein diet were generally less than those of rats fed the higher level of protein, especially at weeks 8 to 10 after birth, corresponding to puberty in human. This difference gradually decreased and became insignificant by the end of the growth period. The effect of protein on the biochemical measures (urinary

755

excretion of creatinine, hydroxiproline, and 17-ketosteroids, and the serum alkaline phosphatase activity) observed in the early period of growth among groups became obscure by the end of the growth period.

Estrus cycle appeared later in the 10% casein-fed rats and seemed to be irregular longer, as compared with that of rats in other groups.

In brief, the growth rate varied directly with the protein intake in the early period, but eventually the differences observed among the groups became insignificant. However, the effect of variance of litters was equal to or greater than that of difference in diet.

The effect of difference in diets on life span was not significant. The effect of variance of litters was relatively great.

An amino acid diet, when properly composed, will support growth and life span closely approaching that obtained with casein.

The number of lesions observed at death did not differ among the groups. The rats fed 10% casein diet showed the largest number of deaths in the early period of life, and those fed 27% casein diet the fewest, notwithastanding that the longevity was almost the same among rats in all the groups. The incidence of mammary tumors seemed to be greater in the groups fed a high protein diet. Protein, even when consumed in a large amount, cannot increase growth above a certain limit that is hereditary in an individual. However, an increased level of dietary protein accelerates growth within this limit during growing period, especially in infancy and at adolescence when growth is active. Even a large amount of protein does not induce growth in an individual in whom growth has ended. There are several causes for differences in the incidence of mammary tumors, such as species difference, difference of morbidity in litters, and nutrition, especially the amount of protein. In adult rats, growth of the body itself has ended. However, if a predisposition to tumors is present, a higher level of dietary protein intake induces tumors. In our experiments, having used female rats only, a low protein diet compared with a high protein diet seemed to have the effect of reducing the incidence of mammary tumors.

PROBLEMS OF NUTRITION IN THE AGED

Dodda B. Rao, M.B.B.S., D.T.M. & H.

Dept. of Geriat. Med. & Chronic Diseases, Oak Forest Hosp.

Oak Forest, Illinois 60452 - U.S.A.

The idea that in a progressive civilized society the State has the responsibility for its vulnerable citizens at all levels, is a relatively new concept.

An adequate nutritions diet and the maintenance of sound health are cardinal requirements in such a society.

Increasingly recognized is the conspicuous lack of appropriate concern and action by both professional and political powers for the many needs of the aged population. The extent of hardship and unmet needs continues to mount, even in an affluent society. Indeed, the "invisibility" of these unfortunate citizens is paving the way for them to emerge as the "new poor". The scientific and therapeutic advances of recent years have had a profound effect on longevity. All over the world the number of people in the late age decades is increasing by millions each year.

The importance of nutrition in the life and health of any human being cannot be overestimated. George R. Minot once said, "Man's future will depend very largely upon what he decides to eat". The quality of the remaining life of the aged person also greatly depends upon what he chooses to eat. The intake of food is one of the greatest variables in life. With the onset of old age, additional factors impinge on this variable and interfere with nutrition.

This article concerns an investigation into some of the socio-economic and dietary aspects of the aged with reference to nutritional status and health. There seems to be a dire need for such studies since there is a striking lack of information. Nutritionists have never been able to define accurately the nutritional needs of

the elderly. Much of the approach has concerned evidence for malnutrition, which may be defined as a disturbance of form or function
based on a lack (or an excess) of calories or of one or more nutrients. (United Kingdom, Department of Health and Social Security,
1970). The resultant pathologic state may be caused by:

(a) Undernutrition: Consumption of an inadequate quantity of
food, and hence a caloric deficit, over a prolonged period.

(b) Overnutrition: Consumption of an excess quantity of food,
and hence a caloric excess, over a prolonged period.

(c) Specific deficiency resulting from a relative or absolute
lack of an individual nutrient.

(d) Imbalance resulting from a disproportion among the essential
nutrients.

In these days of "internationalism", a starving country is a
center of unrest, just as a hungry man is an angry one. The effects
are not self-limiting, but instead lead to social disorder.

The wealth of a nation ultimately is in the health of its people. Aging and nutrition are of international concern. Poverty is
found amidst plenty. The justification for this study lies in the
problems of the increasing millions of dependent senior citizens and
their nutritional requirements.

The lengthened life span has created not only more physical,
psychologic, sociologic and economic issues, but also the problem of
maintaining health and efficiency during the added years. Since,
with the elderly, the chief problem is the maintenance of good health,
the nutrition of this group is of fundamental prophylactic and therapeutic importance. The number of persons reaching the older ages is
expected to increase, thanks to therapeutic and technologic advances,
but this achievement will be in vain unless the quality of life is
preserved. Today's failure to plan in this regard imperils the prospects for the future. Nutritional maintenance is evidently of primary importance in the process. This involves many changes and a
greater appreciation of the problems of the elderly, by the health
professionals in particular and by society in general.

FACTORS LEADING TO IMPROPER
NUTRITION IN THE AGED

Several factors may lead to dietaty inadequacy or imbalance in
the elderly:

1. Limited income may restrict the purchase of adequate amounts of

the right kinds of food, proper cooking facilities, and refrigeration.

2. Loneliness, unhappiness and bereavement can lessen the appetite.

3. Reduced activity, increased fatigue and weakness, or living alone may affect the incentive for eating. Lonely men who are unused to cooking or women who used to cook for family, lose interest in the process. Thus they tend to eat a poorly balanced diet consisting of foods which require little preparation.

4. Elderly persons living in urban areas are particularly prone to social isolation, which leads to mental and physical deterioration.

5. Many factors operate when support from family, friends or community is not available. Such deprivation often leads to apathy, depression and impairment of appetite.

6. Food fads and fallacies and chronic alcoholism can pave the way for poor nutrition.

7. Chronic invalidism, of whatever origin, fosters lack of appetite and a poor nutritional status.

8. Poor dental health may be a serious factor in the poor eating habits of the elderly.

9. Mental disturbances, such as confusion or depression, significantly affect eating habits and the nutritional state.

10. In general, a combination of several of these factors makes the elderly especially vulnerable to malnutrition.

   Most aged people have a life-time accumulation of food habits which may not have altered with changing physiology or with the changing methods of food processing, preservation and preparation.

NUTRITION AND HEALTH OF THE AGED

   The apothegm "You are what you eat", says both too much and too little. It says too much in that diet is not the sole determinant of the body composition. It says too little in that some dietary abnormalities are not ordinarily discernible as changes in major body constituents, but as clinical signs and symptoms or biochemical changes. Nevertheless, the importance of nutrition in the maintenance of health, vigor and enjoyment is well recognized.

   One of the basic nutritional concepts is that all persons,

throughout life, have need for the same nutrients, but in varying
amounts. The necessary amounts of nutrients are influenced by age,
sex, size, activity, and state of health. Advancing years impose
changes and hence the need for "matching food with the mouths", but
all too often this does not happen and the result is malnutrition,
which is a precursor of ill health. The vicious cycle of malnutrition and disease in the aged is summarized in Figure 1.

ASSESSMENT OF MALNUTRITION

     The diagnosis of malnutrition cannot be made from the dietary
history alone. Medical examination, with special knowledge of the
aged, is essential in order to differentiate primary and secondary
causes of malnutrition in the aged. In our studies, comprehensive
clinical evaluation included assessment of the general condition,
mental state, physical activity, and physical illness with special
attention to the skin, hair,nails, eyes, lips, tongue and skeletal
system. Physical measurements included height, weight, skinfold
thickness and x-ray examination

     The importance of clinical and laboratory examinations under
standard conditions should be emphasized.

     Most of the clinical descriptive literature on nutritional de-
ficiencies such as beriberi, scurvy, rickets, or kwashiorkor refers
to special groups of people who have been victims of frank malnutrition or starvation. Elderly people who have stood the test of time
with their diets and dietary habits are less likely to show the overt
manifestations of malnutrition. Indeed, evaluation of the nutritional health of the aged person in the absence of severe malnutrition is difficult[1].

     Some changes attributable to the aging process may mask or over-
lap the recognized signs and symptoms of nutritional deficiencies.
The problem is further complicated by the fact that a number of deficiency signs lack nutrient specificity. For example, glossitis
(a common finding in the aged) can be associated with niacin, folic
acid, vitamin B-12 or riboflavin deficiencies, or with uremia, or
with long-term systemic antibiotic therapy. Another common finding
in the aged is angular stomatitis, a sign associated with riboflavin,
niacin or pyridoxine deficiencies, or with poorly fitting dentures.
The wearing of dentures, poor oral hygiene or peridontal disease may
cause changes in the tongue and gums that cannot be distinguished
from those caused by a specific lack of vitamin B[2]. Follicular
keratitis, dependant edema, frailty, or overweight also may have an
obscure etiology. Most signs of malnutrition are not specific for
lack of any one nutrient. They even can be produced by a combination
of various non-nutritional factors. The etiology is often complex
and some of the interrelationships are still unknown[3].

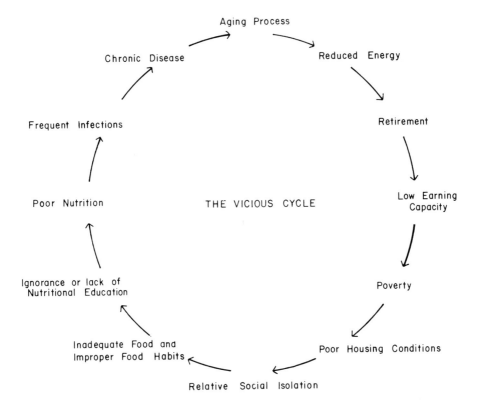

Fig. 1 - Interrelation of health, economics, social conditions and disease in aged.

PUBLIC AND PROFESSIONAL ATTITUDES
TOWARD NUTRITION IN THE AGED

Unfortunately, large segments of the general population do not
have good food habits[4]. This is particularly true of the elderly.
For them, socio-economic and physio-pathologic factors create added
complications. Most problems of the aged are compounded by poor
nutrition. Nutritional deficiencies respond well to treatment, with
dramtic improvement in comfort, strength and efficiency. This, in
turn, enhances psychologic well-being. Total rehabilitation is unat-
tainable without nutritional rehabilitation. With good nutrition,
the person is better equipped physically and emotionally to enjoy
his added years.

The physician, dietitian and other health personnel engaged in
the care of the elderly must be aware of nutritional status and dietary requirements. They should be familiar with the economic, racial,
religious, social and other factors that impede proper nutrition.
However, the vital subject of "clinical nutrition" has been given
very low priority in teaching and communication. The major problem
is lack of knowledge regarding the relationship between food and
health, and the potential of the soil for producing health-promoting
foods. Tradition and commerce have been the main promoters of nutritional practices, although significant scientific advances have been
made in recent decades. This scientific knowledge, usually reported
in technical terms in scientific journals, ought to be translated into
simple language and communicated widely to the public in order to
bridge the gap between knowledge and its application for human welfare.

Our technologic society tends to be oriented toward productivity
and profit. Consequently youth vigor are regarded as of prime importance[5]. The aged and infirm, being unproductive, are accorded low
status and receive low priority in welfare planning. Cautious effort
should be made to avoid such a mistake and there should be legislative
provision to safeguard the welfare of such vulnerable groups.

Poverty of knowledge concerning both nutrition and economics
endangers the health of the aged.

Economic poverty accounts for poor nutrition in many cases, and
in many more there is need for advice on how to spend a limited income
wisely for such things as food, clothing, heating and transportation[6].
In view of the spiralling cost of living, a regular review of old age
pensions and social security benefits is an imperative initial step.
Since the factors involved in malnutrition are numerous, on one solution can be offered. The one humane approach, however, is a direct
look at the problems through scientific research and a realistic application of available knowledge through coordinated effective services and income maintenance that permit dignity and decent living

conditions. Only through this kind of concentrated effort can a reasonable beginning be made in solving the nutritional problems of the
elderly. In the meantime, such welfare services as Meals on Wheels,
invalid and special diets, club meals and home health visits should
be expanded[7]. Nutritional counselling also can be carried out at day
centers, welfare homes and community centers[8]. In view of the limited
physical abilities of the aged, simplicity and practicability should be
of paramount consideration in these programs.

The unmet needs of the aged in the community are numerous, and a
combination of factors may account for malnutrition (Fig. 1). Many of
the factors operate when support from family, friends or community is
not available. Lack of social contact often leads to apathy, depression and disinclination to buy and prepare food, especially if the old
person is recently bereaved and socially isolated. Thus a social situation steadily declines into a state of malnutrition and a subsequent
medical problem. This is not to say that the diet alone must have affected health, as the reserve might be equally true. More often than
not, there is a combination of disease and malnutrition. Thus research in the field of geriatric nutrition is complicated by individual variability and the high incidence of chronic disease[9]. More longitudinal studies are needed to delineate the nutritional status and
the nutritional needs of this age group.

In the meantime there is great need for coordinated planning to
provide the services required. Restaurants might be encouraged to
serve special meals for the elderly at lower cost, and free transportation might be provided by local authorities. Provision might also
be made in public housing for a daily central meal service for both
residents and neighborhood old people.

The elderly vary considerably in their dietary likes and dis-
likes, according to their physical and mental state. Our study and
reports in the literature indicate that as people reach 65 years of
age, their diets are likely to become poorer[10, 11], particularly deficient in some nutritents such as ascorbic acid, calcium and iron,
and characterized by protein-calorie malnutrition[12, 13].

NUTRITIONAL GUIDANCE AND THE QUALIFICATIONS
OF AN "IDEAL" FOOD FOR THE ELDERLY

To help elderly people maintain health, they should receive
education in nutrition relative to their special needs — fewer calories in view of reduced activity and metabolism, but as many minerals, vitamins and proteins as required by younger people. Food
products which are easy to prepare can be a real boon to the elderly,
especially if the products are nutritious, clearly marked and fairly
inexpensive. The general qualifications of an ideal food for the
elderly are: 1) acceptable and readily available, 2) cheap and proc-

essed from local ingredients, 3) easy to prepare quickly, 4) appetizing and amenable to easy mastication and digestion, 5) will keep well even under poor conditions of storage, 6) can be used in liquid form, as a drink or for tube feeding, 7) can be mixed easily with other foods, and 8) is rich in proteins, minerals, essential elements and calories.

IMPACT ON HEALTH SERVICES

In Great Britain, the elderly constitute about 12 per cent of the population but they account for nearly 30 per cent of the cost of the Health Service. Wherever there is an increasing proportion of the aged and infirm in the population, the greater will be the demand on health services. The prevalence of subnutrition enhances the chances of illness. Therefore it is more desirable to promote well-being with proper nutrition than to foster chronic invalidism with its excessive consumption of the resources of health services.

Although overt manifestations of malnutrition and of specific nutrient-deficiency states in the aged are not common, there is ample evidence for the existence of subnutritional states with their varied effects. A poor nutritional state and poor health are associated in many aged, confused[14] and lonely persons.

PRINCIPLES OF SOUND NUTRITION
IN THE AGED

To promote healthy nutritional practices in the aged, the following principles are recommended. Sound nutrition in the elderly differs only slightly from that in younger adults. The daily food intake must be sufficient to supply all essential nutrients and...

1. Fewer calories, in view of reduced physical activity and metabolism. If obesity is present, weight should be reduced. The ideal is to maintain a "desirable weight", i.e., the average weight for a given height and sex at age 25.

2. An increased intake of proteins (1.5 to 2.0 gm per kg of body weight) in order to maintain normal nitrogen balance. About 20-25 per cent of the total calories should be derived from protein.

3. A reduced intake of fats. About 20 per cent of the total calories should be derived from fat.

4. Sufficient carbohydrate to provide the remainder of the required calories.

5. Sufficient vitamins and minerals, particularly iron, calcium and vitamin C, to maintain a normal balance.

6. An adequate water intaкe (6 to 7 glasses a day), so that the urinary output will be at least 1500 ml per twenty-four hours.

7. Enough fluid intake and "roughage" to prevent or treat constipation a common complaint among the aged.

FACTORS IN PLANNING SUCCESSFUL DIETS FOR THE AGED

Successful diets for the aged involve attention to the following:

1. Individual preferences. The upsetting of old habits may be physically and emotionally disturbing.

2. Social, religious, racial and psychologic factors, in addition to life styles.

3. The preparation and presentation of meals. A judicious touch of seasoning should be allowed, and an aperitif need not be denied except for specific medical reasons. Ease of preparation is essential.

4. A wide choice of foods. Ready availability is important. Flexibility in the number of meals. Regular and small feedings four or five times a day (for total daily dietary requirements) are desirable.

BALANCED NUTRITION: BASIC FIVE SOURCES

To achieve the foregoing objectives, and to provide a pattern for balanced nutrition in the aged, it is important to plan a simple guide to good nutrition. The basic five sources of nutrients are:

1. Meat and allied foods: Meat, poultry, eggs and fish. These supply protein fat, iron and vitamins.

2. Dairy products: Whole, skimmed, evaporated or dried milk, and milk products such as cheese, butter and buttermilk. These supply calcium, proteins, fat and vitamins.

3. Vegetables and fruits supply vitamins, minerals and fiber. It is important to include citrus fruit for vitamin C, and dark green or yellow vegetables or fruit for vitamin A.

4. Bread and cereals; Plain or enriched bread, fortified cereals, rice, and such items as macaroni, noodles, biscuits and cakes.

5. Fluids: Water is basic, but coffee, tea, cocoa and fruit juices play an important role.

The foregoing items should be provided in a balanced pattern consisting of breakfast, the main meal (mid-day or at night), afternoon tea, supper, and a bedtime beverage.

The amounts and size of portions should be adjusted to individual requirements. The choice of foods and the method of preparation can be modified according to personal preference.

The essential feature is to ensure that the total daily allowances are consumed within one day. The number of meals can vary, but regularity is of paramount importance.

In all endeavors related to nutrition in the aged, it should be realized that eating should not be transformed into a chore or a dietary experiment. It should be a genuine pleasure at any age, and especially in old age when so many other pleasures of the younger years have faded away.

REFERENCES

[1] Weir DR and Houser HB: Problems in the evaluation of nutritional status in chronic illness, AM J Clin Nutrition 12: 278, 1963.

[2] Chinn AB (ed.): Working with Older People. Vol. IV. Clinical Aspects of Aging US PHS (HEW) Publ. No. 1459, US Govt Printing Office, Washington D.C., 1971, p. 269.

[3] Beaton GH: Interrelationships of nutrition, Proc Nutr Soc 20: 30, 1964.

[4] Rudd JL and Margolin RJ: "Maintenance Therapy for the Geriatric Patient". Charles C Thomas, Publisher, Springfield, Ill., 1968.

[5] Reynolds R: Medical Interns and Chronic Disease and Attitude Study, Doctoral Thesis, Johns Hopkins University, 1970.

[6] Exton-Smith AN and Stanton BR: "Report of an Investigation into the Dietary of Elderly Women Living Alone". Papworth Everard, Pendragon Press, England, 1965.

[7] Rao DB and Kataria MS: Camberwell Nutritional Survey: The Medical Officer, No. 3065, Vol. 117: 207, 1967

[8] Rao DB: Day hospitals and welfare homes in the care of the aged, J Am Geriatrics Soc 19: 781, 1971.

[9] Hoffman AM (ed.): "The Daily Needs and Interests of Older People". Charles C Thomas, Publisher, Springfield; Ill., 1970.

[10] Le Bovit C: The food of older persons living at home, J Am Dietet A 46: 285, 1965.

[11] Fry PC, Fox HM and Linkswiler H: Nutrient intakes of healthy older

women, J Am Dietet A 42: 218, 1963.
[12]Food Consumption and Dietary Levels of Households as Related to the Age of Homemakers, United States — by Region. US Dept. of Agriculture, 1955 Household Food Consumption Survey, Rep. No. 14, US Govt, Printing Office, Washington, D.C., 1959.
[13]Johnson B and Feniak E: Food practices and nutrient intake of elderly home-bound individuals, Canad Nutr Notes 21: 61, 1965.
[14]Rao DB and Kataria MS: Dietary and clinical aspects of a nutritional survey, Brit J. Geriatric Pract. 5: 21, 1968.

NUTRITION OF AGING PEOPLE IN FINLAND

M. Pekkarinen, K. Hasunen, R. Seppanen, and
H. Norja
Dept. of Nutr., Univ. of Helsinki
Helsinki
FINLAND

INTRODUCITON

The number of elderly people is increasing in the more developed countries. In Finland about 9% of people are over 65 years old; this will be 11 per cent by 1980 and 12 per cent by 1990. The proportion of those over 75 will be particularly increased. Because women tend to live longer than men (72 as opposed to 65 years respectively on average) the proportion of elderly women to men will be great. The increasing numbers of elderly people may give rise to numerous problems, many of them economic and social but some also physiological and nutritional.

MATERIAL

The study presented here is part of a multiphasic screening program carried out by the mobile clinic of the Social Insurance Institute of Finland from 1967 through 1972 in urban and rural districts in different parts of Finland. The diet and health status of nearly 10,000 people over 15 years of age were studied, food and nutrient intakes of 1788 persons over 55 years old were studied (Table 1). They were divided into four age groups, and 21 per cent were 65 years or over. In all groups there were more women than men. About half of the men and less than half of the women lived in rural districts.

METHODS

The study made with the diet history interview method. The interview focused on the average food consumption and nutrient intake

TABLE 1

NUMBER OF MEN AND WOMEN OVER 55 YEARS OF AGE
IN DIFFERENT AGE GROUPS

| | AGE, YEARS | | | | |
	55-59	60 64	65-69	70-	TOTAL
Men	319	275	127	124	845
Women	326	303	160	154	943
Total	645	578	287	278	1788

RESULTS AND DISCUSSION

Food Consumption

The mean consumptions of different foods by men and women in age groups are given in the following tables (Tables 2 and 3). A general feature among elderly people is the high use of liquid milk products and butter, the rather liberal use of grain products and potatoes but the very small consumption of vegetables, fruit and berries. With advancing age the consumption of most foods, with the exception of milk by women, was decreased. Because of the greater energy requirement, men consumed more food tham women. When adjusted for energy qualitative differences were not observed in the food consumption between groups and sexes with the exception of vegetables and fruit, the use of which was higher by women of all age groups.

The dietary pattern varied to some extent in different parts of the country depending on the local food habits. Thus in northern Finland reindeer was used more than other meats, rye in eastern Finland more than in the west, and fish in the north more than elsewhere. The use of grain products, butter, milk and potatoes was more abundant in farming than urban areas, while vegetables, fruit and meat were more popular in urban areas. Old people of larger families consumed more of the cheaper foods, especially grain products and potatoes, and less of the more expensive foods such as meat. The diet of old people living alone and that of widowers seemed to be the most monotonous. In the higher social classes the use of meat and fruit was more liberal than in lower classes, obviously due to a better economic situation.

TABLE 2

CONSUMPTION OF DIFFERENT FOODS
(GRAMS PER PERSON PER DAY)
BY MEN IN DIFFERENT AGE GROUPS

| | AGE, YEARS | | | |
	55-59	60-64	65-69	70-
Grain products	330	304	283	270
Potatoes	249	230	201	202
Vegetables	85	82	67	58
Fruit and berries	76	90	84	83
Butter, margarine, oil	57	51	53	47
Liquid milk products	973	942	919	833
Chesse	11	12	8	8
Meat products	132	139	133	110
Fish	41	45	46	33
Egg	27	31	23	24
Sugar	74	74	69	62

Intake of nutrients

The intakes of energy and nutrients by the elderly showed great individual differences. Energy intake decreased with age, and in men more than women (Table 4). The intake was highest in men living in rural areas, clearly due to higher physical activity. In general the mean energy intake was liberal and certainly exceeded the requirements for those age groups.

Obesity has been shown to increase with advancing age. When the prevalence of obesity was studied in this population using a body mass index (Quetelet's index) of over 30 as the criterion, 34 per cent of women and 14 per cent of men aged 60 years or over were very obese. The following table (Table 5) shows some anthropometric data on these obese individuals. It can be seen that women tended to be a little more obese than men. However, the food intake of the very obese people was a little lower than that of those less obese. Obviously obesity had become established much earlier but food retriction was not severe enough for weight loss. Obese people seemed to restrict the use of fats, sugar and grain products especially and had fewer meals per day, a phenomenon also observed in other studies[1, 2, 4].

TABLE 3

CONSUMPTION OF DIFFERENT FOODS
(GRAMS PER PERSON PER DAY)
BY WOMEN IN DIFFERENT AGE GROUPS

| | AGE, YEARS | | | |
	55-59	60-64	65-69	70-
Grain products	251	242	220	219
Potatoes	173	..52	142	145
Vegetables	104	01	90	80
Fruit and berries	140	25	128	124
Butter, margarine, oil	39	40	37	41
Liquid milk products	677	685	700	740
Cheese	10	9	7	6
Meat products	90	84	85	76
Fish	25	26	25	20
Egg	24	22	20	21
Sugar	67	64	56	57

The mean intakes of energy yielding nutrients, proteins and fats
were liberal (Table 6). The percentage contribution of proteins to
total energy was 14 per cent; that of fats in men 30-40 per cent and
in women 36-39 per cent, which are more than in the overall Finnish
population. The highest fat intake was in old people living alone
and divorced persons, being 41-45 per cent of total energy. Dietary
fats were also very saturated, the P/S ratio being between 0.17 and
0.18 in all age groups. The energy share of carbohydrates remained
below 50 per cent. 20 per cent of carbohydrates originated from sucrose in men and 23 per cent in women.

Of the dietary minerals studied (Table 7) calcium intake was generally liberal at 0.5 g/1000 kcal, due to the high milk consumption.
However, 50-55 per cent of men and 40-45 per cent of women obtained
less than this. The intake was lowest among women living alone.

The average intake of iron decreased with age in both men and
women. The daily intake remained below 8 mg in 15 per cent of men
over 70 years, in 15 per cent of women under 65 years, and in 20 per

TABLE 4

MEAN DAILY INTAKE OF ENERGY BY MEN AND WOMEN
IN DIFFERENT AGE GROUPS

		AGE, YEARS			
		55-59	60-64	65-69	70-
Men					
Energy,	kcal	2970	2870	2730	2490
	MJ	12.4	12.0	11.4	10.4
Women					
Energy,	kcal	2190	2120	970	2020
	MJ	9.2	8.9	8.2	8.5

cent of women over 65. Energy-adjusted iron intake was lowest in
divorced persons and women of larger families.

The prevalence of anaemia and iron deficiency seem to be very
common among old people. Takkunen and Seppänen[6] in studies of the
same population found Hb to be under 130 g/1 in 7.9 per cent of men
aged 65 or over and under 120 g/1 in 3.8 per cent of women of the
same age range. Transferrin saturation was below 15 per cent in 6.4
per cent of men and 7.7 per cent of women. Iron deficiency was especially common in the rural population. It seems that dietary
habits are significant in the etiology of iron deficiency among old
people as iron deficient persons consumed less meat products and
more liquid milk products than others.

Though the average intakes of vitamins studied (Table 8) seem
to be adequate and even abundant, some old people may obtain too
little. For example, nearly 20 per cent of men over 65 and 20 per
cent of women over 70 consumed less than 2500 I.U. vitamin A daily.
It is unlikely that these amounts were low enough to cause clinical
signs of vitamin A deficiency.

The diet of these old people contains enough vitamins of the B
group on average. Our earlier studies[3], however, have shown that
thiamine might have been too scanty in some groups of old people
especially before the enriching of white wheat flour with thiamine
had begun in Finland. Also, the results of the transketolase reactivation tests indicated rather low thiamine intakes.

TABLE 5

SOME ANTHROPOMETRIC DATA OF OBESE MEN AND WOMEN
OVER 60 YEARS OF AGE

	MEN	WOMEN
Number	83	218
Height, cm	170.1	154.5
Weight, kg	94.3	79.7
W/H^2 x 10^4	32.5	33.3
Triceps, mm	14.5	27.4
Subscapular, mm	29.9	24.4

TABLE 6

MEAN DAILY INTAKE OF PROTEINS, FATS AND CARBOHYDRATES
(GRAMS PER PERSON PER DAY)
BY MEN AND WOMEN IN DIFFERENT AGE GROUPS

	AGE, YEARS			
	55-59	60-64	65-69	70-
Men				
Proteins	104	102	97	86
Fats	129	125	122	108
Carbohydrates	367	340	326	307
Women				
Proteins	71	71	68	67
Fats	89	87	83	87
Carbohydrates	287	275	248	252

TABLE 7

MEAN DAILY INTAKE OF CALCIUM AND IRON
BY MEN AND WOMEN IN DIFFERENT AGE GROUPS

| | | AGE, YEARS | | | |
		55–59	60–64	65–69	70–
Men					
Calcium,	mg	1480	1440	1380	1250
Iron,	mg	17.4	17.0	15.5	13.9
Women					
Calcium,	mg	1080	1070	1060	1090
Iron,	mg	13.1	12.4	11.3	11.2

Due to the low consumption of vegetables, fruit and berries the intake of vitamin C is rather low especially when considering the great losses during food preparation. Old people living alone, single men, and women of larger families obtained the lowest amounts of vitamin C in their diets. The results of our earlier study[5] are in agreement with this. The very low dietary intakes were also reflected in the plasma levels of vitamin C which were found to be under 0.1 mg/100 ml in 24 per cent of the old persons studied. Slight clinical symptoms of vitamin C deficiency were found in some old persons.

CONCLUSIONS

Though mean values of energy and nutrient intakes seemed to be adequate, about 30 per cent of elderly people obtained less nutrients than generally recommended. Old men members of larger families, especially, may form a nutritional risk group. Obesity and anaemia are the most common nutritional disorders among elderly Finnish people.

REFERENCES

[1]Fabry, P., Fodor, J., Hejl, Z. & Braun, T. 1964. The frequency of meals: its relation to overweight, hypercholesterolaemia and decreased glucosetolerance. Lancet 2:614-615.
[2]Pawan, G.L.S. 1972. Feeding patterns in resistant' obesity. Proc. Nutr. Soc. 31: 90A-91A.
[3]Pekkarinen, M., Koivula, L. & Rissanen, A. 1974. Thiamine intake

TABLE 8

MEAN DAILY INTAKE OF VITAMINS BY MEN
AND WOMEN IN DIFFERENT AGE GROUPS

| | AGE, YEARS | | | |
	55–59	60–64	65–69	70–
Men				
Vitamin A, I.U.	5490	5470	4940	4530
Thiamine, mg	1.9	1.9	1.8	1.6
Riboflavin, mg	3.1	3.1	2.9	2.7
Niacin, mg	17	17	16	14
Vitamin C, mg	80	82	73	64
Women				
Vitamin A, I.U.	6090	6160	5370	5810
Thiamine, mg	1.4	1.3	1.3	1.3
Riboflavin, mg	2.3	2.3	2.3	2.3
Niacin, mg	12	12	11	10
Vitamin C, mg	88	81	78	76

and evaluation of thiamine status among aged people in Finland.
Internat. J. Vit. Nutr. Res. 44: 95–106.

[4]Ries, W. 1973. Feeding behavior in obesity. Proc. Nutr. Soc. 32:
187–193

[5]Roine, P., Koivula, L., Pekkarinen, M. & Rissanen, A. 1974. Vitamin C intake and plasma level among aged people in Finland. Internat. J. Vit. Nutr. Res. 44: 95–106

[6]Takkunen, H. & Seppanen, R. 1978. Iron deficiency in the elderly
population in Finland Scand. J. Soc. Med. In press.

A NEW METHOD IN FOOD SELECTION PROCESS

OF THE ELDERLY IN WESTERN SOCIETY

Magdalena Krondl

Dept. of Nutr. & Food Science, Fac. of Medicine

Univ. of Toronto, Ontario, CANADA

Many factors contribute to the undesirable diet of the elderly. The social factors are of special interest because they are likely to differ with culture. In the West, the nutritional status of elder- lies is affected particularly by social isolation low motivation to prepare meals, and income[1]. In contrast, the seniors in societies with strong family ties are respected can cared for and their eco- comic situation is of less importance[2]. In the Western society many nutrition programs have been introduced[3], but not many of them were designed on research of the target population. Also the effective- ness of the programs is rarely evaluated. Therefore we have attempt- ed to develop a methodology which would provide a data base for a specific nutrition program development and its evaluation. The method includes criteria to assess the presonality profile of elderlies, their food use and food selection determinantes. The data were col- lected in the fall and winter, 1977.

The sample of seniors was drawn from a list of clusters of 2,600 names from sampling frame of 12,000 seniors from Metropolitan Toron- to, Canada. The final number of subjects was 194. Included were those whom it was possible to contact and those who were suitable, as well as willing, to participate in the study. Many who were contact- ed refused co-operation because of preocupation or social withdrawal. The subjects included 85 males and 109 females, aged 65-77, living alone, and second generation Anglophones. Fourteen percent of our sample was still employed and 31% participated in voluntary work. Their socio-economic status was relatively good, the majority fell into the $500 per month income bracket. The average expenditure on food was $21 per week. The majority of the sample completed high school and had some post secondary education. The majority, 76%, rated themselves to be in good or excellent health and their heights and weights were within acceptable range.

The sample was further categorized according to an "activity index". Before developing the "activity" questionnaire, an attempt was made to use life satisfaction index [4], [5], [6] which was designed for use with seniors. The findings did not reflect the level of mental performance and physical exertion, or social interaction. As the activity according to Sherman[7] and Peppers[8] was found to be the most important personality trait of elderly and independent of sex, age, health and living arrangement, it seemed appropriate to measure it. The framework of the instrument designed for this study to measure "activity" was the Canadian Radio and Television Commission's study from Toronto[7]. It included 15 questions on four areas: 1. concern about the world around, 2. self-assistance, 3. physical activity, 4. willingness to associate with others. The answers were scored. The mean activity score of the sample was 49 out of a maximum of 100. The mean value for men was 46 and for women, 51. The difference in activity between men and women was statistically significant $p<0.005$. The "active" were defined as those with scores above the median, while "non-active" had scores below.

In determining the food habits of the elderly the food frequency method was used. The advantage of this method is that it informs more about foods than nutrients. A matrix of 194 foods used by seniors in Ontario, Canada, was first developed. A trained asked the seniors to rate each food according to their frequency of its use. An internal scale registered foods eaten more than once a day, daily, 4 - 6 times per week, 1 - 3 times per week, 2 - 3 times per monthly or less. Foods never eaten and seasonal consumption were recorded separately. Food use frequency scores were calculated for each of the foods using the equation

$$\frac{R_1 S_1 + R_2 S_2 + \ldots R_n S_n}{n}$$

- where R_1 to R_n is equal to the percentage of respondents for each scale rating.

- where S_1 to S_n is equal to the scale rating, n being the maximum rating.

The significance of the scores is that, according to them, the frequency of use of each food can be ranked and that the scores can be correlated with other variables. The scores of the foods included in the matrix were divided into equal thirds, and the foods were divided into high, medium and low categories. From the list of 194 foods, only 16 foods were found in the high use category. These were the foods eaten daily or several times a week by a majority of

the seniors.   Forty-seven foods were included in the medium category
and 131 foods were in the low category. The list of foods in the high
use category is presented in Table 1. Tea, coffee, bread and margarine were the expected items of frequent use by the elderly. The high
use of eggs, carrots, lettuce and orange juice may be more typical
for this sample where the majority were not below the poverty line.
Milk did not appear on this list in contrast with other populations,
e.g. low-income homemakers[10].

In comparing the food habits according to 'activity', two approaches were used. The first was comparing the food use frequency
scores of 16 foods which have been found to be most frequently eaten
by all the seniors. The second was to compare the never-eaten foods
included on the food matrix with 194 foods. The number of never-
eaten foods was expressed as a percentage. With both approaches, it
was shown that there was a significant difference between the active
and non-active at the 5% level. The active subjects had higher frequency scores for all 16 foods examined except for potatoes. The
less active seniors have never eaten a higher percentage of foods.
These findings have supported the assumption that the active seniors
have a more varied diet than the non-active. When the two parameters
of frequency scores and percentages of foods never eaten were compared between the sexes, it was found that women had a more varied diet
than men, which is in disagreement with Reid and Miles[7] who found a
higher variety of foods among men than women.

The main focus of the study was on the food selection process.
It is essential to know why the seniors make their specific food
choices so that their attitudes to foods can be considered in the
nutrition program. For that purpose, we examined from nine different
aspects, how the seniors perceive selected foods. These perceptions
of foods were expressed as food selection determinants. The determinants were: satiety, tolerance, taste, familiarity, prestige, price,
convenience, health belief and nutrition knowledge. The rationale
for the choice of the determinants is described by Krondl and Lau[11].
In principle, the nine variables reflect physiological, cultural,
social and personal motives.

The perception of foods was measured by the use of five-point
semantic differential scales. The scales for[1] perception of satiety
ranged from "very satisfying" to "not satisfaying at all", for[2] perception of tolerance from "not any amount" to "any amount", for[3] perception of taste from "very good" to "very poor", for[4] perception of
familiarity from "eaten since grade one" to "never eaten", for[5] perception of convenience from "very convenient" to "not al all convenient", for[6] perception of price from "very cheap" to "very expansive",
for[7] perception of prestige "would always serve to guests" to "would
never serve to guests", for[8] perception of health belief, from "very
good for health" to "not at all good for health". The knowledge
question asked about the nutrient content of the food under investig-

TABLE 1

FOOD USE FREQUENCY SCORES

FOODS USED MOST FREQUENTLY	FOOD USE SCORE (Range 0 100)
Tea	83
Eggs	77
Whole Wheat Bread	76
Coffee	75
Margarine	74
Carrots	72
Lettuce	72
Orange Juice	68
Hard Cheese	68
Tomatoes	67
Potatoes	67
Chicken	66
Bananas	64
Onions	64
Apples	63
Cold Cereals	62

ation. The limiting nutrients were considered to be calcium, iron, vitamin A, vitamin C, polyunsaturated fats and fibre[12, 13, 14]. Responses to this question were either "agree", "disagree" or "do not know". In the present study, 14 foods were selected for detailed assessment. Their nutritional value was the rationale for their choice. The food selection determinants for the 14 foods were scaled by the seniors during a personal interview. Food selection determinant scores were calculated using a similar methodological principle described for the food use frequency score.

Examples of the food selection determinant scores are shown in Table 2 for skim mild powder and whole wheat bread. Skim milk is seldon used by the elderly and bread is often used by them. It is interesting to notice that the elederly know in both cases the nutritional importance of these two foods judging by the high scores for knowledge. In the case of milk it is interesting to notice much lower scores for tolerance, belief, convenience, prestige, price, taste and mainly familiarity than in bread. We assume that the low scores

TABLE 2

FOOD SELECTION DETERMINANT SCORES

	SKIM MILK POWDER	WHOLE WHEAT BREAD
Knowledge	86	98
Tolerance	65	94
Health Belief	51	83
Convenience	43	83
Prestige	41	83
Price	35	54
Taste	34	82
Familiarity	24	74
Use - frequency	29	76

for milk powder are mainly because this food was not available in the childhood of today's elderly, that is, at their food habit formation stage.

The overall significance of food selection determinants was evident when their ratings were correlated with the food use ratings. Out of the possible 14 correlations, that is for 14 foods, a maximum of 12 were found significant for health belief, taste, price and satiety and the least for knowledge (Table 3). These findings suggest that seniors' choice of most of the studied foods depends on these types of motives. In other words, the elderly eat foods because they believe them to be healthy and that they taste well, not because they know their nutrient content. It must be explained that nutrition knowledge differs from health belief, as health belief is not usually scientifically substantiated. This finding is in accordance with Grotkowski and Sims[1], who found health belief to be an intervening variable between personal attitudes and specific nutrient intake.

The influence of food selection determinants on the use of individual foods is apparent from the results of multiple regression analysis. The results for tea, liver, squash and skim milk powder are shown in Table 4. R^2 for tea indicated that the combined effect of all nine determinants explains only 55% of the variance. In other words, the nine determinants were not the only ones which caused the selection of tea. Possibly other factors such as its availability, or the individual's age, sex, income, culture would explain the re-

TABLE 3

RANKING OF THE FOOD SELECTION DETERMINANTS
IN THE USE OF 14 FOODS

FOOD SELECTION DETERMINANT	NUMBER OF FOODS WITH SIGNIFICANT r's
Health Belief	12
Taste	12
Price	12
Satiety	12
Convenience	7
Tolerance	6
Familiarity	5
Prestige	2
Nutrition Knowledge	2

maining 45%. The partial effects of the food selection determinants were calculated as beta weights. They indicated the relative influence of individual food selection determinants on food choice. In the case of tea, the most important factor was its taste, for eating liver it was health belief and for eating squash it was convenience.

Using the example of skim milk powder, it can be illustrated how this information can be used in designing a nutrition program. Skim milk powder was used little by the elderly and it would be desirable to increase its use. Because the elderly who use it, select it for its perceived healthfulness, convenience, taste and price, these factors should be used for its further promotion among the non-users, rather than stressing the other factors, tolerance, knowledge, prestige, familiarity and satiety. These factors would be of little use in promotion of skim milk powder.

The findings were applied in the manner suggested above in an experimental nutrition program. From the total sample of 194 seniors, 60 attended a seven week session. Active and non-active elderliers were equally represented. The findings (which determinants are important for the food selection of seniors) were used to influence the use of specific foods. The program included classroom discus-

TABLE 4

INFLUENCE OF FOOD SELECTION DETERMINANTS ON THE USE OF
TEA, LIVER, SQUASH AND SKIM MILK POWDER

FOOD	R^2	F	BETA WEIGHTS OF SIGNIFICANT FDS	
Tea	0.55	20.98 ($p < 0.001$)	Taste	0.321
			Satiety	0.295
			Health Belief	0.189
			Tolerance	0.131
Beef Liver	0.22	4.91 ($p < 0.001$)	Health Belief	0.213
			Convenience	0.130
Squash	0.53	19.66 ($p < 0.001$)	Convenience	0.337
			Satiety	0.311
			Familiarity	0.157
Skim Milk Powder	0.35	10.39 ($p < 0.05$)	Price	0.203
			Health Belief	0.196
			Convenience	0.189
			Taste	0.167

sions, demonstrations and food preparation sessions. The food selection determinant rating of foods was registered in the beginning and at the end of the program.

The evaluation of change in food use frequency, which means in food habits, due to the program is planned in a follow-up study seven months after the program to assess the long term effect. A comparison will be made between the seniors who attended the program and the control group.

In conclusion, a method employing instruments to assess an "activity" profile of seniors, food use frequency and nine food selection determinants was described. Variety of food intake appeared to be greater among the females than among the males, and among the active than among the non-active.

On examining motives for choice of the 14 foods, health belief, taste, price and satiety were stronger than convenience, tolerance, familiarity, prestige and nutrition knowledge. Application of food selection motives in a nutrition program was illustrated.

This paper has been prepared in co-operation with D. Lau, M.Sc. The data on "activity" are part of the M.Sc. thesis of Mary Anne Yurkiw.

The research was supported by Health and Welfare, Canada grant No. 606-1464-43.

REFERENCES

[1] Grotkowski, M.I. and L.S. Sims. Nutritional knowledge, attitudes and dietary practices of the elderly. J. Amer. Diet. As. 72: 499-506, 1978.

[2] Pathak, J.D. Our Elderly. Some effects of aging in Indian subjects. Medical Research Centre, Bombay, 1977.

[3] Kohrs, M.B., Hanlan, O. and Eklund, D. Title VII - Nutrition Program for the lderly. 1. Contribution to one day's dietary intake. J. Amer. Diet. As. 72: 487-492, 1978.

[4] Adams, D.L. Analysis of life satisfaction index. Gerontology. 24: 470, 1969.

[5] Wolk, S. and Telleen S. Psychological and sociological correlates of life satisfaction as a function of residential constraint J. Gerontol. 31: 89, 1976.

[6] Neugarten, B.L., R.J. Havighurst, and S.S. Tobin. The measurement of life satisfaction. J. Gerontol. 16: 134, 1961.

[7] Sherman, S.R. Leisure activities in retirement housing. J. Gerontol. 29 (3): 325, 1974.

[8]Peppers, L.G.   Patterns of leisure and adjustment to retirement. The Gerontologist. 16 (5) : 441, 1976.

[9]Canadian Radio and Television Commission - Reaching the Retired: A Survey of the media habits, preferences and needs of senior citizens in Metropolitan Toronto. Information Canada, 1974.

[10]Reaburn, J.A., M. Krondl and D. Lau. Social determinants in food selection. J. Amer. Diet. Assoc. (In print) 1979.

[11]Krondl, M. and D. Lau. Food habit modification as a public health measure. Can. J. Pub. Health. 69 (1) : 39-43, 1978.

[12]Reid, D. and E. Miles. Food habits and nutrient intakes of non--institutionalized senior citizens. Can. J. Publ. Health. 68: 154-158, 1977.

[13]Leichter, J., J.F. Angel, and M. Lee. Nutritional status of a selected group of free-living elderly people in Vancouver. C. M.A. J. 118: 40-43, 1978.

[14]O'Hanlon, P., M.B. Kohrs. Dietary studies of older Americans. Amer. J. Clin. Nutr. 31: 1257-1269, 1978.

AUTHOR INDEX - Volume III

A

Ancona Lopes, F
Astrand, P.

B

Bamji, M.S.
Barna, M.
Blackburn, G.L.
Blomstrand, R.
Brin. M.
Brubacher, G.

C

Campagnoli, M.
Carneiro, L.
Carroll, K.K.
Chandra, R.K.
Chaves, N.
Covian, M.R.
Crawford, M.A.

D

Delbue, C.
Delpeuch, F.
Deutsch, F.

E

Epstein, F.H.

F

Faria Jr., R.
Fashakin, J.B.
Ferro-Luzzi, A.
Foglia, V.G.
Fujiwara, M.

G

Gabr, M.
Gori, G.B.
Gounelle de Pontanel, H.
Grassman, E.

H

Halpern, S.L.
Hassan, M.M.
Hejda. S.
Hermus, D.J.J.
Huffman, S.L.

I

Illingworth, D.R.

J

Jelliffe, D.B.

K

Kallner, A.
Kanamori, M.
Kay, R.M.
Krishnamachari, K.A.V.R.
Kritchevsky, D.
Krondl, M.
Kusnetzoff, J.C.
Kusunoki, T.

L

Layrisse, M.
Lebenthal, E.
Lechtig, A.
Le Magnen, J.
Lewis, B.

Preschool children (continued)
  mean daily energy and
      protein/carbohydrate
      intake of, 668–669
  nutritional status vs.
      exercise in, 657–675
  protein-energy malnutrition
      in, 165–168
Preservatives, allergies to 277
Progesterone, as anabolic agent
      in protein metabolism, 5
Prolactin, thyroid stimulating
      hormone and, 232
Prostaglandin synthesis, long-
      chain fatty acids and,
      25
Protease inhibitors, cancer and,
      506
Protein
  in aging, 755–756
  body growth and, 37
  breast cancer and, 497
  casein and, 379
  cardiovascular disease and,
      379–384
  cholesterol levels and,
      379–384
  in hemoglobin regeneration,
      209
Proteinase inhibitors, in plants
      and animals, 303
Protein-calorie malnutrition,
      see also Malnutrition;
      Protein-energy malnutri-
      tion
  anemia in, 209
  breastfeeding and, 178–179,
      634
  endocrine modifications in,
      229
  incidence of, 177–179
  in infants, 169–173
  iodine deficiency and, 235
  serum T_3 and T_4 concentrates
      in, 232
  in Sudan, 177–184
  thyroid function and, 233
Protein-calorie restriction,
      physical/mental develop-
      ment and, 680

Protein concentrates, essential
      amino acids in, 180, 643
Protein deficiency
  adaptation to, 157–164
  liver protein and, 161
  lysine and, 164
  in New Guinea, 164
Protein diet, see also Diet(s)
  cholesterol turnover and
      metabolism in, 381–383
  coronary heart disease and,
      384
Protein-energy deficiency, vs.
      lipid metabolism in
      rats, 702
Protein-energy malnutrition
  in preschool children,
      165–168, 657–675
  vitamin A deficiency and, 90
Protein hydrolysates, in weight
      reduction, 333
Protein metabolism
  corticosterone and, 7
  in injury, 15
  liver function disorders and,
      521–524
  muscle protein turnover rate
      and, 157
  N-balance in, 4
  in pregnancy, 3–7
  progesterone in, 5
Protein proteinase inhibitor,
      from eggplant exocarps,
      303–306
Protein synthesis, see also
      Protein turnover
  abdominal operations and, 17
  alcohol and, 522–523
  block in, 18
  food intake and, 11, 14
  in humans, 10
  protein gain and, 14
Protein turnover, see also
      Protein synthesis
  end-product method in, 9
  estimation of, 9
  in injury, 15–18
  in malnutrition, 9–13
  obesity and, 13–15
  orthopedic operations and, 17

LOW

ry